Quantitative Fisheries Stock Assessment

The data sets and programs that had been included on a floppy disc with previous printings of this volume are now available on Chapman & Hall's FTP site (ftp://ftp.thomson.com/chapman&hall/hillborn/) for free downloading. The description of the available files in the Preface pertains to the FTP site files as well.

Quantitative Fisheries Stock Assessment

Choice, Dynamics and Uncertainty

Ray Hilborn

and

Carl J. Walters

CHAPMAN & HALL

I(T)P® **International Thomson Publishing**
Thomson Science
New York • Albany • Bonn • Boston • Cincinnati • Detroit • London • Madrid • Melbourne
Mexico City • Pacific Grove • Paris • San Francisco • Singapore • Tokyo • Toronto • Washington

Cover design by Mary McKenna Ridge

Printed in the United States of America

For more information contact:

Chapman & Hall
115 Fifth Avenue
New York, NY 10003

Chapman & Hall
2-6 Boundary Row
London SE1 8HN
England

Thomas Nelson Australia
102 Dodds Street
South Melbourne, 3205
Victoria, Australia

Chapman & Hall GmbH
Postfach 100 263
D-69442 Weinheim
Germany

International Thomson Editores
Campos Eliseos 385, Piso 7
Col. Polanco
11560 Mexico D.F.
Mexico

International Thomson Publishing - Japan
Hirakawacho-cho Kyowa Building, 3F
1-2-1 Hirakawacho-cho
Chiyoda-ku, 102 Tokyo
Japan

International Thomson Publishing Asia
221 Henderson Road #05-10
Henderson Building
Singapore 0315

6 7 8 9 XXX 01 00 99 98 97

Library of Congress Cataloging-in-Publication Data

Hillborn, Raymond, 1947--
Quantitative fisheries stock assessment: choice, dynamics and uncertainty.
I. Title. II. Walters, Carl J.,
ISBN 0-412-02271-0
597.0524

Visit Chapman & Hall on the Internet http://www.chaphall.com/chaphall.html

To order this or any other Chapman & Hall book, please contact **International Thomson Publishing, 7625 Empire Drive, Florence, KY 41042.** Phone (606) 525-6600 or 1-800-842-3636. Fax: (606) 525-7778. E-mail: order@chaphall.com.

For a complete listing of Chapman & Hall titles, send your request to **Chapman & Hall, Dept. BC, 115 Fifth Avenue, New York, NY 10003.**

This book is dedicated to the life and work of
John A. Gulland

CONTENTS

Preface

This book really began in 1980 with our first microcomputer, an Apple II+. The great value of the Apple II+ was that we could take the computer programs we had been building on mainframe and mini-computers, and make them available to the many fisheries biologists who also had Apple II+'s. About 6 months after we got our first Apple, John Glaister came through Vancouver and saw what we were doing and realized that his agency (New South Wales State Fisheries) had the same equipment and could run the same programs. John organized a training course in Australia where we showed about 25 Australian fisheries biologists how to use microcomputers to do many standard fisheries analyses. In the process of organizing this and subsequent courses we developed a series of lecture notes. Over the last 10 years these notes have evolved into the chapters of this book.

A second major factor that shaped this book is a course in stock assessment that we gave for the South Pacific Commission in 1986. In attempting to put together readings for that course, we realized how narrowly most books treated the subject matter: stock assessment was often taught as simply estimating maximum sustainable yield (MSY) or yield-per-recruit from catch data. We just could not see how such assessments are of much value to fisheries people in developing situations as commonly found in countries of the South Pacific. This prompted us to put together a set of notes on the behavior of exploited populations and the behavior of fishing fleets that became chapters 3 and 4 of this book.

In the title of the book we emphasize *Choice, Dynamics and Uncertainty*. These three words symbolize our approach to fisheries management. Fisheries management involves making choices: without choice you can be an observer, but never a manager. Throughout the book we try to emphasize the need to compare options, rather than calculate any single quantity, be it MSY, fishing effort to maximize yield-per-recruit (F_{max}), or whatever. Fisheries systems are dynamic: there is simply little room (or need) for equilibrium or static descriptions. Using computer simulations we emphasize how easy it is to explore the dynamic behavior of fisheries systems, and how different will be your predictions if you move beyond equilibrium views.

Finally, uncertainty pervades fisheries analysis, and we believe that one of the reasons that major fisheries management disasters occur with discouraging regularity is the traditional inability and unwillingness of fisheries management agencies to deal directly and explicitly with uncertainty.

Before you jump into looking at the chapters of the book, let us offer you a few words of guidance and caution. This book is intended as a textbook and reference for courses in fisheries stock assessment and management. It is a beginning quantitative text, designed for "how to do" courses. It is what we would tell you about fisheries management and stock assessment if we had the opportunity to do so.

We cover a broad range of subjects, a few in detail. The book is divided into four sections; (*I*) Introduction, (*II*) Behavior of Fisheries, (*III*) Estimation of Parameters and (*IV*) Managing Fisheries. Most books and manuals deal almost exclusively with the material in section *III*. To us, section *II* is the most important in the book since it gives our view of what fisheries is really about, fish and fishermen and how they behave. Section *IV* also contains much information not normally associated with stock assessment books and manuals. Our chapters on harvest strategies and tactics (Chapter 15), optimization methods (Chapter 16), and adaptive management (Chapter 17) provide our perspective on quantitative methods in formulation and analysis of fisheries options under uncertainty.

In selected areas, particularly estimation of the parameters of stock and recruitment models and biomass dynamics (surplus production) models, we go into considerably more depth than any existing text. To some extent this is a reflection of our own interest. However, we really want to allow readers to understand and do stock assessment, not just run packaged programs that someone else has written. Stock and recruitment and biomass dynamics models are the best places to really understand the various problems with fisheries dynamics models, and how to estimate their parameters.

This book is not an encyclopedic compilation of the literature on different methods as was Ricker's 1975 book, nor is it a full technical review of the state of the art methods now used in fisheries. We do not treat in great detail the refinements of virtual population analysis (VPA), statistical catch-at-age models, and stock-recruitment time filtering methods that have appeared in the last few years. Nor do we devote much effort to derivation and mathematical analysis of models.

Our goals are to 1) provide material sufficient for students and practitioners to know the basics of the different approaches, 2) know their weaknesses, and 3) be able to actually use them on a computer.

Mathematics and Statistics

Most of the sections of this book can be read by anyone with a working knowledge of high-school algebra. In a few places calculus is used, but we have generally concentrated on computer implementation of discrete time models and numerical optimization and estimation procedures. Computer programming will be a more useful skill than calculus. Some of the topics require matrix algebra; if you are not familiar with matrix algebra, you should learn enough of it to understand the notation. Previous courses in statistics will be very helpful, particularly a working understanding of regression and analysis of variance. Some sections may be a little hard to follow for those with little statistical training.

Computers

Anyone who knows us could not imagine us working without some computer programs. We believe that easy and frequent computation is essential for learning about fisheries stock assessment and management. Almost all modern stock assessment methods require computation; the era when you could just take 15 data points and a hand calculator and "do" a stock assessment is long past. While there are a few highly experienced people who can derive many insights just by "eyeballing" a set of data, these people have usually honed their intuition and experience by a lot of formal data analysis and computation.

Dealing realistically with *choice, dynamics and uncertainty* requires computation—quite frankly, if you are not comfortable writing computer programs and playing with numbers, you should not be interested in fisheries management!

This book has a floppy disk with data sets and programs. This disk is a high density (1.2 MB) PC-AT compatible that has a separate directory for each chapter of the book. There are no direct references in the book to programs on the disk, since this would require an even longer book, and this allows for post publication updating of the disk. You should first read the file "README" in the main directory of the disk. The following are features of the attached disk.

1. All programs are written in ASCII Microsoft QuickBasic. This is the language we have found easiest to use, it is inexpensive and can be read by anyone familiar with FORTRAN, PASCAL or C. It provides for complete structured programming, including subroutines, functions and libraries.
2. We provide compact programs that illustrate the essential elements of techniques, rather than large complex "packages" that do everything. We think of these programs as templates: they have a bare minimum of input and output, and present the essentials of the particular algorithm in question. We recommend that you build upon these templates, by adding your own input and output to create your own packages. Don't use someone else's package—build your own. We believe that if you cannot write your own computer program to do an analysis, that you should very likely not be doing the analysis because there is a good chance you will misinterpret the

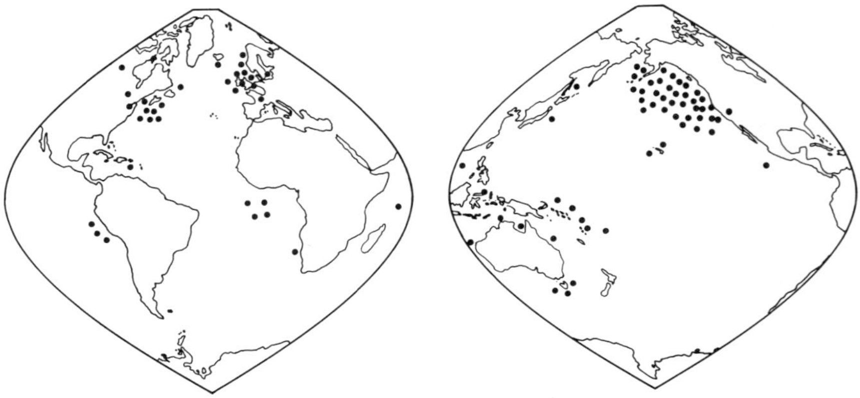

Geographic coverage in this book. Dot on the map represents an example we have used.

results. In some cases we have included packages just to show you what we have done for our own use, and to provide ideas on input and output.

3. As part of the disk, there is a directory called "SUBS" that contains subroutines to do many standard statistical, graphical and numerical tasks. These subroutines are used in many of our programs, and we encourage you to use them in developing your own.
4. The programs to generate most of the figures in the text, and the data sets used as examples, are included on the disk.
5. For most chapters there are problems, and answers or programs to solve the problems on the disk.
6. Another directory entitled "GAMES" includes several management simulation games (Hilborn and Walters 1987a) that we have developed to demonstrate different principles of fisheries management.
7. The contents of the disk are public domain and can be freely copied and used so long as the source of the programs is acknowledged.

Areas of Coverage

This book naturally reflects our experience, and the examples we use come primarily from where we have worked: the West Coast of Canada and the U.S., Australia, the South Pacific and New Zealand. The map above shows the geographic distribution of examples used in the text. While some examples come from all continents, Africa, South America and Asia are underrepresented. We do feel we have a good blend of temperate and tropical examples.

There are two topic areas that most readers will readily identify as underrepresented after a brief examination of the table of contents: tropical stock assessment and analysis of length data. We do not feel that tropical fisheries deserve special treatment: the major problems of tropical fisheries (low budgets and difficulty of determining the age of fish) are in fact typical of many temperate fisheries as well.

Analysis of length data poses special problems for us. It is a very fashionable topic at the moment, but we are very skeptical that much can be learned from length data even in principle (with the best of modern methods). It would take a full length book to review the methods of analysis of length data, and we obviously do not have room for that here. Further, we believe that not enough is known about the reliability most of the new length-based methods to write such a book at present.

There are four other books that we hope will be complementary to this one: they are John Gulland's 1988 version of *Fish Population Dynamics* which has many excellent chapters dealing with particular fisheries or selected topics; Colin Clark's 1985 book *Bioeconomic Modelling and Fisheries Management* which covers models of fish populations and fishermen in much more mathematical detail than possible in here; the 1984 Dahlem Conference Proceeding edited by Robert M. May, which has many wonderful chapters on particular problems and topics; and Ricker's classic 1975 compendium of assessment methods and literature.

The data sets and programs that had been included on a floppy disc with previous printings of this volume are now available on Chapman & Hall's FTP site (ftp://ftp.thomson.com/chapman&hall/hillborn/) for free downloading. The description of the available files in the Preface pertains to the FTP site files as well.

Acknowledgments

We owe our greatest debt to our friends in colleagues we have worked with in various stock assessment courses and projects, particularly at the University of British Columbia, the South Pacific Commission, the University of Washington, and several Australian State and Federal agencies. Don Ludwig, Jon Schnute and Rick Deriso deserve special thanks for the years of discussion and argument they have suffered at our hands.

Reviewing drafts of the chapters was an onerous task, and Ana Parma did a wonderful job, finding many errors and clarifying many sections. Saul Saila, Alejandro Anganuzzi, Miguel Pascual, John Caddy, and Lou Botsford also provided very helpful comments on the first draft of the text. Troy Buckley, Begona Campos, Steven Hare and Mike Ward helped proofread the final version of the manuscript. Marianne Johnson, Ulrike Hilborn, Claribelle Corinado-Herandez and Maxine Davis have all provided assistance in the preparation of the manuscript. Our wives, Ulrike and Sandra, have put up with the long and trying hours we spent putting this book together. The bulk of the final production of this book has been done while the first author was at the University of Washington, and the H. Mason Keeler Professorship has provided financial support. Stacy Waters provided frequent TEXnical assistance. Finally we thank Steven Wozniak and Steven Jobs for inventing and marketing the Apple II, which started it all.

Part I

Introduction

1

Role of Stock Assessment in Fisheries Management

1.1. What is Stock Assessment?

Stock assessment involves the use of various statistical and mathematical calculations to make quantitative predictions about the reactions of fish populations to alternative management choices. Two key words are critical in this thumbnail definition: *quantitative* and *choices*. The basic concern of stock assessment is to go beyond the obvious qualitative predictions that any student of nature should be able to make about natural limits to production, risks of overfishing spawning populations, the need to allow fish to grow to a reasonable size before they are harvested, and so forth. Furthermore, it does not make sense to engage in the risky and often embarrassing business of quantitative prediction in settings where there are no management choices to be made in the first place, except perhaps as an aid to scientific thinking and hypothesis formulation.

It is widely accepted that the fundamental purpose of fisheries management is to ensure sustainable production over time from fish stocks, preferably through regulatory and enhancement actions that promote economic and social well-being of the fishermen and industries that use the production. To achieve this purpose, management authorities must design, justify (politically), and administer (enforce) a collection of restraints on fishing activity. In some systems, the management authority is also empowered to restrain various other economic actors who might impact the ecological basis for fish production (water polluters, etc.), and to engage in activities to "enhance" the basis for production (habitat improvement, hatcheries, etc.).

Practically all management activities, from fishing restraints to habitat improvement, are done as matters of degree rather than as yes or no decisions. Management authorities must make very difficult and quantitative choices about how much development of fishing to encourage or permit, what specific limits to place on catches (times of fishing, sizes of fish, total landings, locations of fishing), how much financial resource to spend on enforcement of regulations versus enhancement of production, and so forth.

The distinction between stock assessment and management

Too often the term stock assessment is synonymous with biological advice about harvest levels. In this view of the world, the biologists assess the status and potential production of the stock and make recommendations about catch levels, efforts, and so on. Any modification of regulations by politicians or fishermen is considered interference in the rightful mission of biologists to determine appropriate management actions.

Such a view fails to recognize the distinction between assessment of biological potential and the decision about how to manage the stock. Once the stock assessment is complete, choice remains. As we see in later chapters, the same biological yield is often possible over a very wide range of fishing efforts. Stock-assessment biologists are not the appropriate people to make such decisions. Similarly, fisheries management decisions often trade off between average yield and variability of yield. The stock assessment should provide estimates of the nature of the tradeoff, but the choice should be made on social and economic grounds. Some of the chapters of this book, particularly the third section, deal with the tools of stock assessment; the fourth section with the tools of decision making. Both are an integral part of fisheries management, but we should not fail to distinguish between them.

Alternative modes of stock assessment

A management authority can go about the difficult business of making choices among quantitative alternatives in three ways. First, it may simply mimic choices made under similar circumstances by other authorities under the assumption that previous decision making has already involved careful evaluation of alternatives. Second, it may make an initial choice that "looks reasonable" on intuitive grounds, then plan to systematically vary the choice while monitoring biological and economic responses, so as to eventually find the best choice by an empirical process of trial and error. Third, it may engage in formal stock assessment, the construction of quantitative models to make the best predictions possible about alternative choices based on whatever data are available to date, and then base its choices on the models while expecting to refine or modify the choices later as more data become available. A combination of the second and third approaches, using a mixture of quantitative modelling and empirical management experimentation, has come to be called "adaptive management" (Walters and Hilborn 1976, Walters 1986).

The simplest way to think about the role of formal stock assessment in fisheries management is as a means to move beyond mimetic or "seat of the pants" policy making, by providing at least some structured use of available data in comparing choices. This is not the same as making the ambitious (and naive) claim that the role of stock assessment is to find the "best possible" or the "optimum" choice of such variables as catch quotas. The choice of policy options is normally, and rightfully, a political, social, and economic decision. The role of stock assessment is to provide the best possible technical support to these decisions. It is reasonable to expect that careful data analysis, using models whose assumptions are clearly specified, will in general result in technical support that is closer to the mark than subjective or intuitive guesses that may involve even worse hidden (subconscious) assumptions.

The choice in fisheries management is not really whether to do stock assessment, but whether to do it well. In the words of John Gulland (1983), perhaps the most broadly experienced stock assessment expert in the world,

> "All those concerned with making policy decisions about fisheries must take into account, to a greater or lesser extent, the condition of the fish stocks and the effect on these stocks of the actions being contemplated."

Any decision choice will somehow "take into account," or make some assumption about, the stock dynamics; thus, in a sense, any choice will necessarily be based on some predictive model, whether this model is explicitly stated or not. Another way to think about the role of stock assessment is as a means to force clear and explicit recognition of what model is being used as the basis for choice. Such clear recognition is a key step toward learning to do better over time.

Where careful assessments are not made, due to distrust of models and available data or to lack of expertise in assessment techniques, management authorities often grasp at (or are forced to use through political pressures) estimates provided by advisors to industry interests. In other words, if you do not provide some answers for policy makers, someone else will. There are always "experts" available to provide whatever estimates will suit the purposes of people who want greater immediate access to fisheries. In the political arenas, where key regulatory decisions and limits are made, debate often focuses on simple summary statistics and simplistic (common sense) models of fish responses; in such arenas, it is common to use deliberate disinformation as a tactic to confuse debate and delay effective action. In the "good old days", before this tactic became widely known, there was a tendency to accept scientific assessment advice uncritically. Today a scientist who tries to wing it (by providing numbers that are not justified through clear, precise, and credible calculations) is liable to find himself either ignored or grossly embarrassed.

The breadth of stock assessment

Stock assessment is sometimes viewed as a rather narrow biological discipline that might be summarized as "the interpretation of commercial catch statistics to estimate potential yields." A well done stock assessment is much more than this. First and foremost, stock assessment involves understanding the *dynamics of fisheries*. This recognizes that fisheries are dynamic entities that will respond over time to management regulations, and to extrinsic factors. Modern stock assessment is not just the task of making static predictions about equilibrium sustainable yields. It should also involve making predictions about the time trends expected in response to policy change and about how policies should be structured in order to deal with the unpredictable changes that will inevitably occur.

Fisheries are also much more than fish catch. Fishermen are an important component of the dynamic system we call a fishery, and stock assessment must take into account how fishermen will respond, and also make predictions about things important to fishermen such as catch per unit effort. Fisheries are considered in crisis when income to fishermen drops below some acceptable level, and we believe that making predictions about how catch per unit effort (and therefore fishermen's income) will change is more important than predicting changes in total catch.

Processing and marketing are often very important components of the fishery system. Recognizing that fishermen are the political center of most fisheries and that returns to fishermen (in money) are the key measure of management success, one simply cannot ignore processing and marketing. For instance, for some biological models, one can show that the average biological yield will be maximized by holding the fish stock at a constant level. This will normally result in rather high year to year variability in catch, often with absolutely no catch in some years. Such variability is quite destructive of processing infrastructure and the maintenance of markets—yet many biologists have held that the "optimum" way to manage these fisheries is to maintain constant stock biomass.

1.2. The Changing Role of Assessment in Fisheries Development

Commercial fisheries are not static systems that can be manipulated and reshaped at will by management. They usually develop initially through a dynamic process that involves several distinct stages. One generalized diagram of these stages is shown in Figure 1.1. First, there is discovery and spread of information about the existence of a potentially valuable stock. This is labeled predevelopment. Second, there follows a period of rapid growth of effort attracted by the success of initial fishermen. Next the fishery reaches full development, where yields are near or perhaps a little above a long-term sustainable level. The rapid development results in declining rates of fishing success as the stock is reduced and more fishermen compete for the remaining fish until either some limit on fishing pressure is established

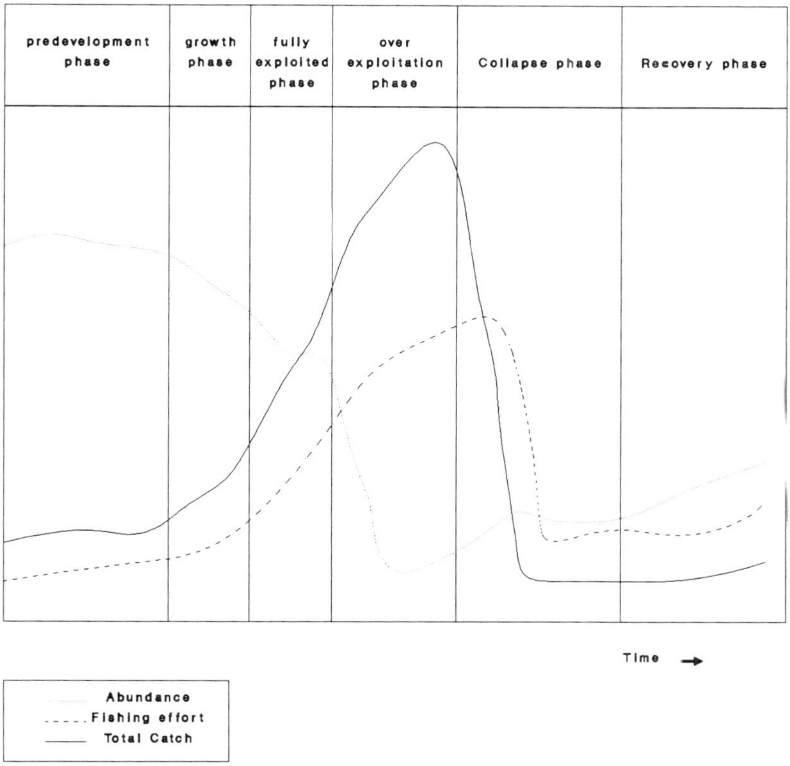

Figure 1.1. Phases of development of uncontrolled fisheries. Redrawn from Csirke and Sharp (1984).

through management or success rates become too low to attract more pressure. The fishery often then enters an overexploitation stage, which is followed by a collapse. If the collapse is not too catastrophic, there is often a period of declining fishing pressure as the less successful fishermen find it no longer worthwhile to pursue the stock. The stock may or may not recover somewhat on its own during this period. On a longer time scale, there may be an evolutionary rhythm in which occasional technological innovations result in increased fishing success and attraction of more fishing pressure and hence a repetition of stages three and four of the initial development, unless fishing effort is carefully managed through each technological transition.

The extent to which the collapse is severe, or the fishery does not collapse at all will depend on the price of the fish product, the delays in investment, the extent to which fishing success declines as abundance declines, and whether regulatory agencies act to reduce effort or catch before a collapse occurs. Figure 1.1 represents only one possible trajectory.

The role of stock assessment is different in each of the stages of fishery development. At first, assessment is critical in setting basic expectations and limits for development and in designing monitoring programs to provide estimates of key population parameters. Later, it can play an important role in "fine tuning" the fishing system for higher yields, in developing plans for stock rehabilitation in cases where the initial development results in overfishing, and in developing strategies for management during technological transitions to more efficient fishing methods.

Setting expectations early in development

Consider the situation management faces when a new fishery begins to develop. Someone (usually entrepreneurial fishermen) has discovered that there is a stock worth pursuing and is beginning to make a profit (or finding unusually good recreational action). Word is beginning to spread about the opportunity, attracting the interest of other fishermen and/or agencies concerned with the promotion of fishery development. It appears that fishing effort and related investment (for example, in processing plants, marketing arrangements, and tourist facilities) is likely to increase rapidly. There is no history of experience with the stock, so little is known about its distribution, total abundance, and productivity.

The most important management (and assessment) question at this point is obvious: what level of fishing pressure should be permitted (or encouraged, or subsidized) as an initial development target, recognizing the risk of eventual economic overcapitalization and/or biological overfishing? On a sustainable basis, is the stock eventually likely to support 10 fishermen, or 100, or 1,000? Notice that at this point in time, even an order-of-magnitude assessment will be of considerable value in development and regulatory planning; there will be time later to obtain more precise assessments, provided the initial development proceeds within reasonable bounds.

Beyond providing rough initial estimates of the stock distribution, stock size, and productivity, an important role of stock assessment early in fishery development is to help define key monitoring requirements that will permit more precise assessments later in the development. It is often not recognized by fishery biologists that even after many years of fishery data are available, some key assessment calculations will still depend heavily on data that can only be gathered early in the fishery's development. For example, estimation of potential yields, optimum sizes of fish to harvest, and most refined methods for stock size estimation all depend on having a reasonable estimate of the average natural mortality rate. Usually the only time this rate can be measured is early in fishery development, when it is possible to examine the relative abundance of animals of different ages before the population age structure is much distorted by fishing. Later in development, the age structure carries information only about the combined effects of fishing and natural mortality. There is no reliable way to separate the combined effects (and hence measure the mortality rate due to fishing) unless the early data are available for comparison, the fishing mortality rate can be measured directly, or the fishing mortality changes greatly. Estimating fishing mortality directly is difficult because you would need to know the catch and total stock size directly or have an unusually good tagging program.

These considerations apply primarily to large single unit stocks. If the fish resource consists of many spatially isolated subunits, then one can (in theory) overexploit some, while underexploiting others. Unfortunately, the normal tendency of fishing fleets is to make fishing pressure uniform over all subunits.

Two biological characteristics of the fish species can provide very useful information very early in development. Knowledge about the longevity of the species can provide useful guidelines about the potential for large unsustainable yields. Fish that live a long time normally provide very large yields at the beginning of a fishery because of fishing down of the older age classes. Secondly, knowledge of the behavior of the species can provide useful information about the danger of severe stock collapse. Pelagic schooling fish are well known to collapse because they re-form in schools that provide easy targets for fishing nets. Species that do not form large aggregations are normally more resistant to dramatic collapse.

Improving information as development proceeds

Increasing catches during fishery development are generally accompanied by (1) a decrease in fish density in areas where fishing effort is initially concentrated, (2) movement of effort into less preferred fishing areas, and eventually (3) a decrease in indices of fishing success, such as catch per unit of fishing effort (CPUE). Changes in catch, CPUE, and other indices of fishing impact (e.g., age composition, average fish size) can be used to provide estimates of the stock size, and to measure rates of "surplus" production (excess of reproduction and growth over natural mortality) as related to the changing stock size. Pelagic schooling fish typically do not show declining catch rates as development proceeds, because the schooling assures that fish are found in high densities even when total abundance is greatly reduced. Such fisheries are very difficult to manage because the fishermen and managers get little sense that the stock is declining unless careful analysis of the spatial structure of the stock is made.

A key role of stock assessment during development is to provide regular updating and "feedback" of population parameters and estimated potentials into the management decision process. In particular, systematic and regular assessments may provide good early warnings of overfishing and help to prevent severe overcapitalization of the fishing industry.

It is important to understand that there is no substitute for the experience gained during development for estimating rates of surplus production and potential yields. The surplus of reproduction and growth over mortality that a given stock can produce is a result of the particular quantitative values that several key rates assume in the particular environment where the stock resides. These rates can differ drastically from stock to stock, even within a single species over a narrow geographic range. Rough bounds on productivity can be established from general biological information about the species (growth, longevity, etc.), but there are no precise quantitative "laws" or principles that might be used as predictive substitutes for the role of stock assessment in helping to "learn as you go."

One of the fundamental tenets of traditional fisheries theory is that there is a repeatable relationship between fishing effort and average catch as shown in Figure 1.2 (a similar figure is the first figure to appear in Gulland 1983 and in Clark 1988). Yield increases as fishing effort increases up to some point, at which point yield begins to decline with further increases in fishing effort. There may be year-to-year fluctuations about the average relationship, but the following must be true: (1) in the absence of fishing effort there will be no catch, (2) at very high levels of fishing effort the stock will be fished to such low numbers that the remaining fish will not be able to produce a large surplus, and (3) the maximum average yield is therefore somewhere in between no effort and very high effort. If the fishing gear is inefficient it may not be possible to fish the stock very hard, but in general we accept the above three propositions as a starting point in the analysis of most populations.

Unfortunately, the apparently obvious nature of Figure 1.2 has pointed stock assessment in the wrong direction. Once we accept Figure 1.2, it is easy to believe that the purpose of stock assessment is to estimate the level of fishing effort that will generate the maximum average yield and to estimate what that average yield will be. Indeed, a discouragingly large proportion of fisheries stock assessments concentrate on precisely and only these two questions: what is the optimum effort and what is the maximum sustainable yield (MSY)? *These are the wrong questions.*

A simple minded view of stock assessment as development proceeds is that the thing to do is to monitor fishing effort as it increases slowly and gradually, while monitoring yields so as to eventually be able to plot a relationship like Figure 1.2. According to this view, once the fishery has reached the top of the curve (and yield begins to drop), you know you have found MSY. What could be simpler! Unfortunately, quite a few things are simpler. You cannot find the top of a curve as in Figure 1.2 without going beyond

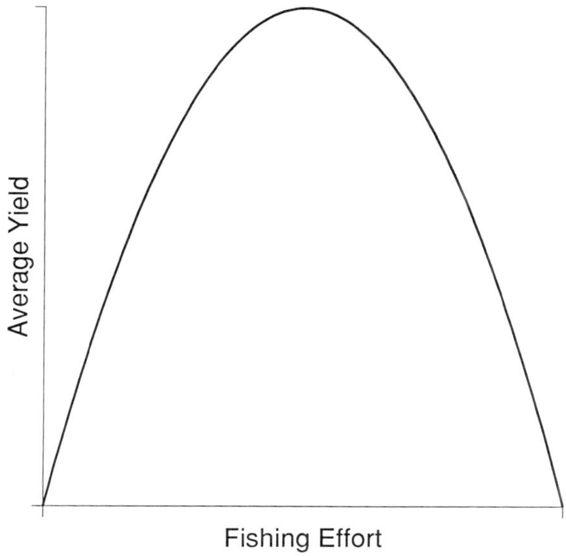

Figure 1.2. The assumed relationship between fishing effort and average catch. Perhaps the most commonly printed illustration in fisheries textbooks, and the most dangerous.

the top. The noisier the data are, the farther beyond the top you have to go before you are sure that you have actually found it.

Figure 1.3 shows the yield-effort relationship for yellowfin tuna (*Thunnus albacares*) in the eastern Atlantic Ocean, from Hunter et al. (1986). In 1975, the International Commission for the Conservation of Atlantic Tunas (ICCAT) analyzed the data from 1964 to 1973 estimated the sustainable yield as approximately 50,000 tons, and estimated the optimum effort as approximately 60,000 fishing days. They believed they had reached the top of the yield-effort curve.

However, ICCAT was politically unable to constrain effort at the "optimum," the fishing effort continued to increase, and by 1983 the yield was over 100,000 tons. It was clear that the 1975 estimate of 50,000 tons was a "false summit," and that the real top was at higher efforts and higher yields. The second curve fit in Figure 1.3 gives the 1985 estimate of the optimum effort and maximum sustainable yield. The key question is "Did they find the top in 1985?"

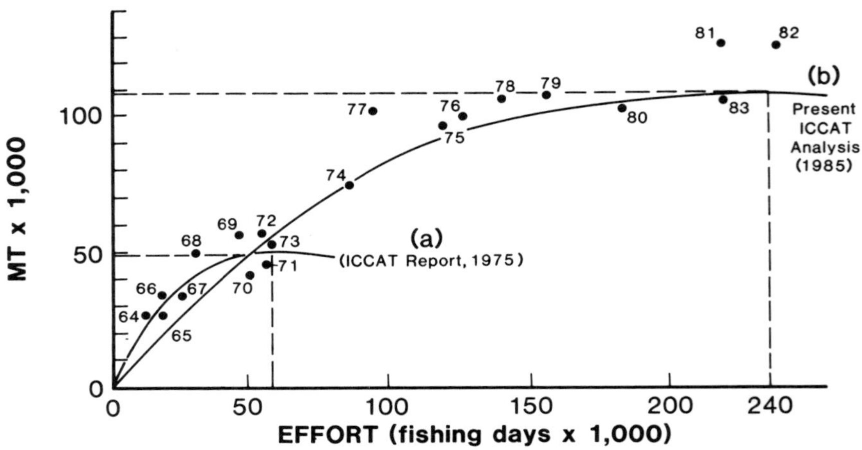

Figure 1.3. Yield-effort relationship for eastern Atlantic yellowfin tuna. From Hunter et al. 1986. Reproduced by permission of the Food and Agriculture Organization of the United Nations.

Principle: You cannot determine the potential yield from a fish stock without overexploiting it.

The yellowfin example illustrates the problem of finding the top: you simply cannot do it without going past it, and perhaps quite a bit past it. In subsequent chapters we discuss some of the biological and economic reasons why you must exceed the maximum considerably. For our current purposes it is sufficient to accept that one must go well past the top to find the best effort and MSY. This means that once the MSY and optimum effort have been found, the next step is to reduce the fishing effort down to the optimum.

Principle: The hardest thing to do in fisheries management is reduce fishing pressure.

Reducing fishing effort is the hardest thing to do in fisheries! It involves either driving fishermen from the fishery or reducing everyone's catch. Neither of these is politically practical or socially acceptable. By the time the optimum has been found, the catch per unit effort is quite a bit lower than it was when the fishery started. Note that the 1975 ICCAT optimum catch was roughly 1 ton per day; at the 1985 "optimum" the catch per day was less than 0.5 ton. This illustrates the point that by the time you know fishing pressure should be reduced, the fishermen's returns per day are low. In fact, by then fishermen are normally just breaking even. Remember that we still do not know if the ICCAT 1985 level is the true optimum, and we need to increase effort another 20% to 30% to make sure!

If we follow the simple stock assessment prescription of gradually increasing effort until we detect we have passed the top of the yield-effort curve, we must then somehow reduce fishing pressure by 20% to 30% at a time when fishermen are in tight financial conditions. *This is a prescription for disaster*.

Thus, we argue that you cannot predict the MSY at early stages of development and that, once you have found it, it may be too late to do much good. It would be very nice if we could predict MSY and optimum effort prior to reaching the top, but this is an unobtainable dream. Rather, we must concentrate stock assessment efforts during the development phase on assuring that we (1) detect the top as rapidly as possible and (2) build mechanisms into the fishery so that it will be possible to reduce effort when the time is necessary. These mechanisms can include biological tactics, such as setting aside spatial refuges that are unfished during development, as well as economic tactics such as imposing high taxation rates that can be reduced later to compensate for lowered catches.

Again, these difficulties are much less severe if the fish resource consists of discrete subunits and there is contrasting fishing effort in the subunits. As a general rule, the more independent spatial units are available, the easier is the assessment.

Fine tuning and rehabilitation planning in ''fully developed'' fisheries

The world fisheries catch went through a period of remarkable (6% per annum) growth during the 1950s and 1960s. With the collapse of the Peruvian anchoveta (*Engraulis ringens*) fishery in the early 1970s the growth flattened out, but by the 1980s it had begun to rise again (Figure 1.4). Some major fisheries that sustained the growth spurt in the 1950s and 1960s, such as Peru's anchoveta fishery, have subsequently collapsed. Although there is some potential to develop new fisheries in offshore locations (krill in the Antarctic, deep water pelagic fishes), the main opportunities now for increasing fishery yields around the world are through "fine tuning" regulations on stocks that are fully exploited, and through rehabilitation (rebuilding) programs for stocks that have been overexploited. Stock assessment will play a key role in defining these opportunities and in planning how to make use of them.

The alternative to better management of stocks is the high technology world of aquaculture and genetic engineering. Aquaculture is undoubtedly going to continue to grow; however, it should not be viewed as a substitute for good management of natural resources.

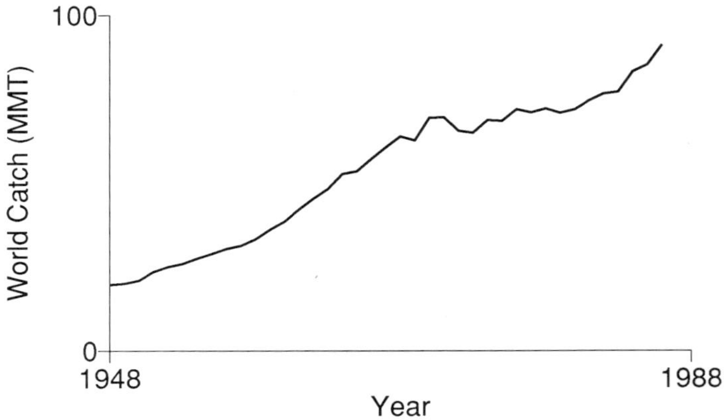

Figure 1.4. History of world fisheries catches. Data from Fisheries and Agricultural Organization of the United Nations FAO *Yearbook of Fishery Statistics* (1980–1985).

Many fisheries around the world have been stabilized or limited by regulatory policies that were developed piecemeal and without good data as the fisheries grew. In many cases the current, relatively stable regime is likely to be sustainable, but is far from optimum in terms of regulations on where, when, and how the fish are taken. In particular, many fisheries focus on fish that would produce higher yields if they were allowed to grow more (and often, move offshore). Other fisheries use gear that is wasteful of the target species (kills adults or juveniles without capturing all of the kill) or of other species that are taken as "by-catch" or "discards." The role of stock assessment in such cases is to provide a coherent framework of calculations for putting together data on fish growth, movement, mortality, and vulnerability to fishing. This framework of calculations (the assessment model) can then be used to systematically search for better policy options.

Perhaps 60% of the world's major fish stocks are now overexploited, in the sense that stock sizes have been driven to lower levels than would produce the largest annual biological surplus or net economic value. The 60% is a very rough guess. Reviews we have seen of the status of stocks in Canada, the U.S., Australia and New Zealand indicate few if any underexploited stocks and a large proportion of stocks thought to be overexploited. Rebuilding these stocks to more productive levels will involve difficult choices to reduce catches in the short term, so as to produce more in the long term. A key role of stock assessment is to quantify the choices as precisely as possible: how long will rebuilding take if only a small catch reduction is acceptable, and how much can this painful time for fishermen be shortened if a larger reduction is accepted? Stock rebuilding programs will only be politically "saleable" if these questions can be answered with calculations that are credible to all parties involved in decision making.

Some fisheries agencies are attempting to buy their way out of overfishing situations without reducing harvest rates, by instead trying to artificially enhance the productivity of stocks through technologies such as fish hatcheries and provision of artificial habitats. A key role of stock assessment in these situations is to help measure whether the production enhancement is in fact working as planned and whether it is having any deleterious side effects on those parts of the stock (and other stocks) that are not directly enhanced. The assessment work may well discover for example that the enhancement activities are successful enough to stimulate increased fishing pressure, which then compounds the overfishing problem on unenhanced portions of the stock so as to cause a further loss in natural production. That further loss in natural production may more than make up for any gains from enhanced parts of the stock. An example of this happening is the fishery for large and highly prized chinook salmon *(Oncorhynchus tshawytscha)* in the North Pacific Ocean (Figure 1.5). American and Canadian fisheries agencies tried to keep up with growing demand for chinooks by releasing more fish from hatcheries, and at first they were very successful. But the natural stocks kept declining, even faster in some areas, and in recent years the catch has declined dramatically. There is now a major United States/Canada treaty initiative to rebuild the chinook stocks, and fisheries have been cut back in many areas of the Pacific coast.

1.3. Stock Assessment and Cooperative Management

A traditional view of stock assessment is as a collection of analyses aimed at estimating stock size and productivity from statistics gathered from the commercial (or sport) fishing process. In this view, the fishery is seen as the primary and most economic sampling device for getting information about the stock. Typically total catch statistics are gathered, along with measures of fishing effort (the area or volume of water searched by fishermen in their "sampling"); the catch per effort is then often assumed to be proportional to the actual stock size. In some fisheries, catch and effort statistics are recorded in detailed logbooks to permit spatial mapping of relative fish abundance, at least over whatever area is fished enough to give reasonable sample sizes. The catch is usually sampled for other characteristics of the fish, such as size and age composition.

Unfortunately, the pursuit of fish for sport or economic gain usually results in a highly unrandom and nonrepresentative sampling pattern in time, space, and characteristics of fish sampled. For instance, fishermen can often concentrate their effort on areas of relatively high fish density, so that catch per effort can remain high over time, even while the total stock size is re-

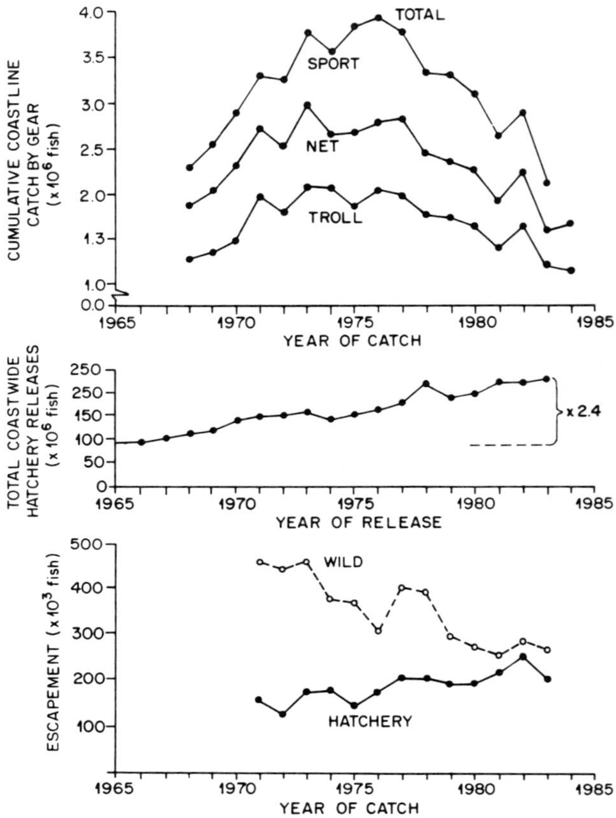

Figure 1.5. Chinook salmon catches, hatchery releases, and relative abundance of hatchery and natural spawners in the Northeastern Pacific. Is hatchery production helping to sustain the fishery? From Walters and Riddell (1986).

duced substantially (e.g., as fewer areas of high density are left). Alternatively, catch per effort can decline rapidly as a few local concentrations of fish are depleted, but with little change in the total stock size if only a small percentage of the fish are in the local concentrations. The net effect of ignoring such distortions in stock assessment calculations can be absolutely disastrous: gross overestimates of stock size and potential yield, missed opportunities for fishery expansion, and wasteful regulations on the sizes of fish to be harvested and the times and places of harvest.

An important role of stock assessment is to help identify whether catch and effort statistics are likely to give a misleading picture of stock trends and health, and hence whether some more systematic and expensive sampling program may be worthwhile. The design of better sampling programs involves challenges that go beyond where to locate sampling stations and what to measure; a key issue becomes who should gather the samples and under what economic incentive system.

Recognition of dangers in using commercial and sport sampling data alone has led many agencies to invest in research sampling programs involving agency-owned or chartered fishing vessels. These programs are generally very expensive to develop and maintain, and a single research vessel can collect only a tiny number of samples compared to a whole commercial fishing fleet.

An alternative to agency research vessels or direct charters is to provide incentives for fishermen to work cooperatively with the management agency by spending part of their time "fishing for information," using standardized fishing procedures and gear on informative sampling locations (transects, grids, preset stations). Such methods have been used in Australia. An obvious incentive for such cooperation is to allow increased catches by cooperating fishermen; stock assessment techniques can help decide how large a catch increase would be "safe" or worth accepting in terms of the value of improved information. Stock assessment may also help to define other incentives for cooperation, such as preferential access to particularly lucrative times and places of fishing.

For many fisheries, it will probably never be worthwhile to engage in refined data collection, stock assessment, and regulation unless some type of cooperative data-gathering system can be established with the fishermen. In such cases, stock assessment will either continue to operate in a twilight of potentially misleading data from normal fishing activity or else take a leading role in defining imaginative schemes to make cooperation worthwhile for both fishermen and the management agency.

Even where very accurate sampling and survey programs have provided a good picture of past changes and the current status of a stock, it is often not possible to predict how that stock will respond to new management initiatives (or further development of fishing pressure) whose effects have never been seen before. A classic role of stock assessment has been to provide some reasonable extrapolation (or best prediction based on available information) about such circumstances. In providing such extrapolations, it is easy to overlook the fact that even perfect data on how a complex system has behaved over a limited range of historical circumstances can be misleading about how it will behave outside that limited range, even when a very reasonable biological model is available. A wiser role for stock assessment when faced with policy questions that involve gross extrapolations would be to provide assistance in the design of management experiments to test the extrapolations in a relatively safe and economically productive man-

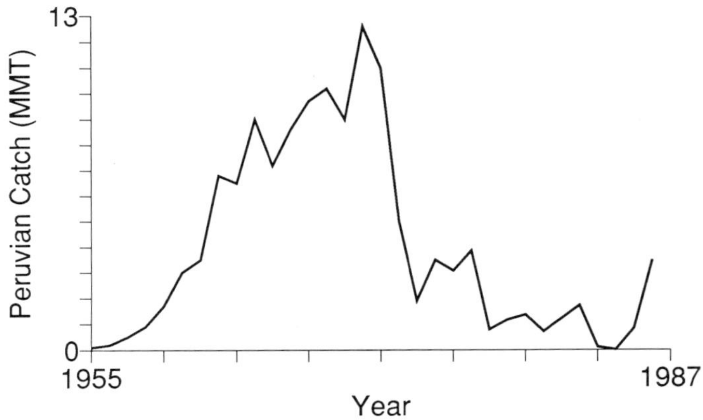

Figure 1.6. Catch history of Peru's anchoveta fishery. Data from FAO Yearbook of Fisheries Statistics (1980–1985).

ner. Such experiments would be especially powerful tools for improving management if used in conjunction with cooperative data-gathering programs involving fishermen (for an example of that approach, see Walters and Collie 1989).

1.4. Two Great Failures in Fisheries Management

To illustrate some of the points made earlier in this chapter, let us briefly examine the history of two of the most infamous fisheries, the Peruvian anchoveta fishery and the North Sea herring fishery.

The Peruvian anchoveta fishery

Figure 1.6 shows the catch history of Peru's fishery for anchoveta. The fishery became a major producer of fish meal in the early 1950s, stimulated somewhat by the collapse of the California sardine (*Sardinops sagax*) fishery and the movement of processing and harvesting equipment from California to Peru. By the mid 1960s, the anchoveta fishery was the largest fishery in the world. However, in the mid 1960s there was a dip in catch associated with an *El Niño* event, and a few biologists began to issue warnings about the future of the fishery. By the late 1960s, there was more widespread concern about the potential for overfishing, and numerous overseas experts were called in to assess the stock. As usual, the experts could not agree, and produced numerous estimates of sustainable yield ranging from 7 million to over 10 million tons per year. A consensus of sorts was established at a level of 9.5 million tons. However, the economic pressures for growth were very strong, and the fishery continued to catch more than what all but the most optimistic of biologists felt was sustainable.

These arguments became somewhat academic in 1972–1973, when a second *El Niño* oceanic condition (warm surface water, reduced upwelling) apparently had two effects: it initially concentrated the fish close inshore where they were highly vulnerable to the fishing boats, and then it caused poor juvenile survival for the offspring of the remaining spawners. The net result was a general recruitment failure. All the biologists agreed that a major if not total reduction in fishing was necessary to let the stock rebuild. The Peruvian government was unable or unwilling to reduce fishing pressure, and high exploitation rates continued for several years until the stock was reduced to such low levels that it became economically unimportant. The stock apparently started to recover in the late 1970s, but was hit again by a strong *El Niño* event in 1982–1983. There remains no general agreement about the relative importance of the *El Niño* events and continued exploitation as causes of collapse in this fishery, but almost everyone agrees that both are to some extent responsible.

For the purposes of this book, we want to ask "What can we learn about the role of stock assessment from this example?" First and foremost, the stock assessment work prior to the 1972–1983 *El Niño* concentrated almost exclusively on trying to predict the MSY; the government of Peru wanted to know what level of harvest was sustainable, and the stock assessment experts tried to provide a number. This is probably the greatest failing of fisheries scientists. When decision makers ask the wrong question, try to convince them to ask a better question instead of providing them a silly answer that will eventually lead them even further astray. In the 1960s, it was widely recognized that species such as the Peruvian anchoveta are prone to major fluctuations; after all, the California sardine had recently collapsed. At the time, it was felt that if we just kept the catches low enough it might be possible to avoid such collapses.

The stock assessment biologists should have emphasized more forcefully that they could not predict the sustainable yield with any reliability (perhaps plus or minus 50%). They should have told the Peruvian government that the time would come when the stock would decline, they should have insisted on helping to work out a management plan for that contingency.

It is easy to make these pronouncements in retrospect; in the late 1960s, fisheries scientists did not have as much appreciation of the natural variability of fish stocks as we do today, nor had they seen enough cases of collapse to issue any warning with great confidence. However, we no longer have those excuses today. The role of stock assessment is not to make best guesses at MSY, but rather to help design a fishery management system that can respond to the types of variability we see in nature.

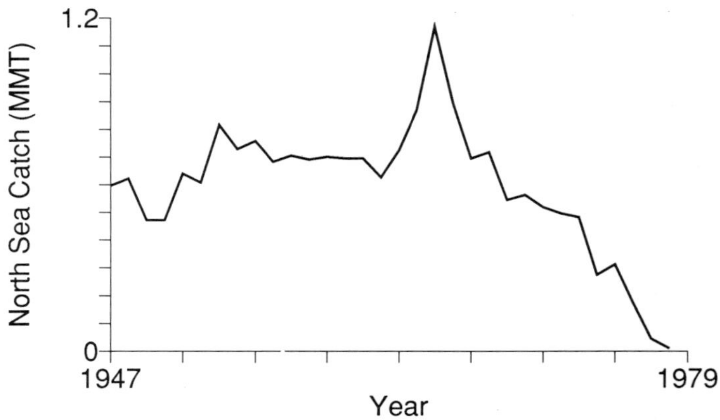

Figure 1.7. Catch history of North Sea herring. Data from Saville and Bailey (1980).

North Sea herring

Another great disaster in fishery management involved the stocks of herring in the North Sea (*Clupea harengus*). Figure 1.7 shows the catches from 1947 to 1978. The fishery had maintained yields of between 300,000 and 1,000,000 tons between 1903 and 1965, but beginning in the late 1960s the fishery began a serious decline.

As the stocks declined, biologists started to recognize that overfishing was occurring, but this recognition was slow in coming. After all, the stocks had a history of substantial natural variation; who would want to raise an outcry when a natural recovery might be just around the corner? Saville and Bailey (1980) said, "the advice given to the regulatory bodies on management . . . has tended to be much too optimistic in respect of total allowable catches." In the herring case, the overoptimism was due at least in part to the use of virtual population analysis (VPA), which depends upon estimates of current fishing mortality rates to provide a "reconstruction" of historical stock trends. As it turned out, estimates of fishing mortality for North Sea herring were consistently too low, leading to overestimates of stock biomass and therefore overoptimism about the current state of the stock.

The North Sea herring experience, along with a number of other fisheries with similar problems in using VPA, have provided fisheries scientists with quite a bit more skepticism about our ability to detect stock trends, as they develop, by the analysis of catch data. The blame for the collapse of the North Sea herring does not fall solely on the stock assessment biologists. They were slow in detecting the problem, but, once it was detected, the political decision makers were even slower in heeding their advice. Saetersdal (1980) provides a review of the management advice offered and the actual management taken. By 1970, the stock assessment biologists were recommending major cuts in fishing pressure, and, by 1974, the managers had agreed to a total quota of 494,000 tons (which was larger than the total stock at the time). However, it was not until 1977 that effective regulation was enforced. Between the biological recommendation for reduction in fishing pressure in 1970 and the implementation of effective regulation in 1977, the stocks had dropped to less than 200,000 tons.

The North Sea herring fishery illustrates two of the major problems in fisheries stock assessment. Biologically, we cannot usually detect overfishing until it has already become quite severe. This is the principle cited earlier. The North Sea herring illustrates how the point remains valid even when quite detailed data and assessment procedures are used instead of just catch and effort data. Socially and economically, it is difficult to effectively reduce fishing pressure even after we recognize that such reduction is necessary.

1.5. Summary and Critique

If fisheries science is to be successful we must learn from and avoid the mistakes of the past. We must recognize that stock assessment involves understanding and making predictions about the response of fishery systems to alternative management actions. We must help managers make *choices* about *dynamic* fishery systems in the face of *uncertainty*.

This is a difficult task. However, we are fortunate that our predecessors have made lots of mistakes, and in many cases they have documented these mistakes, so that we can learn from them. We cannot avoid making mistakes, but we do our predecessors a great disservice if we do not take advantage of what has been learned and try not to repeat the same mistakes.

In the context of this chapter, building on historical experience involves a recognition that stock assessment does not consist of making static predictions about optimum efforts and sustainable yields, but concerns the assessment of time trajectories of fish and fishermen in response to management and other changes. It also involves a recognition that stock assessment biologists must educate managers and decision makers to ask appropriate questions and to think of the dynamic response of fisheries to change.

Bibliographic Notes

This chapter has provided our view of the role of stock assessment. Other views can be found in Gulland (1974, 1983), McHugh (1984), Rothschild (1986), and Royce (1987). Details of the Peruvian anchoveta fishery can be found in Pauly and Tsukayama (1987) and Glantz (1979). Saville and Bailey (1980) and Saetersdal (1980) discuss the North Sea herring.

2

Objectives of Fisheries Management

2.1. Why Objectives?

The fisheries of the world are managed for the benefit of man. Certainly some fish are protected for reasons of conservation, but by definition a fishery means harvesting fish. With the exception of overzealous government press agents, few people argue that fisheries take place for the good of the fish.

Given that we are managing a fishery for the benefit of mankind, how do we evaluate how well we are doing? To answer this question, we must first endeavor to define what we really want to achieve with management — we must be clear about what we mean by the benefit of mankind.

This discussion is about fisheries management and not fisheries biology. While a fisheries biologist may bury his head in the sand and ignore management objectives, fishery managers cannot ignore the reason for their existence. All fisheries management must at some point decide which of several alternative management decisions will cause the most desirable outcome. There are two stages to such decisions. First, for each possible action, possible outcomes and their probabilities must be assessed. This is the traditional realm of stock assessment and modelling. We estimate that 99.9% of effort by fisheries management staff is devoted to such evaluation activities.

The second stage is to decide which of the possible actions is best, in recognition of the outcomes that it may produce. Usually, a decision maker is presented with possible actions and outcomes and asked to exercise his craft without any precise guidance about what criteria to use in weighing the alternatives. This occasionally works, but the method is far from perfect. If the analyst is ignorant of the decision maker's objectives, he is not likely to present the full information required for a wise decision. Let us assume that two harvesting plans are being considered, for each of which which the biologist might present his prediction of average catch. Higher average catch will also be associated with higher variability, and it may well be that high variation in catch is undesirable to the decision maker or the users. Yet it may never occur to a biologist to present the variance as well as the average of catch to the decision maker.

Moreover, if an analyst does not know the decision maker's objectives, he may not even consider management options that might be very useful. For example, few biologists would even consider rapidly "mining" a very unproductive stock down to low levels, and banking the proceeds to earn interest at a higher rate than the stock could produce if left at high levels. Yet in some cases, there may be compelling social or economic reasons to adopt such an extreme policy. Unless the biologist understands these reasons well enough to go to the decision maker with sound counterarguments, the biologist invites having the advice ignored. Going to the decision maker with eyes open to the possibility of rapidly depleting the stock, the biologist might at least devise a compromise strategy that provides for high immediate returns but with monitoring and shutdown provisions that insure long run safety for the stock.

Fisheries management objectives have received only modest attention in the literature. Gulland (1983) devotes 1 page to the topic. Alverson and Paulik (1973) provide a good review of some of the issues. The traditional biological assumption of maximum sustained yield (MSY) has been repeatedly challenged and was the subject of an American Fisheries Society symposium (Roedel 1975). Everyone recognized that MSY is not generally the sole objective in fisheries management and that other objectives, such as economic or recreational value, need to be considered. Unfortunately, the general literature on objectives has largely taken a very static view of fisheries, as though you could always count on fisheries being near equilibrium and the only questions were about how this equilibrium should look. There is a growing literature (Clark 1985, Mangel 1985, Mendelssohn 1982) on how to deal with objectives related to temporal change (short-term vs. long-term catches, variability of catches, tradeoffs involved in stock rebuilding programs), but this literature is mostly highly mathematical and therefore difficult for general readers to follow.

In this chapter we will not discuss whether fisheries should be managed for biological, economical, or recreational value, nor will we make any broad assertions about how to deal with changes over time. We will, however, be concerned with how to find out what the objectives are and how to express their implications more clearly. We will discuss their structure and show how to deal with tradeoffs between different users and tradeoffs in benefits over time.

2.2. Alternative Management Objectives

We like to group the various and sometimes conflicting objectives of fisheries management into four general areas: biological, economic, recreational, and social. There may be some fisheries where only one of these is important, but we suspect that, in most cases, management is concerned with two or more general objectives and the critical problem is how to trade off between them. But before discussing any tradeoffs, let us look at the features of each general type of objective.

Biological

The traditional default objective for biologists has been maximum sustained yield (MSY). Under MSY, we attempt to maximize the average weight of catch over an infinite time horizon. Year-to-year variability is not considered, and only average catch is seen to matter. A variation, particularly in recreational fisheries, is to use MSY in numbers rather than weight. There is no intrinsic biological justification for MSY except a general notion that more catch is better, but the doctrine of MSY has long been present in fisheries management considerations. Larkin (1977) said in reference to MSY,

> The basic idea was enshrined in national policy documents, incorporated in international treaties, and, in effect, became synonymous in most people's minds with sound management. Most fishery managers and politicians engaged in a steady dialog of explaining why they had to compromise a bit on MSY for "social reasons" but, in so doing, they sounded apologetic. They knew they were sinning.

The major benefit of MSY is that it is simple; the fisheries textbooks provide methods for determining it, and a biologist can use it as an operational principle with a clear conscience. Unfortunately, MSY is really not as simple as it was once thought, and in the same paper cited above, Larkin lists a number of reasons why there probably is no such thing as maximum sustainable biological yield. Fish stocks simply vary too much, and are too complex, for it to be practical to identify and obtain a true maximum sustainable yield.

A number of variations on the basic theme of MSY have been suggested as "safer" biological objectives for management. For example, if the fishing rate that will give MSY is predicted to be, say, F_{MSY}, then one widely supported alternative is the socalled $F_{0.1}$ policy. $F_{0.1}$ is just the fishing rate where the slope of the yield-per-recruit line is 10% of the slope of yield-per-recruit at the origin. This is a rather arcane and arbitrary excuse for setting F lower than F_{MSY} but has evolved into something of a world standard. The trouble with policies like $F_{0.1}$ is that we have no objective way to decide what is safe enough from a biological perspective; who is to say whether it is worth giving up 10%, or 20%, or even 50% of the projected MSY?

Indicators

The normal indicator of biological yield would be annual weight or numbers of fish caught. Average size or distribution of sizes in the catch is sometimes taken as an indicator of population status or risk of overfishing (too many small fish may mean not enough are managing to spawn). Average or maximum age of fish in the catch may serve a similar role. A big question is how to capture risk of stock collapse in evaluating a policy. The most common indicator for risk is how low the population is expected to go, or what percentage of years the stock will be below some specified level.

Economic

Economists have long objected to MSY because maximization of the weight of fish harvested is an arbitrary measure that may bear little relationship to how the fishery contributes to society. They argue (Crutchfield 1965, 1979) that the purpose of a fishery is to produce income rather than fish, and the costs of catching the fish must be considered. A simple economic objective is to maximize the net profit from the fishery, essentially maximization of the difference between the landed value and the harvesting costs. Harvesting costs can be broken into fixed costs (moorage, depreciation, insurance, etc.) and variable costs (fuel, food, crew expenses, etc.). In general, fixed costs do not depend on how much a vessel fishes, whereas variable costs do. Figure 2.1 shows a simple relationship between the landed value, fixed costs, and variable costs, plotted against harvest rate.

Landed value is assumed proportional to landings, so the landed value less total costs are the profits (the height of the shaded area of Figure 2.1). Although MSY and landed value may peak at harvest rate A, profit is maximized at a lower harvest rate B. Profit is always maximized at a lower harvest rate (and therefore a larger stock size) and a lower effort than MSY.

The economic rationale for profit maximization is that the harvesting of fish requires society's resources, such as fuel, steel, electronics, and labor, and that the additional social resources required to increase the harvest rate from point A to point B on Figure 2.1 do not justify the slight increase in catch. The resources could presumably be put to better use in other productive activities. In countries such as Iceland, where fish exports provide one of the major sources of foreign exchange, it is sometimes argued that total landed value may be more appropriate than profits, because these countries can use domestically produced resources that are not scarce, labor in particular, to provide fish as a medium of foreign exchange. If price is independent of harvest, maximization of total landed value becomes equiva-

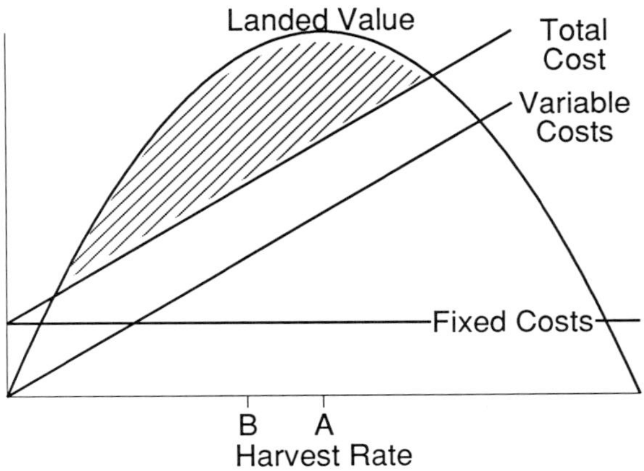

Figure 2.1. The general relationship between landed value, fixed costs, variable costs, and profits in a fishery. Point A is the effort that maximizes biological production and landed value, and effort B maximizes profit. The height of the shaded area indicates profits.

lent to MSY, but, if price decreases with the volume harvested, the optimal effort will be lower than MSY.

Indicators

An indicator of economic performance would be the resource rent, that is, the value of catch less costs of harvesting.

Recreation

Many of the world's fisheries are important for recreation. The freshwater fisheries of North America are a classic example; however, recreational fisheries can be found in most countries of the world. The many indicators of recreational benefits can be grouped roughly as those based on recreational catch and those based on recreational effort.

Depending upon the fishery, the total recreational landing of fish in numbers may be an appropriate indicator of success. In other fisheries, the size of the fish might be particularly important; a few trophy-sized fish may be much more valuable than dozens of smaller fish. The success per day fishing may be a major indication of recreational benefits, again measured in either numbers, weight, or trophy-sized fish. Big-game management faces similar problems, and many jurisdictions have adopted regulations that attempt to maximize biomass yield in some hunting areas and number of trophies in other areas.

Economists suggest that the benefits from a recreational fishery depend on how many people go fishing and how often. Thus, the best indicator may be total effort. Economists often go one step further and attempt to place a value on a recreational day's fishing effort, usually in the form of dollars per day (Cauvin 1980).

Indicators

Indicators of recreational benefits include the estimated value of the recreational fishing effort (dollars per day fished × number of days effort), the amount of recreational fishing effort, and the amount or size of recreational catch.

Social objectives

Fisheries often produce benefits for regions that are economically disadvantaged. On the West Coast of North America, native Indians depend on fishing for a major portion of their food. In many regions of the world, fishing provides one of the few sources of employment in isolated villages, even though quite often economic and possibly biological efficiency would be much higher if the fish were caught by a few large vessels instead of thousands of small boats. Fisheries are often maintained as an explicit social policy and the key indicator for the decision maker might be the number of jobs involved or total fishing effort. If number of jobs is an important indicator of fishery performance, a good policy might be to use lots of effort while regulating it to be highly inefficient.

In many countries, social issues dominate fisheries policy; large inefficient fleets could be replaced by much smaller fleets that cost much less to operate, and the realized rent could be transferred to those eliminated from the fishery to provide them with a higher income than they derived from fishing. Such economic arguments have not been particularly successful at convincing fishermen or politicians to restructure fisheries. Clearly, the maintenance of traditional community structure and lifestyles is a dominant objective in realized fisheries policy.

Indicators

Social indicators of fishery performance include total employment, income distribution to rural communities, and maintenance of traditional rural/urban living patterns.

2.3. Marginal Value and Risk

MSY is a convenient objective to use as a benchmark in considering alternative objectives. In the previous section we looked at alternative indicators and found that we might wish to maximize many things other than average catch. In this section, we look at a single indicator and see how our objective is related to the level of that indicator. The MSY objective implies that twice as much catch is twice as good or, generally, that the relationship between catch and happiness is a straight line. But what if we view MSY as an average over time rather than a fixed value for each year. Then if we apply the same reasoning about twice as much making us twice as happy, we should be just as happy to take the catches in any pattern over time.

We can also think of alternative outcomes as probabilistic outcomes of decisions. Suppose we have two possible management actions, where action A will produce an annual yield of 10 tons per year every year for sure, whereas action B might produce 20 tons per year but might also produce 0 tons per year. If the two possible outcomes of B (20 or 0) have equal probabilities of 0.5 each, then the expected value of B is $0.5 \times 20 + 0.5 \times 0 = 10$. Given MSY as the objective, A and B are equally preferable because they have the same expected value.

This decision choice is similar to one we might make between taking a salaried job which paid \$30,000 per year and a commission based job in which we had a 50% chance of making \$45,000 and a 50% chance of making \$15,000. Some of us might prefer the salary income, others would prefer to gamble and go on a commission basis. These "risk takers" may actually be adjusting the odds in their minds and believe that given their motivation and skills the chance that they will do well and make \$45,000 is greater than 50%.

A major component of whether a sequence of (*10, 10, 10, 10*) is equal in value to a sequence of (*20, 0, 20, 0*) depends on how much each additional fish in the catch adds to our satisfaction. We need to understand our marginal value for additional fish. The plots in Figure 2.2 will help us do so.

MSY assumes that each additional fish is equally valuable, as shown in Figure 2.2 a. Figure 2.2 b shows a case where the marginal value decreases as catch increases. A subsistence food fishery, for instance, might have little use for catches over the food requirements of the community and display such a curve. This type of curve is frequently called "risk-averse," because a sequence of (10, 10, 10, 10) would be greatly preferred to (20, 0, 20, 0), because a catch of 20 is not twice as good as a catch of 10. A risk averse decision maker would generally prefer an assured yield and not gamble on a risky management action that might increase catches significantly. Risk-averse objectives show a decreasing marginal value of catch as catch in-

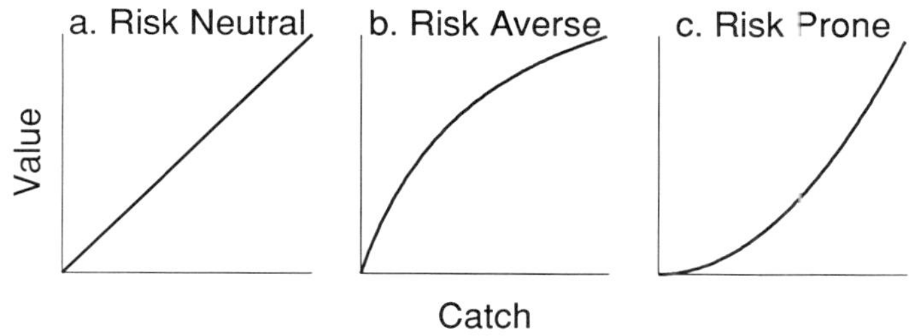

Figure 2.2. Three types of curves represent the value for different levels of catch. **a** A curve where the marginal value of each fish is constant or "risk neutral." **b** A curve with decreasing marginal value as catch increases, also called "risk-averse." **c** Increasing marginal value with catch or "risk prone" objectives.

creases. Figure 2.2 c shows a curve where the marginal value increases with catch. Such a curve might be appropriate whenever it is inefficient or undesirable to catch small numbers of fish. If there are minimum cannery runs, or marketing difficulties with small supply, it might be preferred to have a sequence of (20, 0,20, 0) rather than (10, 10, 10, 10).

There is an extensive literature on the marginal valuation of attributes (Keeney and Raiffa 1976), but for our purposes, it is most important to recognize that it is possible for different people to have different attitudes toward the marginal value of increasing catch or any other indicator of fisheries value. Different attitudes toward marginal value will lead to different rankings of management alternatives, and the standard objectives used in fisheries analysis, MSY and maximization of profits, both assume a risk-neutral or constant marginal value as shown in Figure 2.2 a. We will show later that real decision makers usually have a more risk-averse objective, and this difference between the assumed objective of the analyst and the actual objective of the decision makers may account for much of the frustration of biologists and economists who see their advice consistently rejected.

What may appear to be a highly risk-averse attitude toward catch may really be a risk-neutral attitude toward a more relevant indicator such as company profits, the probability of staying in business, or political pressure. Market and processing limitations will almost always produce decreasing marginal value of catch for purely economic reasons, because high harvests may depress markets and processing or storage limitations may greatly reduce the product value in years of very high harvests. Similarly, it may be very important to have a minimum level of catch in all years in order to maintain customer loyalty. When processors limit the catch of fishermen in exceptionally good years, we see one indication of economic reality.

2.4. Discounting

Wildlife biologists know that a bird in the hand is worth two in the bush, but is a fish in the hold worth two in the water? The answer is probably yes, because there is considerable uncertainty about the ability to catch those two elusive fish in the water. We discuss questions of uncertainty elsewhere, but for the purpose of this section we may well ask if a fish in the hold today is worth more than a fish (of the same size) in the hold next year. From an economic perspective the answer is surely yes. We can sell the fish today, put the money in the bank, and have it plus interest next year. We also assume that prices are constant or at least uncertain with some expected value, and we are not considering the dynamics of fish recruitment and growth.

The fact that commodities normally exchanged for money can earn interest, and that many commodities not normally exchanged for money can be enjoyed, leads to the conclusion that a commodity today is worth more than an equal commodity next year. This is called discounting and should not be confused with problems in valuation caused by inflation. Governments normally require discounting of future costs and benefits for their projects; in Canada a 10% discount rate (in constant dollars) is required for federal projects. This says that a fish caught next year is worth roughly 90% of a fish caught this year, or that economic or recreational benefits will be worth 10% less next year than they are today. The normal discount formula is given as:

$$V_i = V_t \times \frac{1}{(1 + d)^i} \tag{2.4.1}$$

where V_i is the value of a commodity i years in the future, V_t is the value of a commodity today and d is the discount rate. Table 2.1 shows the relative value of goods produced at different times in the future.

A 10% discount rate rapidly diminishes the value of any future yield. The impacts of such discounting are extreme on any project that does not produce very short term benefits. The implications for normal stock management are not so severe; discounting usually only lowers the optimum stock size a little because we now require 1.1 fish next year to make us willing to give up one fish this year. However, the implications are more serious for very unproductive stocks such as whales; if the stock left in the water will only grow 5% by next year, whereas you could earn 10% on the money from those fish if you took them this year, should you be concerned about leaving any stock in the water?

Discounting does affect stock management plans significantly when considering stock rebuilding alternatives, even for productive stocks. If we must forgo catch for a number of years in order to increase average catch in the

Table 2.1. Relative value of goods produced at different times in the future given several discount rates. Value of a good today is 1.0.

	Discount rate			
Years	0.03	0.05	0.10	0.20
1	0.97	0.95	0.91	0.83
2	0.94	0.91	0.83	0.69
3	0.92	0.86	0.75	0.58
4	0.89	0.82	0.68	0.48
5	0.86	0.78	0.62	0.40
6	0.84	0.75	0.56	0.33
7	0.81	0.71	0.51	0.28
8	0.79	0.68	0.47	0.23
9	0.77	0.64	0.42	0.19
10	0.74	0.61	0.39	0.16
15	0.64	0.48	0.24	0.06
20	0.55	0.38	0.15	0.03
50	0.23	0.09	0.01	0.00
100	0.05	0.01	0.00	0.00

future, high discount rates will demand that the increases in future average catch be very large in order to justify reduced catch in the near term.

If, however, the implications of high discount rates seem surprising, consider the implications of no discounting. Let us assume we could increase the yield of a stock 1% by a one-time investment of a large sum of money, and we do our value calculations for this investment without any discounting. Imagine that a salmon passage over a falls could be built at a cost of 1 million dollars and there would be no future maintenance expenses. Then, even if the passageway only produced 10 extra fish per year, the future value of 10 fish every year forever would be more than 1 million dollars; it would in fact be infinite. To avoid this problem of nonsense benefits, we either have to compare specific investment alternatives in terms of per-year benefits (e.g., look for a passageway or other investment that gives 20 fish instead of just 10) or else use discounting as a general indication of the vast variety of alternative ways that we could spend the 1 million dollars.

This is an exaggerated example, but it is obvious that any economic costs and benefits of a fisheries management plan should be discounted at some rate: perhaps the traditional difference between interest and inflation of 2% to 5%. Why do governments and economists insist upon a 10% rate, so much more than the usual interest rate? The reason is uncertainty. We must recognize that there is uncertainty about the future, and it is quite possible that the world will not be the same 20 years from now. The extra margin above the interest rate is an allowance for the unknowable future. Whether 10% is more appropriate than 6% or 8% is beyond our understanding and is best left to economists.

The dominant role that a discount rate can play in the valuation of development projects is shown in a study by Fox and Herfindahl (1964) who reevaluated 178 water resource development projects by the United States Army Corps of Engineers. Together, the projects had investment costs of over $3 billion and all had benefit-cost ratios of greater than 1.0 at a discount rate of 2.6%. Fox and Herfindahl reevaluated the benefit-cost ratios at 4, 6, and 8% and found that the benefit-cost ratios dropped below 1.0 in 9%, 65%, and 80%, respectively, of the projects. The projects required a very low discount rate (2% to 3%) to be profitable because they required substantial cash investments in early years and did not produce benefits until many years later. It is ironic that we tend to associate high discount rates with a short term rip-off, no-concern-for-the-future attitude, yet the biggest habitat destruction agency in North America requires low discount rates to justify its projects.

Salmon hatcheries in the Pacific Northwest are also extremely sensitive to discount rates. Coho salmon hatcheries, which produce fish that return at age 3, are going to look better than chinook hatcheries, which produce fish which return at age 4 and 5, unless each chinook is valued far more highly than each coho. Any fishery development plan that requires substantial front-end costs will only be economically viable if the returns occur soon after the investment, or discount rates are low.

There is substantial disagreement among economists about discounting of nonmonetary benefits such as jobs or recreational days. When discounting jobs, for instance, a project which produces many jobs in the near future is worth more than a project that produces fewer jobs on a permanent basis. This difference has led some to advocate two discount rates, a monetary discount rate to be applied to monetary costs and benefits, and a "social" discount rate, usually lower than the monetary rate, to be applied to nonmonetary costs and benefits.

The use of any discount rate leads to a paradox, because any discounting will put no value on a resource 100 or 200 years hence, and yet deep in our heritage of resource management is the concept of perpetual conservation. Few decision makers or politicians would publicly accept a management policy that will lead to the destruction of the fishery a hundred years hence.

Different individuals will have different discount rates. A fisherman who has a $10,000 interest payment due on his boat will undoubtedly have a different perspective on the tradeoff between catch today and catch in the future than a fisherman whose boat is paid off and whose children are about to enter the fishery or go off to university.

Finally, discount rates can play a dominant role in the conservation of a stock. Economic returns from a fishery (or any renewable resource) are maximized by harvesting the stock to extinction if the stock's intrinsic growth rate is less than the discount rate (Clark 1976), and if the stock can be harvested economically at low numbers. Most marine mammals and many long-lived fish have annual growth rates close to or below the 10% per annum discount rate. Should they be harvested to extinction simply to make some extra money? Clearly society says no. This remains something of a paradox in the theory of renewable resource management.

2.5. Multiple-Factor Tradeoffs

Most fisheries produce many benefits, and decision makers must balance the relative importance of each. The simplest method is to convert all benefits into a common unit such as dollars, but, in practice, this is often quite difficult. If we are trading off between recreational catch and commercial catch and convert both to a dollar equivalent, we will produce the highest value by assigning all the catch to the higher valued user, which is usually recreational. It is true that when recreational and commercial uses conflict, the commercial user is usually driven out, but there are many fisheries where both coexist. Is this inefficient allocation, or is the simple dollar conversion masking more complex real objectives?

The use of dollars as a currency for comparison is too simplistic and yet the alternative is to face a frighteningly large body of literature which has accumulated on the subject of tradeoffs between multiple objectives. However, there are some relatively straightforward aspects of the problem which we can illustrate in this section. Readers wishing to explore the subject in more detail should examine some of the references given in the bibliographic notes of this chapter.

The "Fisheries Act" is the major piece of legislation under which fish are managed in Canada. Along with other legislation and treaties, it has been interpreted as implying an order of priority for allocation of fish:

1. Conservation of stocks
2. Native subsistence fisheries
3. Commercial and sport fisheries

Strictly interpreted, this priority scheme suggests that there should be no fishing until spawning requirements are met. Once spawning requirements are met, native subsistence fisheries should be permitted until their needed catch is reached, and then the commercial and sport fisheries should be opened. Perhaps this is the way the world should work, but it is certainly not the way the world does work. In practice the Canadian government tries to do a much more complex balancing act, pretending to meet the priorities while in fact responding to political pressures from the user groups in complicated and sometimes unpredictable ways.

We prefer to think of objectives as a score card; the more points a manager

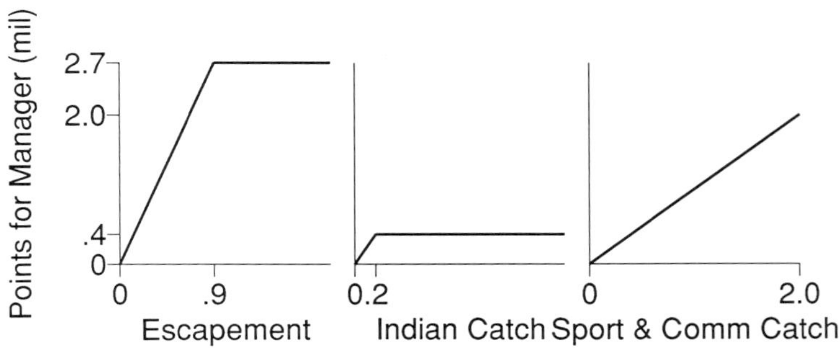

Figure 2.3. Graphical representation of initial score card for Skeena River management.

accumulates, the better he is doing. We could, then, translate the dictates of Canadian fisheries priorities as follows. Let us assume that we are considering the management of Skeena River sockeye salmon, where the needed escapement is 900,000, the Indian food requirement is 200,000, and a multitude of sport and commercial fishermen are eager to catch any extra fish. How can we construct a score card that will be realistic and operable? One possible score card might look like this: 3 points for each fish in the escapement until 900,000 fish escape, and no additional points for fish over 900,000 in escapement; 2 points for each fish in the Indian food fishery up to 200,000 fish, and no additional points for Indian food catch over 200,00 fish; 1 point for each fish in sport or commercial catch.

This score card insures that management will strive to put the first 900,000 fish into escapement, the next 200,000 fish to the Indians, and any remaining fish to sport and commercial catch. The points allocated to each use are unimportant so long as the points are highest for escapement and lowest for sport and commercial catch.

Although such a score card might nominally meet the requirements of the priority list, it does not provide total guidance to a manager. How does he trade-off between sport and commercial catch? If a total run of 900,000 came in, his score would be maximized by assigning all the fish to escapement and leaving no fish for the Indians; however, allowing an Indian harvest of 100,000 and an escapement of 800,000 would hardly endanger the long-term health of the stock and would certainly be beneficial for the Indians. Does this mean the scoring method is wrong, or was it wrong in the first place to think in terms of a simplistic priority ordering of users?

The score card proposed above can be graphically illustrated and is shown in Figure 2.3. It would be used to evaluate a fisheries management plan and would enable us to consider the tradeoff between escapement and other indicators in the long term. Let us assume that an escapement of 900,000 would normally produce a total return of 1,800,000 which would allow for

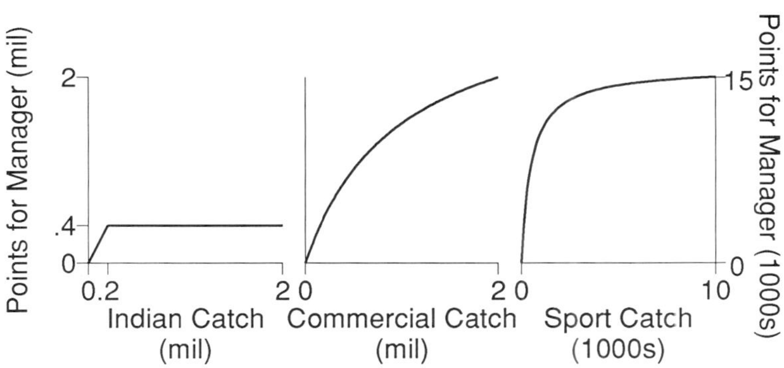

Figure 2.4. A second possible score card for Skeena River allocation.

200,000 Indian catch and 700,000 sport and commercial catch. If we had a poor run of only 900,000 and allocated all the run to escapement, we would get no points for Indians or commercial and recreational catch, but we would then have the expected return be 1,800,000, which would provide 200,000 Indian points and 700,000 sport and commercial points.

If we assigned some of the poor return of 900,000 to the Indian catch this year, we would be taking away from our expected return next time, but we would be getting 2 points for each fish we allocated to the Indians this year and losing 1 point for each fish we lose next generation in the sport and commercial catch. We therefore ought to allocate escapement to Indian catch until we expect to lose two fish next time for each fish not spawning this time.

We should actually take escapement out of the score card and use stock recruitment dynamics to help us evaluate future runs and thus future points. We need to realize that we are not at all interested in escapement for its own sake, but very much so in how it will produce fish, and therefore catch, in the future. Given a decreasing rate of production with increasing escapement, the points assigned for escapement should decrease as escapement increases.

By eliminating the escapement from the score card, it is now simpler, and we have resolved some of the problems associated with trading catch for escapement. We have not resolved how to allocate between sport and commercial fishermen. There is currently a limited demand for sportfishing of sockeye on the Skeena River, as well as limited cannery and harvesting capacity for commercial fish. A more realistic point system is shown in Figure 2.4. The Indian points remain the same, assuring Indians the highest priority on catch; the sport fishermen get roughly 1.5 points per fish, but with marginal values steeply declining at approximately 10,000 catch; and the commercial fishery is assigned roughly 1 point per fish with a slight marginal decrease in value at higher catches.

The score card of Figure 2.4 is quite functional and it could be used to evaluate any specific fishing plan; but where did it come from? We made it up, of course, using some guiding principles from Canadian laws. We believe it a serious mistake that real objectives are rarely quantified, since you cannot know how well you are doing unless you know what you are trying to do. Everyone associated with Skeena River management probably has some intrinsic objectives that can be put into score card form. Sport fishermen presumably believe that sport catch should have the highest priority; Indians believe that they have the highest priority, and so on.

What score card should managers use? There is no simple answer. The manager's score card (or objective) could be developed from his perception of the nature of the job, it could be dictated to him by his supervisors, or it could, in principle, be developed by a negotiation process between user groups and management.

Assuming that we want to develop a score card, or an objective function as it is often called, how do we do it? One of the most commonly used methods involves posing a set of questions to the individual about tradeoffs between alternative outcomes. This is generally done for each indicator (Indian, sport, commercial catch) to establish the level of risk aversion or changes in marginal value for each indicator. The second step is to ask questions about the tradeoffs between levels of indicators to establish their relative value. The relative value for each indicator is then converted to a number between zero and one where zero represents the worst possible outcome and one the best. Figure 2.4 would simply be reduced so that each point ranges from 0.0 to 1.0, assuming that the maximum commercial catch we anticipate is 2,000,000. Normally, we would then decide the relative value of the points for each attribute so that the sum of the weights is 1.0. Thus if all attributes were at their best, the total score would be 1.0; if all attributes were at their worst, the score would be 0.0. The objective function would be written

$$P_T = w_i P_i + w_s P_s + w_c P_c \tag{2.5.1}$$

where P_T is total points, w_i is the relative weight for Indian catch, P_i is the total points for Indian catch, w_s and w_c are the weights for sports and commercial catch, and P_s and P_c are the total points for sports and commercial catch.

If the Ps are 0 at their worst and 1.0 at their best and if the ws sum to 1.0, then we have a nice score card system which is 1 at best and 0 at worst. We can easily calculate from Figure 2.4 that

$$w_i = \frac{400{,}000}{400{,}000 + 15{,}000 + 2{,}000{,}000} = 0.166. \tag{2.5.2}$$

Similarly $w_s = 0.006$ and $w_c = 0.828$.

The details of the questioning procedure used to derive a person's objective functions are given in Keeney and Raiffa (1976), but our experience has been that the most important feature of discussing objectives is to get some firm objective written down as a starting point. Once this is done, the decision maker's intuitive evaluation of an outcome can be compared to that calculated from his written objective.

An alternative to the interview procedure is direct assessment. This is actually another form of interview, but instead of building objectives one at a time, the questions deal with comparison of packages of indicators. Thus the interviewer would ask, "Do you prefer 0 Indian catch, 120,000 sport catch, and 1,000,000 commercial catch, or 200,000 Indian catch, 10,000 sport catch and 500,000 commercial catch?" By asking dozens of such questions, one can begin to deduce something about tradeoffs between different attributes. Both techniques depend upon the interviewee actually knowing his objectives and responding appropriately, as well as on the interviewer, whose skill and experience can make or break such an assessment.

A third alternative method is to look at what decision makers actually do and determine their objectives from their actions. Even though this method will produce general ideas about the objectives of management, as will be seen in the next section, it is less likely to help us formulate a usable score card. In the absence of legislated objectives, we must look at actual decision makers and see what objectives their decisions reflect.

2.6. Inferring Objectives from Management Behavior

Instead of asking abstract hypothetical questions about manager's objectives, we can look at how managers behave and infer their objectives. One must be cautious with this approach since any particular behavior can usually be explained by several alternative hypotheses.

Figure 2.5 shows the historical harvest rates on Skeena River sockeye salmon plotted against total run size. The solid line shows what a fixed escapement policy would look like: no harvest at all until the optimum escapement (800,000) is reached and then an increasing harvest rate. The actual harvest rates used from 1950 to 1968 show higher harvest rates than the fixed escapement required for MSY when stocks are low, and a somewhat lower harvest rate than optimal for MSY at high stock sizes.

We could infer from this information that the real objective was a "risk-averse" curve where the managers always want to provide some fishing. We could, in theory, actually calculate which type of objective is most consistent with the manager's actual behavior and then suggest that this was his objective. The problem with such an inference is that the behavior shown in Figure 2.5 is consistent with a fixed escapement policy allowing for impre-

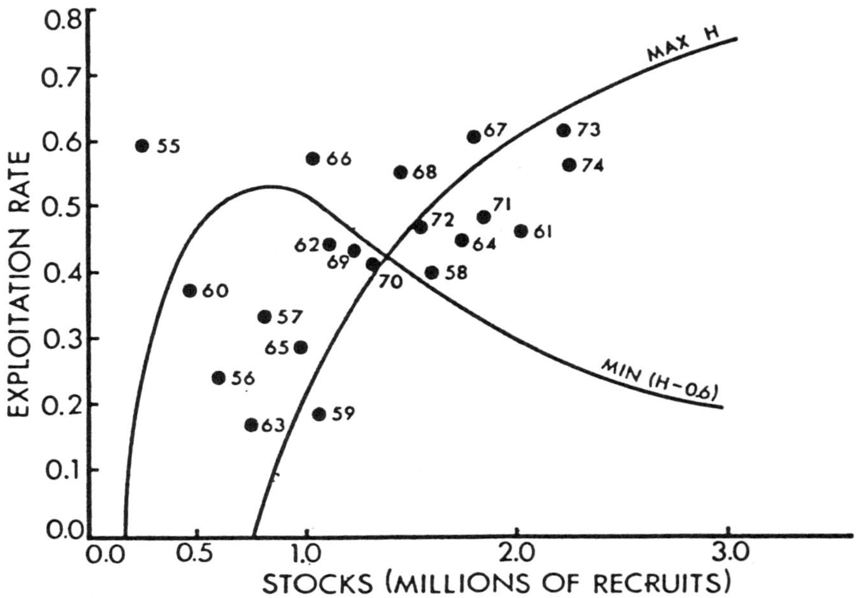

Figure 2.5. Harvest rate plotted against run size and a line denoting the maximum sustained yield (Max H) objective. From Walters. (1975). Can. J. Fish. Aquat. Sci. 32: Fig. 7.

cision in the manager's ability to regulate the harvest rates. Since the total return is not known accurately until the run of fish has largely gone past the commercial fishery, the managers often do not know that the run is weak and should not be fished until much of it has passed the fishery. Similarly, if a run is very large, the manager will not know it until the run is partly past the fishery and then a limited number of vessels or cannery capacity may limit ability to catch the fish.

Discussions with the managers on the Skeena River during the period shown in Figure 2.5 indicate that both mechanisms were operating; some risk aversion and some uncontrollability of the fishery. In this case it is not easy to infer management objectives.

Another interesting attempt to infer objectives is the management of Pacific halibut by the International Pacific Halibut Commission (IPHC) in the late 1970s and early 1980s. Halibut stocks were depleted in the late 1960s and early 1970s because of a rapid growth of incidental catch by crab and groundfish fisheries and continued exploitation by the traditional longline fishery. Once the extent and consequences of the incidental catch were understood, it appeared that halibut stocks were greatly depressed below their optimum size. Management by MSY would have suggested a complete

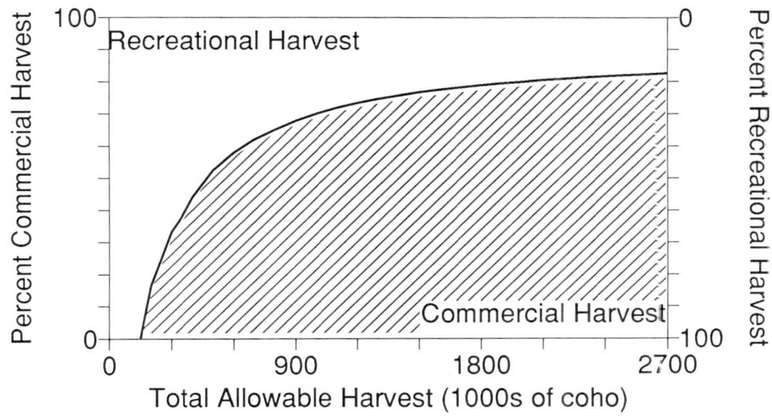

Figure 2.6. Proposed allocation of coho salmon south of Cape Falcon, Oregon. Data from Salmon Plan Development Team (1988).

closure of the fishery until the stock was rebuilt. Since halibut are slow growing, such a period of rebuilding would likely have taken a number of years. Under MSY this would have meant a cessation of fishing and resultant destruction of the existing halibut longline fishery. The IPHC chose to estimate the surplus production each year and allocate 75% to catch and 25% to rebuilding the stock. This was a deliberate and explicit move to maintain the existing fishery while opting for a slower rebuilding of the stock. MSY was neither acceptable nor appropriate in the eyes of the IPHC.

A good example of implicit objectives that illustrates a decreasing marginal value for one user group is given in Figure 2.6, which shows the proposed allocation of coho salmon between recreational and commercial fishermen for a major fishery in Oregon. Note that at small stock sizes the recreational anglers receive all of the fish, but as the run size increases the commercial fishing share increases. This reflects an implicitly higher valuation of recreationally caught fish at low recreational catches, and the decreasing marginal value of recreational catch.

When Canada took control of its cod fisheries on the East Coast by declaring 200 mile jurisdiction, the stocks had been depleted by unrestricted access. Canada embarked on a rebuilding plan similar to that of the IPHC; a portion of the surplus production was allocated to catch and a portion to stock rebuilding. Again, the large number of fishermen and shore workers militated against a narrow-minded biological objective.

There are well-documented cases of long-term fishing moratoria that appear roughly consistent with MSY. The collapse of the British Columbia herring fishery in the late 1960s was followed by a total ban on fishing for several years until the stocks had substantially recovered. Commercial harvesting of North Pacific fur seals was halted for several decades while the stock recovered. In both of these cases, the industry as well as the stocks had crashed, and the moratoria did not prevent anyone from making a living because there was no living to be made from the greatly reduced stocks. It is quite possible that all of the examples given in this section illustrate a real economic objective.

When there are existing economically viable vessels and processors, a modestly risk-averse objective that permits some continued fishing is appropriate. However, when the industry is dead, first priority is assigned to rebuilding the stocks.

2.7. Summary and Critique

In the preceding sections, we have attempted to discuss the major issues involved in objectives for fisheries management. We have not actually provided direct advice on what to do. We do not believe that a fishery manager will usually receive explicit instructions on the management objective; politicians generally prefer much less rigorous statements that can be interpreted as they see fit when political circumstances change.

On the other hand, it is definitely not the stock-assessment biologist's or manager's job to actually make decisions about management objectives. At best, the technical advisors should work with the decision makers to clarify the implications of different objectives. We would recommend the following general approach when presenting evaluations of alternative fishery management plans.

1. Present time series of different indicators, including biological, economic, recreational, and social objectives
2. Attempt to determine the shape of the marginal value curves for each of the indicators, in particular to identify those where year-to-year constancy is highly important
3. Apply government specified discount rates to all monetary expenditures and outputs from the fishery and provide standard net present value and cost benefit statistics on monetary items
4. From the above three items, attempt to construct a quantitative scorecard of management performance, similar to that shown in Figure 2.4, and present the numerical "score" as another indicator of performance, not as a substitute for the presentation of a broader range of indicators.

Unresolved problems in fisheries management objectives

There are currently two major conflicts in the determination of fisheries management objectives. The first has to do with the shape of the marginal value curves for specific indicators. This shape determines the general form the optimal fishing plan will take (Deriso 1985). If value increases proportionally with catch, fishing plans that generate high average yields with high year-to-year variability will look good. If marginal values decrease at higher outputs, fishing plans that provide more stable outputs will be preferred. A few studies (Powers and Lackey 1976, Keeney 1977, Hilborn and Walters 1977) have addressed these issues directly.

The second, and more important, major conflict is between economic and social performance. In most of the world, the fishing fleets are much larger than they need to be to maximize either economic or biological yield. Yet to reduce the fishing fleets would be to reduce employment. The salmon fisheries are perhaps the extreme example. A typical salmon fishery, such as the Skeena River sockeye fishery, might involve several hundred boats, each worth $20,000 to $150,000, and employ perhaps a thousand people during the harvest. The fishery could be replaced by a few dozen people operating a few vessels or perhaps some form of fish traps. The profits from the "efficient" fishery could be used to pay those displaced from the fishery more money than they earned while fishing, and the national economy would be much better off because fewer foreign imports would be required to run the vessels.

The major issue would appear to be one of employment: fishermen who have a rewarding way of life would be turned into glorified welfare recipients—and many of them would presumably migrate to the major urban centers, which would lead to the decay of many rural communities. This would be amplified because of secondary employment in boat supply and maintenance that would be lost.

An alternative would be to pay the fishermen to operate their boats for tourist sightseeing during the same time they used to fish. Tourists would be provided with free boat trips, the fishermen would still be employed for the same amount of time in a rewarding job with a desirable lifestyle, total costs would be reduced at least through reduced imports in fishing nets and electronic fishing gear. Everyone would seem to be better off, and yet to our knowledge such a proposal has never been seriously considered. We have no answer to this question, and we will leave it as an item for further thought.

It could be argued that once we have excess fishing fleets (that is, more fishermen and boats than are needed for a biological or economic optimum), it is very difficult to reduce them because of the loss of employment. The argument would continue that managers do not want to get to a situation of excess employment, but because of the dynamics of exploitation, and in particular the unsustainable yields during fishery development, we usually end up with too many fishermen. Given a choice, decision makers will not trade off economic viability for increased employment.

As an illustration of the dilemma between economic efficiency and employment, consider the Tasmanian abalone fishery (Prince 1989). This fishery has had strict limited entry of 120 abalone divers since 1969. Since 1984 there has been an individual transferable quota system of 28 tons per diver. Only the licensed diver can harvest abalone and each diver normally employs a deck hand who assists with boat handling and remains on the surface to assist the diver while in the water. In 1987, the 28 ton individual quota had a landed value of roughly $350,000 (Australian dollars), and the divers net income was usually between $250,000 and $300,000 (excluding the cost of the license). Divers who had entered the fishery early did very well indeed, and even those who had recently entered normally paid off their license in a few years. The license value had risen to $850,000.

Each diver worked about 100 days per year to obtain his 28 tons, and made a rather handsome income by almost any standard. Over time we would expect the license price to rise so that new entrants did not do quite as well. The annual license fee was $20,000, which paid for the cost of administering the abalone license system and paid for the biological work conducted by the State of Tasmania.

Assuming the 28 tons per diver was sustainable, there is an obvious societal decision to be made about how many divers to permit. The number of licenses could be doubled to allow each diver 14 tons per year which would have expanded the employment opportunity. Financially the fishery could probably support as many as 4 times as many divers, each making a considerably more modest income than the current 120 divers. The societal benefits of more divers would have been more employment (more divers, more deck hands, more boat repairs, etc.) and more equitable income distribution (not many individuals earn $300,000 in rural Tasmania). Social costs of having more divers would have included less income to the government (since the $300,000 incomes placed the divers in a very high marginal tax bracket), increased management enforcement difficulties, and increased imports of boat related items such as fuel and electronics. Which future for the Tasmanian abalone fishery would be better; a few rich fishermen providing a lot of income to the government, or lots of not so rich fishermen, providing more of a drain on the national balance of payments? This is not an easy question.

Are formal objectives actually necessary?

Few fisheries operate under any sort of rigorous formal objectives and our discussion of methods for the analysis of objectives is not intended to make the case that legislators should formally define objectives. There are certainly dangers from not having formal objectives—in the absence of an explicit management objective the political power of interest groups will normally be used to enhance their particular view of how the fishery should be run.

Formal objectives can be used by fishery managers to remove them from the political power of interest groups. A major rebuilding of salmon runs in Washington State occurred after court-mandated priorities for allocation between escapement, Indian catch, and non-Indian catch. Formal objectives are required for most types of optimization; an analyst cannot tell if a new option he comes up with produces "better" results than old options unless he knows what "better" is. This problem is discussed in the chapter on optimization.

There are undoubtedly some dangers from formal objectives: it is quite possible for lawmakers to mandate MSY or equally antiquated objectives, which then become a millstone to the managers who must carry out the law. Our general view is that even though management agencies should devote some effort to clarification of objectives, it is probably unwise to set down formal rigorous objectives that are not subject to frequent review.

Bibliographic Notes

The question of objectives in fisheries management is almost totally ignored; Alverson and Paulik (1973) present one of the best papers. On the suitability of MSY, Larkin (1977) and Sissenwine (1978) are the best individual papers, with diverse points of view presented in the symposium edited by Roedel (1975). Crutchfield (1965) discusses economic objectives and Cauvin (1980) recreational ones. Pikitch (1988) considers objectives for multispecies fisheries. The best book on the formal aspects of objectives, particularly multifactor tradeoffs is Keeney and Raiffa (1976). For detailed examples of how these methods can be applied see Keeney (1977) and Hilborn and Walters (1977).

Part II

Behavior of Fisheries

3

Behavior of Exploited Populations

3.1. Introduction

The essential biological feature of any fishery is the dynamics of the fish population; the analysis of population dynamics involves trying to make predictions about the birth, death, growth, and movement processes of the fish. The purpose of this chapter is to discuss what is known about fish population dynamics and to establish some preliminary models about how fish stocks respond to harvesting. We begin by exploring how populations behave, first in their unexploited state and then when they are harvested. This requires an analysis of historical and geological records, as well as an examination of modern fisheries. We then discuss some basic ecological theory about the dynamic behavior of populations to provide a framework for later discussion about specific elements in the life history of fish.

The next stage is to develop quantitative models of fish populations: first, simple models that consider only the biomass of the population, and then explicit age-structured models that include more detailed population processes, such as growth and recruitment. Next, the distribution of fish stocks in space is discussed at some length, because it has major implications to exploitation. Finally, we consider the impact of external variables such as predation, food supply, and environmental conditions.

3.2. Types of Observed Behavior

To understand how populations will respond to exploitation, we need to appreciate how they behave when unexploited.

Behavior of unexploited populations

Lessons from other taxonomic groups

Most fish have the unfortunate property of being quite difficult to observe; nearly all of our understanding of fish populations comes from exploited stocks. A brief look at animal populations other than fish may prove instructive. If we survey the animal kingdom, we can find examples of all sorts of behavior—some populations are rather stable on a scale of tens of years, many more fluctuate dramatically. Counts of breeding pairs of birds are among the more stable of animal numbers, while many insects show fluctuations of several orders of magnitude on a year-to-year basis. Figure 3.1 shows time trends in abundance for herons in two areas of Great Britain, snowshoe hares and lynx in northern Canada, and two species of moth.

The heron population appears to be modestly stable, whereas the hare and moth populations show large fluctuations. There are few long-term histories of any type of animal population, and we cannot in any way think of the examples in Figure 3.1 as a random sample. Populations that fluctuate wildly have been more interesting for biologists and have been studied more often than less variable populations. Probably the most important thing fisheries scientists can learn from looking at other groups of animals is that constancy is an exception rather than the rule, and we should expect that fish populations will change in abundance with or without fishing.

Long-term records of fish populations

Anoxic sediments. There do exist a few long-term records of unexploited fish populations. The best data come from a series of marine cores from anoxic sediments, showing historical patterns of deposition of fish scales from the water column above. Anoxic sediments preserve the scales, so the pattern of deposition (number of scales per centimeter) is felt to reflect the historical pattern of abundance. Soutar and Isaacs (1969) developed this technique off the coast of California while exploring the long-term history of sardine (*Sardinops sagax*) and anchovy (*Engraulis mordax*). Figure 3.2 shows their reconstruction of sardine and anchovy abundance. The pattern of periodic "outbreaks" and disappearances has many implications. The California sardine fishery, which developed in the early 1900s, had collapsed by the 1950s, and great reductions in fishing pressure have not resulted in the rehabilitation of that stock. The historical record that Soutar and Isaacs produced indicates sardines have historically disappeared long before commercial fishing began, and perhaps "explains" why sardines have not recovered.

This same technique has now been applied in two other areas: off the coast of Peru (De Vries and Pearcy 1982), and the coast of Namibia (Shackleton 1987). Both are areas where major fisheries for anchovy and sardine-like fish have taken place, and results in both areas indicate a similar pattern to that found in California—periodic outbreaks and disappearances are common.

Historical data. What we know about the abundance of fish populations

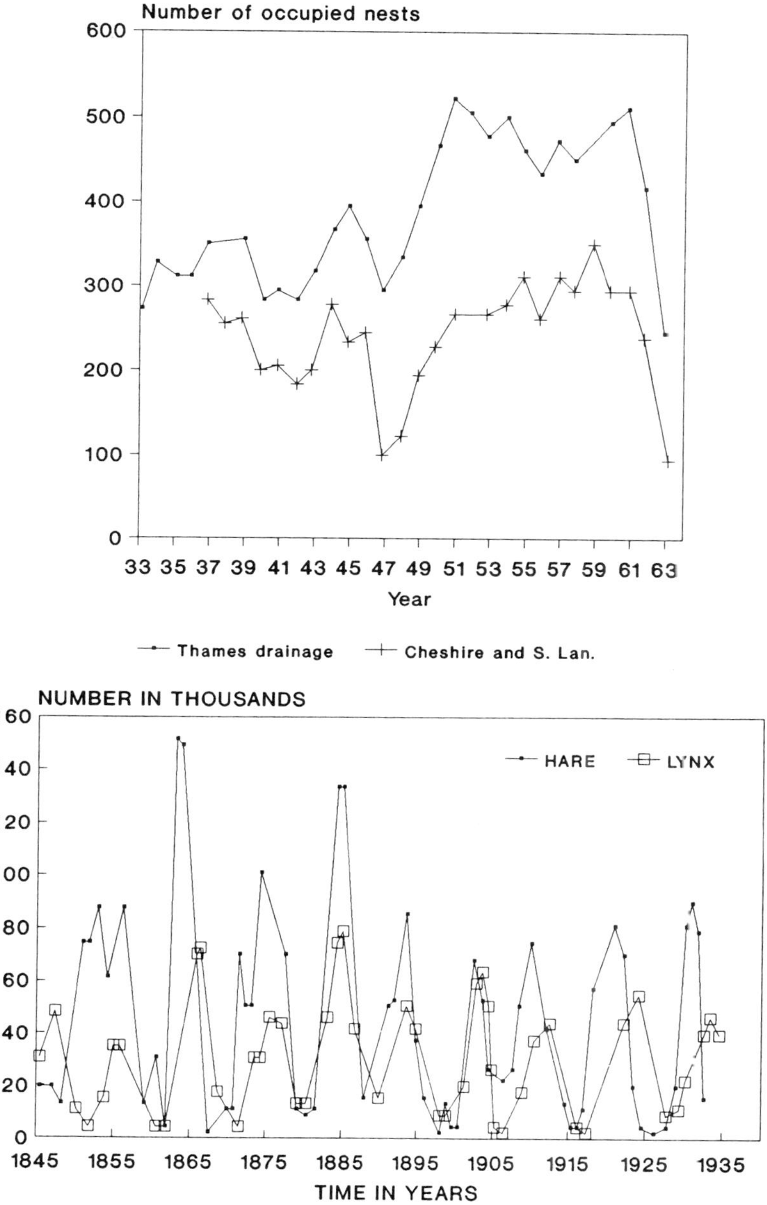

Figure 3.1. Time trends in the abundance of herons in two areas of great Britain between 1934 and 1966 (redrawn from Lack 1954), snowshoe hare and lynx of northern Canada (redrawn from MacLulich 1937), and two species of moth, the pine moth (redrawn from Varley 1949) and the larch budmoth (redrawn from Baltensweiler 1964).

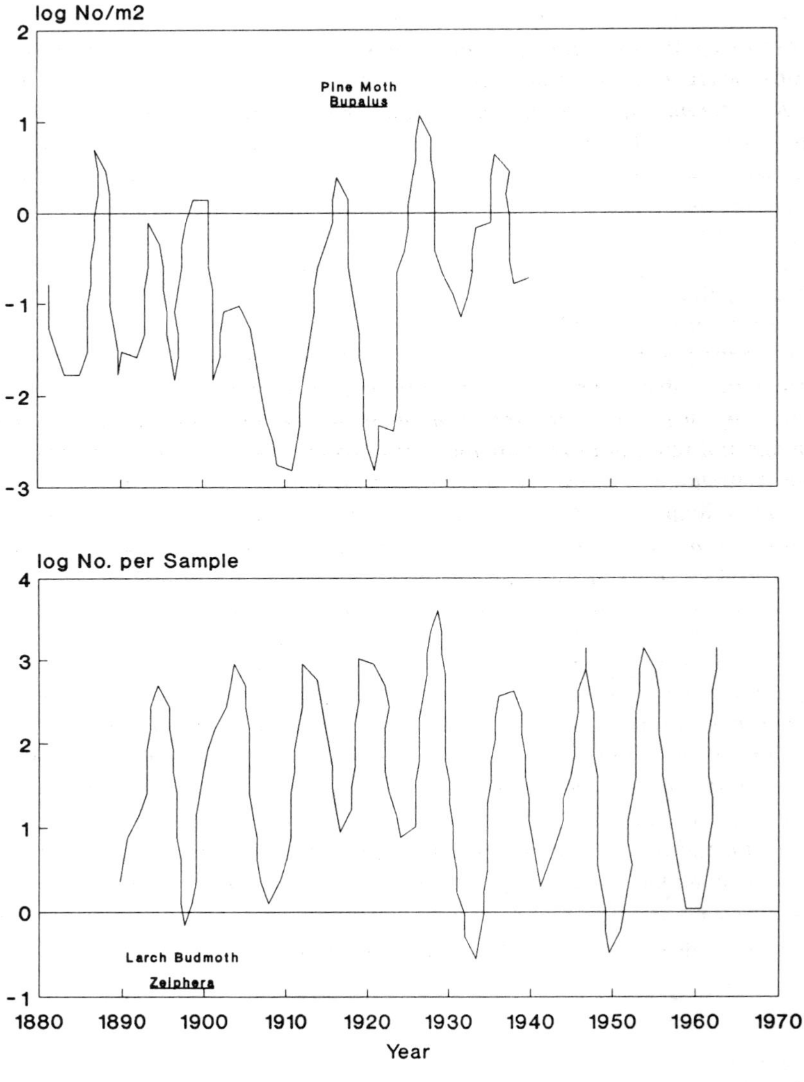

Figure 3.1. Continued

comes primarily from the analysis of catch records. This means that observations on truly unexploited populations are almost nonexistent except those obtained by indirect methods, such as the scale records described in the previous section. Nevertheless, we can establish some time-series of lightly exploited populations that should prove indicative. Cushing (1982) reviewed the histories of the Swedish and Norwegian herring fisheries and showed that there appears to have been a periodic alternation in abundance between

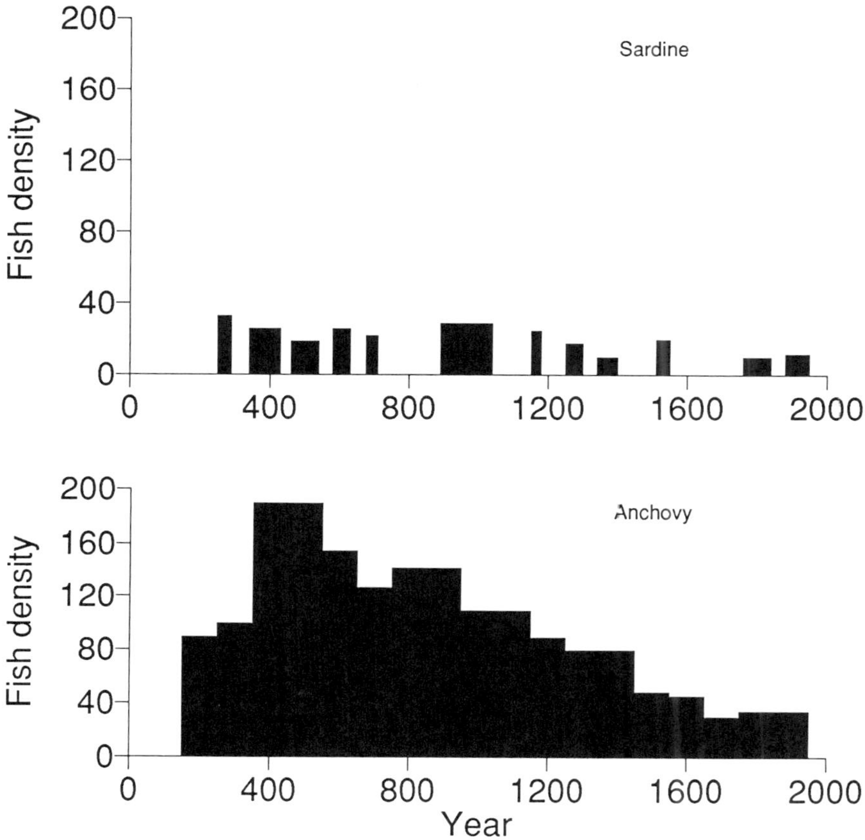

Figure 3.2. Historical patterns of anchovy and sardine abundance from anoxic sediment cores. Data from Soutar and Isaacs (1969).

the Norwegian stock outside the Baltic and the Swedish stock inside the Baltic. Periods of good catches in one fishery are generally periods of poor catches in the other, as shown in Figure 3.3. The catches over the 500 years Cushing considered are low compared to those of the 20th century, so we have some justification in considering the stocks lightly exploited during most of this time. The pattern, therefore, is rather similar to the one revealed by long term scale records; periods of high abundance alternating with periods of scarcity.

Behavior of exploited populations

Exploited populations may behave in quite a few different ways. Caddy and Gulland (1983) surveyed the history of a number of fisheries from

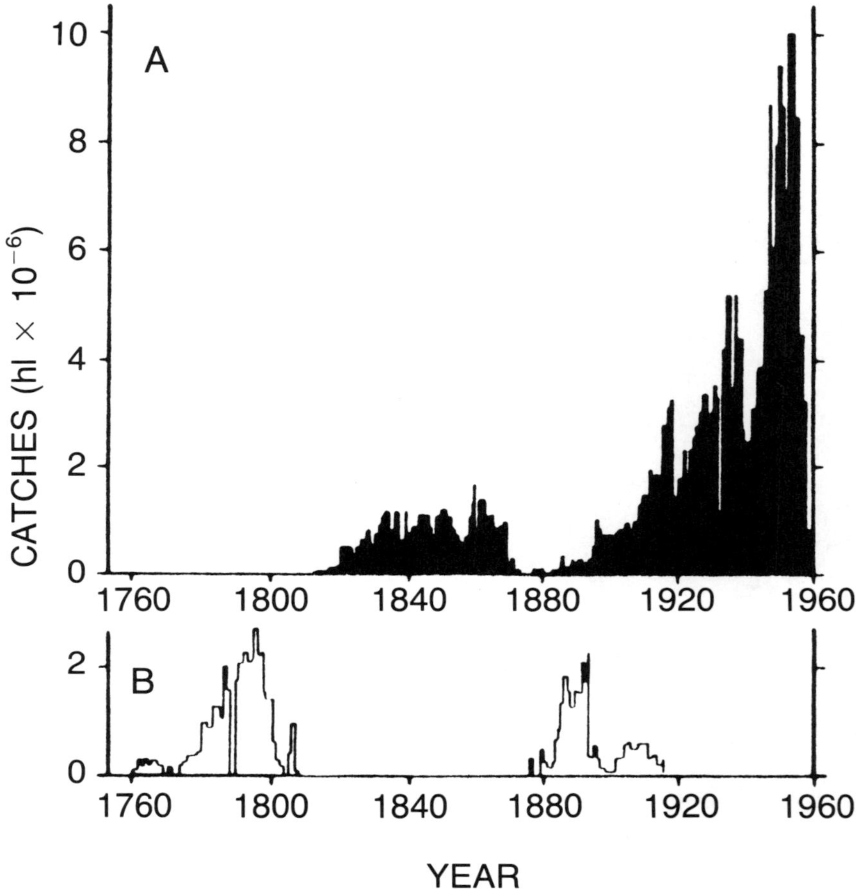

Figure 3.3. History of Swedish (a) and Norwegian (b) herring catches. Redrawn from Cushing (1982), Figure 26.

throughout the world, and classified the stocks into four groups. Their categories are as follows:

1 *Steady state*. Stocks where a more or less steady yield seems to be sustainable over reasonably long periods of time
2 *Cyclical*. Stocks that show strong cyclical yields, with periods of high catch regularly followed by periods of low catch
3 *Irregular*. Stocks that show irregular periods of high abundance, without the consistency in alternation between abundance and scarcity shown in cyclical stocks

4 *Spasmodic*. Stocks that have produced major yields and then collapsed without any major recovery.

Certainly these classifications are not definitive and many alternative schemes could be invented, but the major point is that most stocks do not appear to be capable of producing a steady sustainable yield; steady yield appears in fact to be the exception rather than the rule. This observation does not bode well for much of the methodology of stock assessment and the "theory of fishing," where there has been a preoccupation with equilibrium calculations. Figure 3.4 shows an example of each type of behavior.

This point is not clearly appreciated in fisheries—most fisheries management is built on the assumption that more or less sustainable yields are possible. As we examine the body of fisheries management theory later in this book, we will find that it has very little to say about how best to manage cyclical, irregular, or spasmodic stocks.

Thompson-Burkenroad debate

Ever since the efficiency of fishing gear has become sufficient to catch a significant proportion of some fish populations, there has been an ongoing debate about the importance of fishing in determining fish abundance. As early whalers depleted the Greenland bowhead stock, they invariably believed that poor catches reflected a change in the migratory pattern of the whale or adverse ice conditions rather than reduced abundance due to harvesting. Some who realized that many females and calves were being killed expressed concern that perhaps too many whales were being taken, but, because poor years were often followed by good years, it was easy to believe that the fluctuation in catch was due to natural events rather than fishing. Only after the fishery had gone on for over 200 years did it become widely accepted that the demise of the Greenland whale was due to fishing (Nansen 1924), and even today this is challenged in some quarters (Vibe 1967).

The same sequence of arguments can be found in almost all major fisheries; fishermen and some biologists tend to ascribe fluctuations to natural events rather than fishing, until the stock is greatly depleted. Others, mainly biologists, have tended to ascribe stock declines mainly to fishing, and stock recoveries to the effects of management. The classic 20th century debate about these alternative explanations took place with reference to Pacific halibut (*Hippoglossus stenolepis*) between W. F. Thompson (Thompson et al. 1931, Thompson 1950), and M. D. Burkenroad (1948, 1950, 1951, 1953) and has been reviewed by Skud (1975). Thompson believed that the decline in halibut abundance in the 1920s was due to overfishing and that the decline was reversed by decreasing the fishing pressure on the stock. Burkenroad argued that the decline could equally well be explained by natural fluctua-

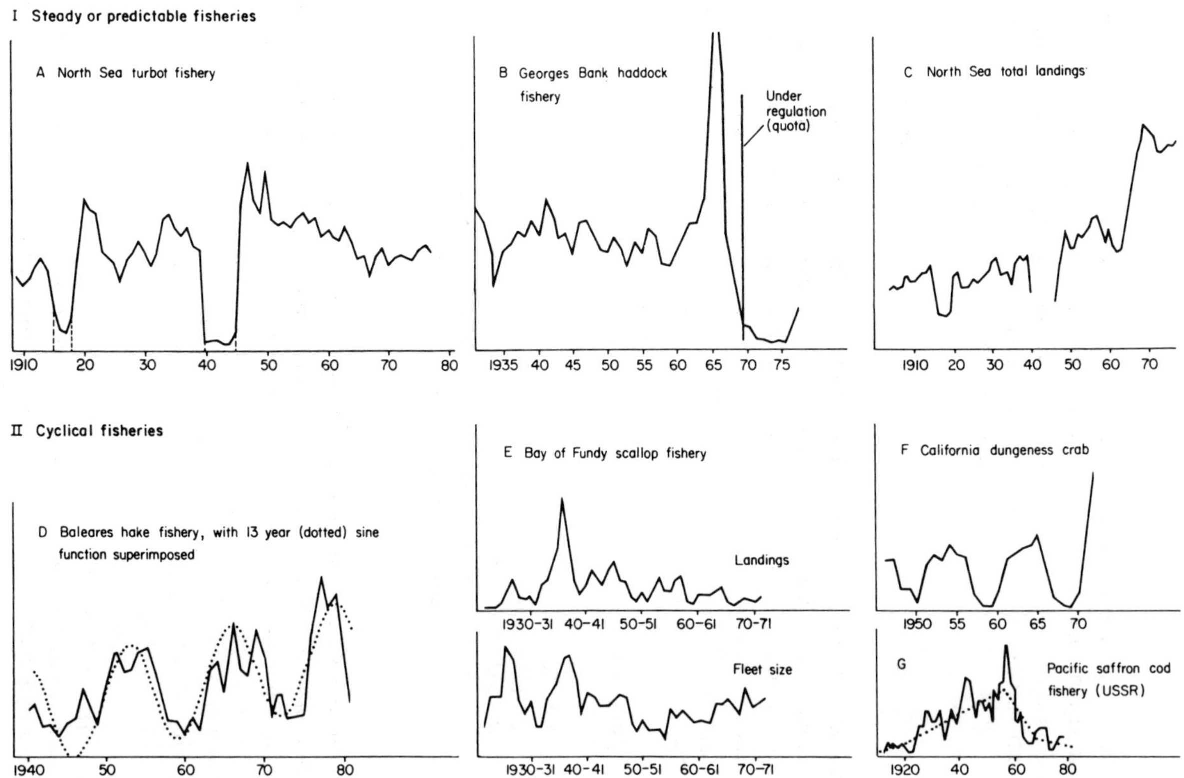
I Steady or predictable fisheries
A North Sea turbot fishery
1910
20
30
40
50
60
70
80
B Georges Bank haddock fishery
Under regulation (quota)
1935
40
45
50
55
60
65
70
75
C North Sea total landings
1910
20
30
40
50
60
70
II Cyclical fisheries
D Baleares hake fishery, with 13 year (dotted) sine function superimposed
1940
50
60
70
80
E Bay of Fundy scallop fishery
Landings
1930-31
40-41
50-51
60-61
70-71
Fleet size
1930-31
40-41
50-51
60-61
70-71
F California dungeness crab
1950
55
60
65
70
G
Pacific saffron cod fishery (USSR)
1920
40
60
80

Figure 3.4.

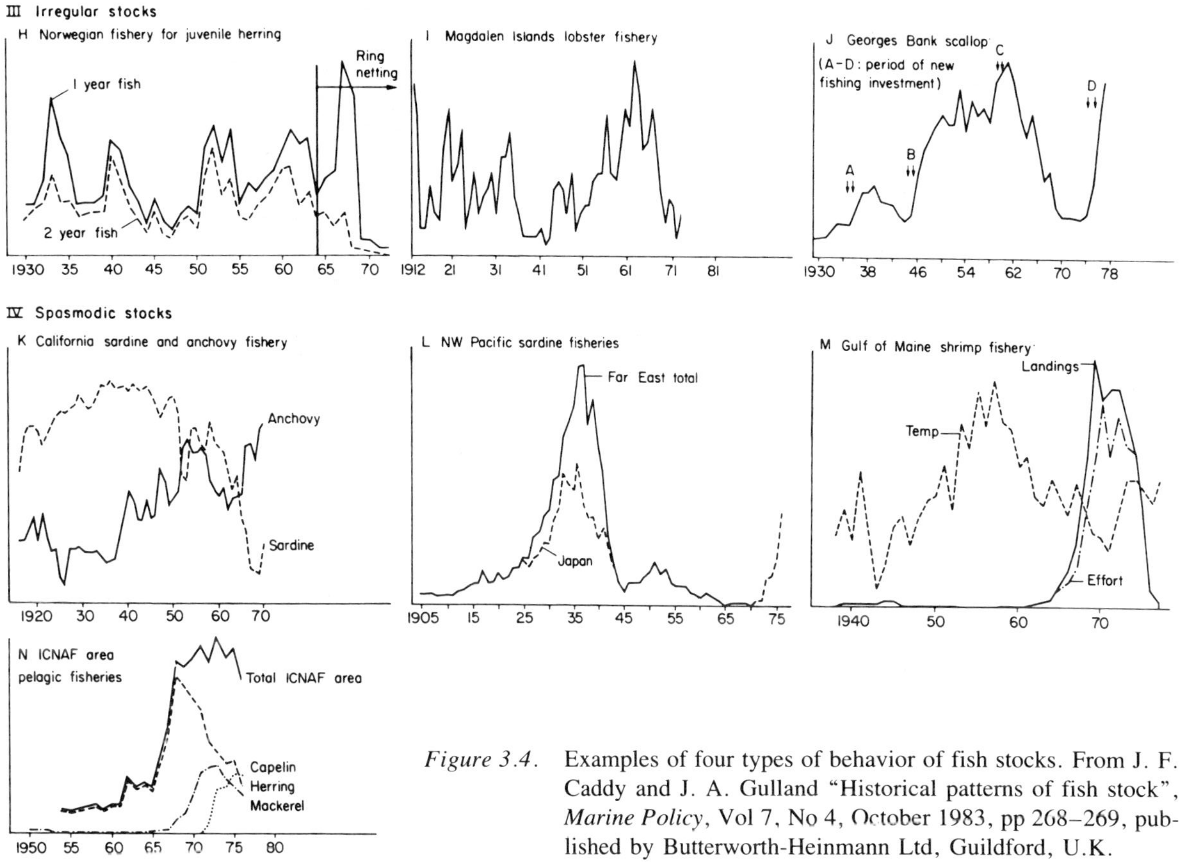

Figure 3.4. Examples of four types of behavior of fish stocks. From J. F. Caddy and J. A. Gulland "Historical patterns of fish stock", *Marine Policy*, Vol 7, No 4, October 1983, pp 268–269, published by Butterworth-Heinmann Ltd, Guildford, U.K.

tions in oceanographic conditions for juvenile halibut survival; he pointed out that management responses will almost inevitably involve reduced fishing during declines, and these responses will likely occur just in time to allow the managers to take credit for natural recoveries that would have occurred anyway.

At the time the debate raged (1940s and 1950s), Thompson emerged as the winner, and by the mid 1950s the importance of fishing pressure in determining fish abundance was very well established. It was so well established in fact, that almost our entire body of fisheries theory is built upon the assumption that changes in abundance are due to fishing mortality balanced against the natural innate growth and reproductive processes of the fish.

However, in the last decade the debate has been revived as many clear cases of fluctuation due to "natural" factors such as temperature, upwelling, competition, and predation have come to light. The historical records from anoxic sediments have provided strong evidence that environmental factors and "natural" change can be as important as fishing mortality. The debate is perhaps more mature now—we question the relative importance of environmental conditions and fishing—but the essence of the debate remains the same. Furthermore, as Walters and Collie (1989) note, we will likely be very slow about resolving the debate, because it will continue to be prudent for managers to respond to declines by assuming that fishing may be responsible, so we will continue to get data where the effects of management and environment are statistically confounded (both effects present in the data).

As a historical note, the Thompson-Burkenroad debate with respect to Pacific halibut remains unresolved. Despite what is arguably the best fisheries data set in the world, one can explain the history of the Pacific halibut stock equally well as changes due to environment or changes due to fishing. Figure 3.5 shows that the recruitment has been quite cyclic. In fact, Thompson was almost surely wrong in his original argument about overfishing, since the largest recruitments of halibut have come from the *smallest* adult biomass levels. But now the debate is even more interesting: should adult biomass levels be kept low in hopes that large recruitments will be consistently produced (through some mechanism such as reduced cannibalism), or should stock recovery be promoted so as to buffer against the effects of a cyclic decline in juvenile survival due to environmental factors?

3.3. Principles of Dynamic Behavior

As ecologists began to describe and model the behavior of natural populations, they needed a vocabulary appropriate for dynamic systems. As it happened, electrical engineers had such a vocabulary, as well as models that have proved good elementary starting points for ecological analyses. This

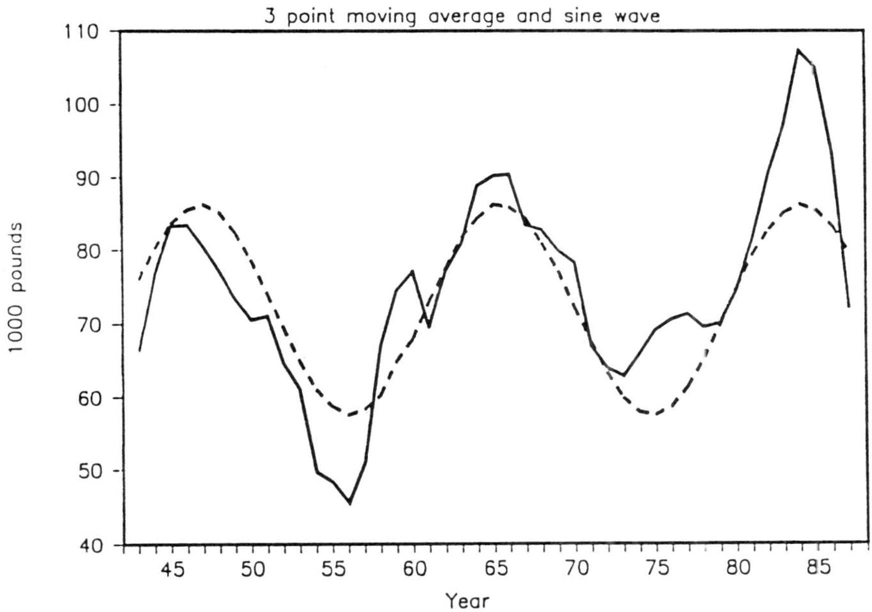

Figure 3.5. The history of recruitment to the Pacific halibut fishery. The sine curve shows the cyclic nature of recruitment. Biomass of 3-year-old individuals is the measure of recruitment. From Deriso (personal communication).

vocabulary and modelling approach is now widely used in ecology (and fisheries biology), and understanding its basic concepts is a necessary foundation for understanding the dynamics of both real fisheries and models of fisheries.

Stability

Stability is a term often used in fisheries and is used in many conflicting ways. In an engineering sense, a system is stable if it tends toward a constant state when unperturbed and returns back to the same constant state when perturbed. To illustrate principles of dynamic behavior, we are going to explore the behavior of a very simple fishery simulation model. In our model, we will consider the biomass of the fish stock (B) and the size of the fishing fleet (E). The biomass will behave according to the simple logistic model, and the fishing fleet will grow when catch per boat is above a critical level and shrink when the catch per boat is below the critical level (U^*). The equations for this model are

$$B_{t+1} = B_t + rB_t\left(1 - \frac{B_t}{k}\right) - C_t$$
$$C_t = B_t E_t q$$
$$U_t = \frac{C_t}{E_t}$$
$$E_{t+1} = E_t(1 + s(U_t - U^*)) \qquad (3.3.1)$$

where U_t is the catch per boat, C_t is the total catch, r is the intrinsic rate of growth of the fish stock, k is the "carrying capacity" or unfished equilibrium stock size, q is the efficiency of each fishing boat, and s is the responsiveness of the fleet size E_t to U_t. If s is small, then the fleet will grow (or shrink) slowly, and, if s is large, the fleet will respond rapidly to the difference between U_t and U^*. U^* is the catch rate where fishermen have zero profits. If U_t is greater than U^* the fishery is profitable and more fishermen will enter. If U_t is less than U^* fishermen will leave the fishery. For the illustrations given below, we will use $r = 0.4$, $k = 1{,}000$, $q = 0.01$, and $U^* = 5$. We will vary s, depending upon what type of behavior we wish to illustrate.

If we begin a simulation with the stock in an unfished state ($B_t = k = 1{,}000$), a fleet of 5 boats, and the response parameter $s = 0.05$, we get the time pattern shown in Figure 3.6 from time 1 to time 50. The population declines to 500 and the fleet increases to 20, with a slight overshoot along the way. At time 50, we add 20 boats, which causes the biomass to initially drop and the fleet size to decline. After another 50 years, the biomass and fleet size return to the stable equilibrium of 500 units biomass and 20 boats. On the right side of Figure 3.6 is a graph showing fleet size versus biomass. The trajectory for the first 50 years of simulation shows biomass declining and fleet increasing. After the perturbation, the fleet size declines and the biomass declines and then recovers.

Such a system is called stable, because if perturbed it will return to equilibrium. Most of the fish population models we consider in this book are stable; if the stock is fished, the biomass declines, but, if fishing is halted, the stock will return to its unfished state.

Instability

Instability is the description of a system that does not return to its original state when perturbed, but instead leads to the extinction of some element of the system, either the biomass or the fleet in this example. We can make our simple biomass-fleet model unstable by increasing the response parameter s to a value of 0.3. This means that the fleet will respond very rapidly to catches per boats that differ from $U^* = 5$. If $U_t = 6$, then the fleet would

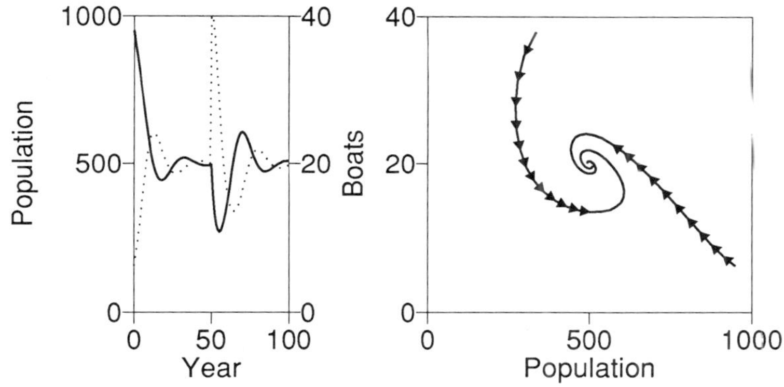

Figure 3.6. Behavior of a stable population when perturbed. The left hand graph shows number of boats and fish. In year 50, the number of boats is increased to 40. The right-hand graph shows the number of boats plotted against fish-population size.

grow by 30% in 1 year. This leads to the behavior shown in Figure 3.7. The system is started near the equilibrium, but large fluctuations result, which cause the biomass and fleet to eventually go to zero.

Cyclicity

A cyclic system is never in a true steady state but is instead always changing. Cyclicity is displayed by our model when the response parameter $s = 0.22$. The type of cycle shown in Figure 3.8 is called a stable-limit cycle because if perturbed the system tends to return to its original cyclic pattern.

Boundaries and multiple stable states

A system that will always return to its original state regardless of how much it is perturbed is called globally stable. The alternative is local stability, where the system will return if pushed to certain limits, but if pushed beyond those limits will not return. Implicit in the concept of local stability is the existence of boundaries, which are states beyond which the system will fail to return. Consider two competing species, such as sardines and anchovy in California. Imagine that in the undisturbed state, sardines are abundant and anchovy are rare, and if slightly perturbed the system will return to that condition. However, if pushed too far the system will flip and become one of low sardine abundance and high anchovy abundance. This provides a potential model for explaining the history of the California sar-

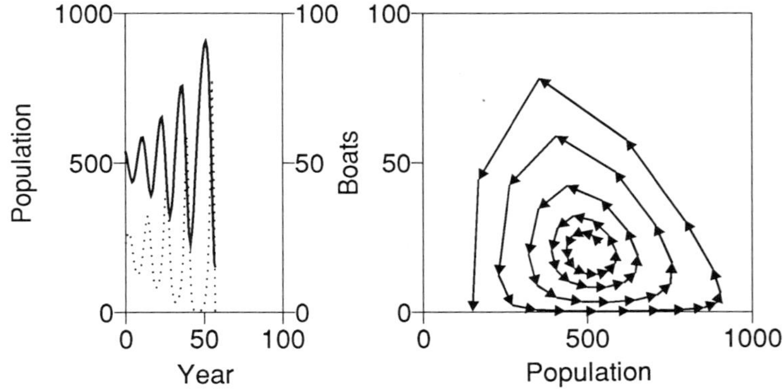

Figure 3.7. Behavior of an unstable population. When disturbed, the system does not return to equilibrium.

dine fishery. Skud (1982) offers such an explanation for the flip between sardines and anchovy in California and mackerel (*Scomber scombrus*) and herring (*Clupea harengus*) in New England. Skud suggests that the flip is caused by major environmental changes.

The area inside particular boundaries can be thought of as a region of stability or domain of attraction. From a practical perspective, we may not care if a fish population is truly stable inside this region; so long as it remains in roughly the same region, cyclic behavior or random variability within this region may be perfectly acceptable to the fishermen or managers. Although multiple equilibrium behavior caused by competition is theoretically attractive, the only real evidence for it is rather circumstantial. Perhaps more evidence will accumulate in the next few decades.

Predator thresholds

Boundaries can be thought of as thresholds: to go beyond the threshold is to enter into a region of new behavior. One particular kind of threshold occurs when fish populations are faced with a predator whose abundance does not respond rapidly to changes in the abundance of the species being preyed upon. For instance, if the predator spends most of the year feeding on one prey species, but also has a secondary prey species, the abundance of the predator may be quite independent of that of secondary prey, yet the predation impact on the secondary prey can be substantial. If the predators are efficient at finding and capturing prey, then the number eaten will be more or less constant. As the prey abundance declines, the mortality rate caused by the predators increases, until at some point the mortality rate is higher than the intrinsic ability of the prey to reproduce themselves. At this point, all other things being equal, the prey population will decline toward

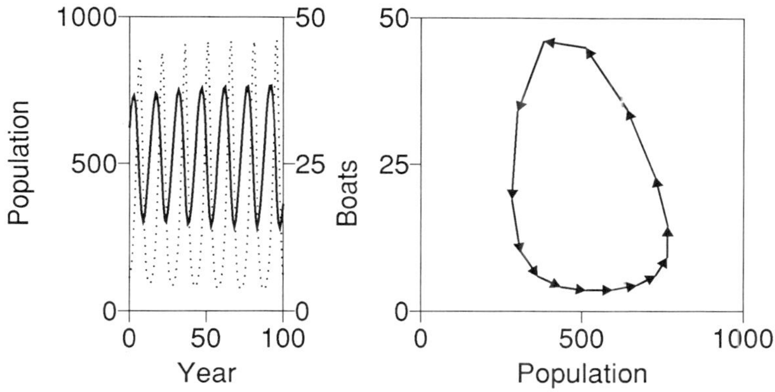

Figure 3.8. A cyclic fishery.

extinction. This phenomenon is called depensatory mortality and its graphical representation in Figure 3.9 shows that below some density, the prey will be unable to sustain the predation pressure and their growth rate will be less than 1.0.

As an example, consider a modification of our previous model of fish biomass and a fishing fleet. Let us assume that the fleet size is held constant through some combination of social programs, economic incentives, or regulatory measures. Let us also assume that the amount of fish each boat catches increases as biomass increases up to a maximum amount that the boat can handle in a year considering time and hold capacity constraints. The relationship between prey abundance and amount consumed per predator (boat) is called the functional response. To represent this response as a curve that is steepest at low fish density and increases to a maximum, we will use the Michaelis-Menten or Holling disc (Holling 1959) equation for U_t. The overall model is then

$$B_{t+1} = B_t + rB_t\left(1 - \frac{B_t}{k}\right) - C_t$$

$$U_t = \frac{aB_t}{b + B_t}$$

$$C_t = U_tE \qquad (3.3.2)$$

If we have a constant fleet size $E = 10$ boats, and use the values $a = 10$ and $b = 100$, we obtain the behavior shown in Figure 3.10. So long as the biomass stays above roughly 200, the system is stable, and will always return to an equilibrium population of approximately 900. If the biomass

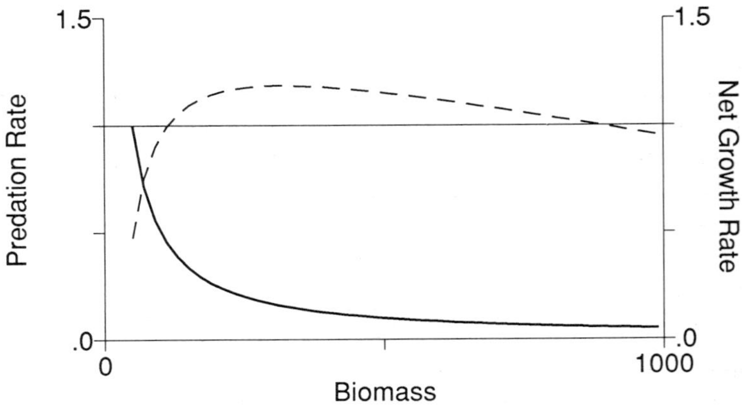

Figure 3.9. Mortality rate inflicted by a predator with constant abundance (solid line), and prey net growth rate (dashed line) plotted against prey biomass. When prey biomass is very low, the predation rate is high and the net growth rate is less than 1.0 (the thin horizontal line).

somehow drops below a critical level near 200, the population will decline toward zero.

It is felt that such predator thresholds may be quite common in nature (Peterman and Gatto 1978); however they have been difficult to document. The best examples deal with the most efficient of predators—man. Figure 3.11 shows the fishing mortality rate on Norwegian herring as the fishery declined in the 1960s and 1970s. The catch remained high while abundance declined, and the stock was driven very close to extinction. The fishing fleet acted as a predator with a near-constant consumption rate until the stock abundance was very low.

A variation on the theme of Figure 3.11 can occur where each fisherman exerts less effort as the stock declines. In this case, the functional response will be what is called type III, and it will have a sigmoid shape. Instead of being purely depensatory, fishing will instead act in a "compensatory" fashion at low stock sizes: whenever the stock declines further there will be less effort, but whenever it increases there will be more effort. This behavior will tend to trap the stock at a low level, rather than drive it to extinction. The Peru anchoveta stock may have been caught in such a situation. Predator pits have attracted considerable attention because of their implications for fisheries management. If there are predators other than man that produce "natural" predator thresholds, then the fish-fisherman-predator system may be stable under low fishing pressure, but may collapse when fished below a critical stock level, and, even if fishing is stopped, the stock may not recover.

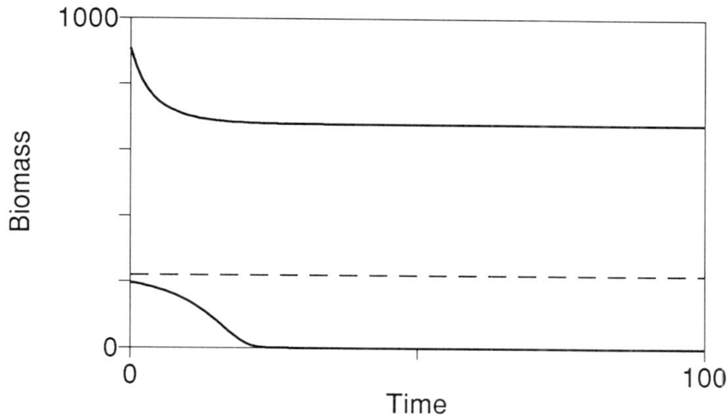

Figure 3.10. Multiple equilibrium behavior with a predator threshold. The dashed line is the biomass threshold of 200. If the stock stays above 200, the equilibrium is at approximately 900, but, if the stock is reduced below 200, it collapses.

Resilience

Resilience is another term like stability that has a technical usage that is slightly different from the dictionary definition. C. S. Holling (1973) noted that most natural systems are characterized by change, and that man's attempts to stabilize nature for his own purposes often lead to short-term constancy and long-term disaster. His perception was that change is necessary to maintain the character of many natural systems. Natural systems, from this perspective, are not stable, but do exhibit changes within certain bounds or regions of stability. A system with a large region of desirable behavior is called *resilient*. Man's attempts to reduce local fluctuations often cause the boundaries to shrink, so the region of desirable behavior becomes smaller until it is so small that management actions cannot contain the system within this region, and it flips to another, much less desirable condition.

Figure 3.12 illustrates this concept using forest-fire prevention as an example. In the natural state, fuels build up until sufficient for a low-heat fire, and then such a fire will occur, rarely killing the mature trees. This is the loop labeled "natural fires." Fire prevention inhibits the small fires, until so much fuel accumulates that a fire able to kill most of the mature trees occurs. This is the loop labeled "tree-killing fires." The management action of fire prevention, in attempting to stabilize the system, reduces its resilience and leads to catastrophe. Resilience is, therefore, the ability of a system to utilize, profit from, and absorb natural variation.

The concept of resilience has many important implications for fisheries

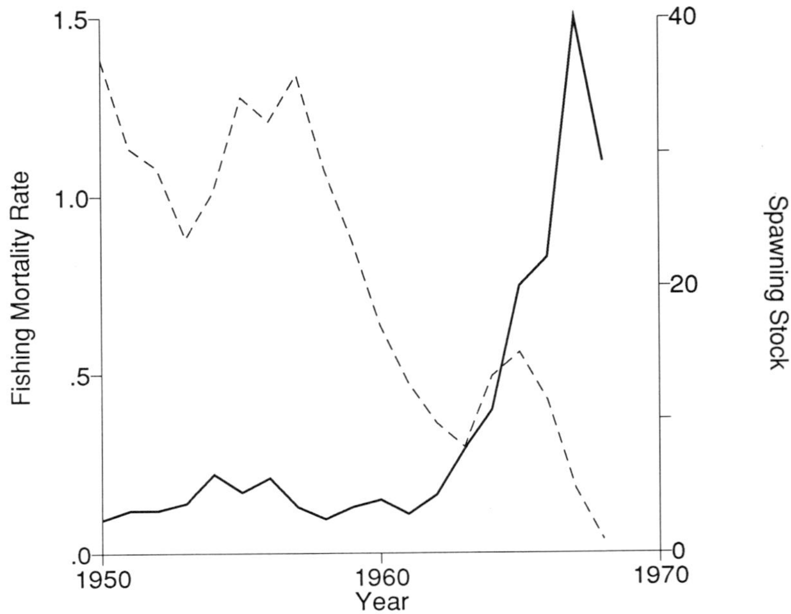

Figure 3.11. Exploitation rates (solid line) and spawning biomass (dashed line) of the Norwegian herring fishery. Data from Dragesund et al. (1980 table 3).

management. If it is true that natural variation is important for the long-term maintenance of system behavior, we might not want to try to stabilize population size about some hypothetical optimum level. Rather, we might want to let population size fluctuate naturally. Fish-fisherman interactions may be naturally resilient; fishermen may learn to absorb and even utilize the year-to-year fluctuation in yield by adopting social and economic mechanisms to buffer themselves from year to year changes in income. Management actions designed to stabilize yield may make the fishermen less resilient to income fluctuation, leading to bankruptcy and other disruptions when a major change in stock abundance occurs.

Myths of dynamic behavior

Now that we have seen how some real populations behave and examined the terms and concepts used to characterize such behavior, we can now consider several alternative "world views" or, as we prefer to call them, "myths." We use the term myths to indicate that they are gross oversimplifications of reality, but they do provide a reference point in describing an otherwise impossibly complex world. We do not intend the term myth to mean something that is intrinsically untrue. Each of these myths is a construct that

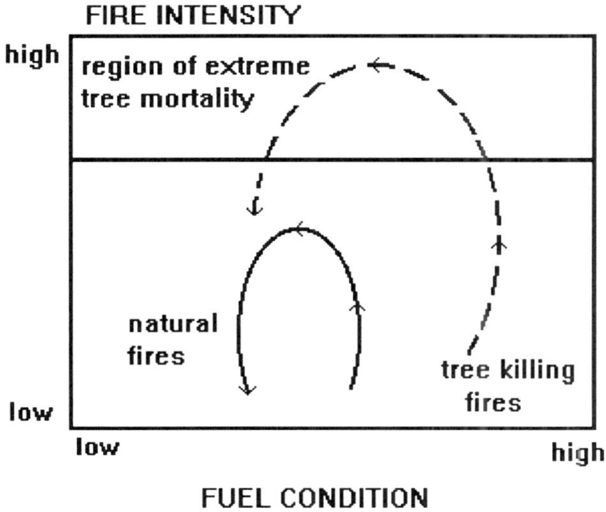

Figure 3.12. Dynamic behavior of a forest fire-fuel system. In the unmanaged forest, frequent small fires take place but are not intense enough to inflict severe tree mortality (solid line). When small fires are suppressed, fuel accumulates, and, when fires do occur, they inflict severe tree mortality (dashed line). Redrawn from Holling (1980).

enables us to characterize how we believe nature responds to fisheries management, and a framework for the interpretation of the data we may observe.

The balance of nature

One myth of population dynamics is closely tied to a rather naive ecological myth that in the absence of man nature exists in some sort of balance, so that on average the abundance of species is roughly constant over time. There might be year-to-year changes of a minor magnitude, but in general the underlying structure of the ecosystem is unchanging on a scale of decades. This myth is tied to a view of ecosystems as globally stable; if perturbed they return to their natural equilibrium. Even though few practicing ecologists or fisheries scientists would agree that this myth has much to do with reality, almost all fisheries theory is based on it! Almost all fisheries models used for the analysis of data and evaluation of management options are built on the assumption of historical equilibrium and/or equilibrium as a desirable goal of management. This is probably the single greatest indictment of fisheries management as practiced today.

Randomness

A second myth is that the world is mostly random; species may come and go, food web connections may appear and disappear, whole communities may disappear for decades, all driven by external climatic and environmental forces. The classic ecological debate in the 1950s over density dependence revolved around this issue. Some argued that most of the fluctuation in abundance seen in nature was due to random, usually weather-related events (the Andrewartha and Birch school). The Thompson-Burkenroad debate discussed earlier revolved around the issue of how much of the fluctuation in Pacific halibut abundance was due to environment, as opposed to fishing.

Spatial balance

A compromise myth about ecosystem behavior is that the world may be quite random with reference to any single spatial location or small area, but that there is a "balance" in the larger spatial scale, achieved either by statistical averaging over many spatial sites, or by the dynamics of spatial connectedness. This myth is based on ecological observations that things may come and go at any particular location, but that at a larger spatial scale there is more constancy.

If one accepted this myth, the types of models and the types of fisheries policies we would have to adopt would be much different than those we use today. Rather than setting fisheries regulations (quotas, seasons, etc.) for very large areas, we might attempt to regulate for spatial diversity, allowing for some areas to be at high densities, some to be at low densities, and so forth. Most fisheries regulations, coupled with fishermen's behavior as described in the next chapter, will assure a minimum of spatial heterogeneity—the regulations and fishermen tend to homogenize any spatial variability that exists. If the long-term persistence of fish ecosystems depends on spatial variability, such policies will be disastrous.

Multiple stable states

The final myth is that ecosystems have a number of intrinsic possible conditions or states, and they periodically bounce between these states depending upon environmental conditions or other external perturbations (such as fishing). Things are not random, in the sense that there are internal relationships between the species that determine what states are possible, but randomness may have an important role in determining what state is achieved at any time.

Accepting this myth would require fishery managers to be much more careful than if they accept the balance of nature myth. If fishery systems may not recover from overfishing, managers would need to be aware of limits and be much more cautious in formulating management plans. This myth poses many technical difficulties; can we detect limits without exceeding them, and if we have passed a limit, can we recover? Concerns reflecting this myth are now appearing in fisheries management. In some cases managers are now including risk of collapse estimates as one indicator to examine in choosing regulations, but the implications of this myth are so frightening that most managers simply prefer to ignore it.

3.4. Representations of Population Dynamics

Having surveyed some general principles of how populations behave, we now look at some of the many alternative models used to describe population dynamics. A confusing variety of mathematical models and equations are used to represent fish population dynamics. In part, this state of affairs is a reflection of the history of development of the field of stock assessment, where various lines of thinking and mathematical notation have developed in parallel at different research centers around the world. But more importantly, there is in practice no single ideal model or way of representing the dynamics of exploited populations; such a model would have to cope with an impossible variety of arbitrary definitions of the exploited stock, with a great variety of biological life history patterns, and with data ranging from very simple and aggregated to very complex and detailed. In practice, we find it most useful to be flexible about defining populations and to have a variety of models (representations) available for use when different qualities and quantities of data are available.

What is a population?

Much fisheries management is based on the presumption that the most important biological entity or element or unit of monitoring and regulation is the single species unit stock. A unit stock is an arbitrary collection of populations of fish that is large enough to be essentially self-reproducing (abundance changes are not dominated by immigration and emigration), with members of the collection showing similar patterns of growth, migration, and dispersal. The unit should not be so large as to contain many genetically distinct races or subpopulations within it, although it may be smaller than any genetically distinct racial unit of the species (as measured through morphologic or biochemical indicators).

Often it is obvious how to define unit stocks through clear physical boundaries and distinctive patterns of migration and spawning. In other cases, the species may exhibit complex geographic variability, with clinal variation perhaps, but no obvious boundaries between subpopulations that might respond differently to management. In some cases, it has been found that there is highly localized genetic variability among fish from different spawning areas and times of spawning; in such cases, trying to manage each genetic unit separately might involve impossibly cumbersome and costly programs of monitoring, specification of regulations, and enforcement of regulations.

It is important to recognize that the idea of a unit stock is a very slippery one. It is convenient to think of a unit stock as a homogeneous collection of fish that are all subject to the same opportunities for growth and reproduction and the same risks of natural and fishing mortality. If such a unit actually existed in nature, it would be very easy to monitor and to model for predictive purposes; we would need only to define the growth, reproduction, and mortality risks for a typical fish in the collection, and then multiply these by the number of fish to predict overall population behavior.

In fact, any natural stock of fish that is large enough to be of management interest will consist of a highly heterogeneous collection of individuals with different sizes, ages, growth rates, movement patterns, reproductive abilities, behaviors in response to fishing gear, and risks of mortality. Nor will these attributes of individual variation even be independent of one another; for example, larger individuals are likely to have lower risks of natural mortality due to predation. We might classify individuals according to attributes such as age and size, hoping that individuals within each class will behave more homogeneously (and predictably), but there will still remain substantial variability within any class that is of manageable size in terms of monitoring (sampling) and prediction.

Biologists sometimes talk about a fish stock as though it were a coordinated biological entity or superorganism subject to natural selection and evolution. They speak about populations "struggling to survive" in the face of fishing, "pushing back" against environmental and fishing stresses, or "displaying compensatory responses" to stresses. These images, analogies, and metaphors for thinking about population dynamics can be grossly misleading. Although natural selection can in principle operate on groups or populations under particular circumstances, there is generally no reason to expect that such circumstances apply to the larger and arbitrarily defined collections of individuals that we call unit stocks. To the extent that unit stocks do exhibit predictable behavior of statistical measures, such as total stock biomass or numbers, it is foolish to hope that this behavior will be explained or predicted by reference to models for how individual organisms survive.

It is also convenient to talk about populations as having certain fixed parameters from which some aspects of population "behavior" can be predicted. The changes in time and space that we call population behavior are in fact the statistical results of a very large number of interactions among individual animals and between individuals and their environment. These statistical results would be fully predictable from average parameters of survival, reproduction, movement, and growth only if all the individuals were identical to one another, were not subject to natural selection and evolution, and were all subject to the same environmental conditions. Obviously, real biological systems are not so simple; any so-called parameter is likely to exhibit at least some change over time, and vary from place to place for different stocks of the same species.

The fact that there are really no fixed biological parameters brings us to a key point; we know that any model assuming fixed parameters will be "wrong" in a strict sense, that is, it will make numerical predictions that are not precisely correct. Therefore, the issue in choosing a model for prediction is not whether it is right or wrong. Rather, the practical issue is deciding which simplifying conveniences are good enough, or which are the best among several imperfect choices, for making predictions given whatever limited historical information is available.

The simplest models used in fisheries look only at the biomass of the stock, and are called biomass dynamic or production models (see, for example, Figure 3.13). There are four major directions in which these models are often extended. Most commonly, explicit age structure is added to the models. The other three directions for extension are (1) adding the dynamics of the fishing fleets, processing, and marketing, (2) adding interactions with other species and the environment, and (3) adding explicit spatial representation of stock structure.

In later sections of this chapter and in the next chapter on fleet dynamics, we explore these models by beginning with a basic biomass dynamics model and then adding the above components.

Representations of temporal change

Consider an arbitrary collection of fish that are being treated as a unit stock. This collection has a number of quantitative attributes that measure its potential to produce immediate yields. These attributes include the total number of fish, density of fish per unit area or volume over the areas where they are harvested, the total biomass of fish, the biomass per unit area, the mean size (weight or length) of fish, and so on. Stock-assessment models attempt to make various assertions about how one or more of these attributes will behave over time in response to fishing. The assertions (or predictions) can be expressed in three different ways:

1 As rates of change in attribute values
2 As attribute values after some fixed time period, in relation to the current values

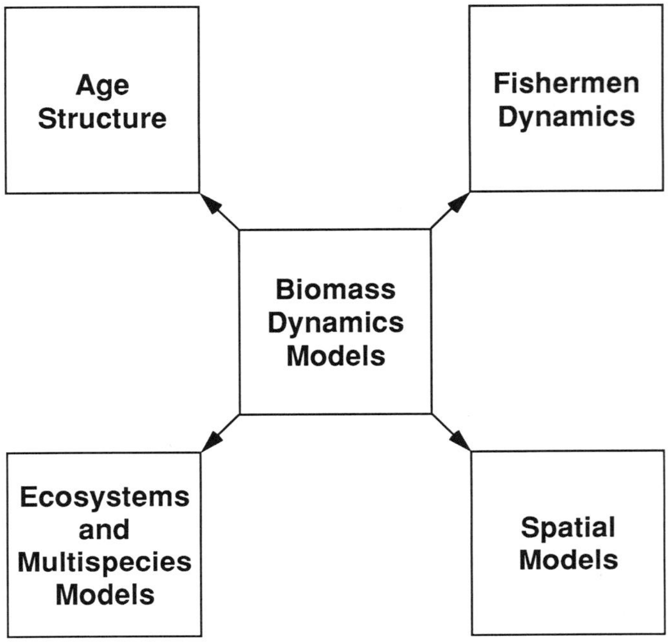

Figure 3.13. Biomass dynamics as the basic core of fisheries models and four directions of possible elaboration.

3 As equilibrium or long-term average values of the attributes, toward which the values will tend to move over time.

Assertions of type (1) are known as differential equation models; in developing them, we pretend that the attributes are continuous quantities that are subject to inflows and outflows analogous to the amount of water in a container. Assertions of type (2) are called *difference equation* models; they attempt to predict the net result of processes operating over a fixed time period, without looking at the detailed patterns of events within the period. By linking together a sequence of type (2) predictions (where the result of each prediction is used as the current value for the next prediction, repeatedly), we generate what is called a *simulation* of longer term behavior. Simulations can also be done using differential equation models. Assertions of type (3) attempt to avoid the obvious risks of short-term prediction associated with assertions of types (1) and (2) in favor of making simpler (and hopefully more robust) predictions about directions of change and eventual values of attributes.

Differential equation models can be solved over time (integrated) to give difference equations. This can result in the proliferation of symbols and what appears to be more complex and realistic predictions, when in fact very simple assumptions are being made. Consider for example the simple type (1) model for change in the number N of fish in relation to a fishing loss rate F:

$$\frac{dN}{dt} = -FN \tag{3.4.1}$$

which says that N will decrease faster when N is large and more slowly as N falls, even if the fishing pressure remains the same. This differential equation has the solution

$$N_1 = N_0 e^{-F} \tag{3.4.2}$$

where N_0 is the initial number and N_1 is the number after one unit of time. An equivalent and simpler way to write this model is as

$$N_1 = N_0 s \tag{3.4.3}$$

where s is the survival rate over the unit of time. Notice that the differential equation solution, with its e^{-F} term, looks more complicated and realistic. This is often the case with type (1) model solutions; they tend to impress people who are inexperienced with mathematics, just because they are visually more elaborate. The real key to evaluating the two alternative predictions for N_1 from N_0 is not which equation is more impressive looking, but rather how we are to assign reasonable values to e^{-F} or its simpler equivalent s.

Early theoretical studies in fish-population dynamics were usually based on differential equation models, such as Schaefer's (1954) model based on logistic population growth and Beverton and Holt's (1957) dynamic-pool models for yield in relation to recruitment, growth, and mortality. These models could be solved mathematically to give predictions with a relatively small amount of computational effort, at a time when numerical calculations were burdensome and time consuming. Today, most models are developed in terms of difference equations and make predictions over discrete time steps; these equations are intuitively easier to formulate and to understand, and are easily solved for numerical predictions using computers.

Equilibrium or average predictions are developed by first formulating a model for change over time, then solving this model for its equilibrium values. For example, if population size next year is described as "a survival rate times population size this year plus new recruits," we can write

$$N_{t+1} = sN_t + R \tag{3.4.4}$$

and for fixed survival rate s and recruitment R the predicted equilibrium will occur when $N_{t+1} = N_t = N$. Substituting N for N_{t+1} and N_t in the above equation and solving for N gives

$$N = \frac{R}{(1 - s)}. \tag{3.4.5}$$

This simple and useful equation predicts that the population size at equilibrium will be the recruitment rate R divided by the total mortality rate $1 - s$, and the fraction of recruits R/N at equilibrium will simply be the mortality rate $1 - s$. Many elaborate looking equations of stock assessment involve exactly such simple assumptions about survival and recruitment (and growth) and such simple algebraic manipulation, but are given an appearance of greater realism through the use of more complex symbols and notation.

An important trick in stock assessment modelling (and in understanding other people's models) is to first get a general "balance equation" for each quantity to be predicted, then work out how to further elaborate the terms or components of this equation in relation to variables of interest. For example, prediction of numerical change usually involves the balance equation given above ($N_{t+1} = sN_t + R$); simple models are obtained by treating the components s and R as constant (e.g., $R = 0$ in a closed cohort of fish), and more complex predictions are obtained by making s and R vary in relation to factors such as fishing effort and parental stock size. It is useful to think in terms of an hierarchy of calculations: the balance equation is the top level in the hierarchy, at the next level down are the components s and R, and at the next level are equations used to predict s and R from other factors.

Representations of changes in time are relatively easy to formulate once the modeller has defined precisely what variables are to be predicted. Most apparent differences in models for temporal prediction are just notational differences that arise from whether the modeller began his analysis with differential or difference equations to represent the basic components of change. The situation concerning population structure is not so simple; therefore a great variety of representations are possible.

Representations of population structure

Population models vary greatly in how they carve up the total stock size into smaller, presumably more homogeneous and predictable, groups of fish. At one extreme are general models for surplus production, which look only at the total, aggregate biomass of the stock. At the opposite extreme are models that keep a separate account of the number of animals in many size and age groups, spatial locations, genetic races, and so forth. The most common breakdown of total population size is into the number of animals by age class, where age is associated with differences in size, growth rate, fecundity, and vulnerability to fishing. Less common are models that divide the population by size structure; with variable growth rates, age cannot be used as a simple surrogate for size. Still less common are models that divide the population into different spatial subpopulations or areas.

Two factors determine whether it will affect overall population predictions to do the calculations separately for a collection of structural units such as age classes: (1) variation in parameters among units and (2) variation in relative contributions by different units over time. Consider for example a situation in which the population is divided into a number of units, each with N_i animals (i may be age, for example) subject to an annual survival rate s_i. The total number of surviving animals in this population after 1 year will be predicted to be

$$N_{t+1} = \sum_i s_i N_i \tag{3.4.6}$$

where $N_t = \Sigma_i N_i$. Notice here that N_i can be written as the total N_t times the proportion P_i that are in class i ($P_i = N_i/N_t$). This means that N_{t+1} can be predicted as the total N_t times the sum of the s_i times P_i:

$$N_{t+1} = N_t \sum_i s_i P_i \tag{3.4.7}$$

The term $\Sigma_i\ s_i P_i$ is a weighted average survival rate, say $\bar{s}$, with s_i being weighted by the proportion P_i of type i animals in N_t. Thus the overall prediction can be written more simply as

$$N_{t+1} = N_t \bar{s} \tag{3.4.8}$$

Notice that if the s_i are assumed constant over time, this overall prediction will be identical to the more complicated sum of component predictions, provided the P_i are also constant over time. Thus, it is not just differences in s_i values among the groups that determine whether a more detailed representation will be more accurate; for the more detailed model to do any better it is also necessary that the population composition (P_i values) is changing in time or in relation to policy changes.

It has long been recognized that population age compositions can be dramatically altered by fishing, with associated changes in average fecundities and body sizes per fish. The so-called yield-per-recruit and dynamic-pool models attempt to account for such changes, as do most fish-population simulation models. The main policy concerns in such modelling are finding the best size of fish to harvest (preventing "growth overfishing") and measuring the impact of different harvest policies on total reproductive effort by the stock (preventing "recruitment overfishing").

Less well recognized (at least by population modellers) are the changes in spatial structure that often accompany fishery development. Typically, fishing is first directed at the most accessible and concentrated groups of fish. Often these groups are overharvested, and the fishermen move on to less accessible and more dispersed parts of the stock. Recoveries in the more accessible groups may be prevented by the fishermen switching back onto these groups whenever they show any sign of recovery. Better models of spatial structure can help one interpret changes in fishery statistics (catches, catches per unit effort) during development, and assist in the development of policy options for redistributing fishing effort so as not to waste (through overfishing) the potential production from any part of the spatial structure.

Finding the best representation: matching models to data

Most biologists appear to assume that more is better when it comes to detail in population modelling, as though taking account of more aspects of population structure in space and time will always lead to more accurate predictions of population change. This intuitive notion is dangerously misleading in two important regards: (1) more detailed predictions can in fact be less accurate due to cumulative errors when many inaccurate parameter estimates are used; and (2) gathering progressively more detailed information can become very expensive, consuming financial resources that management agencies might well put to better use in improving the collection of certain basic statistics and in other management activities, such as enforcement of regulations.

Choosing the best model or representation for dealing with any particular policy question first involves being very careful about what is meant by "best." The stock assessment literature is filled with examples where biologists have presented measures of goodness of fit (correlation between model and data, etc.) to justify the use of particular models. But the ability of a model to fit past data says absolutely nothing about whether it will make correct policy predictions! It is quite possible to construct a model that fits past data very well, but whose individual parameter values are so poorly determined that the parameters most important to policy are in fact very badly estimated. For example, the estimate of optimum fishing effort (obviously a key policy parameter) from biomass dynamics models is very sensitive to model parameters that measure the rate of population growth when population size is small (intrinsic rate of increase, maximum productivity per spawner when spawning stock is reduced, etc.), but is insensitive or invariant to parameters that measure how large the stock will be when it is not fished (carrying capacity parameters). If the data used for estimating population parameters are mainly from a historical period of high stock sizes, the "unimportant" carrying capacity parameters will be well estimated, whereas the growth rate parameters (and hence optimum effort) will be very poorly determined even if the model fits the data very well.

It is often assumed that the best model is one that will forecast short-term catches or stock sizes most accurately. But on inspection, this assumption is foolish. In many cases, such forecasts are not even needed, because there is an in-season regulatory system that can provide good stock assessments and regulatory changes within each fishing season without having to rely on accurate preseason predictions. In other cases, the most accurate short-term forecast can be obtained by using a statistical time-series model that predicts future values from some linear combination of past values; although the structure of time-series models is good for capturing patterns (trends, cycles, etc.) as they develop over time, such models are inappropriate or misleading if used to predict the effects of substantial alterations in regulations because population dynamics are nonlinear, especially in relation to recruitment relationships.

When the stock-assessment scientist has decided precisely what policy variables or management factors to make predictions about (or has been charged with making recommendations about), then the scientist can begin to make intelligent comparisons of alternative models. Usually these comparisons involve deliberately building (or selecting) several models, fitting the models to available data, and using methods, such as the Monte Carlo simulation, to decide which alternative is most likely to give the most accurate result. The key point is that a good assessment scientist will not become wedded to any particular model or method of analysis; deliberately and objectively a range of alternatives must be explored. There is a useful adage that "the truth often lies at the intersection of competing lies," and to use this adage in stock assessment means deliberately comparing a range of alternative models.

It is important in comparing alternative representations to not be fooled by the apparent differences among alternatives created by variations in mathematical notation or presentation. For example, many different equations are used to describe the average relationship between recruitment and spawning stock size. When recruitment overfishing is present, so that recruitment rate is clearly an increasing function of spawning stock, it does not really matter what equation is used to describe this increase; the key policy prescription (that spawning stock should be allowed to increase) will be the same for all models. When the data do not show clear evidence of overfishing (so recruitment looks like a shotgun scatter when plotted against spawning stock), no equation will allow a safe extrapolation about how low the stock can be driven before a relationship does become apparent.

General biological knowledge about a stock (including distribution, longevity, and growth) is not a reliable guide to the best model representation for a particular policy purpose. Ludwig and Walters (1985, 1989) have recently shown that the best model choice can even be one that is known to be unrealistically simple. Ludwig and Walters generated simulated data sets using an age-structured model (so they knew the "correct" model for the entire data set), and then showed that there are conditions under which the simulated population could be better managed by fitting data from it to a simpler biomass dynamics model than by fitting to the model that was used to generate the data in the first place. Generally, they concluded that simpler models are likely to outperform the correct (biologically more complex and realistic) model in situations where there has been little informative variation in harvest patterns over time; in these cases, parameter estimation performance for the simpler models was much better than for the realistic model. Fitting the realistic model involved trying to extract too much information from the data, so that all parameters ended up being poorly estimated for it.

Some theoretical work has been done on optimal model complexity. As more and more variables are added to a model, the model's ability to fit a data set will naturally improve, but the number of assumptions that must be made (parameters) increases even faster. O'Neill (1973) considered the problem of optimal model size in terms of measurement error, systematic bias, and resulting model inaccuracy, and suggested there was an optimal model size. Walters (1986) considered the same problem in terms of parameter uncertainty and prediction error and came to a similar conclusion. The relationships they suggest are shown in Figure 3.14.

The problem of choosing among alternative models is an important research area in stock assessment. In this text, our emphasis is on trying to provide the reader with a variety of choices and tools so that he or she can make intelligent comparisons and choose without having to rely on any one assessment model or technique.

3.5. Biomass Dynamics Models

Now that we have seen how natural populations behave and examined the general types of behavior seen in some simple models, we begin a review of the kinds of behavior found in the most commonly used fisheries models. The simplest fisheries population models are those that consider only a single indicator of population size, usually the biomass. These models ignore age structure and do not explicitly consider growth, recruitment, or differential vulnerability to fishing gear. These models are called biomass dynamics models because they deal primarily with the total biomass from the fish stock, rather than characteristics of age and size. Biomass dynamics

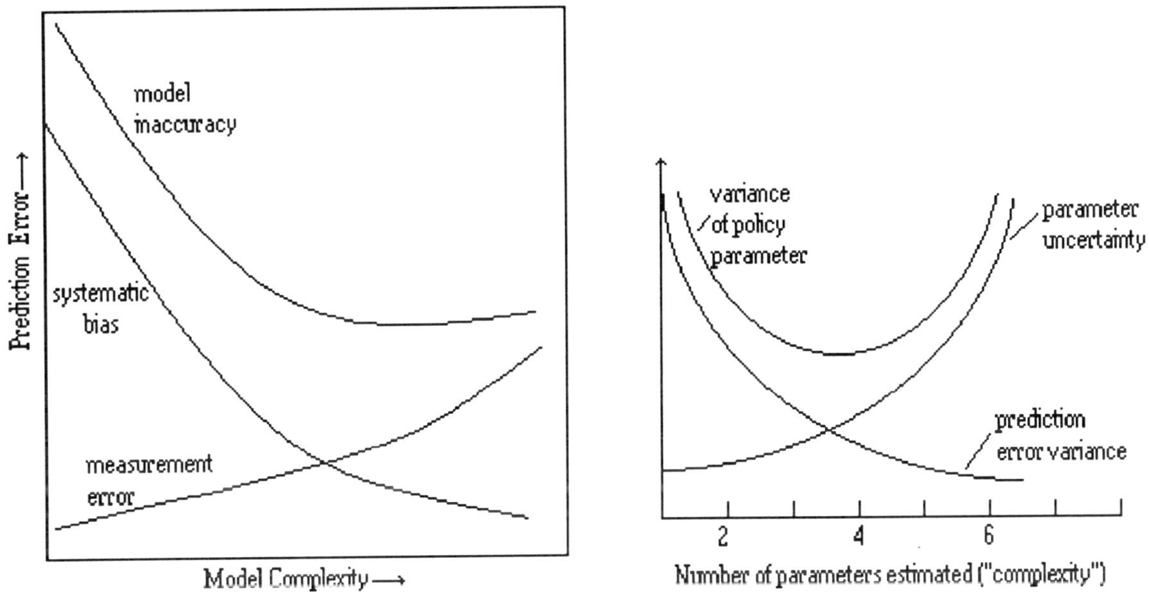

Figure 3.14. Determinants of optimal model size. Redrawn from O'Neill (1973) and Walters (1986).

models are variations on the traditional logistic models of ecology, and are quite widely used in fisheries management. Chapter 8 is devoted to estimation and use of biomass dynamics models, but in this chapter we look only at their behavior.

Logistic

The simplest population model commonly used in fisheries is the logistic growth model extended to include catch:

$$\frac{dB}{dt} = rB\left(1 - \frac{B}{k}\right) - C \tag{3.5.1}$$

where B is the biomass of the stock, r is the intrinsic rate of growth, k is the biomass the stock would tend toward if unfished, and C is the catch rate.

Normally the catch is assumed to be proportional to fishing effort and to stock size, which results in the model proposed by Schaefer (1954):

$$\frac{dB}{dt} = rB\left(1 - \frac{B}{k}\right) - qEB \tag{3.5.2}$$

where E is the fishing effort and q is a parameter describing the efficiency of the fishing gear (think of q as the proportion of the stock B taken by one unit of effort). The equilibrium properties of the model are

$$\begin{aligned} B^* &= k\left(1 - \frac{q}{r}E\right) \\ \left(\frac{C}{E}\right)^* &= kq\left(1 - \frac{q}{r}E\right) \\ C^* &= qkE\left(1 - \frac{q}{r}E\right) \end{aligned} \tag{3.5.3}$$

Figure 3.15 introduces a type of *behavior portrait* that we will use in this section to examine the equilibrium and dynamic behavior of fisheries models. Figure 3.15a shows the equilibrium yield-effort relationship (solid line), and a dynamic trajectory of yield-effort as effort increases at 15% per year starting with low effort and a virgin stock (arrows). The equilibrium relationship shows yield increasing with effort up to a maximum equilibrium yield (maximum sustainable yield or MSY). As effort increases beyond this point yield decreases until the fishing mortality rate is equal to the intrinsic rate of growth r, beyond which the stock is fished to extinction.

The dynamic trajectory in Figure 3.15a shows the same general pattern, except that yield peaks at a higher effort level and at a higher yield. As effort increases, short-term unsustainable yields can be obtained. This has

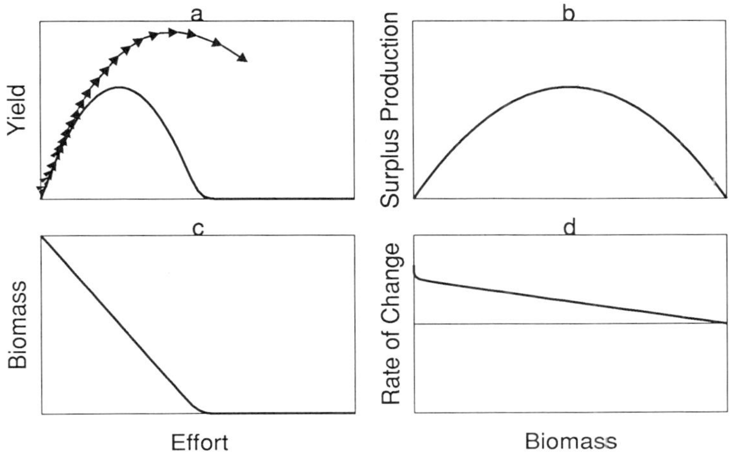

Figure 3.15. Equilibrium and dynamic relationships for the Schaefer model.

deluded many fisheries managers (and fisheries biologists) into believing that the sustainable yields from a fishery are higher than they truly are. One of the most persistent errors made by fisheries biologists is to try to estimate yields by assuming that the data come from a stock at equilibrium. Fish stocks are rarely at equilibrium and falsely assuming equilibrium usually results in overestimation of yield.

Figure 3.15b shows the equilibrium relationship between biomass and surplus production. Surplus production is defined as the catch plus the net change in biomass over some finite time period, generally 1 year. If the stock is at equilibrium, then surplus production is simply catch. If the stock is growing, surplus production is catch plus net growth of the population. As we can see in Figure 3.15b, surplus production is a symmetric curve, which is low at low biomass and also at high biomass.

Figure 3.15c shows the equilibrium relationship between biomass and fishing effort. Biomass decreases linearly with increasing effort. Figure 3.15d shows the rate of change versus biomass. The rate of change is relatively how quickly the population would change in the absence of harvesting. At high biomass, the rate of change is close to 1.0, but as biomass decreases, the rate of change increases, and in fact approaches $1 + r$ as biomass approaches zero.

Pella and Tomlinson

The Schaefer model of Equation 3.5.2 can be rewritten as

$$\frac{dB}{dt} = rB - \frac{r}{k}B^2 - qBE \tag{3.5.4}$$

Pella and Tomlinson (1969) suggested that the Schaefer model could be modified by raising B to an arbitrary power m to provide considerably more flexibility in the shape the surplus production vs. stock size relation might take. The Pella and Tomlinson model is therefore

$$\frac{dB}{dt} = rB - \frac{r}{k}B^m - qEB \tag{3.5.5}$$

where m is the power. When $m \neq 2$, then k is no longer the unfished equilibrium population size.

Figure 3.16 shows the portrait of equilibrium and dynamic behavior when $m = 4$. The effect of changing the power from 2 to 4 is to skew the surplus production curve to the right, so that maximum surplus production is obtained very near the unfished biomass (Figure 3.16b). If $m < 2$ then the surplus production curve is skewed left. Also note that the yield-effort curve (Figure 3.16a) and biomass-effort (Figure 3.16c) also show a rapid and somewhat sudden dropoff at higher efforts.

3.6. Age-Structured Models

The basic processes associated with age-structured models (Walters 1969) are straightforward and widely accepted; namely

$$\begin{pmatrix}\text{number alive}\\ \text{at age } a\\ \text{at time } t\end{pmatrix} = \begin{pmatrix}\text{number alive}\\ \text{at age } a-1\\ \text{at time } t-1\end{pmatrix} - \begin{pmatrix}\text{natural}\\ \text{mortality}\end{pmatrix} - \begin{pmatrix}\text{fishing}\\ \text{mortality}\end{pmatrix} - \begin{pmatrix}\text{fishing}\\ \text{induced}\\ \text{mortality}\end{pmatrix} \tag{3.6.1}$$

which simply says that for a closed population, with no immigration and emigration, the only three sources of loss are natural mortality (predation, disease, and other vagaries of nature), fishing mortality (catch), and fishing-induced mortality (fish killed by the fishing gear but not counted as catch).

If we add the following assumption for fishing mortality,

$$\begin{pmatrix}\text{fishing}\\ \text{mortality}\end{pmatrix} = \begin{pmatrix}\text{number}\\ \text{alive}\end{pmatrix} \times \begin{pmatrix}\text{fishing}\\ \text{removal rate}\end{pmatrix} \times \begin{pmatrix}\text{relative}\\ \text{vulnerability}\\ \text{to gear}\end{pmatrix} \tag{3.6.2}$$

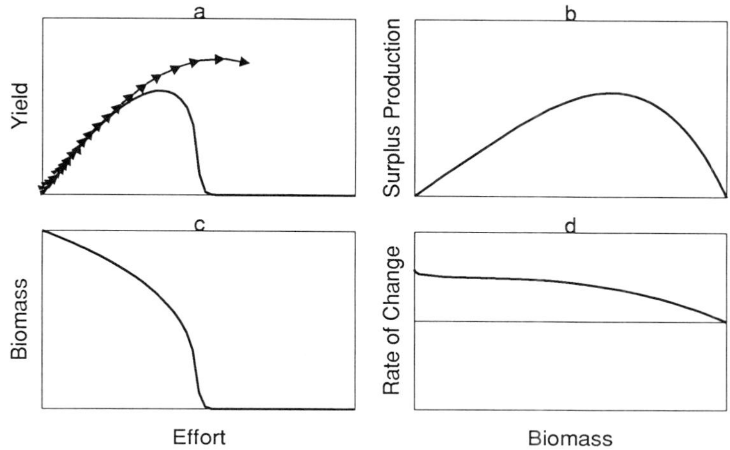

Figure 3.16. Portrait of behavior of the Pella and Tomlinson model.

and a similar assumption for fishing-induced mortality, we obtain the general set of age structured difference equations

$$N_{t+1,a+1} = N_{t,a}(1 - m) - C_{t,a} - G_{t,a}$$

$$C_{t,a} = N_{t,a} f_t v_a$$

$$G_{t,a} = N_{t,a} f_t g_a$$

$$B_t = \sum_a N_{t,a} w_a \qquad (3.6.3)$$

where m is a natural mortality rate, C is the catch, G is the fishing induced mortality, f_t is a fishing mortality rate for fully vulnerable individuals, v_a is an age specific vulnerability to fishing mortality, and g_a is an age-specific vulnerability to fishing-induced mortality.

If we also allow for age-specific fecundities (e_a), then the number of eggs produced by each cohort is

$$E_{t,a} = N_{t,a} e_a \qquad (3.6.4)$$

and the total egg production E_t is the sum over all ages of the eggs produced. The next assumption usually made is that there is an average relationship between total egg production in 1 year and the recruitment of 1 year olds the next year:

$$N_{t+1,1} = R(E_t). \qquad (3.6.5)$$

The two common stock-recruitment curves used are those of Ricker (1975) and Beverton and Holt (1957). The commonly used forms of these two curves are

$$N_{t+1,1} = (aE_t)/(b + E_t)$$
$$N_{t+1,1} = aE_t e^{-bE_t}. \qquad (3.6.6)$$

Basic processes and behavior

Going beyond simple biomass dynamics models enables us to capture many features of population dynamics associated with the time lags of age structure, and age-specific growth and vulnerability to fishing. In this section, we start with the simplest of age-structured models and examine its behavior and then gradually add new components to see how they contribute to the potential richness of the model behavior.

Natural mortality

The simplest age-structured model assumes constant recruitment and age-independent weight and vulnerability to fishing. We track the population in terms of total numbers (since all individuals weigh the same). The basic equations are

$$N_{t+1,a+1} = N_{t,a}(1 - m) - C_{t,a}$$
$$C_{t,a} = N_{t,a}f_t$$
$$N_{t+1,1} = R \qquad (3.6.7)$$

where $N_{t,a}$ is the number of individuals of age a alive at the beginning of time t, m is a natural mortality rate, $C_{t,a}$ is the number of individuals of age a captured during time t, f_t is an exploitation rate at time t, and R is the annual recruitment.

Figure 3.17 shows how this model behaves with $R = 10$ and $m = 0.2$. Figure 3.17a shows the equilibrium relationship between fishing effort and sustainable yield (solid line) and the transient relationship between fishing effort and yield as effort is increased at 15% per year (dashed line). Yield increases as effort increases. Since recruitment is constant and there is no growth, then the best biological yield will be obtained by fishing as hard as possible to capture the fish before any of them have a chance to die of natural mortality.

Figure 3.17b shows the equilibrium relationship between biomass and surplus production. Surplus production is highest at the lowest stock sizes, since at low stock size there is less natural mortality. Figure 3.17c shows the equilibrium relationship between biomass and fishing effort. As effort increases the stock size is reduced.

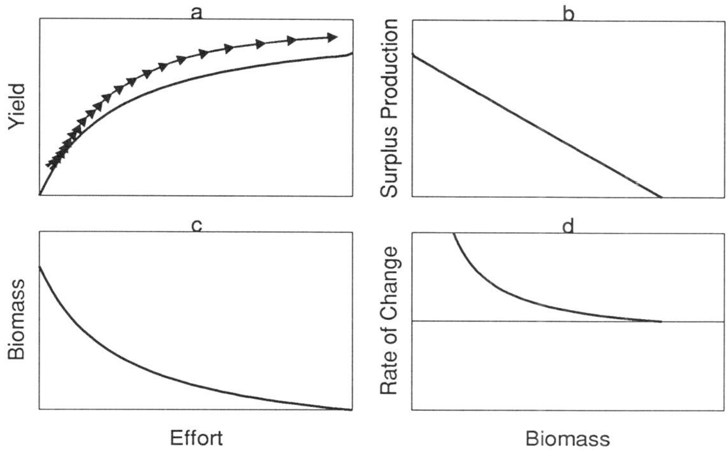

Figure 3.17. The equilibrium and dynamic behavior of a simple age-structured model with age-independent vulnerability to fishing gear and weight and constant recruitment.

Figure 3.17d shows the relationship between biomass and the rate of change. Rate of change is defined as (biomass next year + catch this year)/biomass this year. When unfished, the rate of change is 1.0, the biomass does not change, and there is no yield. As the stock is fished harder, the denominator is decreasing while the yield is going up.

Growth

The world is more complex than pictured above, and almost all fish gain weight as they age. If we allow for a weight-age relationship, w_a, that is time and stock size invariable, then our age-structured model becomes:

$$N_{t+1,a+1} = N_{t,a}(1 - m) - \frac{C_{t,a}}{w_a}$$

$$C_{t,a} = N_{t,a} f_t w_a$$

$$B_t = \sum_a N_{t,a} w_a$$

$$N_{t+1,1} = R \tag{3.6.8}$$

where $C_{t,a}$ is now catch measured in weight units, and B_t is the total biomass of the stock. Table 3.1 shows a weight vs. age relationship, and Figure 3.18 shows the portrait for this model of fish population dynamics.

The major change from the previous portrait is that yield now decreases

Table 3.1. Weight-age relationship used for Figure 3.18.

Age	1	2	3	4	5	6	7	8	9	10
Weight	0.8	1.5	2.5	3.3	4.0	4.3	4.4	4.5	4.5	4.5

with increasing fishing effort due to *growth overfishing*. As fishing effort increases, fewer individuals survive long enough to make it into the heavier age classes. Also note in Figure 3.18a that surplus production again drops off at lower biomass because at small biomass the population consists mostly of small individuals who are not gaining as much weight. Even if small individuals grow at a very high rate, if they are very small, the total biomass gain of the population is low. Note, too, that the trajectory of yield-effort for a developing fishery with 15% annual growth in effort (dashed line in Figure 3.18b) now diverges more from the equilibrium than it did in the previous graph. If one were to make the mistake of assuming that the yields obtained during a development period were equilibrium sustainable yields (an assumption often made—see chapter on biomass dynamics models) one would overestimate the sustainable yield and the corresponding level of fishing effort.

Differential vulnerability to fishing gear

Another common property of fisheries is that young fish are less vulnerable to fishing gear than older fish. As shown in Table 3.2, age 1 fish are

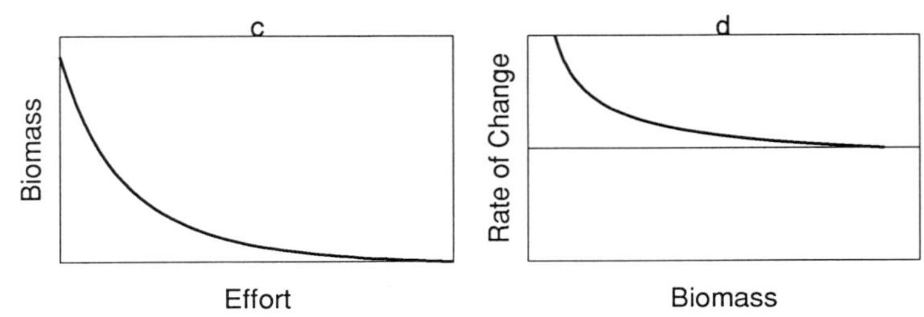

Figure 3.18. The equilibrium and dynamic behavior of a simple age-structured model with age-independent vulnerability to fishing gear and constant recruitment, but with age-specific body weights.

Table 3.2. Weight-age relationship used for Figure 3.19.

Age	Weight	Vulnerability
1	0.8	0.0
2	1.5	0.2
3	2.5	0.5
4	3.3	0.7
5	4.0	1.0
6	4.3	1.0
7	4.4	1.0
8	4.5	1.0
9	4.5	1.0
10	4.5	1.0

completely invulnerable to the gear, age 2 fish are only 20% vulnerable, age 3 are 50%, age 4 are 70%, and only at age 5 are fish fully vulnerable. Our model now becomes

$$N_{t+1,a+1} = N_{t,a}(1 - m) - \frac{C_{t,a}}{w_a}$$

$$C_{t,a} = N_{t,a} f_t v_a w_a$$

$$B_t = \sum_a N_{t,a} w_a v_a$$

$$N_{t+1,1} = R \qquad (3.6.9)$$

where we have now introduced v_a as the age specific vulnerability to the gear, and now B_t is the exploitable biomass. This means that the young, faster growing fish of ages 1–4 are reasonably invulnerable. We see in Figure 3.19a that yield increases with fishing effort. We cannot cause growth overfishing of this stock because of the vulnerability schedule. Therefore, this portrait looks much more like Figure 3.17; the harder you fish the more you get. Further the stock can never be fished to extinction because recruitment is constant — it would be a nice world, wouldn't it.

Recruitment

Tomorrow's recruitment depends upon the reproductive output of today's fish. If we fish a stock to extinction, there will be no stock tomorrow. Whereas there is considerable disagreement among fisheries scientists regarding the importance of the relationship between spawning stock and subsequent recruitment *over the range of historically observed stock sizes*, we know that in many fisheries you can fish hard enough to cause a reduction in future

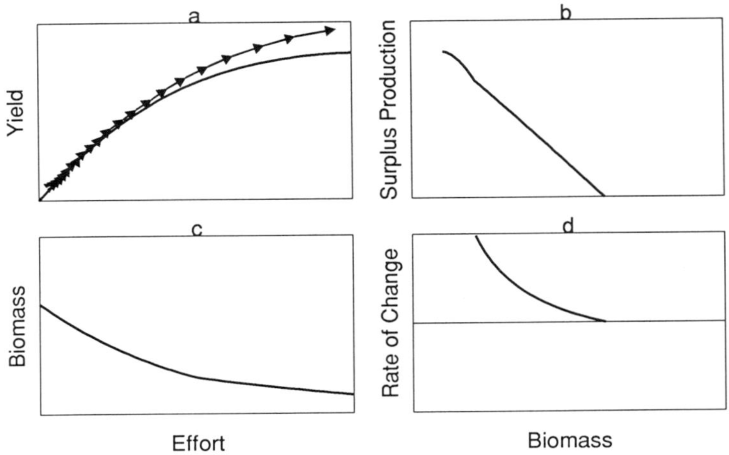

Figure 3.19. The equilibrium and dynamic behavior of a simple age-structured model with constant recruitment, age-specific weights, and age-specific vulnerability to fishing gear.

recruitment. Chapter 7 is devoted entirely to this topic. We can easily incorporate a stock-recruitment relationship into our analysis as follows

$$
\begin{aligned}
N_{t+1,a+1} &= N_{t,a}(1 - m) - \frac{C_{t,a}}{w_a} \\
C_{t,a} &= N_{t,a} f_t v_a w_a \\
B_t &= \sum_a N_{t,a} w_a v_a \\
E_{t,a} &= N_{t,a} e_a \\
E_t &= \sum_a E_{t,a} \\
R_t &= (aE_t)/(b + E_t) \\
N_{t+1,1} &= R_t. \qquad (3.6.10)
\end{aligned}
$$

We have added an age-specific fecundity (e_a) to calculate the total number of eggs produced (E_t), and used the Beverton-Holt stock-recruitment relationship to predict R_t from these eggs. The age-specific fecundities used are given in Table 3.3. Note that we have assumed that the fish do not become reproductive until age 4 and that by age 6 egg output is proportional to body weight. The Beverton-Holt relationship is discussed more thoroughly in Chapter 7, but for the purposes of this chapter it is sufficient to note that, as the total egg production increases, so does recruitment, but at a diminishing rate, so the total recruitment approaches a maximum level (the pa-

Table 3.3. Weight, vulnerability, and fecundity schedules used for Figure 3.21.

Age	Weight	Vulnerability	Fecundity
1	0.8	0.0	0.0
2	1.5	0.2	0.0
3	2.5	0.5	0.0
4	3.3	0.7	2.0
5	4.0	1.0	3.0
6	4.3	1.0	4.3
7	4.4	1.0	4.4
8	4.5	1.0	4.5
9	4.5	1.0	4.5
10	4.5	1.0	4.5

rameter a). Figure 3.20 shows an example of the relationship between egg production and total recruitment.

Adding stock and recruitment to the age-structured model makes a great difference in the behavior of the population. The portrait in Figure 3.21 illustrates this difference. Figure 3.21a shows that increasing effort initially results in an increased yield, but too much effort reduces the total yield, potentially to zero. It also shows how different the fishery development yield curve is from the equilibrium relationship. The assumption that the initial yield-effort relationship is the same as that achievable at equilibrium is a very serious error, and is particularly dangerous when there is a strong link between spawning stock and subsequent recruitment. Panels b, c, and d of

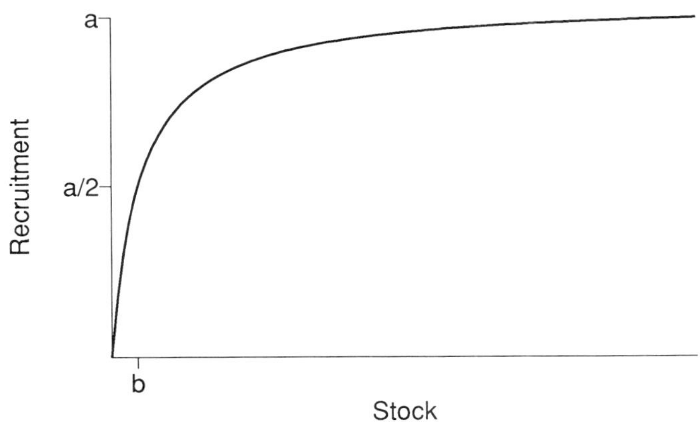

Figure 3.20. A Beverton-Holt stock-recruitment relationship. Note that maximum recruitment is a, and that, the recruitment is $a/2$ at an egg production of b.

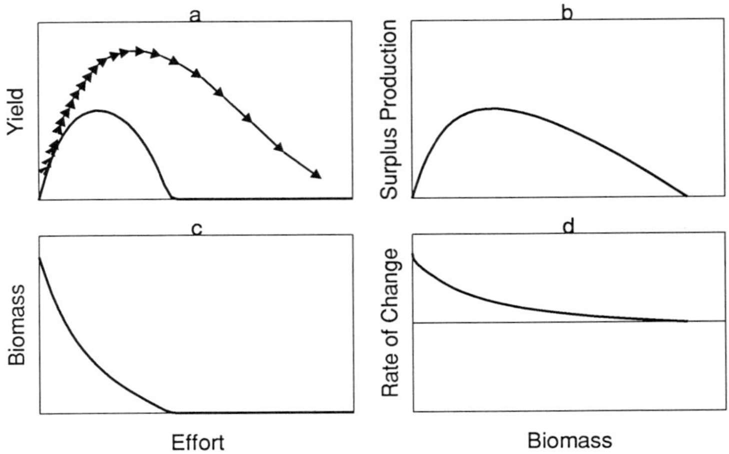

Figure 3.21. Portrait with stock and recruitment added to the model.

Figure 3.21 are reasonably straightforward and look remarkably like the equilibrium relationships obtained from the biomass dynamics models of Section 3.5.

Getz (1980) and Lawson and Hilborn (1985) presented explicit solutions for the equilibrium conditions of Equation 3.6.10. We will not present them here for two reasons. First they are a bit long and would take up considerable space. More importantly, because the solutions are for equilibrium conditions, we are afraid they would encourage readers to do calculations and models in terms of equilibrium — and this we want to discourage.

One drawback of the way that the recruitment curve is formulated in Equation 3.6.10 is that any change in the parameters a and b results in a different unfished equilibrium. Thus we cannot lower parameter b to see if making the stock-recruitment curve steeper affects the behavior, without also changing the unfished stock size. A convenient way to solve this problem is to redefine a and b as follows (Mace and Doonan, 1988):

$$
\begin{aligned}
a &= \frac{E^*}{R^*}\left(1 - \frac{z - 0.2}{0.8z}\right) \\
b &= \frac{z - 0.2}{0.8\, zR^*} \\
R &= \frac{E_t}{a + bE_t}
\end{aligned}
\tag{3.6.11}
$$

where B^* is the unfished egg production, R^* is the recruitment when $B =$

B^* and z is the steepness of the stock-recruitment curve, defined so that $R_t = zR^*$ when $E_t = 0.2\ E^*$. If $z = 0.5$, then, at 20% of virgin fecundity, recruitment is 50% of what it was at virgin fecundity.

This method of reparameterizing the stock-recruitment curve lets us adjust the steepness of the curve (z) without affecting the unfished behavior of the stock. We can use this to look at the impact of different z values on the sustainable and unsustainable yields. The overall potential yield of a simple age-structured model depends primarily on the natural mortality rate and the steepness of the stock-recruitment curve. Figure 3.22 shows contours of maximum average yield as they depend upon the annual survival rate and the steepness of the stock-recruitment curve. These contours are for one specific selectivity and growth relationship, but the results are quite general. The unfished equilibrium biomass is 1,000. This figure was obtained by searching over different survival and steepness values for the harvest rate that maximized average long-term catch. As the survival rate decreases and the steepness of the stock-recruitment curve increases, the average sustainable yield increases.

The unsustainable yield depends very much on survival and "steepness." Figure 3.23 shows the ratio of the unsustainable yield to the sustainable yield. The unsustainable yield is defined as the difference between the virgin biomass and the biomass at the harvest rate that produces the highest long-term average yield. Thus if the virgin biomass was 1000, and the biomass at optimum harvest rate is 500, then the unsustainable yield is 500. The unsustainable yield is then divided by the sustainable yield to produce the ratios shown in Figure 3.23. If the sustainable yield had been 100 and the unsustainable yield 500, then the ratio would be 5. Note that for annual survival rates of 0.7 or greater, the unsustainable yield is usually 10 times the sustainable yield.

3.7. Spatial Distributions

Thus far, we have considered fish populations as a single, spatially homogeneous entity. We have taken the biomass dynamics models of Figure 3.13 and added age structure. This is the traditional approach, yet many, if not most, fisheries are not spatially homogeneous. Densities are higher in some areas than others, the fish themselves may be quite mobile or completely sedentary. For most fisheries, we must consider the spatial structure of the population, because this will directly affect how the stock responds to exploitation.

Our approach in this section is to briefly describe the life histories and fisheries of several species and demonstrate how to model the biology of these species.

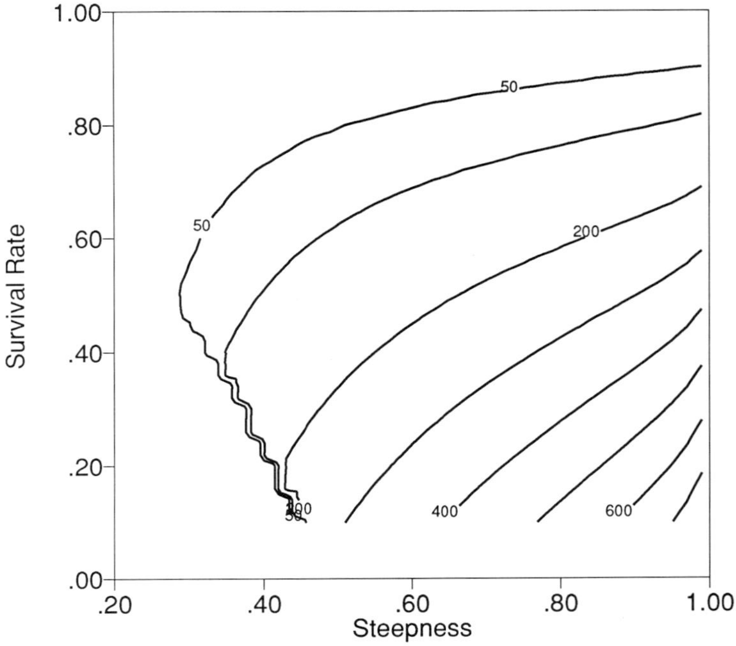

Figure 3.22. Maximum average yield that can be sustained with a constant harvest rate policy for different combinations of survival and stock-recruitment steepness. Unfished equilibrium biomass is 1,000.

Tasmanian abalone

The abalone (*Haliotis rubra*) population in Tasmania is the antithesis of the homogeneous stock (Prince 1989). Abalone are a cold-water species that are harvested around the entire perimeter of Tasmania, more than 1,000 kilometers of coastline. Abalone are reasonably long lived in Tasmania (annual survival rates of adults are about 80%), and adults are sedentary, normally moving only tens of meters in their entire life, although a few individuals may move hundreds of meters. The larvae are pelagic for a few days, but recent work indicates that effective dispersal of larvae also operates on a scale of tens of meters. Thus, the effective unit of an abalone stock is an isolated reef a few hundred meters on a side. This is consistent with observations that individual reefs can be fished to very low numbers and not recover, despite the presence of large populations on nearby reefs.

Such short-term movement at all life history stages means that the Tasmanian abalone fishery operates on thousands of individual, independent stocks of abalone. This poses statistical problems for fisheries managers, because it would be quite difficult to do stock assessment and regulate each

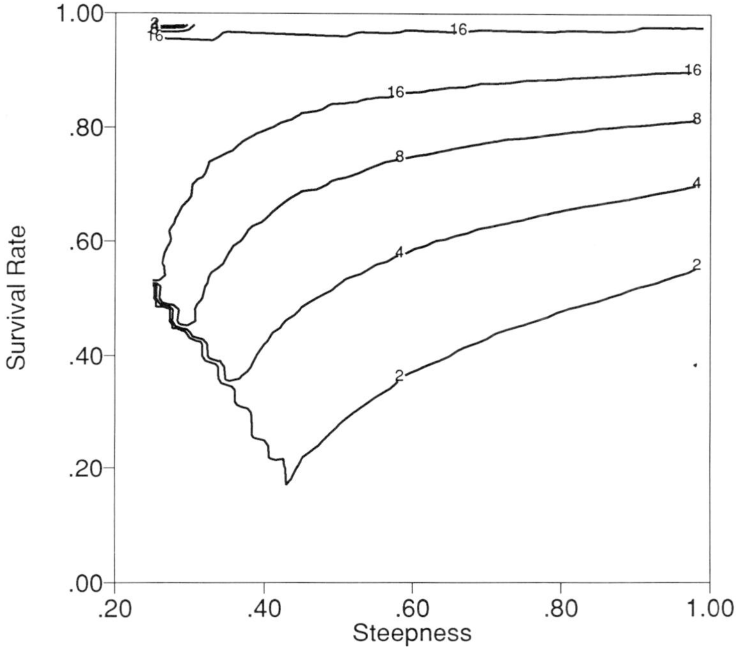

Figure 3.23. Contours of the ratio of unsustainable yield to sustainable yield as it depends upon the survival rate and steepness of the stock-recruitment curve.

individual reef. However, it is quite easy to model; we simply use our equations for a regular population model, whether a production model or a full age-structured model, and add another subscript for area. We then need to add area-dependent survival rates, growth rates, vulnerabilities, and fishing mortality rates, but the basic equations are quite simple:

$$
\begin{aligned}
N_{t+1,a+1,i} &= N_{t,a,i}(1 - m_i) - C_{t,a,i} \\
C_{t,a,i} &= N_{t,a,i} f_{t,i}\, v_{a,i} \\
B_{t,i} &= \sum_a N_{t,a,i} w_{a,i} v_{a,i} \\
E_{t,a,i} &= N_{t,a,i} e_{a,i} \\
E_{t,i} &= \sum_a E_{t,a,i} \\
R_{t,i} &= f(E_{t,i}) \\
N_{t+1,1,i} &= R_{t,i}
\end{aligned}
\tag{3.7.1}
$$

We have simply added the i subscript to designate the stock or site and replaced the specific Beverton-Holt stock-recruitment function by a generalized stock-recruitment function f. We assume there is no exchange of individuals between sites for all age classes.

The dynamics of this model, in each site, are exactly the same as seen in the discussion of age-structured models earlier. In the next chapter, we consider spatial allocation of fishing effort, which produces some results that are important in our understanding of trends in catch and catch per effort.

Skipjack tuna in the western Pacific

Skipjack tuna (*Katsuwonus pelamis*) are a pelagic species that live over most of the tropical oceans of the world. They produce many extremely small eggs, with females often spawning every day for extended periods. The eggs drift in the surface waters, and, as the juvenile fish develop, they remain near the surface. Kearney (1982) has described skipjack as the piscine equivalent of phytoplankton. They are everywhere, breed rapidly and have a very high natural mortality rate. In the western Pacific, about 160,000 skipjack were tagged from 1977 to 1980, and individual marked fish were recovered thousands of kilometers from where they were tagged (Kearney 1982). Most fish were recovered reasonably close to where they were tagged.

If we break the ocean into regions, using 200 mile Exclusive Economic Zones (EEZS) for instance, we can describe skipjack movement as a table of probabilities of movement from one region to the next. We might say, for example, that skipjack have a 10% chance of moving from Papua New Guinea to the Solomon Islands in a given month. This type of simple movement model can be written (in a biomass dynamics model framework) as:

$$\frac{dB_i}{dt} = rB_i\left(1 - \frac{B_i}{k}\right) - C_i + \sum_j p_{j,i}B_j - \sum_j p_{i,j}B_i \qquad (3.7.2)$$

where again the subscript i represents the site. The parameter $p_{j,i}$ is the movement rate from area j to area i. Table 3.4 shows the rates of movement estimated among 7 regions of the western Pacific.

This basic approach to modelling is more simply stated as a set of discrete time equations

$$\begin{pmatrix}\text{Biomass}\\ \text{next time}\\ \text{area } i\end{pmatrix} = \begin{pmatrix}\text{Biomass}\\ \text{this time}\\ \text{area } i\end{pmatrix} + \begin{pmatrix}\text{surplus}\\ \text{production}\end{pmatrix} - \begin{pmatrix}\text{catch}\end{pmatrix} + \begin{pmatrix}\text{immigration}\\ \text{from}\\ \text{other areas}\end{pmatrix} - \begin{pmatrix}\text{emigration}\end{pmatrix} \qquad (3.7.3)$$

Table 3.4. Estimated fishing mortalities, catchabilities, and movement probabilities for the seven fishery model.

	PAL	YAP	INT	TRK	PON	PNG	SOL
f or q	0.0049	0.0043	0.0046	0.0054	0.0065	0.00012	0.00003

From	Probability of movement			To				
	PAL	YAP	INT	TRK	PON	PNG	SOL	OTH
PAL	0.677	0.0047	0.107	0	0	0	0	0.168
YAP	0.008	0.707	0.032	0.041	0	0	0	0.211
INT	0.008	0.041	0.808	0.100	0.00*	0.034	0	0.008
TRK	0	0.066	0.027	0.635	0.084	0	0	0.187
PON	0	0	0	0.082	0.683	0.00*	0	0.235
PNG	0	0	0.052	0	0.00*	0.600	0.013	0.336
SOL	0	0	0	0	0	0.024	0.873	0.124

Fishery codes are PAL, Palau; YAP, Yap; INT, International waters between the Federated States of Micronesia (YAP, TRK, PON), Palau, PNG, and Indonesia; TRK, Truk; PON, Ponape; PNG, Papau New Guinea; SOL, Solomon Islands; OTH, loss of tags to other areas and natural mortality. For all countries except PNG and SOL the fishing mortality is estimated (effort assumed constant at 1.0; for PNG and SOL the catchability coefficient q is estimated since effort data were available). If a 0 appears in the movement probability table, it means that these countries are not adjacent and no direct exchange rate was estimated unless an * appears, in which case it means the best estimate was that no exchange occurs between countries. From Hilborn (1990).

which can be written

$$B_{t+1,i} = B_{t,i} + B_{t,i}r\left(1 - \frac{B_{t,i}}{k}\right) - qE_{t,i}B_{t,i} + \sum_j p_{j,i}B_{t,j} - \sum_j p_{i,j}B_{t,i}. \tag{3.7.4}$$

The behavior of this model when unexploited is quite simple; the biomass in each area comes to an equilibrium dictated by the r and k values and the exchange rates with other areas. When fished, the behavior depends upon the spatial structure of the fishery. The most important feature is that any area can sustain higher yields if adjacent areas are unfished. The immigration from other areas can sustain an overexploited fishery in any one area in a way that will not occur in spatially isolated fisheries.

Figure 3.24 shows yield plotted against fishing mortality for the International zone in two cases: one in which movement occurs and one in which no movement occurs (values taken from Table 3.24). In both cases we assumed there was no fishing in other countries. Note that yield is much higher when movement occurs and that it does not decrease with increasing fishing mortality because immigration from other areas effectively swamps the local

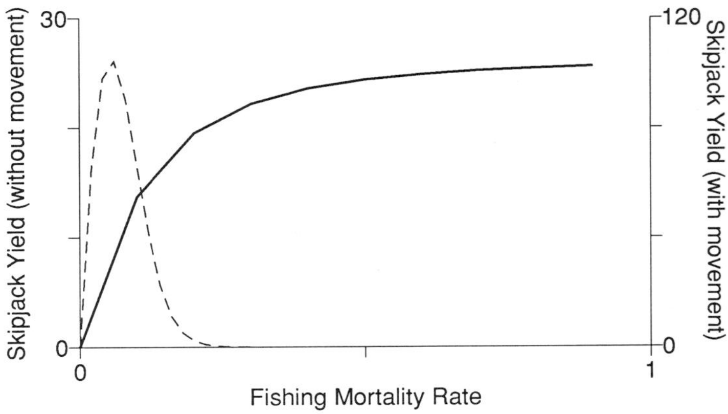

Figure 3.24. Yield-fishing relationship for the International waters with and without exchange of tuna among areas.

dynamics. For this simulation an r value of 0.1 was used. The movement rates into the international zone are 0.11 from Palau, 0.03 from Yap, 0.03 from Truk, and 0.05 from Papua New Guinea, which total to 0.21. This is much larger than the potential for production from the area itself. If, however, the Solomon Islands or Palau had been chosen (where there is little movement into the area), there would have been little difference in yield. Note that, because we assumed no fishing in other countries, the impact of movement will be maximized.

Pacific salmon

Pacific salmon have a truly migratory life history. They spawn in the rivers of North America and Asia and spend from a few months to a few years in freshwater. They then embark on a migration down the streams and rivers to the ocean and spend from a few months to several years in the ocean. The key characteristic of this migration is that at a specific point in the life history, the spatial distribution of the stock is reasonably predictable; we know where they will be.

Migratory life histories are very common in fisheries. The Arcto-Scandinavian herring undergoes an annual migration from Norway to Iceland and back. Grey whales migrate to feed in the North Pacific in the summer, and they winter primarily along the coast of Mexico. The North Pacific fur seal returns each summer to specific islands to breed and spends much of the rest of the year in the open ocean feeding in reasonably predictable locations. Many marine fish species undergo spawning migrations or have other biological characteristics such as buoyant eggs, such that the eggs and juveniles are carried into inshore areas where they settle and begin to grow; as they

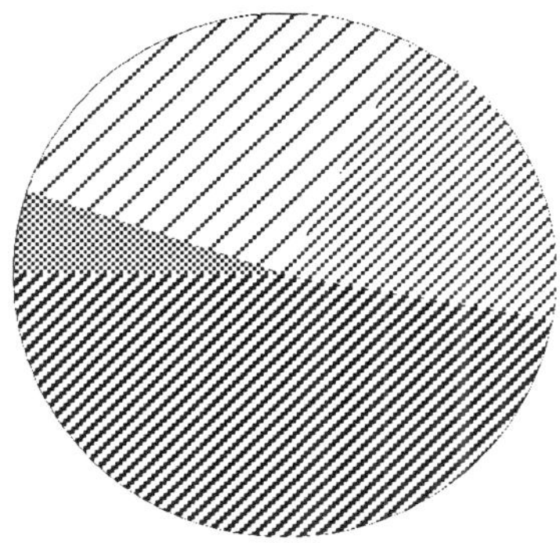

Figure 3.25. Distribution of the ocean catch of 4-year-old chinook salmon from the Columbia River.

grow, they most often begin to move progressively offshore into deeper water.

At a superficial level, migratory patterns are easily described; at a certain time of the year, the fish will be located in a certain spot. However, when we look at a smaller spatial scale, we must ask some important questions about the distribution of the fish. For instance, Figure 3.25 shows the distribution of catch of chinook salmon (*Oncorhynchus tshawytscha*) from the Columbia River at 4 years of age. Roughly half of the catch comes from Alaska, the rest is more or less evenly split between southern and northern British Columbia with a small catch from Washington and Oregon. If we assume that catches are proportional to abundance, we can model such a distribution as a fixed set of "residence proportions."

$$N_{a,t,i} = N_{a,t}\, p_{a,t,i} \tag{3.7.5}$$

where N_{ati} is the number of fish from area a alive at time t in area i, $N_{a,t}$ is the total number of fish alive from area a at time t and p_{ati} is the residence proportion. This type of migration model is conceptually the simplest, and is often used for modelling of Pacific salmon. However, it makes three major assumptions: (1) the ocean distribution is not affected by year to year changes in the environment, (2) the ocean distribution is not affected by density, (3) the ocean distribution is not affected by exploitation patterns, and (4) fish do not move between areas during the time period when catch is taken.

We know that assumption (1) is false, as the ocean distribution of some species of salmon has been shown to be affected by temperature. Blackbourn

(1987) has demonstrated that sockeye salmon will be found farther north in warm years and farther south in cold years. It is reasonably easy to model this, simply by adding another subscript to p, for different types of years.

There is no evidence yet to show that the distribution of salmon is affected by density. Density-dependent distribution has been shown to be the case for anchovy (next section), and the same methods used to model anchovy could be used for salmon. The residence proportion of Equation 3.7.5 does not make much biological sense when one considers the realities of fish movement and exploitation. Equation 3.7.5 says that the fish magically appear in a certain distribution, regardless of the previous exploitation history. This would require a rather magical ability of the fish to sense the distribution of all other salmon and adjust themselves accordingly. This type of mechanism is quite possible in a small spatial area where fish can detect the presence of others, but it is more difficult to believe for the open ocean. In particular, imagine that a very intense fishery is mounted in northern British Columbia so that all fish there are caught. It would be impossible for any fish to get to Alaska; yet under Equation 3.7.5, a certain proportion are supposed to be there!

A functionally more realistic movement model would be that of Equation 3.7.2 used for skipjack tuna, with time-specific movement rates. Under any constant exploitation regime, a set of residence proportions would emerge, but these would change if exploitation patterns changed.

California Anchovy

California anchovy display what can be called density-dependent spatial distribution in their spawning and larval distribution. As the abundance of the stock increases, the total area covered also increases. Figure 3.26 shows the distribution of larvae in two years.

MacCall (1990) has proposed a bowl analogy which suggests that the fish go to the bottom of the bowl at low abundances and that as abundance increases the bowl fills. In other words, the area covered by the stock expands more slowly than the abundance of the stock.

This type of spatial distribution can be found in many schooling fishes. Many herring stocks have demonstrated a reduction in range, rather than density, as the stock abundance was reduced. Schooling and territoriality are obvious biological mechanisms that could lead to density-dependent spatial distributions. The consequences of density-dependent spatial distributions are rather bad for fisheries management, since it usually results in increasing fishing mortality as stock size decreases. This in turn has often led to poor assessments of the trend in stock abundance and eventual overexploitation. The mechanism for this is discussed in the next chapter.

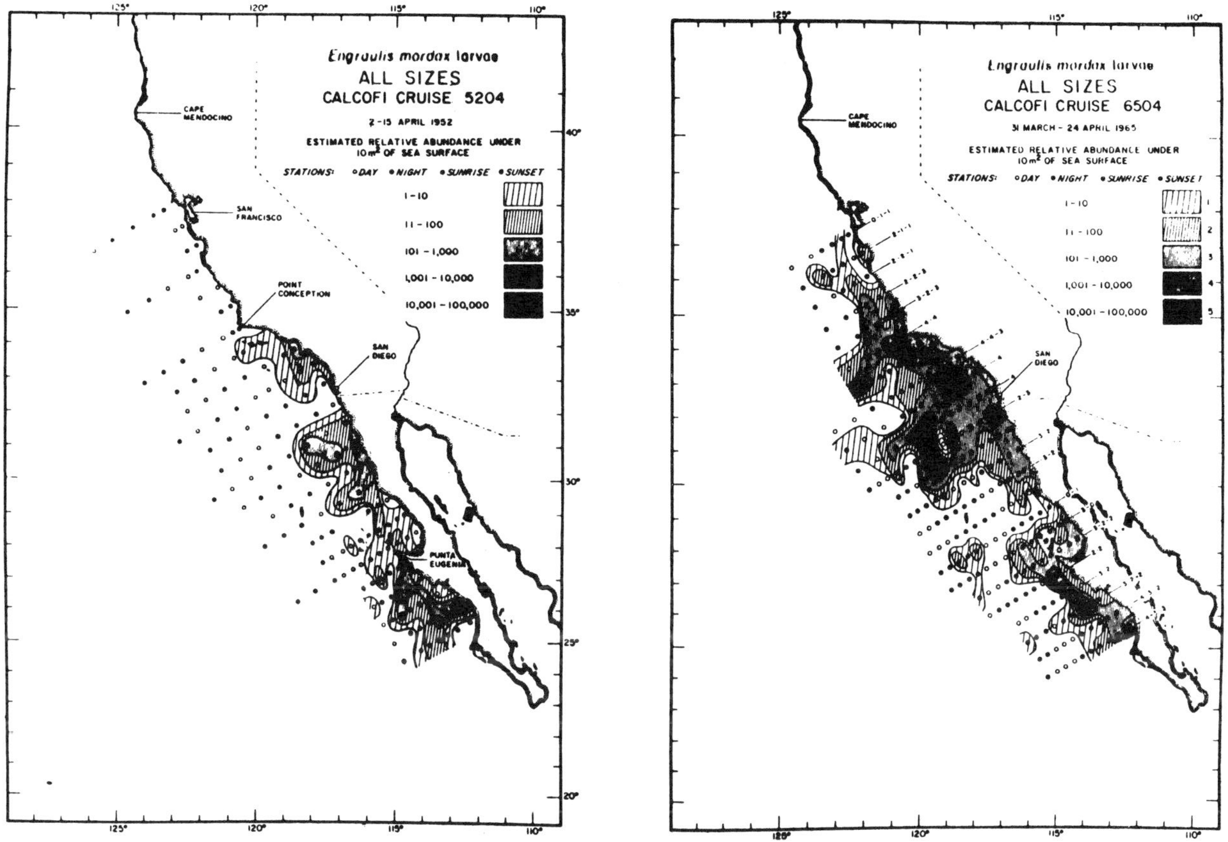

Figure 3.26. Spatial distribution of the larvae of California anchovy in 1952 (left) and 1965 (right). In the left-hand graph the total abundance was low, and the stock covered a small area. In the right-hand graph the stock was much more abundant and covered a larger area. From Lasker, R, and A. MacCall (1983).

Pacific halibut

Pacific halibut are found along the northeastern Pacific from Oregon to Alaska. The adults are quite sedentary and rarely move more than a few tens of kilometers, however the eggs and larvae drift north along the coast, so that most 1- and 2-year-old halibut are found in the Bering Sea and near the Aleutian Islands. The juvenile halibut make their way south, apparently between ages 2 and 7, and are sedentary adults (age 9–11) by the time they become vulnerable to the commercial longline gear. Their migration pattern is shown in Figure 3.27. Recruitment to a stock (the adult population in an area) will depend primarily upon immigration of young fish, rather than production from the local "stock." This type of life history is common among sedentary species, particularly invertebrates, that have a mobile egg, larval, or juvenile stage.

3.8. Species Interactions

All of the models we have discussed thus far, whether biomass dynamics models, age-structured models, or spatially explicit models, have one key thing in common; they consider only a single stock, with no explicit interaction with its predators, competitors or other environmental variables. This is the tradition in fisheries! By the mid 1950s modern fisheries science had taken its basic form, and the basic models and prescriptions were in place. The biomass dynamics models of Schaefer (1954, 1957) and the age-structured models that include vulnerability schedules, stock-recruitment relationship, etc., were codified in Beverton and Holt's (1957) classic textbook. Ricker's (1958) handbook provided a final reference to cement the growing consensus on what fisheries science involved.

There were two key elements to this consensus; first was the acceptance of MSY as a reasonable (not necessarily the optimal) objective for fisheries management. Second, was the acceptance that reducing fishing pressure was the solution to fisheries problems. If a stock was collapsing, the solution was to reduce effort. This became a firmly entrenched dogma that remains today as the core of fisheries textbooks. It is predicated on a single-stock view of the world where interactions with other species are not key controlling factors and on the Thompson side of the Thompson-Burkenroad debate, that fishing and not environment determines fish abundance.

This body of fisheries science is self-reinforcing because it contains an objective (MSY), an analytic framework (biomass dynamics models, age-structured models), and a prescription for declining catches (reduce fishing pressure). However, this edifice is collapsing under the weight of evidence that in many cases the single-stock view of the world is not sufficient.

The evidence that is crushing the single stock paradigm is varied. It in-

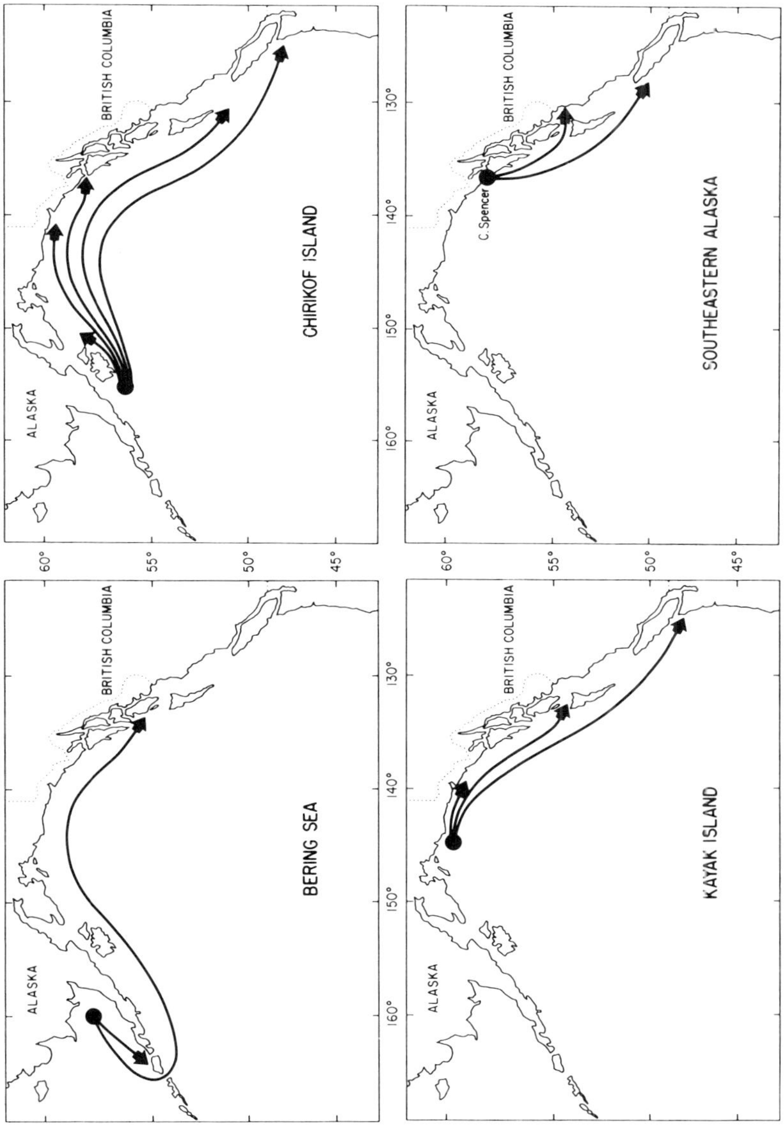

Figure 3.27. Migration pattern of Pacific halibut. From Skud (1977).

cludes the continuation of the Thompson-Burkenroad debate and many related cases where there appear to be very clear environmental correlates with fish abundance, normally via effects on recruitment. The long-term histories from anoxic sediments are damaging, because the single-stock paradigm is built upon a globally stable single-species model; these models cannot accommodate stocks that have major fluctuations without fishing pressure. Finally, the single-stock paradigm is assaulted by growing evidence of species interaction. Fish stocks do not exist in a vacuum; fish must eat something, they must compete with other species for what they eat, and other species may eat them.

Although the single-stock edifice is collapsing, no new paradigm has yet risen to replace it. We know how to model species interactions, and these basic approaches are described subsequently in this section. Unfortunately, data collection and parameter estimation problems normally make it very difficult to use these models. This is discussed later in Chapter 14 entitled "Multispecies Approaches." In the rest of this section, we will introduce the common types of models used to describe species interaction.

Competition

Competition can take place either because one species consumes some resource (food, space) that the other species needs or because one species prevents access to this resource but does not necessarily consume it. There are two common ways of modelling such interaction. One method involves explicitly keeping track of the resource in demand (prey for instance) and accounting for competition via consumption by the competing species. This approach falls under the classification of *food web models* discussed below.

The second, and more common approach, is to consider the effect that each individual or unit of biomass of a species has on other species. This is the model of Lotka and Volterra (Lotka 1925) and is written in difference equation form as

$$B_{1,t+1} = B_{1,t} + r_1 B_{1,t}\left(\frac{k_1 - B_{1,t} - a_{2,1}B_{2,t}}{k_1}\right) - C_{1,t}$$

$$B_{2,t+1} = B_{2,t} + r_2 B_{2,t}\left(\frac{k_2 - B_{2,t} - a_{1,2}B_{1,t}}{k_2}\right) - C_{2,t} \qquad (3.8.1)$$

where $B_{1,t}$ is the biomass of species 1 at time t, the r, k, and C's are the traditional biomass dynamics model parameters and catches, and the new parameter $a_{i,j}$ is the effect one individual of species i has on species j. If the a values are 1, then each individual of species 2 reduces the surplus production of species 1 as much as an individual of species 1 reduces the surplus production of species 1.

This model is easily expanded to include any number of species, and, even though it has not been used a great deal in fisheries, it has seen considerable application in modelling plant community interactions. Its use in fisheries has been primarily limited by the difficulty of estimating the a coefficients. It is hard enough to estimate the r and k values from actual data, and it is effectively impossible to estimate the a values for most data sets. This problem is discussed in more detail in Chapter 14.

Predation

The most common approach to modelling predation is to assume that the consumption rate per predator is proportional to the abundance of the prey. This is based on a simple random search model of predation, assuming that each time a predator detects a prey it instantaneously captures it and begins searching again. More realistic predation models recognize that each predator has an intrinsic limit to how much it can consume per unit time, either due to handling time, gut capacity, or intrinsic energy demand. This type of model, formulated by Holling (1959) and Ivlev (1961), can be used in a predator prey model as follows

$$
\begin{aligned}
V_t &= \frac{bB_t}{c + B_t} \\
B_{t+1} &= B_t + rB_t\left(\frac{k - B_t}{k}\right) - V_tP_t \\
P_{t+1} &= P_ts + dV_tP_t - eP_t^2 \qquad (3.8.2)
\end{aligned}
$$

where V_t is the amount of prey biomass eaten per predator at time t, b and c are parameters of the prey consumed per predator versus prey biomass relationship, B_t is the prey biomass at time t, r and k are the normal parameters of the biomass dynamics model, s is a survival rate for the predators, d is the conversion efficiency, which determines how many units of predator are produced per unit of prey biomass consumed, and e is a parameter that affects the density dependent self regulation of the predators.

Complex food webs

Using the basic building blocks of single-stock population dynamics, competition and predation, we can construct food-web models that display very interesting behavior. In most cases, the predators will be preying upon more than one species, and a modification of the functional response curve is needed to allow for this biological fact. This relationship is generally called the *multiprey functional response* and a commonly used form is

$$V_{t,i} = \frac{N_{t,i} a p_i}{1 + \Sigma_j N_{t,j} h_j a p_j} \tag{3.8.3}$$

where $V_{t,i}$ is the number of prey species i eaten per predator at time t, $N_{t,i}$ is the density of prey i at time t, a is the area searched per predator during time t, p_i is the probability that if an individual of prey species i was in the area searched, it was chased and successfully captured, and h_j is the time it takes a predator between the detection of a member of species j and when it resumes searching.

In fact, these complex food web models are of little if any use in stock assessment. Thus far, it has proven too difficult to obtain reliable parameter estimates for the models and too difficult to establish the significant interactions among species. A central problem in the estimation is that it is not enough to just show that one species eats another, or even to estimate how much they eat. One must also show that the mortality due to being eaten is "additive" to other mortality agents, i.e., that the eaters are not selectively taking individuals that have higher probabilities of mortality in the first place.

3.9. Summary and Critique

This chapter introduces most of the fundamental biology that serves as the basis for the rest of the book. The two basic fisheries concepts of *growth overfishing* and *recruitment overfishing* are probably familiar to most readers. Appreciation of the *dynamic* behavior of fish populations and fish population models is critical to our view of fisheries stock assessment and management. Widespread access to microcomputers permits anyone involved in stock assessment to examine dynamic behavior; there simply is no excuse to look only at simple equilibrium relationships. Understanding *unsustainable yields*, which are obtainable when fishing pressure begins or is increased, is critical to appreciating the expected pattern of yield as a fishery develops.

One of the glaring gaps in current fisheries theory is that 99% of fisheries models involve simple age-structure effects but not other structural features of populations. These models almost always exhibit very stable, frankly boring behavior, just as do biomass dynamics models. Only by adding very extreme recruitment assumptions can we sometimes generate cyclic behavior. The world does not appear to be so stable or constant. Our brief review of unexploited populations, Caddy and Gulland's (1983) review of fish populations, and the long-term histories from anoxic sediments all indicate that large-scale fluctuations are common. The types of models we have presented just do not and cannot display such rich behavior.

It is not hard to build models that behave in interesting ways. The problem is that generating such interesting behavior requires further assumptions about predator pits, or interesting species interactions, or interactions with abiotic changes. As we shall see later, we simply cannot distinguish between all of the alternative models. For any particular fish stock, we can usually fit its behavior with a simple biomass dynamics model, so why do we need to fit a more complex model? Unfortunately, the simple models generally will eventually fail, and our stock will stand revealed as cyclic, spasmodic, or irregular. Two interesting challenges for future fisheries work are to design management strategies for these types of population behavior and to find ways of identifying the behavior early enough to be useful.

Modelling is easy; it is not difficult to write down a number of alternative models of population dynamics and get them running on a microcomputer. No model is ever correct, but some models are more useful for some applications than others, and every model omits some components of a system. As computers became widely available, many people felt that bigger, more complete models were better. Experience has shown this is not generally true, and for specific applications, such as formulating fishing plans, very simple models may be better than more complex, more realistic ones. This is a very exciting area of current research; we cannot yet say what level of model complexity is best.

Bibliographic Notes

It is surprising how difficult it is to obtain data on the dynamics of unexploited populations. The same examples given in Section 3.2 can be found in almost every ecology textbook, as there simply are not a lot of examples to choose from. One of the best sources of further reading on the dynamics of exploited populations is *Exploitation of Marine Communities* edited by R. M. May (1984). This conference proceeding has a number of excellent papers and discussions about single species and multispecies fish communities and models.

The behavior of exploited populations is discussed in Caddy and Gulland (1983), and some interesting long-term histories can be found in Cushing (1982). Steele and Henderson (1984) considered some problems of modelling the long-term fluctuations often seen in fish stocks. The history of the Thompson-Burkenroad debate is reviewed by Skud (1975). The concepts of stability are discussed in more detail by Holling (1973, 1978). A more mathematical approach can be found in May (1973). Walters (1986) discusses approaches to modelling in more detail. Garcia (1984) considers some consequences of many fish stocks not behaving in a nice stable fashion.

4

The Dynamics of Fishing Fleets

4.1. Introduction

Stock assessment is often viewed as a purely biological and statistical subject, except in so far as the fishing process must be carefully examined in order to develop standardized measures of fishing effort that can be used in abundance and mortality estimation. This narrow focus has led to management strategies and regulatory schemes that ignore the dynamic responses of fishermen to changes in stock size and to management itself. These responses can dampen or even reverse the intended effects of regulation and create socioeconomic problems that make future regulation for biological conservation progressively more difficult to implement. Thus it is foolish to study only the prey in the predator-prey system that is a developing fishery; it is equally important to monitor and understand basic processes that determine the dynamics of the predator — the fishermen.

In this section we drop the old assumption that fishermen and fishing effort are "policy variables" to be controlled at will by management regulations. Instead, suppose that, like natural predators, they (1) require incentives to go out at all, (2) respond to incentives by expanding their activities, (3) evolve more efficient tactics over time by processes of individual learning and longer term selection (competition) among individuals and gear types, (4) respond to declining opportunities by switching to other fisheries or nonfishing activities, and (5) respond to regulatory measures just as they respond to other obstacles that limit their success.

In making the supposition that fishermen have dynamics analogous to natural predators, we are taking a competitive and market-oriented view of the economics of fishing; this is more realistic for most fisheries than to assume that fishing is a communal, cooperative, and altruistic process. We discuss that alternative viewpoint in a later section on integration of data collection and fishery development.

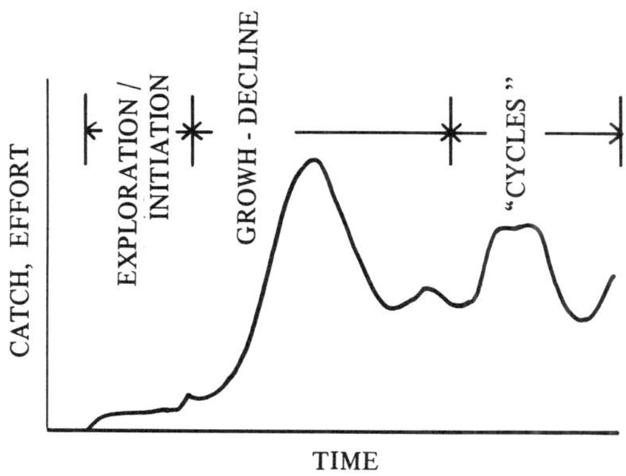

Figure 4.1. A typical pattern of fishery development

4.2. Patterns of Fishery Development

A key step in understanding any dynamic system is to examine its "transient behavior" when it is far from a balanced or equilibrium state. The changes that occur during fishery development offer a key source of information about both stock and fisherman dynamics.

The economic development of a fishery has three basic stages: (1) initiation through exploratory activity and special incentives of various kinds; (2) growth and later decline to bionomic equilibrium; and (3) later development cycles involving episodes of innovation and consolidation (Fig 4.1). Each of these stages involves different key factors and processes.

The initiation of development

Obviously a fishery will not develop until someone discovers a potentially valuable stock of fish; exploratory activities that result in such discoveries are undertaken by most management agencies and by curious and "hungry" fishermen. However, what happens next is very poorly understood, even in the industrialized and well-studied economies. Sometimes a quite valuable resource will be left untouched for many years, even when the appropriate fishing technology, markets, and financing are readily available; then development will suddenly take off into a rapid growth period.

Initiation of development appears to require four basic conditions: (1) the availability of a harvesting technology that is considered physically safe and obviously profitable in the sense that exploratory fishing with it produces catches that are comfortably more valuable than the operating costs to take those catches; (2) the presence of a market that will immediately accept and reward the exploratory fishermen and that contains processing/transport/marketing facilities adequate to handle larger catches; (3) availability of low-risk financing (i.e., low-interest-rate loans, grants, etc.) to cover both initial capital (equipment) and initial operating (gas, maintenance, labor, etc.) costs; and (4) perhaps most important, some daring and/or economically desperate individuals who are willing to try putting the other three conditions together as a gamble, before there is enough direct human experience to make most potential fishermen believe that the conditions really do hold.

Much economic and development planning work has concentrated on helping to establish the first three conditions above (abundance, market, financing). Yet some experience suggests that the key factor may be the fourth, individual risk taking. This factor is in the realm of psychology and sociology, and has not been well studied in fisheries.

Growth and decline to bionomic equilibrium

After some poorly defined point in time when it becomes apparent to a wider community of potential fishermen that a new fishery is worth trying in comparison to their current activities, there will be a rapid period of recruitment of new fishermen. Growth during this period may be fuelled by (1) continued financial subsidies of various sorts, (2) reinvestment of profits (including sharing of experience and labor) from earlier good catches, (3) transfer of gear (boats, nets, etc.) from other fisheries at relatively low capital cost, and (4) rapid learning about fishing techniques and distribution of the fish.

This growth inevitably results in declining catch rates for the individual fishermen, due both to depletion in the accumulated stock of fish and also, in many cases, to direct competition among units of gear for the best fishing spots. However, this decline in catch rates is a noisy signal or warning to potential new fishermen, and there is inevitably some time lag from when an investment decision is made based on past catch rates and when the new investors can actually start fishing (Figure 4.2). Thus it is practically inevitable that growth will continue (new fishermen start going out) until well past the time when catch rates have fallen low enough not to attract new fishermen (cannot meet loan payments and operating costs). In other words, the population of fishermen will peak at a level that is not sustainable.

Then there is generally a decline in fishing activity, marked by (1) established fishermen spending less time at sea, (2) new fishermen (and some experienced ones) dropping out of the fishery to seek other activities, (3) a desperate search for new fishing grounds and more efficient fishing methods, and (4) strong pressure on management agencies to develop technological fixes (artificial propagation, hatcheries, etc.) to help sustain the industry. It is important to recognize that this painful period of desperation

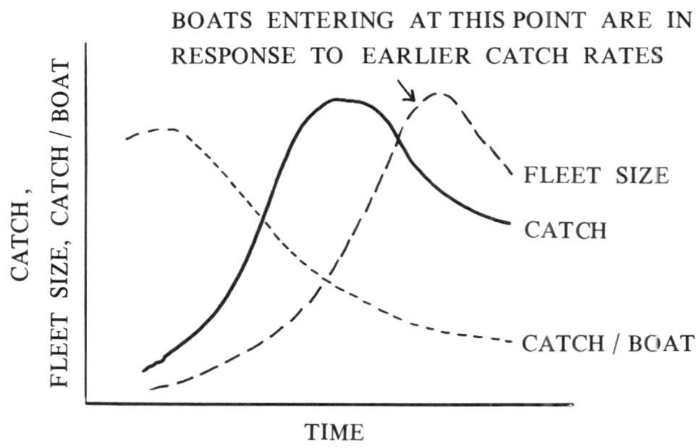

Figure 4.2. During fishery development, boats may enter the fleet until well after catch rates are too low to attract further growth, due to time lags between investment decisions and actual entry.

and decline is a practically inevitable feature of fishery development and generally cannot be prevented by providing better early predictions of catch rate declines or by providing economic assistance during and after the peak period.

In some fisheries, the decline will be followed by establishment of a temporary "bionomic equilibrium," with low but fairly stable catch rates per fisherman. On average, these catch rates will barely be adequate to meet operating and capital replacement costs; below average fishermen will be dropping out of the fishery but these losses will be balanced by recruitment of new fishermen who see a chance of doing better than average.

A complete biological collapse of the target stock in single species fisheries is rare, so that a bionomic equilibrium is never attained. In multispecies fisheries (most of the world's fisheries), the larger and more valued species may continue to be depleted toward extinction, but the fishery will tend to stabilize by moving onto progressively less valued species.

Innovation periods and development cycles

It is important to recognize that any bionomic equilibrium will be a temporary phenomenon, set by the catch rates and reinvestment patterns associated with a particular fishing technology and knowledge about the distribution of the stock. As a period of frustration (high competition, low profitability), any bionomic equilibrium will provide incentives for innovation in fishing technologies and search for new grounds.

When successful innovations/discoveries occur, the fishery is reset into a condition where growth and investment are again favored. Just as in the original development pattern, noisy catch rate signals and time delays in investment response will lead to temporary overdevelopment. There will then be a decline, perhaps not as violent as after the original development peak. A new period of temporary equilibrium will then ensue, unless the new technology is so efficient that it grossly depletes the stock during the period of temporary overdevelopment.

Effect of catch regulation on the stages of development

None of the above discussions of development stages said anything about changes in the total catch; we spoke only of changes in catch rates per fisherman. This should provide a clue about the effect of regulations on total catch: such regulations act exactly the same as biological depletion, except that declines in catch rate per fisherman after a total catch limit is imposed will be due to increasing competition among fishermen for the limited total catch available, rather than declining abundance, if the catch regulation is successful as a conservation measure. The increasing competition will rarely take the form of direct interference or "contest" competition among the fishermen; instead the fishermen will see it in terms of increasingly harsh regulatory tactics (shorter fishing seasons, gear restrictions, individual quotas or limits, etc.) that are put in place to achieve the overall limit.

Figure 4.3 shows the historical length of the fishing season for Pacific halibut *Hippoglossus stenolepis*. The United States has not imposed any form of limited-entry, so the United States fishery now takes place in 2 or 3 very brief (8–24 hour) fisheries. The Canadian fishery lasts considerably longer because of limited-entry. Short fishing seasons have a number of undesirable consequences. It becomes very important for an individual fishermen to be in the right place and have all of his gear working. A mechanical breakdown or bad choice of location can mean a lost season. When the fishing season is only several days long, lives are often lost because fishermen will fish continuously through the whole fishery opening and will risk weather conditions that would be considered too dangerous with a longer season.

In other words, total catch regulations do nothing to prevent the dynamic changes associated with economic overdevelopment, decline to bionomic equilibrium, and innovation associated with ways to improve individual catch rates as shares of whatever limited total is available. Total catch limits do not prevent economic pain.

4.3. Basic Principles of Fishing Fleet Behavior

In this section we review some general principles about how fishermen behave. Our analysis is divided into four parts: (1) determinants of fleet size

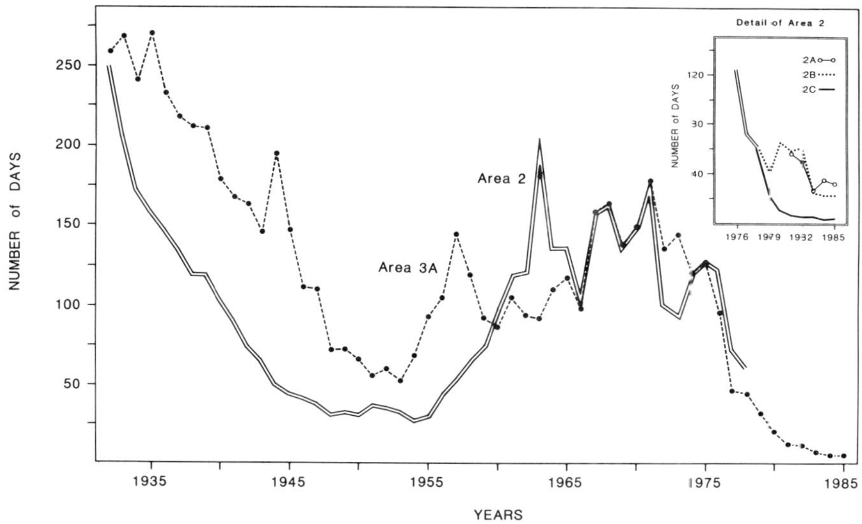

Figure 4.3. The evolution of the fishing season in the International Pacific Halibut fishery. From IHPC (1987).

(investment and disinvestment from the fishery), (2) determinants of when and where to fish, (3) determinants of catching power, and (4) determinants of discarding.

These are the four essential elements in fishermen dynamics and provide us with a framework for asking questions and analyzing data about how fishermen behave. The purpose of this analysis is straightforward: we want to understand how fishermen will respond to changes in biological and economic conditions and how they will respond to regulation.

Investment and Growth

The first step in understanding how a particular fishery works is to determine how many vessels there are in the fishery. This may be straightforward in some cases where vessels have only a single use. Other fisheries are much more complex: fishermen may use the same gear and vessels for many species, or may make minor modifications of gear to fish different species. Most artisanal fisheries would fall into this category, it will be much more difficult to define who is "in" the fishery and who is not. There are two basic processes involved in determining the size of a fishing fleet: loss of vessels and addition of vessels.

Loss of vessels

Vessels may disappear from a fishery for four basic reasons: (1) sinking or accidental losses; (2) intentional scrapping; (3) movement to another fishery, perhaps with some modification of gear; and (4) vessels may simply tie up and not fish if fishing is not profitable.

Sinking and scrapping. Sinkings can be thought of as the fishing equivalent of natural mortality; fishing places vessels in dangerous conditions and loss of vessels is a common occurrence. Aranson (1984) examined losses from the Icelandic demersal fleet and found a constant loss rate of a few percent per year until vessels reached about 20 years of age, when loss rate increased rapidly. It was assumed that the losses after age 20 represented scrappings and the few percent per year were sinkings.

Unintentional sinkings should presumably be a more or less constant factor in a fishery, whereas scrappings should depend upon economic conditions. Scrapping of vessels can take two forms; vessels may simply be written off and abandoned or they may be replaced by new vessels. When incomes are poor we would expect a vessel to be abandoned sooner than when incomes are good. However, one can expect that scrapping for replacement by newer vessels, as might occur under license limitation, might occur faster when economic times are good. To our knowledge no one has ever presented data on this aspect of vessel loss.

Exit to other fisheries and tie up. When there are alternative fisheries in which a vessel can engage, we would expect vessels to move to the alternative fisheries when economic times are hard. One can imagine a threshold difference in expected economic value between alternative fisheries that would have to be surpassed before vessels shifted between fisheries, and this threshold might depend upon the costs of conversion between fisheries.

For instance, the poor prices for skipjack and yellowfin tuna in the early 1980s made many vessel operations unprofitable. A significant portion of the United States seine fleet tied up, and some of the vessels were converted to trawling for pollock in Alaska, an expensive undertaking. Many longline and pole-and-line vessels were also idle, but since there are really no alternative fisheries for them, they were typically either scrapped or tied up.

The collapse of the Alaska crab fishery in the 1980s is a similar example, where boats were converted to almost any possible other use, including pollock fishing and serving as oil drilling tenders.

Entry of new vessels

Entry of new vessels is the essence of a developing fishery. Key questions in understanding why a fishery develops as it does are (1) where the vessels actually come from and (2) who brings them into the fishery.

Vessel construction. Construction of new vessels is an obvious indication of a developing fishery. Often new construction is a sign that the fishery has reached a stage of technological maturity that requires a specific vessel design adapted for the particular fishery. In the United States tuna fishery, which has been well documented, the seine boats that originally replaced the pole-and-line boats were seiners converted from other fisheries, particularly the sardine fishery. As fishermen gained more experience in the fishery, they began to build new boats especially suited for tuna.

In 1987, orange roughy (*Hoplostethus atlanticus*) (a highly valued fish) was found in commercial abundance in southeastern Australia. Because they were found in an area subject to a limited-entry permit based on tonnages and in deep water where only large boats could fish, many small vessels' licenses were bought in order to construct new large vessels capable of fishing for orange roughy. This had the inadvertent effect of reducing the small inshore trawl fleet that supplied much of the fresh fish market to Melbourne and Sydney.

Another similar boom in vessel construction (on a much larger scale) took place as United States fishermen took over the freezer trawler harvest and processing of Alaskan pollock. Many new large processing boats were constructed especially for this fishery.

Ideally one would have a register of all vessels in the fishery, and their age and origin. This could then be used to determine how the number of new vessels being built for the fishery depended upon the catch rates or profitability. Lane (1988) presents a unique analysis of actual investment in fishing vessels in the British Columbia troller fleet. Utilizing income tax returns from 100 vessels he was able to determine the investment over a 10 year period in licenses (it is a limited-entry fishery), vessels, gear and electronics, and other equipment. His results are shown in Figure 4.4. Despite the rare opportunity to actually look at financial records of individual fishermen, this does not provide a complete picture of investment in the fishery. Investment in vessels, for instance, may mean buying someone else's boat, or it may mean new construction. It would be very interesting to know how many new vessels were added to the fleet.

Lane also provides a model of investment by individual fishermen, which appears to explain the observed investment patterns. Lane's model is rather complex; however, its basic assumption is that each fisherman is attempting to maximize his expected profitability while avoiding high-risk situations. This type of model has not yet seen very broad application, but is one of the few attempts to look in depth at the investment decision making process.

Figure 4.5 shows the history of the North Pacific pelagic fur seal (*Callorhinus ursinus*) fishery (Wilen 1976), plotting fleet size against the seal herd size. As the fishery developed, the herd was driven down and more boats entered, until the early 1890s, when the herd was reduced enough that

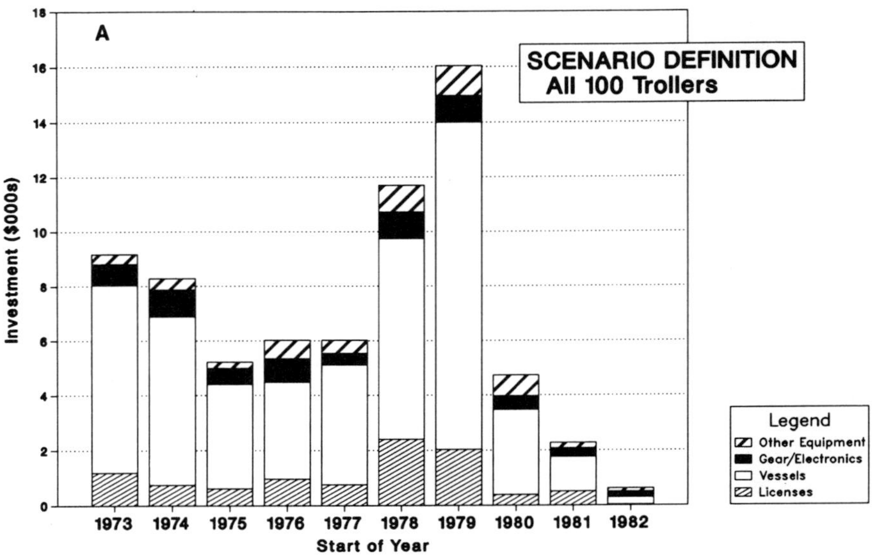

Figure 4.4. Investment by 100 trollers from 1973 to 1982. From Lane. 1988. Can. J. Fish. Aquat. Sci. 45: figure 3.

the fishery was only marginally profitable. Most vessels then left the fishery, primarily to fish for halibut.

Entry from other fisheries. The second major source of growth in a fishing fleet is entry from other fisheries, and this is how most fisheries get started. Gear is modified from some other use, or exactly the same gear is used. The development of the South Pacific tuna fishery relied on techniques developed by Japanese and American fishermen in other waters. We would expect that vessel movement into a fishery would be related both to the profitability in the fishery, and the profitability in the "old" fishery. The movement of United States seiners into the Western Pacific in 1984–1985 is generally attributed to higher catch rates in the Western Pacific than in the Eastern Pacific. Similarly the French and Spanish seine fleets moved into the Indian Ocean in 1985 from the Atlantic.

Simple models of investment

To understand how a fishery is going to develop we need to understand what drives the movement into the fishery of vessels from other fisheries and the construction of new vessels. When vessels are easily exchanged between fisheries we can imagine a simple movement model based on relative catch rates — vessels will simply move into a fishery that has higher

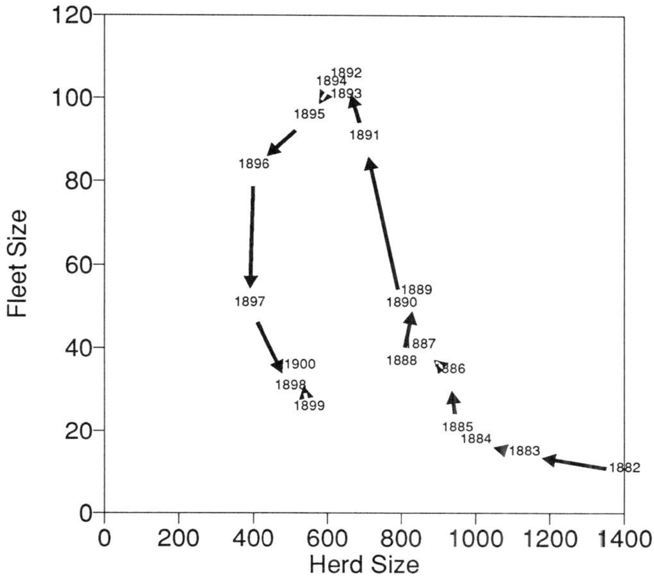

Figure 4.5. Herd size plotted against number of vessels for the North Pacific fur seal fishery. Redrawn from Wilen (1976).

catch rates until the catch rates (or profits) are roughly the same in all alternative fisheries.

However, when interchange between fisheries is difficult, either because of technology or political limitations, we want to understand what governs the development of new vessels. This question breaks into two major parts: (1) whether the entry of new vessels is a result of past profits or anticipation of new profits and (2) whether the investment comes from inside or outside the fishery.

Does investment come from realized or anticipated profits? The simplest models of fishery development assume that profits realized in a fishery are reinvested in the fishery and that all growth in the fishery comes from past profits. This is essentially an economic analog of a predator population that takes surplus energy and uses it for reproduction.

We commonly observe that, when a fishery is profitable, there is a large investment in new vessels or gear, and, in many cases, this investment appears to be much greater than the profits being realized. It is important to understand if fishermen are investing because of what they have earned or because of what they think they will earn. The answer is probably both, but there are few data to make a convincing case.

Does investment come from inside or outside the fishery? A second and related question is where investment comes from. One view is that investment almost always comes from people already involved in the fishery. A second view is that when a fishery is profitable many people from other fisheries or even not involved in fisheries will invest. Again there has been little formal analysis of this, but in many fisheries it is well documented that after initial development much of the capital does come from people not previously involved in the fishery. The question then becomes how much capital (new vessels) will move into the fishery. This presumably depends upon investors' perceptions of potential size of the resource, rate of profits and so on.

Time lags in investment

A key feature to understanding development rates, particularly in determining whether there will be a large overshoot in vessel building, is the time lag required in vessel construction. If it takes 2–4 years to design, plan, and build a vessel, then surges in profitability will likely be followed 2–4 years later by an influx of new vessels, even if the fishery is no longer highly profitable. Artisanal and small-scale fisheries may face similar problems due to time lags in funding from aid agencies; it may take several years to get funding for a fishery development project, and, by the time the money has arrived, the fishery may not be as attractive as it was when the proposals were developed.

Implications of diversity in investment structure

An important determinant of how a fishery responds to variation in catch rate, prices and costs is the diversity of investment structure. If a fishery consists of vessels that are owned outright, then the fishermen can weather fairly dramatic fluctuations in incomes; they need only meet operating costs and crew expenses. However, if interest payments on the vessel are significant, then the operation may go bankrupt after only a much smaller drop in income.

The diversity of the investment pattern is equally important. If there is a lot of difference in the interest payments of individual vessels, then a drop in prices or catch rates may cause the most heavily mortgaged or least efficient vessels to drop out, leaving the remaining vessels more fish to share. This is a stable situation in which drops in income affect only the "least efficient" producers very seriously, and their exit from the fishery is everyone else's gain.

Alternatively, as often happens in fisheries that change rapidly, almost all vessels are financed similarly, either by private or public investment. Therefore, if one vessel is not able to meet its costs, many others also will be unable to meet costs, and a large number of fishermen will go broke together. The buffering effect of the least efficient producers is lost when there is little diversity in the investment structure.

Notice that buffering effects have two causes: diversity in costs so that as incomes decrease some people must drop out before others and impact of some vessels leaving the fishery on fish available to remaining vessels (ideally the loss of some vessels would mean higher catches for the others). Three mechanisms can provide higher catch rates for remaining fishermen:

(1) Under a total allowable catch, there would be more time available for each remaining fisherman to fish before the quota is reached

(2) There may be less direct interference competition among fishermen

(3) There may be a direct increase in fish abundance because of reduced fishing pressure.

Investment in items other than vessels

We discussed earlier the "tragedy of the commons" and the tendency of unregulated fisheries to go to bionomic equilibrium where all vessels just meet costs. We tend to speak of fishery investment in terms of vessels, and early attempts (discussed below) at regulating overinvestment concentrated on restricting number of vessels.

Experience has shown that there are other places for investment to take place, either in gear or in licenses. Overinvestment is socially undesirable, particularly in fisheries where fishing equipment is imported, because money that could be used locally for domestic consumption is instead sent overseas to buy fishing equipment that does not increase total catch. Overinvestment need not take the form of fishing vessels; overinvestment can also take place in fishing gear and fishing licenses.

Investment in gear. A fishing vessel consists of much more than a hull. Motor, hydraulics, nets, and electronics are all essential elements of a fishing operation. Many technical advances in catch methods are related to these attributes rather than hull size or tonnage. Therefore, when considering changes in the fishery, we must always be aware of the many components of gear and not concentrate on tonnages. Attempts to limit overinvestment by regulating hull numbers or size have always been met by investment being transferred to gear.

Investment in licenses. When licensing is introduced into any fishery, and there are a limited number of licenses to go around, transferable licenses will acquire a value. This represents a legitimate form of fishery investment that can also represent overcapitalization if the value placed on the licenses is too high. From a national income perspective excess license value is not nearly as undesirable as overinvestment in machinery. This is because excess license values simply transfer cash between individuals and do not mean that unnecessary fuel, nets, and so on are being imported.

Common phases in fishery investment and regulations

The recognition that fisheries tended to move toward a point of biological overexploitation and economic overcapitalization has led most countries to introduce fishery regulations. These attempts have only been partially successful, but they have been highly informative. In the sections below we outline the four stages of increasing regulation that have been found in various fisheries. No fishery comes to mind that has gone through all four stages; rather they represent the collective experience of many fisheries.

Unregulated. In an unregulated state, a fishery will move toward a point where there are no excess profits to be made; that is, if anyone is making good money someone else will come into the fishery. This may or may not lead to biological overexploitation; more often than not it does.

Biological regulation. The first attempts at fishery regulation were concerned with biological overexploitation. This led to restrictions on gear and season closures, both designed to increase future yields. When successful, this biological regulation improved the total value of catch but did not affect the economics of the individual fishermen over the long term, because the incentive to keep investing until profits were gone remained. The fisheries continued to have too many vessels.

Limited entry. The second stage in regulation was to restrict the number of vessels allowed to participate in the fishery, generally by issuing a fixed number of licenses that could be transferred by sale. This had two effects. As expected, the licenses themselves became a valuable commodity. Fishermen also found an amazing number of ways to circumvent the limitation. If the number of hulls was limited, vessels became longer. If length was limited, vessels became wider. If tonnage was limited, horsepower, hydraulics, and electronics were improved. In general, license limitation has been only marginally successful at preventing overinvestment in fishing gear.

Individual Transferable Quotas. The current state of the art in trying to prevent overcapitalization is Individual Transferable Quotas (ITQ) which give the holder the right to catch a fixed number of fish, using whatever level of investment he wishes (Moloney and Pearse 1979, Clark 1980). This technique has been used with varying degrees of success in Canada and Australia, and is now being attempted on a massive scale in New Zealand. The theory is that ITQs remove any incentive for overinvestment in gear, because the holder cannot increase his catch by a better boat — his catch is determined by the quota he owns.

The natural and expected result of ITQs is that the quota rights themselves become very valuable. In the Tasmanian and South Australian abalone fisheries, license values are on the order of $500,000. Thus the system has effectively transferred overcapitalization from equipment to licenses. This is generally thought to be good, but there are some concerns.

First, new investors in the fishery are heavily mortgaged, and the license values appear to be based on historical rates of return. It seems likely that any drop in price or allowable catches (the quotas are often a percentage of a total allowable catch) will drive many recent investors bankrupt. This may well result in strong political pressure for increased quotas or subsidies, which, given the history of fisheries politics, may result in overexploitation.

A second undesirable feature of ITQs is that they create enormous incentives for fishermen to sell fish illegally. ITQs were first attempted in the Bay of Fundy herring fishery in Eastern Canada; there, large-scale illegal sales essentially destroyed the system. Another side effect is that, if there is any difference in price in size of fish or type of fish, there will be a strong incentive for fishermen to discard low-value items and keep only high value items (a practice known as high grading). Thus the ITQ must be in units that exactly correspond to value.

Determination of effort

If we know how many vessels are potentially able to operate in a fishery in given year, the next question is where and how often will they fish. In the simplest case there is a single fishing location with only one type of gear or species target. In more complex cases, the fishermen must make a choice between alternative fishing areas or species targets.

Single area

When there is only a single fishing area, the question becomes how many of the fishermen will go fishing. Figure 4.6 shows the relationship between the number of vessels operating and the income they obtained in the previous week for the British Columbia gillnet salmon fishery. Note that at very low catch rates few individuals fish, but at higher catch rates most everyone fishes. There are four components that are thought to affect the shape of this curve.

Differential costs among fishermen. Given perfect information and an objective of maximizing profits, each fisherman would fish when he could meet his operating costs. Thus the distribution in Figure 4.6 could reflect just the variation of operating costs from fisherman to fisherman. This does seem unlikely to be the sole explanation, because some fishing takes place at such low catch rates that no one could possibly meet costs.

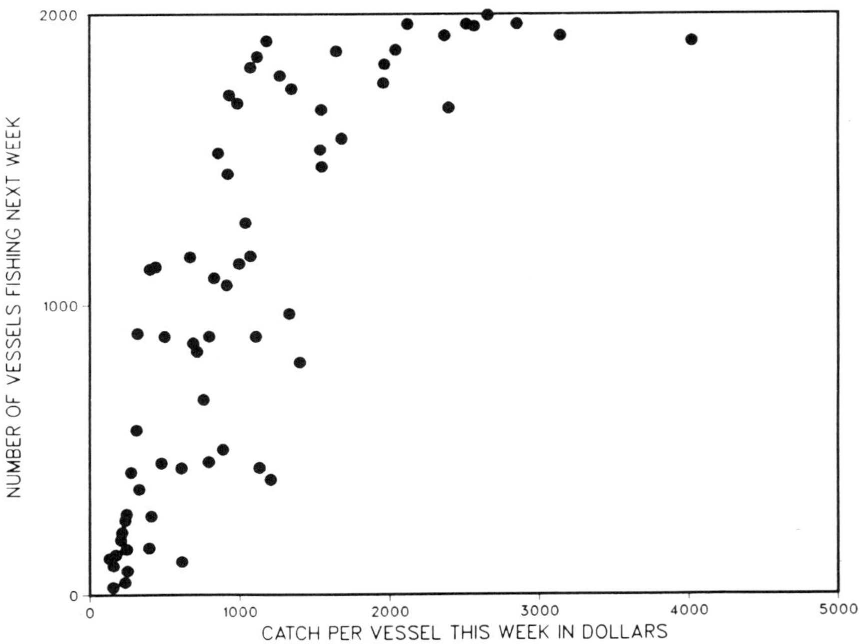

Figure 4.6. The relationship between landed value in a week and fishing effort the subsequent week in the British Columbia salmon gillnet fishery. From Millington 1984.

Information uncertainty. Fishermen do not know what they are going to catch unless they go fishing. Thus some fishing at very low catch rates is fishing for information. Some people have to go out to see what can be caught.

Gear testing. Another factor that affects effort in some fisheries is gear testing. In the salmon gillnet data shown above, many fishermen will fish early in the season when catch rates are poor simply to get their gear and crew working well together. It is recognized they will not meet their costs, but the experience gained will pay off when catch rates are higher later in the season.

Fishing for food or income. A last factor that will affect how many fishermen go fishing at any time is their actual objective. Few fishermen fish only to maximize profits; others fish to get enough food for the year or enough income. Thus we often find that fishermen will quit fishing once they have met their personal "target" for food or income levels, even when there are still plenty of fish available. Another factor in some fisheries is the need for fishermen to accumulate a certain number of days of fishing to qualify for unemployment insurance. It is not unknown in British Columbia for a fishery to be opened when there are no fish available simply to assure that fishermen get enough fishing time to qualify for benefits.

Several areas or species

When fishermen have a choice of places to go within a fishery, we would like to know what determines how many fishermen will fish in each area. The simplest theory suggests that if operating costs are the same in each area and there is perfect information about catch rates, the effort will be distributed so that catch rates are the same in all areas (Gordon 1953). This theory also predicts that if the catch rate of fish in one area increases, vessels should move into that area until the catch rate goes down to the level found in other areas. The implications of this for real fisheries are that if all other costs are the same, fishermen will tend to fish hardest closest to home.

Figure 4.7 shows how the number of vessels fishing in one area in British Columbia increased when abundance increased. The increase in number of vessels was sufficient to quickly equalize the catch rate among areas.

There are a number of reasons why the above theory may be too simple. When we examine actual fisheries data we rarely find that the catch rates are the same in all areas. Below we consider some of the reasons for variation among areas.

Uncertainty. Fishermen rarely have perfect information, and often go to great lengths to keep information about good areas to themselves. This means that catch rates in an area may be higher than other areas, but because of poor information other fishermen may be unaware of it.

Differential costs in each area. Generally the costs of fishing differ among areas. The fuel costs to travel to different areas will be related to distance, some areas may be more prone to gear loss than others, and so on. We would expect that the catch rate would need to be higher in areas with higher costs, and therefore we would expect to find some general relationship between catch rate and distance from port.

Costs of movement. Even when operating costs within all areas are the same, and information is perfect, there will usually be a cost to movement, so that fishermen will be willing to stay in an area with a lower catch rate until the expected increase in catch by movement will pay for the cost of moving. Time lost while traveling may be a significant cost of movement.

Differential costs for different fishermen — home ports. Fishermen may have different costs of operation in different fisheries. If several home ports are involved, then the cost of travelling to an area will differ for different fishermen, and differences in catch rates or incomes may persist simply because it is not worth the travel for the fishermen from the further ports. Similarly there may be slightly different vessel configurations so that some boats can more profitably operate in one area than another.

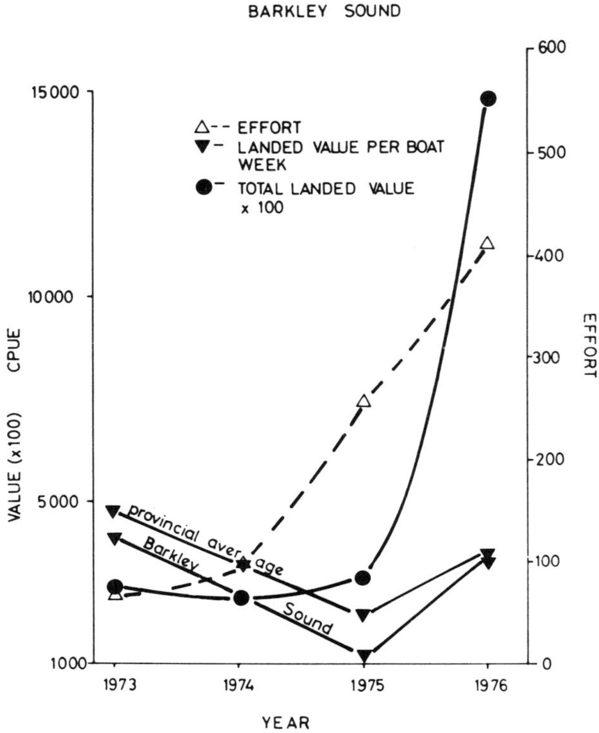

Figure 4.7. The increase in number of vessels fishing in Barkeley Sound, British Columbia, when abundance of fish increased. From Hilborn and Ledbetter (1979). J. Fish. Res. Board Can. 36: Figure 8.

Differential desirability. There are many factors in fishing other than income and catch rates. Most fishermen prefer to be close to home, and protected or safe waters are preferred to exposed waters. Thus we might expect that distance from home, comfort, and risk all combine to make some places more desirable than others. We would expect the catch rates to be higher in the less desirable areas.

Differential skill or knowledge. Similar to differential vessel characteristics is differential skill. Fishermen often specialize in local knowledge about currents, fish behavior and so forth. Thus catch rates may be high in one place, but do not represent the expected catch rates of fishermen fishing elsewhere. Fishermen may fail to move into an area with a high catch rate simply because they know they would not experience the same rates if they moved there.

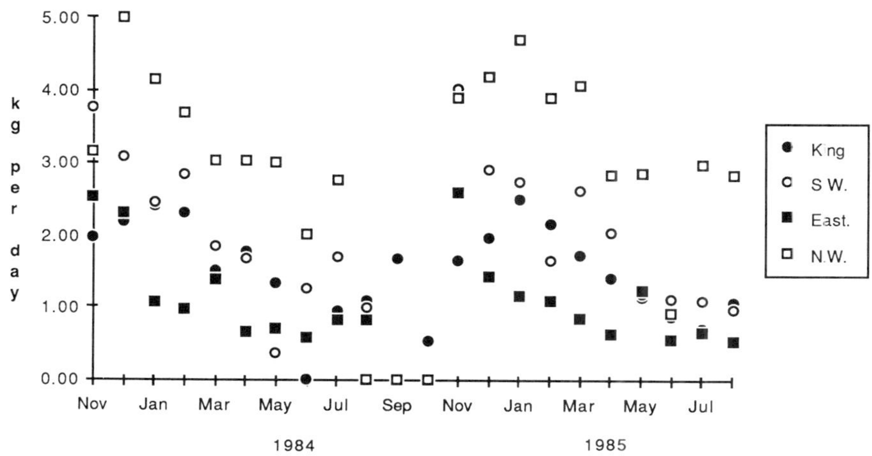

Figure 4.8. Catch rates in four areas of the Tasmanian rock lobster fishery. N.W. is the remote exposed northwestern portion of Tasmania. S.W. is the southwestern portion of Tasmania which is also exposed to poor weather but closer to home. King is King Island, home port to some of the fishermen. East is eastern Tasmania, near the home ports of most fishermen and on the protected side of the island. From Hilborn and Kennedy (1991).

Hilborn and Kennedy (1991) examined the spatial distribution of catch rates in the Tasmanian rock lobster (*Jasus novaehollandiae*) fishery. They found that catch rates were much higher in areas that are remote (far from home), and subject to poor and often dangerous weather. Figure 4.8 shows the distribution of catch rates (kilograms per pot day) for five areas, ranging from the east coast, which is sheltered and close to port, to the northwest coast, which is exposed and far from most fishermens' homes. This graph shows that the west coast areas have much higher catch rates than the other areas, and the ranking of the catch rates by area stays near constant throughout the whole season. This is despite considerable differences in the availability of the lobsters seasonally in different areas. The movement of fishermen among areas and the differential costs of fishing (both monetary and in terms of potential time lost due to poor weather) determine the relative catch rates by area.

Catching power

In the previous two sections, we discussed the factors that determine how many vessels will be present in a fishery and how many will be fishing in a given location at a given time. In this section, we discuss what determines how many fish they actually catch.

The catch of a vessel is determined by three factors: (1) how often it fishes, (2) abundance where it fishes, and (3) the crew's skill relative to other vessels fishing in the same place. Figure 4.9 shows the breakdown of the importance of these factors in two fisheries: a salmon sport fishery and a salmon purse seine fishery in British Columbia. The sport fishery is probably similar to many artisanal fisheries: people go fishing when they have the time, the need, or feel like it, and catch is dominantly determined by time spent. In the seine fishery all three factors are approximately equal in importance.

The topic of catching power is traditionally associated with only the third question: given boats that fish the same amount of time in the same place, what determines how many fish they will catch.

Time budgets

The starting point for analysis of individual catching power is to develop time budgets for vessels. One generally finds that vessels spend a significant portion of their total time available for fishing in other activities than setting the net or putting their hooks in the water. It is useful to think of the total time as consisting of four experimental components:

$$\text{total time} = \text{travel time} + \text{search time} + \text{setting time} + \text{handling time}. \tag{4.3.1}$$

1. *Travel time.* Time spent steaming to and from the fishing grounds is frequently a major component. The long-distance pole-and-line and longline fisheries of Japan operating in the South Pacific are examples of this. A vessel may be able to increase its catch rate simply by being able to steam faster to the fishing grounds.
2. *Search time.* Much time spent by fishing vessels can be classified as searching for the best places to deploy gear. For example, pole-and-line and seine vessels fishing for tuna devote most of their days to searching. If a vessel can increase its search efficiency, either by moving faster or increasing its detection radius by higher towers or aircraft, its catching power will increase.
3. *Setting/shooting time.* Once a likely spot is found, the next time component is the time required to set the gear and bring in the fish. More efficient net manipulation, lower probability of bad sets, and fewer difficulties in bringing in the net can all contribute to increased catching power.
4. *Handling time* Once the fish are on board they must be processed and put in the hold. If this activity is incompatible with the resumption of search and fishing, then it is classified as handling time and can provide a limit to the catching power.

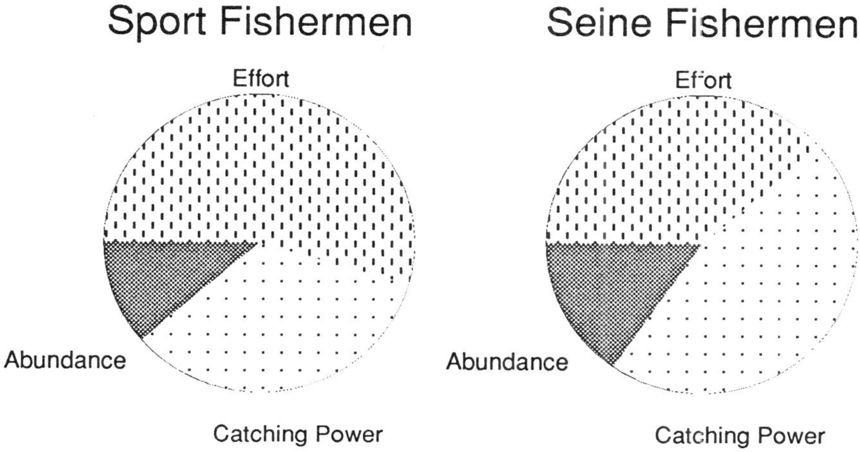

Figure 4.9. Importance of fishing effort, fishing area, and fishermen's skill in two fisheries.

Time budgets in tuna seiners. Tuna purse seining can be used as an example of how time budgets can contribute to an understanding of catching power. If the total time available for fishing is partitioned into the four components above, then the time for actually fishing is total time available less time spent steaming to the grounds. We will call this total fishing time and refer to it as T.

Next consider the search process. Suppose the vessel moves at a speed s (kilometers per hour) and can detect fish at some width w (kilometers) on either side of the vessel. If the vessel has high towers or employs an aircraft, then w will be bigger. For simplicity, assume that any school closer than w km from the vessel will be seen, and any school beyond w will be missed. In reality, there is a probability of detection that decreases with distance from the boat, but this can be translated into an effective sweep width w. The vessel therefore searches an area $A = s \times w \times 2\ km^2$ per hour of search time.

Now assume that once a school is seen, there is a probability p that the vessel will make a set on it. Further assume that each set and handling time takes h hours per set, and that each set takes an average catch of c tons. Finally assume that the density of fish in number of schools per km^2 is D.

Suppose now that we want to predict the number of sets (S) that the average boat will make per fishing time T, and the catch rate $C = Sc$. To construct this prediction, we first note that the total time T can be written as $T = T_h + T_s$, where T_h is total time spent setting and handling, and T_s is total time spent searching. Now, according to the above assumptions and definitions, T_h can be written as $T_h = hS$, that is, the time per set times the number of sets. Next, note that the number of sets will be related to search time and density by the equation $S = pADT_s$, provided schools are distributed at random within the area searched (i.e., S is encounter rate per hour times hours searched, encounter rate per hour is AD). Rearranging this equation results in the equivalent expression $T_s = S/(pAD)$. At this point, we have now expressed the two components of T, T_h and T_s, in terms of the search and handling parameters. Substituting these expressions into the time budget results in the equation

$$T = T_h + T_s$$

$$T = hS + \frac{S}{pAD} \tag{4.3.2}$$

We can now solve this equation to predict S as a function of T, h, p, a, and D, and to predict catch rate C as S times c.

The catch rate in tons per total fishing hour (C) can finally be written as

$$C = \frac{TpADc}{1 + hpAD} \tag{4.3.3}$$

This relationship is often referred to in the ecological literature as Holling's Disc equation, and in the previous paragraph we have simply repeated C. S. Holling's (1959) derivation for natural predators. Note that the ratio C/T will not be a good index of abundance if h is significant.

We can use this equation to relate vessel characteristics to catch rate. If a vessel moves faster or has a higher detection width, it will catch more fish. If a vessel reduces its handling time or increases its total time available for fishing, it should do better, and if it takes more fish per set (c) or school, it should do better.

Other search fisheries. Similar formulations of the search process can be used to examine the catching power in other fisheries. Tuna pole-and-line fisheries are almost exactly analogous to tuna seine fisheries: the search speed, search width, total time available, and handling time have similar meanings.

Individual divers searching for sessile invertebrates also conform nicely to the same model. The search rate is the speed of coverage of the bottom, the search width is the visibility of the diver. The probability of detection, p, is the chance a diver will actually see a target in the area searched. This can depend upon bottom conditions, color, diver skill, and so on. The handling time is how long it takes the diver to physically remove and process each invertebrate or fish taken.

Area swept methods. Other types of fisheries that are similar in structure to the above model are those where a gear is physically passed through or over the fishing ground. Trawl fisheries are the classic examples. One calculates the speed the net is towed, the volume or area swept by the net mouth, the proportion of fish in the area swept that will be caught by the net, and the handling time associated with each unit of fish caught; then one can predict the catch. The major limitation to this method is the assumption about random distribution of fishing effort in relation to the fish.

Hook and line. Hook and line fisheries can also be analyzed with the above search/time budget model. For troll fisheries, the speed of the lure being pulled through the water is the search speed. The search width becomes the distance from which a fish will detect the lure, rather than a characteristic of the fisherman. p becomes the probability that a fish who detects the lure will attack the lure and be hooked. The handling time is the time required to land and process the fish and get the hook back in the water. If the fish movement speed is not small in relation to the speed of the lure being trolled, then a modification of the lure speed needs to be made to include the fact that the fish may swim into the reactive distance as well as the lure being pulled past the fish.

For longline fisheries the structure is similar to that above, except now the fish swimming speed is the search speed. We think of it as the fish sweeping out a volume of water and encountering the lure. In most longline fisheries, some saturation of gear takes place between the time when the gear is set and when it is checked. This requires modifications in the formula.

Trap fisheries — gillnets. Trap fisheries are common for lobster, crabs, and in some cases prawns and other invertebrates, or even fish. Each trap can be thought of as having an effective detection radius, in which individual targets will be attracted to and caught by the gear. The model needs some modification, however, because the individual gear can saturate. The more individuals are in a pot, trap, or gillnet, the less likely it is that subsequent individuals will be caught.

Catching power of individual vessels

The above time-based models are ways of looking at catching power from a mechanistic viewpoint. The alternative is a strictly statistical one, to look at measured catch rates of fishermen in the same place and see what is related to their performance.

Vessel characteristics. There is a large body of literature relating vessel characteristics such as size, tonnage, or speed to catch rates. In some cases, there seem to be good correlations; in other cases, the correlations are rather poor. As a general rule, in a fishery where there are large differences in vessel size and the volume swept by the gear is quite different for different vessels (such as in a trawl fishery), we would expect to be able to explain catch rates by vessel characteristics.

Skipper and crew. In some fisheries, vessel characteristics do not appear to be very important, and differences in catch rate seem to be related to the skill of the fishermen. This is not surprising, but unfortunately there are few data sources that have been examined to compare the relative importance of fishermen and gear (Pálsson and Durrenberger 1982, Hilborn and Ledbetter 1985).

Catching power and vessel standardization

The efficiency of fishing fleets rarely stays constant; fishermen engage in a constant search for more efficient fishing methods and technologies, and this normally involves changes in fishing gear as well as knowledge.

Standardization for different vessel classes. Fishing vessels are different, and some vessels are more efficient than others. A common problem faced by fishery managers is trying to understand how the efficiency of the fishing fleet has changed over time. Indeed, agencies that attempt to use biomass dynamics models, or use some measure of fishing effort in virtual population analysis (VPA) (Chapter 10) or catch-at-age analysis spend considerable energy trying to calibrate effort changes to produce a time-invariant estimate of standardized fishing effort. The fishery for tuna in the eastern tropical Pacific is a typical example. Early commercial catch data were obtained from pole-and-line boats; later purse seining began with small vessels, which evolved over time into large purse seiners over 1,000 tons. The agency responsible for assessment of these stocks, the Inter-American Tropical Tuna Commission (IATTC) has attempted to maintain a consistent measure of fishing effort despite the enormous changes that have taken place in the fishing gear. The method we recommend for standardization of vessel classes was first used by Allen and Punsly (1984), working for the IATTC.

If we think of fishing vessels as discrete classes, we can think of the catch rate obtained as the product of the abundance in a particular year times the efficiency of the vessel class

$$U_{ti} = A_t q_i \tag{4.3.4}$$

where U is the catch rate, A is the abundance of fish, and q is the efficiency. The subscript t refers to time and i refers to vessel class. We can write this as a statistical model as follows:

$$U_{ti} = U_{11} \times \alpha_t \times \beta_i \times \epsilon_{ti} \tag{4.3.5}$$

where U_{11} is the catch rate obtained by the first vessel class in the first time period, α_t is a factor that is the abundance in year t relative to year 1, β_i is

Table 4.1. Catch rate (tons per hour) for three classes of vessels in four different years.

Year	Class I	Class II	Class III
1	0.63	0.85	1.28
2	0.46	0.65	1.09
3	0.35	0.66	1.01
4	0.43	0.48	0.84

the efficiency of vessel class i relative to vessel class 1, and ϵ_{ti} is a factor that accounts for the deviation between the observed U_{ti} and the expected value for t and i.

Taking logarithms of both sides, we obtain the linear statistical model

$$\log(U_{ti}) = \log(U_{11}) + \log(\alpha_t) + \log(\beta_i) + \epsilon_{ti} \tag{4.3.6}$$

We can estimate the values of $\log(U_{11})$, α and β using the Generalized Linear Model (GLM) (Nelder and Mead 1975).

For instance, consider the catch rate data given in Table 4.1. Note that in general class II vessels have a higher catch rate than do class I vessels, and that class III vessels have the highest catch rate of all. The catch rate appears to have been declining over time. If we fit these data using the GLM of Equation 4.3.6, we obtain the parameter estimates shown in Table 4.2.

These results show that the base catch rate, for time 1 and class I is estimated at 0.595 (tons per hour), and that in year 2 the catch rate was estimated to be 88.3% of the rate in year 1, and declined further to 70.% of the year 1 rate in year 3, and 63.4% of the year 1 rate in year 4. The class II vessels are estimated to be 1.41 times more efficient than class I vessels, and class III vessels are estimated to be 2.28 times more efficient than class I vessels.

We have presented only estimated catch rate data, not actual catch and effort data; and it is quite possible, for instance, that the average catch rate, if unadjusted for the vessel class mix, might have changed in a much different pattern. The need for calibration of catch rate data has long been recognized and as mentioned previously has been a routine part of many agencies work. Although there are many ways of doing such standardization, the use of GLMs is by far the most useful procedure, because it provides a framework for incorporating many factors in addition to vessel class and time.

Incorporating vessel attributes. A good example of the power of GLMs is their ability to transcend the vessel classification schemes necessary to divide fishing boats into discrete vessel classes. Vessels often do not fall into discrete classes, but rather exhibit a continuous range of tonnage and horsepower, as well as discrete items such as specific types of nets, electronic

Table 4.2. Parameters estimated from the data of Table 4.1.

Parameter	log value	value
U_{11}	−0.518	0.595
α_2	−0.244	0.883
α_3	−0.355	0.701
α_4	−0.456	0.634
β_2	0.347	1.41
β_3	0.825	2.28

gear, and search assistance such as search vessels or helicopters. For instance a model that could include vessel tonnage, horsepower, and the presence or absence of a piece of equipment *i* (for instance a color hydroacoustic sounder) can be written as

$$\log(U_{ti}) = \log(U_{11}) + \log(\alpha_t) + \beta_1 \log(T_i) + \beta_2 \log(H_i) + \gamma E_i + \epsilon_{ti} \quad (4.3.7)$$

where now U_{ti} is the catch rate of the *i*th vessel in the *t*th year, α_t as before is a scale factor for the catch rate in year *t* relative to year 1, β_1 is the log of the increase in catch rate expected for each unit of tonnage, T_i is the tonnage of the *i*th vessel, β_2 is the log of the increase in catch rate expected for each unit of horsepower, H_i is the horsepower of the *i*th vessel, γ is the log of the increase in catch rate expected if the vessel is equipped with the color hydroacoustic device, and E_i takes the value 0 if the vessel does not have a color hydroacoustic device and the value 1 if it does have the device. ϵ_{ti} is the residual for the *i*th vessel in the *t*th year.

In this model, we have discrete treatment effects such as the time effect and the color hydroacoustic effect, and continuous effects such as tonnage and horsepower. If we simply code tonnage and horsepower as their actual values for each vessel, then interpretation of U_{11} is a little awkward since it is the catch rate estimated for a vessel of zero tonnage and zero horsepower. In practice, interpretation is easier if the tonnage and horsepower are given as the difference between some base level (the standardized vessel), and the value for the particular vessel. For instance, if many vessels in the fleet are 100 tons with 500 horsepower motors, if we subtract 100 tons from the vessel tonnage and 500 horsepower from the vessel horsepower, then the U_{11} will be the catch rate in time 1 for a vessel of 100 tons and 500 horsepower with no color hydroacoustic device.

The GLM approach provides a useful framework for determining what vessel attributes do contribute to catching power. Just as one would do in a multiple regression, you can include or exclude different vessel characteristics to see which ones provide significant increases in predictive ability.

Incorporating spatial differences. Most fisheries have discrete areas for

Table 4.3. Data for three vessel classes in 4 years in three areas.

Year	Class I	Class II	Class III
		Area 1	
1	0.63	0.85	1.28
2	0.46	0.65	1.09
3	0.35	0.66	1.01
4	0.43	0.48	0.84
		Area 2	
1	0.60	0.81	1.10
2	0.52	0.68	0.79
3	0.47	0.54	0.84
4	0.39	0.44	0.79
		Area 3	
1	0.38	0.64	1.45
2	0.32	0.51	1.09
3	0.31	0.54	1.07
4	0.33	0.52	0.90

collection of catch data, and the distribution of fish abundance and fishing effort will usually differ in space. Spatial effects can be added to the GLM model quite easily:

$$\log(U_{tik}) = \log(U_{111}) + \log(\alpha_t) + \log(\beta_i) + \log(\gamma_k) + \epsilon_{tik} \qquad (4.3.8)$$

where everything is the same as in Equation 4.3.6, except that γ_k now represents the average abundance differential in area k. Table 4.3 shows some data for three fleets, four years, and three areas. The results of the GLM analysis are given in Table 4.4. Again we see a declining abundance from year 1 to year 4, and increasing catching power with vessel class. The area effects γ show that area 2 has about 79% the catch rate of area 1, and area 3 has 59% of area 1 catch rate.

However, it is in the ability to analyze interaction between factors that the GLM framework is particularly powerful. In some fisheries, particular classes of vessels are better able to fish in some areas; small vessels may be more efficient close inshore, local fishermen may be more effective in their home waters and so on. In the GLM framework, we examine the interaction between area and vessel class: if there is such an interaction, then we say that vessel's catching power differs by area. The model can be written

$$\log(U_{tik}) = \log(U_{111}) + \log(\alpha_t) + \log(\beta_i) + \log(\gamma_k) + \log(\beta\gamma_{ik}) + \epsilon_{tik} \qquad (4.3.9)$$

Table 4.5 shows the results of this analysis. There is a significant area/class interaction, and most specifically the class III vessels seem to be par-

Table 4.4. Parameters estimated from GLM for time, fleet, area analysis.

Estimated log value	value	Parameter
−0.5774	0.56	U_{111}
−0.1908	0.83	α_2
−0.2921	0.75	α_3
−0.3351	0.72	α_4
0.431	1.54	β_2
0.9543	2.60	β_3
−0.1646	0.85	γ_2
−0.2478	0.78	γ_3

ticularly effective in area 3, where their catch rate is 1.77 times what would have been expected in area 3 for class III vessels.

Another type of interaction that may occur in such data, but not in the data used in this example is area/year interaction. This type of interaction indicates that the abundance trends in the different areas are not the same; some areas may be increasing or decreasing faster than other areas, or going in different directions.

Considering individual vessels. The efficiency of individual vessels may not be related to specific attributes of the vessel, but rather to the skill of the skipper and crew or to some intangible factors associated with the vessel. One can use the GLM framework to ask the simple question "Are vessels different?" by formulating the model as follows:

$$\log(U_{tik}) = \log(U_{111}) + \log(\alpha_t) + \log(V_i) + \log(\gamma_k) + \epsilon_{tik} \quad (4.3.10)$$

where i now refers to individual vessels, and V_i is a vessel effect which simply says how powerful vessel i is relative to vessel 1. In an examination of the British Columbia salmon purse seine fleet, we found that individual vessel attributes such as length and tonnage could not explain more than a few percent of the variation in catch between vessels, but that individual vessel differences explained 18%. This difference is felt to be related to the skill of the fishermen. We also showed that there was a significant interaction between vessel and area and indication that vessels are specialized by area; some vessels fish better in some waters and more poorly in other waters. Discussions with fishermen suggested that there were specialized fishermen, and that their knowledge of local fishing conditions was important in the relative catching rate.

Estimating abundance trends. There are two major reasons for doing the type of analysis discussed above: (1) to understand what determines the catching power of individual vessels and (2) to correct for changing fleet composition to provide a consistent index of abundance based on catch rate data. Many people become obsessed with obtaining a standardized fishing

Table 4.5. Results of model with area/vessel class interaction.

Estimated log value	value	parameter
−0.4578	0.63	U_{11}
−0.1908	0.83	α_2
−0.2921	0.75	α_3
−0.3351	0.72	α_4
0.3146	1.37	β_2
0.7118	2.04	β_3
−0.2401	0.79	γ_2
−0.5312	0.59	γ_3
0.07219	1.07	$\beta\gamma_{22}$
0.277	1.32	$\beta\gamma_{23}$
0.1543	1.17	$\beta\gamma_{32}$
0.5734	1.77	$\beta\gamma_{33}$

effort, in order to calculate a standardized catch per effort. This is unnecessary; calculation of a standardized index of abundance does not require ever calculating standardized effort.

Emerging from the GLM analysis is a *time effect*, the α present in all of the above equations. This is standardized abundance. To state it simply, you put all your catch data into a big GLM analysis, and the time effect that emerges is your best estimate of the abundance trends. If you are dealing with spatially discrete stocks, then the model should be fit with area/time interactions and the combination of time and area/time effects is the area-specific abundance trends.

Where it goes wrong. GLM provides a very powerful, consistent method for examining the effects of vessel differences and area differences to determine abundance trends. However, there are a great many ways such analysis can go wrong. First and foremost, if catch rate by a specific vessel type in a specific area is not proportional to abundance, then the abundance trend estimated from GLM will not be proportional to abundance! You can't avoid this basic problem (see Chapter 6).

Equally problematic is that if there has been any change that is unrelated to quantifiable effects then the GLM will not be able to tell that this change has occurred, and will instead ascribe the change to changing abundance. For instance, if vessel efficiency has been growing due to increasing skill, it will appear that abundance has been increasing and catching power has remained constant. It is only if there are between vessel differences in catching power that are related to a measurable characteristic (tonnage, horsepower, etc.), that the GLM will be able to distinguish between abundance changes and efficiency changes.

The important message of this section is that GLM is the appropriate way to analyze catching power information and the best tool available to calculate standardized catch rates.

Discarding

Once fishermen have fish on the deck, we need to know what is actually kept. This is particularly important in developing fisheries, where species or size groups may be caught and discarded during early stages of the fishery, only to be kept and marketed at later stages (Saila 1983). It is also important because it appears that almost all fish discarded at sea die (Saila 1983). The previously discarded species and size classes represent losses that would not be documented in landing statistics, but are nonetheless real removals from the populations.

It is very difficult to obtain data on discarding, but there are several factors that should influence it:

1. *Markets*. Fishermen are most unlikely to keep fish they cannot sell or eat themselves.
2. *Hold capacity*. If a vessel is on an extended trip and fills its hold capacity, it seems likely that the fishermen would discard less valuable species and only keep larger fish that are more valuable per unit of hold capacity occupied.
3. *Regulations*. If there are regulations on size or species, fishermen who catch controlled fish may simply discard them, regardless of existing laws.

Pikitch (1987) and Pikitch et al. (1988) have studied discarding in a multispecies trawl fishery subject to single-species quotas. Her method was to place observers on board vessels, and, because there was no legal penalty for discarding, there is no reason to believe that the fishermen's behavior was modified by the presence of observers on board. Table 4.6 shows the percentage of each species discarded during periods of trip limits and total prohibition. Not surprisingly when trip limits were large, few fish were discarded, but more were discarded as trip limits were reduced. When a species was totally prohibited, fishermen discarded all they caught of the prohibited species.

Figure 4.10 shows the reasons fishermen gave for discarding. The major reason for three of the species, Pacific ocean perch (POP), widow rockfish, and yellowtail was regulation; fishermen discarded because they were subject to trip limits or landing prohibition. Some discarding took place because the fish were too small to be marketed, and some were discarded for *high grade*, that is smaller, less valuable fish were discarded so the vessels could retain larger, higher value fish of the same species.

Table 4.6. Percentage of the catch (in weight) discarded during different trip limit regimes. From Pikitch (1987)

Species	Gear Type	Regulation	% Discard
Widow rockfish	BGT	30,000 lb per trip	5.7
		3,000 lb per trip	52.3
	MT	Prohibited	100.0
Sablefish	BGT	5,000 lb per trip	
		22 in. length	15.3
		6–12,000 lb per trip	20.4
		Prohibited	100.0
Pacific ocean perch	BGT	10,000 lb per trip	0.6
		5,000 lb per trip	7.6
	ST	10,000 lb per trip	37.9
		5,000 lb per trip	95.7
Sebastes complex	BGT	25,000 lb or more	12.4
		20,000 lb or less	42.5
	MT	all	8.4
	ST	all	46.1
Yellowtail rockfish	BGT	less than 10,000 lb	9.1
		10,000 lb or greater	6.1
	MT	all	0.1
	ST	all	0.4

(BGT = bottom groundfish trawl, MT = midwater Trawl, ST = shrimp trawl.)

4.4. Fleet Dynamics Models

Having discussed the components of fleet dynamics, we can now examine how some simple models of fleet dynamics behave. Just as we explored the behavior of different models of exploited populations in the previous chapter, we will now explore different models of fleets and fish in this section. We will not attempt to construct a series of detailed component models of the different parts of fleet dynamics but will instead introduce some rather simple generalized models, and then discuss one particular model in detail.

Basic Fleet model

The simplest model of a fishing fleet considers only a single homogeneous fleet and a single fish stock. The following model was first presented by Smith (1969). It considers a fish stock modelled by the Schaefer model:

$$\frac{dB}{dt} = rB\left(1 - \frac{B}{k}\right) - C \tag{4.4.1}$$

where B is the stock biomass, r is an intrinsic rate of growth, k is an unfished equilibrium stock size, and C is the catch rate. The simplest assumption for

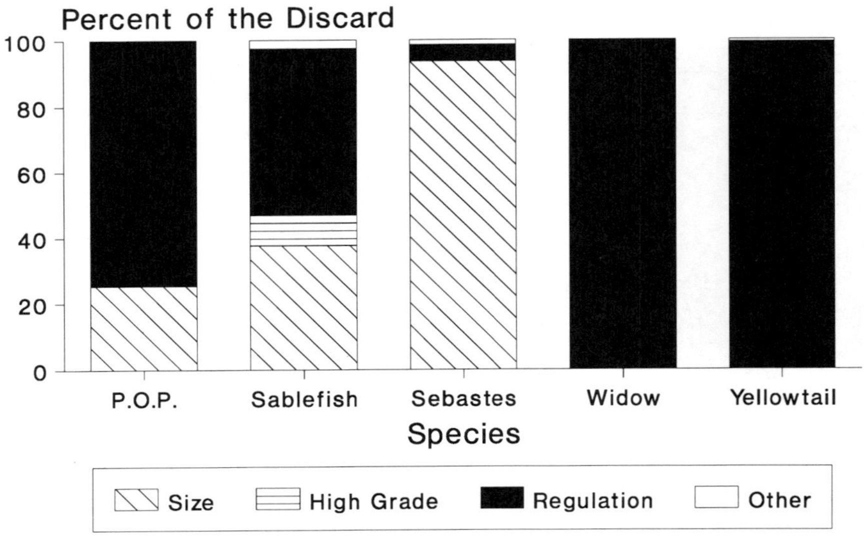

Figure 4.10. Reasons for discarding in the Oregon multispecies trawl fishery. From Pikitch, Erickson and Wallace. 1988. P.O.P. = Pacific ocean perch, Sebastes is a species mix of the genus *Sebastes*, Widow = widow rockfish, and Yellowtail = yellowtail rockfish.

catch rate is that it is proportional to the number of vessels in the fishing fleet V so that

$$C = qVB \tag{4.4.2}$$

where q is the efficiency of the vessels (catchability coefficient).

If we allow the following parameters, $r = 0.4$, $k = 10{,}000$, $q = 0.01$, then we have a stock that when unfished would have a size of 10,000 (tons for instance), which could potentially increase by as much as 40% per year, and would have a sustainable yield of 1,000, and where each boat could take 1% of the stock each year. The optimum fleet size for MSY would be 20 vessels.

Let us assume that each vessel has fixed costs of \$25,000 per year, and variable costs of \$25,000 per year. Further assume that the price of a new vessel is \$150,000, and that fish are worth \$900 per ton. The profits of an individual boat are therefore the total tons captured times the price of fish, minus fixed and variable costs. Let us assume that the change in fleet size is governed by a simple rule, where the profits from the fishery are invested in new boats, and if losses take place, boats withdraw from the fishery. This can be written as

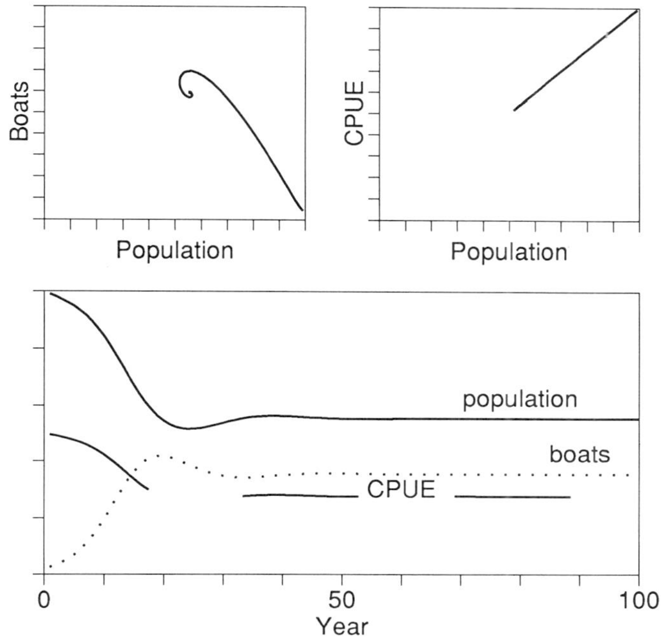

Figure 4.11. Behavior of a simple model of investment.

$$\frac{dV}{dt} = \frac{\text{profits per boat}}{\text{cost of a new boat}} \tag{4.4.3}$$

We are, of course, assuming that vessels come in infinitely small units, and that the decision to invest and the arrival of the vessels on the fishing grounds take place instantaneously. Similarly, when profits are negative, the vessels disappear in infinitely divisible units and the disappearance takes place instantaneously.

In the context of our previous discussion of the dynamics of fishing fleets, we are assuming that there is no leverage on investment, the only investment comes from profits made in this fishery — no mortgages, no speculators etc. We are also assuming that all vessels are identical and that catch rate is directly proportional to abundance. There is no discarding.

Figure 4.11 shows the behavior of this model when we start with a single vessel. The fleet builds up from 1 boat to roughly 18 boats, and the population decreases to roughly half of the virgin stock size. Profits of course decrease to zero. There is a very minor overshoot where the fleet overbuilds slightly, profits drop negative, and some disinvestment takes place. In general, however, this scenario is rather attractive: the system undergoes no

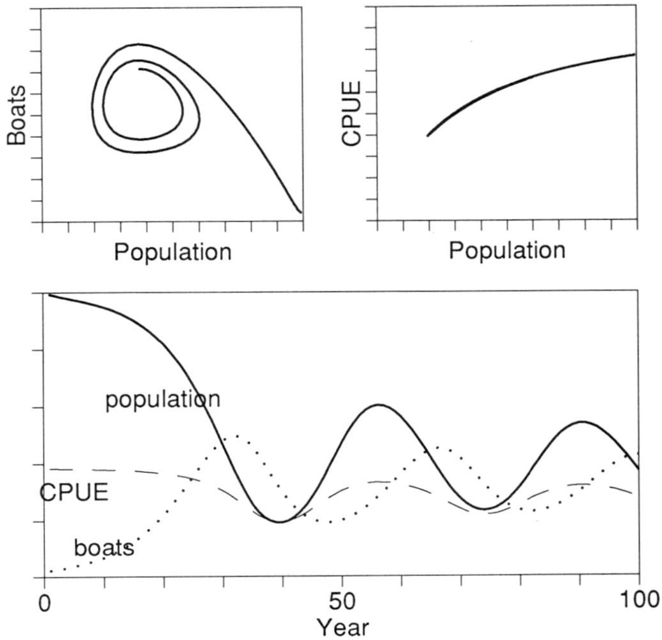

Figure 4.12. Behavior of simple fleet dynamic model when catch rate is not proportional to abundance.

major crash, the stock ends up roughly near MSY, and fishery managers would be reasonably content.

Adding a functional response to fish abundance

We can alter the behavior of this system by assuming that the CPUE is not directly proportional to abundance, but that instead the CPUE is governed by the following equation:

$$\text{CPUE} = \frac{100 * B}{3000 + B} \tag{4.4.4}$$

Under this relationship (the Holling disc equation mentioned in the previous section), the CPUE at virgin stock size will be slightly lower than before, but will not drop as quickly when abundance drops. The behavior of this model is shown in Figure 4.12. Note that now the system cycles with some damping of the cycles. There is considerable overshoot of investment, profits become quite negative, disinvestment occurs, the stock starts to rebuild, and so forth. This is caused by the relationship between CPUE and abundance. In this case, the CPUE stays high enough that the fishery is profitable down to a lower stock size, and at this lower stock size the surplus production is lower, so that the stock continues to decline until fishing effort is decreased.

The behavior of any fleet-fish system depends critically upon the costs, price, and relationship between CPUE and abundance. If price were to increase, or costs decrease, then the fishery would still be profitable at lower stock sizes. A more general and detailed model was presented by Getz et al. (1985) where they considered the possibility of diversity in the fishing fleet in both type of vessels and fishing strategies.

Spatially separated stocks

Many fisheries operate on spatially discrete stocks, and almost all fisheries operate in a number of spatial areas. Caddy (1975) described the Georges Bank scallop (*Placopecten magellanicus*) fishery, where scallop are found in reasonably discrete beds and fishermen search for and allocate their fishing effort among the beds. Figure 4.13 shows the distribution of 18 Canadian scallop boats in 1970. Caddy modelled the fishing pattern of the fleet, as well as the biology of recruitment and growth of the scallops, and investigated the difference between this spatially discrete model and normal biomass dynamics models. Caddy found that the yield-effort relationship was greatly influenced by how the fleet allocated its fishing effort. These relationships are shown in Figure 4.14.

A generalization of the type of spatial model used by Caddy was proposed by Hilborn and Walters (1987). The Hilborn and Walters model assumes that there are a number of discrete stocks, and the fishing fleet is homogeneous. The model assumes that the fishermen know the catch rates, costs, price etc. available in each area and within each year and that effort first goes to the spatial areas where the rate of return is the highest. Rate of return may be catch rate or profitability, whatever the fishermen are trying to maximize. As the fishing year develops, the areas that started out with the highest rate of return may be depleted so that other areas are better, and effort then moves to those areas. Each unit of effort goes to the area where the rate of return is the highest. So long as the effort is allocated in small units, the fishing effort will always be allocated to the best area, and if there is enough effort, the rate of return should tend to be the same in all areas.

We can model such a fishery by considering a set of 10 discrete stocks (although scallops presumably exchange larvae), each governed by a discrete Schaefer model:

$$B_{i,t+1} = B_{i,t} + r_i B_{i,t}\left(1 - \frac{B_{i,t}}{k_i}\right) - C_{i,t} \qquad (4.4.5)$$

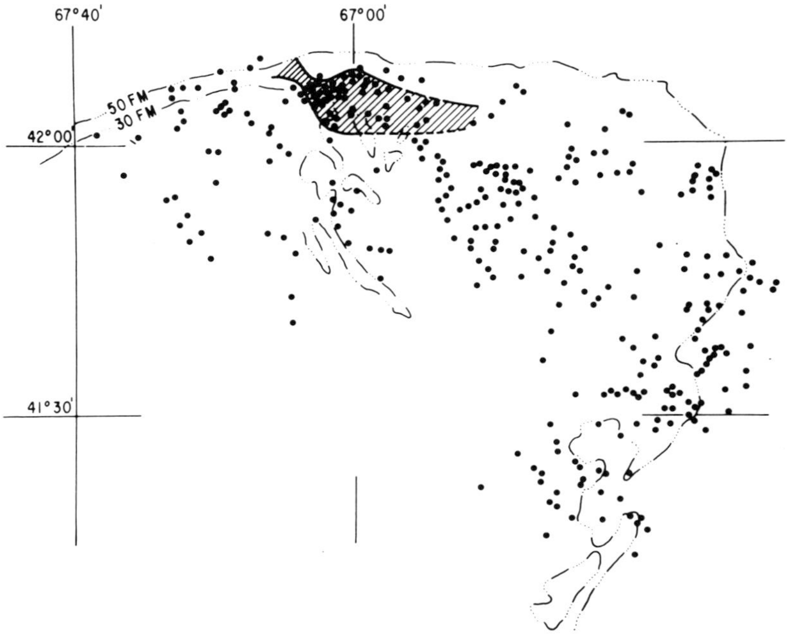

Figure 4.13. Positions fished by a sample of 18 Canadian boats during the first half of 1970. The extent of high-density recruitment of scallops is shaded. From Caddy (1975). J. Fish. Res. Board Can. 32: Figure 3.

where subscripting is now by area i and time t. Within each year, effort is divided into many small units, which are sequentially allocated to the area with the best CPUE. For instance, if we have 1,000 boat/days of effort, we might break them into 100 units of 10 boat days. We then loop over these 100 units of effort, each one determining which area has the best CPUE, and allocating the fishing effort to that area. The following calculations are performed within the year for these 100 units of effort. As effort is allocated to the best area, its biomass will be reduced, and some other area may become the best area.

$$
\begin{aligned}
\mathrm{CPUE}_{\mathrm{best}} &= B_{best} q_{best} \\
C_{best} &= C_{best} + \mathrm{CPUE}_{best} \times e \\
E_{best} &= E_{best} + e \\
B_{best} &= B_{best} - \mathrm{CPUE}_{best} \times e
\end{aligned} \tag{4.4.6}
$$

At the end of this loop over effort units, we have cumulative catch and effort over each area where fishing will take place. To keep the model sim-

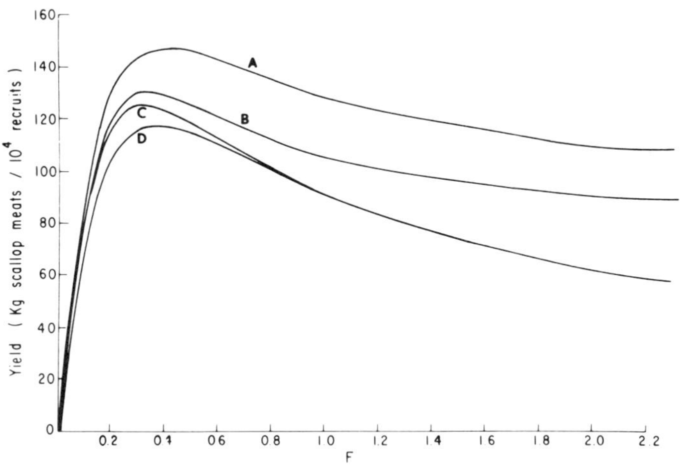

Figure 4.14. A comparison of yield per recruit curves for three models of effort allocation. A: dynamic pool model, B: effort allocated proportionally to CPUE, C: effort allocated by a complex spatial model of fishermens' choice. From Caddy (1975). J. Fish. Res. Board Can.: 32. Figure 13.

ilar to the single-fleet/single-area model described above, we assume that there are 10 areas, each with a k of 1,000. The q in each area is 0.1 instead of 0.01. If we make every area identical, then the model behaves almost the same as the single fleet/single area model (Equation 4.4.1). However, there are many interesting variations we can explore by making the areas not identical.

For instance, instead of assuming that the costs of fishing in all areas are \$25,000 per year, we can make the costs \$15,000 in some areas (designated areas 5 and 6), and have them increase up to \$35,000 in other areas (designated 1 and 10). This pattern gives us low-cost fishing areas in the middle, with higher cost fishing areas at the extremes. Figure 4.15 shows the pattern of development for this fishery model; note that in general things look pretty much the same as in the single-fleet/single-stock model, except the upper left-hand panel now shows the spatial distribution of CPUE. Now we have lower CPUE in the middle of the range and high CPUE on the edges of the range. Because fishermen are trying to maximize rate of return and the costs of fishing are lower in the middle of the range, the CPUE must be higher on the edges of the range to compensate for the higher costs.

A more interesting case is shown in Figure 4.16. Here we assume that the costs are the same in all areas, but the catchabilities are different, and that the catchability q is higher on the edges of the range and lower in the

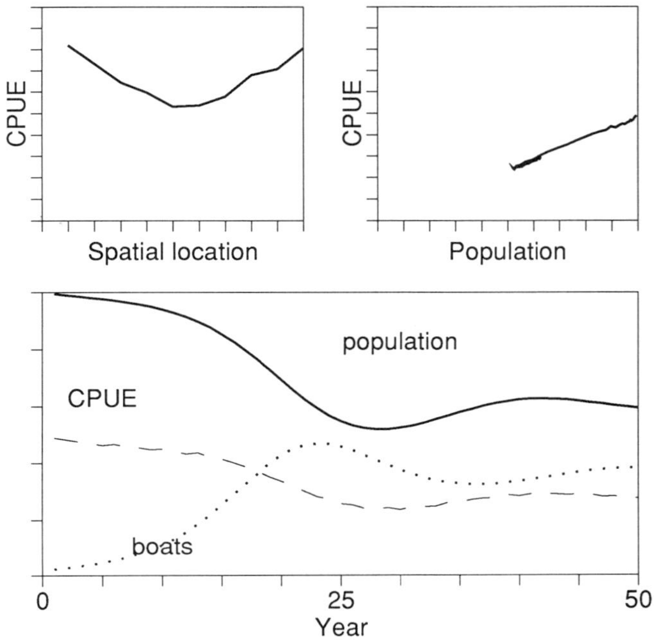

Figure 4.15. The temporal and spatial pattern that develops when costs are lower close to the fishing port.

middle. This means that effort will go first to the edges of the range, where CPUE will be higher, and only after these areas are fished down, will effort go to the middle. We have used *q's* low enough in the middle of the range that it is never profitable to fish them. The fishery development pattern is different in this case. Most significantly, the fishery never exploits the total population as intensely; the stock is never fished down to MSY. The central areas with low *q* are never fished at all, and the adjacent areas are lightly fished. The spatial pattern in CPUE is the same as the previous example, but this time the low CPUEs in the central area are due to low *q's*. Perhaps the most interesting feature is the relationship between total CPUE and the total population size shown in the upper right-hand panel. Note that CPUE actually declines a little faster than abundance. This is because the initial effort goes to a few of the areas, where CPUE declines quickly, whereas total abundance is less affected. This phenomenon we call *hyperdepletion* and is discussed in some detail in Chapter 5.

Allen and McGlade (1986) present a spatial multispecies model. Their assumptions differ slightly from those presented above; they do not assume perfect information of catching opportunities but consider in more detail information exchange, costs of fuel, and travel times.

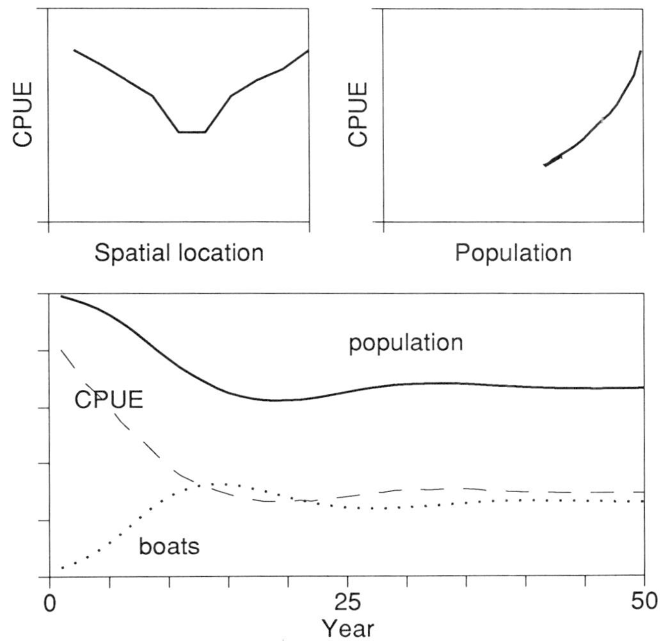

Figure 4.16. Spatial and temporal pattern that develops when catchability is low in middle spatial areas

There are an infinite number of variations one can make on spatial models and such exploration provides a potentially fertile area of research. Another variation on the spatial model theme was a model of tuna seiner and dolphin school movements described by Kleiber and Edwards (1988). This model simulates dolphin schools undergoing a modified random walk, pursued by seiners (fishing for tuna associated with the dolphins), which search for concentrations of dolphin schools. Although this model is computationally very intense, because it tracks individual vessels and dolphin schools on a time step that can be as short as a few hours, it is an approach that will undoubtedly be used more extensively as computational power becomes more available.

A model of the Peruvian anchoveta fishery

Fisheries are dynamic systems that involve fish, fishermen, managers, and the assorted biological, economic, and social factors that affect them. All of the models and data we have considered earlier in this chapter examine parts of this big picture but never really look at the entire system. This section describes a simulation game that we developed for teaching students about the difficulties of managing both the biological and economic development of a fishery. It uses the Peru anchoveta (*Engraulis ringens*) fishery as a case example and tries to simulate three main aspects of that fishery: (1) dynamic changes in the anchoveta stock in response to fishing, environmental variation (*El Niño* events), and changes in abundance of competitors (sardines); (2) dynamics of fishing fleet size and fishing effort in response to changes in the anchoveta stock; and (3) changes in employment and Peruvian national debt associated with the fishery. For a major review of the anchoveta fishery and data sources used in the model, see Pauly and Tsukayama (1987).

A key purpose of the game is to demonstrate how management of a fishery can become progressively more difficult with fewer and fewer viable policy options as development proceeds. Essentially, the game player must try to manage a way through a squeeze between biological and economic forces. On the biological side of the squeeze, increasing harvest rates result in decreased stock size and increased vulnerability to environmental variation, so that the annual catch that can be safely allowed decreases over time toward a relatively low sustainable level. On the economic side of the squeeze, increasing fleet size through investment and borrowing results in a progressively larger debt, so the catch level necessary to earn enough to even service this debt (make interest payments) will increase. Unless the game player is careful in both biological and economic management, he soon finds himself in a nasty situation where he must accept either biological overexploitation or pathological growth in debt (and eventual bankruptcy), or sometimes both.

The following subsections describe how we modelled each of the three major parts of the system. Note how each submodel is relatively crude, since we wanted to produce a game that could be played quickly on microcomputers, while forcing the game player to look at the system as a whole rather than at either the biological or economic details.

Anchoveta production dynamics and ecosystem interactions

The biomass dynamics of anchoveta and sardines are each represented by a Deriso (1980) delay difference model of the form

$$B_{t+1} = gS_t - hS_{t-1}\frac{S_t}{B_t} + R_{t+1} \qquad (4.4.7)$$

where B is stock biomass, S is postharvest biomass, g and h are constants representing the combined effects of natural mortality and growth (see Chapter 9), and R is biomass of new (1-year-old) recruits. Recruitment biomass

Table 4.7. Parameter values used in anchoveta simulation game.

Anchoveta biomass dynamics	Anchoveta catch
a = 1.5	q_1 = 0.01
b = .075	q_2 = 14
c = 0.4	q_3 = 0.1
d = 3.0	
g = 0.9	
h = 0.2	
Sardine biomass dynamics	Bird dynamics
a = 0.1	s = 0.85
b = .033	b_1 = 0.014
c = 0.9	b_2 = 0.0000001
d = 3.0	
g = 0.9	
h = 0.2	
Investment and borrowing	Fishing effort
r = 0.9	fm = 100
k = 200000	er = 25
p = 100000	
i = 0.1	
Balance of payments and employment	
m = 0.75	
m_1 = 10	
m_2 = 1000000	

R for each species is assumed to depend on parental stock size, ocean surface temperature (index of *El Niño* events), and biomass of the other (competing) species:

$$R_t = S_{t-1}exp[a - b[S_{t-1} + cS'_{t-1}] + d(e - T_t)] \qquad (4.4.8)$$

where a and b are Ricker stock-recruit parameters, c is a competition coefficient, S' is the spawning biomass of the competing species, d is a temperature-effect parameter, and T is an ocean-surface temperature anomaly ($T = e$ for average ocean temperature, $T < e$ for cold and productive upwelling conditions, and $T > e$ during warm *El Niño* events).

The a, b, g, and h parameters are chosen (Table 4.7) so as to give a longevity of approximately 5 years in the absence of fishing, an unfished anchoveta stock size of approximately 20 million tons when sardines are low, and an unfished sardine stock size of approximately 1 million tons when anchoveta are abundant. The c parameters are chosen so that anchoveta will "dominate" sardines (keep sardine biomass down to 1 million tons) in the unfished situation, but allow the sardines to increase to a 25 million ton maximum in the absence of the anchoveta; the c parameter for sardine effect on anchoveta is chosen so that anchoveta recruitment rate per spawner will be depressed by about 30% (relative to unfished equilibrium) if the sardines build up to their 25 million ton maximum during a period of low anchoveta abundance.

The ocean-temperature index T is generated as $T = 9r^{5.0} + 14$ where r is a uniform random number between 0 and 1.0. Raising r to the power 5.0 has the effect of making lower values of T (closer to the minimum of 14) more probable, with very high (*El Niño*) values (maximum 9 + 14) occurring only about once every 10–20 years. The e (average temperature) parameter is taken to be 14, and the d parameter is chosen so as to make recruitment drop to about 15% of its average value for high-temperature extreme years. As modelled in the above equation for R, the temperature effect essentially moves the stock-recruitment curve up and down as the temperature index varies.

The anchoveta catch C_t and spawning biomass S_t for each year are calculated from a simple catch equation:

$$C_t = B_t[1 - e^{-q_t E_t}] \qquad (4.4.9)$$
$$S_t = B_t - C_t$$

where B is biomass before harvest, q_t is a catchability coefficient for year t, and E_t is the fishing effort in year t. The catchability coefficient q_t is assumed to depend on stock size and ocean temperature:

$$q_t = q_1 \frac{(T/q_2)}{1 + q_3 B_t} \qquad (4.4.10)$$

Here q_1 is a base fraction of the stock size taken by one unit of fishing effort when the stock size is low and when ocean surface temperature equals q_2, and q_3 is a parameter that measures how rapidly catchability decreases as B_t increases due to gear saturation. The form of the above equation results in catch per effort becoming independent of B_t for high biomasses. Catchability is made proportional to ocean temperature to reflect how *El Niño* events concentrate the fish into a narrow band near shore as the events develop. The model ignores the possibility that extreme events may reduce catchability by pushing the fish into deeper water where they are more difficult to find and less vulnerable to seines. The program is structured so that the user can reduce or increase q_1 by a multiplicative factor to represent regulations involving gear restriction, or technological advances in gear efficiency.

To display how fishery development may have ecosystem influences extending beyond the fish stocks, the model includes a simple representation of population changes in fish-feeding birds. Both *El Niño* events and fishing appear to have a dramatic effect on sea birds off Peru (Tovar et al. 1987). To represent this effect, we assumed that the birds are mainly influenced through the effects of food shortage on reproduction rather than survival. The bird population is modelled as

$$N_t = sN_{t-1} + \frac{N_{t-1} b_1 S_{t-1}}{[1 + b_2 N_{t-1}]} \qquad (4.4.11)$$

Here N_t is population size, sN_{t-1} is survivors from last year's population, and the reproductive rate term $b_1 S_{t-1}/[1 + b_2 N_{t-1}]$ represents the joint effect of anchoveta abundance (S) and bird population size. Reproductive rate is assumed to be directly proportional to anchoveta spawning stock, and to decline with increasing bird population size for any fixed level of anchoveta abundance.

Fleet Dynamics and International Debt. The fishing fleet is assumed to consist in any year of F_t seine vessels. The number of vessels in the fleet varies from year to year according to the depreciation-investment equation

$$\mathrm{F}_{t+1} = rF_t + \frac{I_t}{k} - R_t \qquad (4.4.12)$$

where $1 - r$ is the fraction of vessels lost to physical depreciation per year, I_t is money invested in new boats per year, k is the cost for each new vessel added to the fleet, and R_t is the number of boats deliberately retired from the fleet by policies such as buy-backs or license removals.

The money invested in new boats each year, I_t, is assumed to come from two sources: (1) borrowing by the industry or government from the international banking system, as a deliberate policy choice by the model user (no borrowing occurs except when the user stops the model to take out a single "loan") and (2) reinvestment of earnings by the fishing industry (or government). At the start of the simulation, it is assumed that the initial fleet size has been established through borrowing, so that there is an initial debt to the international banking system.

The investment computation for each year begins with a calculation of income for the industry, G_t:

$$G_t = p_t(1 - lt)C_t + L_t \qquad (4.4.13)$$

where p_t is landed price per million tons of catch, lt is a landings tax (set by the user) assumed to be taken from the landed value $p_t C_t$ at the point of landing and used elsewhere in the economy (i.e., not available to the industry or government for fishery development and debt servicing), and L_t is the amount borrowed (set by the user) in year t from the international banking system.

Income for the industry, G_t is then reduced by operating costs, assumed to be proportional to fishing effort E_t, and if the result is negative then the negative value is added to the international debt (i.e., model "borrows" to pay for operating costs if poor operating decisions are made in any year). If there is a positive balance after paying operating costs, the model applies the remaining balance first to debt interest payments (interest rate i times current debt) then to repay a proportion (set by user) of the debt principal. If the debt interest payments cannot be met, the unpaid interest charge is added to the debt principal for the next simulation year. Finally, if there is any money left over from G_t after paying operating costs, debt interest, and debt principal, then this remaining net income is assumed to be reinvested in fleet development (i.e., $I_t = G_t$ minus all costs).

Early in the simulated development, catch per effort will be high and the international debt will be low, so that I_t will be positive and the fishing fleet will grow (unless very high landing taxes are set by the game player). Fleet growth can be further stimulated by the user by borrowing. Later in development, as stock and catch per effort decline, the model will first stop building new boats. Then if earnings continue to decline, it will stop making payments on the debt principal and eventually on the debt interest. It may even be forced to borrow to meet operating costs, though fishing effort will generally decrease enough under very low stock conditions to make that unnecessary (see next section). When earnings are low enough so that the debt interest payments are not met, the debt will begin to grow geometrically, by a factor equal to 1.0 plus the interest rate (i.e., unpaid interest is added each year to the debt principal, so that the principal for next year, and hence the interest owing next year, will increase each year).

Fishing Effort, Employment, and Balance of Trade. Based on the fleet size F_t present in any year, the model calculates a maximum fishing effort $E_t^{\max}$ (boat days) equal to F_t times a maximum days fishing per boat, fm. However, it is assumed that fishing effort is sensitive to anchoveta abundance B_t, because of short-term (daily, weekly) information-gathering processes by the fishermen. There is assumed to be a distribution of information and "optimism" about fishing success among the fishermen, so that (1) very little effort will be expended when the stock is very low, (2) much fishing effort will be "turned on" as stock size varies through an intermediate range, and (3) effort will approach $E_t^{\max}$ if the stock size is very large. To represent this statistical relationship between total effort (by many fishermen) and stock size, we use the equation

$$E_t = E_t^{\max} \frac{S_{t-1}^2}{[er + S_{t-1}^2]} \tag{4.4.14}$$

where S_{t-1} is the anchoveta spawning stock from last year and er is the value of spawning stock squared at which half of the maximum possible effort will be expended. By squaring S in the above relationship, we generate a sigmoid relationship between effort and the abundance index S; the square root of er is the spawning stock size at the inflection point (point of steepest increase) of the relationship.

Employment in the industry is assumed to come primarily from three sources: fishing, vessel repair and maintenance, and catch processing. Fishing and vessel repair employment are assumed to be proportional to fishing effort, whereas catch processing employment is assumed to be proportional to the catch. For any year, the total number of jobs J_t in the industry are then calculated as

$$J_t = m_1 E_t + m_2 C_t \tag{4.4.15}$$

where m_1 is the number of jobs generated in fishing and vessel repair/maintenance per unit of fishing effort and m_2 is the number of jobs per million tons of anchoveta caught.

As a final indicator of economic impact of the fishing industry, the model calculates a rough index of the overall contribution of the fishery to the national balance of trade. This balance index T_t may be positive or negative and is estimated as

$$T_t = pC_t - D_t - mI_t \tag{4.4.16}$$

where pC_t is the total export earnings from the catch, D_t is total payments to international banks for debt interest and principal, and mI_t represents the share of investment in new vessels that is spent on imports of foreign equipment and technology.

Policy Options and Scenarios. The model is programmed so that the user can interrupt the simulation at any time after a simulation starting year of 1960 (when the anchoveta fishery began to grow) to make changes in harvest and economic management parameters. The user can construct bioeconomic development scenarios by using combinations of biological and economic regulations.

On the biological (harvest regulation) side of management, the user can

1. Limit gear efficiency (relative value of q_1)
2. Limit fishing effort per year (without any catch limit)
3. Set a fixed catch quota (which also limits effort)—when the user sets both an effort limit and a quota, the program checks each year to see which limit would produce a lower catch, and takes this lower value to be the actual catch

On the economic development side of management, the user can

1. Borrow money in any year for boat building (loan)

2. Set the landing tax rate (proportion of gross income)
3. Set a maximum on the number of new boats per year
4. Reduce the fleet size in any one year by a percentage
5. Set a target percentage for annual reduction in the debt principal (reschedule loan repayments).

Borrowing stimulates fleet development, whereas the other measures either slow development or limit the maximum fleet size that will be permitted (if maximum boat building is set to zero) or profitable (if landing tax or debt reduction target is high).

Note that there is no direct connection in the model between the biological and economic regulations. If the user sets a catch quota or effort limit, the program does not automatically limit borrowing or boat building. Thus the fishing industry may become "overcapitalized" even under a quota, with too many boats competing for the quota, if it remains profitable to build boats even after the quota is imposed.

More than 200 students have played the anchoveta game in fourth year and graduate courses in applied ecology at the University of British Columbia. Generally the most successful players have been those who take three key steps early in each simulation (by simulation year 5, or 1964): (1) set a conservative catch quota of 5–7 million tons (mt); (2) set a high debt repayment rate; and (3) use landing taxes and/or boat building limits to prevent rapid growth in the fleet. It is interesting to note that these steps are precisely the opposite of the policies actually followed in Peru (and most developing fisheries): they assumed very optimistic sustainable catches of 10–12 mt, borrowed heavily for development, and took no significant steps to discourage overinvestment in fishing vessels.

4.5. Fisheries Development Paths

As fisheries develop, there are many changes that take place in stock size, catches, catch rates, and vessel profitability. One convenient way to summarize the development pattern in a fishery is as pictured below in Figure 4.17, which looks at the relationship between stock size and fishing fleet size. Pictured are three possible development trajectories all beginning with an unfished stock and an undeveloped fleet.

Overexploitation and collapse

In the solid trajectory of Figure 4.17 the fishery develops rapidly, so that the fleet size exceeds the optimum fleet for sustained biological yield (the x in the middle of the graph). The fleet drives the stock down to low abundance and then the fleet collapses because the catch rate is so low that every-

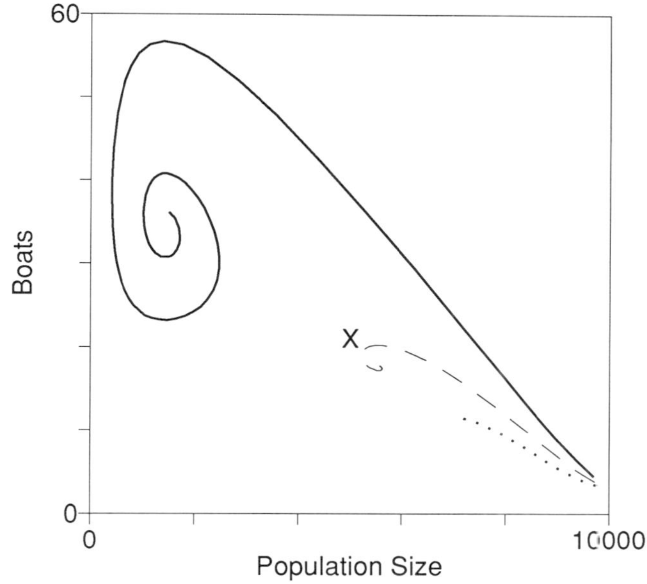

Figure 4.17. Three possible trajectories for fishery development.

one loses money. This is then followed by a period of stock rebuilding and eventually some fleet redevelopment, although the stock is shown to never recover to the biological optimum size.

Such a trajectory would be typical of many herring and anchovy fisheries (North Sea herring, California sardine, Peruvian anchoveta) which were extremely profitable to pursue even at low stock sizes (due to schooling). A common theme in these fisheries is that catch rates did not decline as abundance declined, so that fishing remained profitable until stock size was very low.

Development to biological optimum

The dashed trajectory shows a fishery developing up to the biological optimum stock and fleet size. The fishery may develop much more slowly and never overdevelop to the point where the fleet becomes larger than is sustainable. Such a trajectory is very lucky and unlikely to ever occur.

Incomplete development

The dotted trajectory shows a fishery that never does develop to biological optimum yield. This would occur when the harvest technology is reasonably

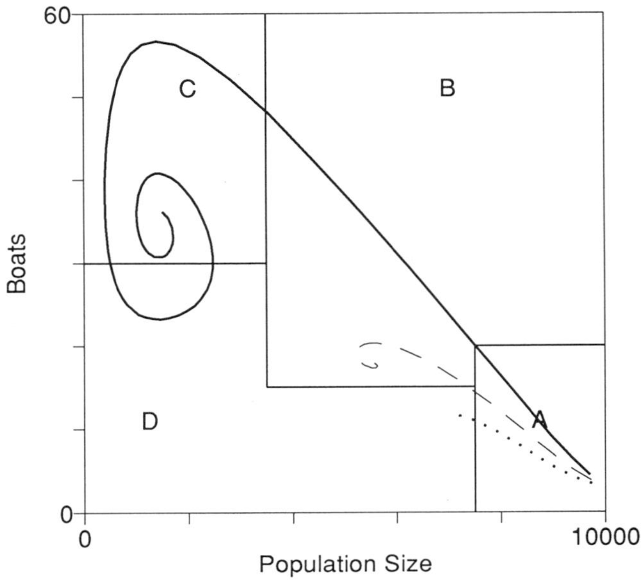

Figure 4.18. Four regions where management and stock assessment issues are different.

inefficient, or the price is so low, or markets saturated enough so that it just does not pay to reduce the stock size any further. Global fisheries for skipjack tuna can be thought of as in this class. At current prices, the stocks of skipjack are far from being overexploited.

Regions in fishery trajectories

Figure 4.18 shows how the stock size/fleet size graph of Figure 4.17 can be partitioned into four regions that represent different stages of the fishery. In region A, the fishery is just beginning, fleet size is small, stock is still largely untouched. Some fisheries will not develop beyond this stage for economic reasons, but generally it is an early and short-lived stage, during which foundations for good management of the fishery must be laid.

Region B can best be called a mature fishery, where the stock is roughly fully exploited, and the fleet is either nearly optimal or too large. This may be a stable, long-term condition as in trajectory 1 above, or just a state the fishery passes through on its way to overexploitation.

Region C is an overexploited fishery that is collapsing. Stock sizes have declined well below optimum, the fishermen are losing money, and everyone is unhappy. Many fisheries never reach this state, either because eco-

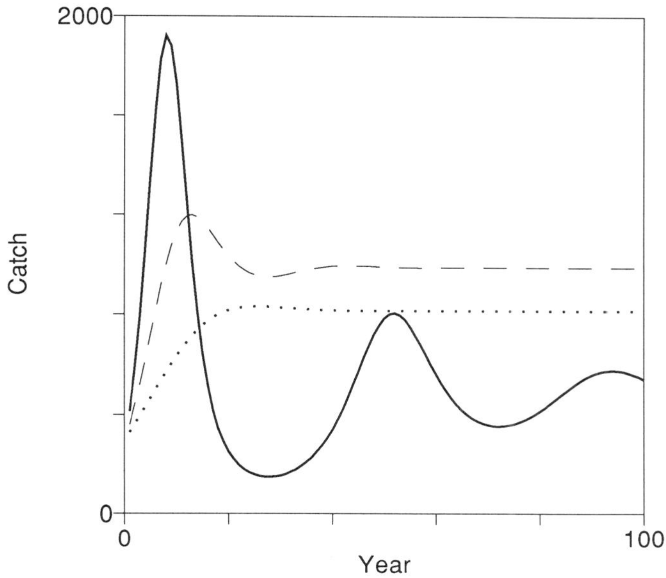

Figure 4.19. Catch trends in three fishery development trajectories.

nomics and vulnerability of the fish prevent overexploitation, or because management actions are taken to prevent stock collapse.

Region D is a recovery period after overexploitation and collapse. Many fishermen have left the fishery, and the stock is gradually recovering. Generally, the stock will never recover to its former abundance, but the fishery will be reestablished at a level well below the previous biological optimum.

4.6. Responses during Fishery Development

During each of the trajectories above, there are changes that take place in catch, abundance, catch rates and profitability. These will have important implications for the fishermen as well as for potential fishery managers.

Catch

Figure 4.19 shows the catch trends for the three trajectories described in Figure 4.17. For trajectory 1, the catch builds up rapidly, then collapses, and finally recovers. When recovery occurs, the catch comes nowhere near its peak. Trajectory 2 looks similar but the collapse is quite modest. However, the sustained catch is below the maximum catch. If the fishery develops rapidly, the early catches may be much higher than sustainable. In

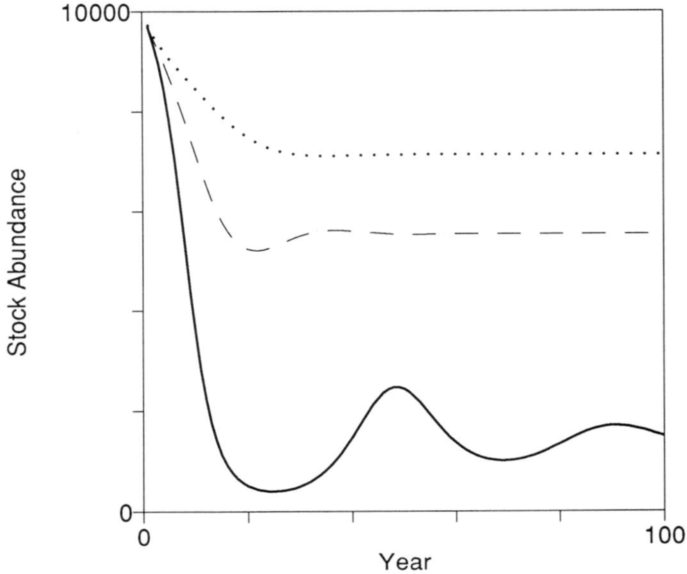

Figure 4.20. Abundance trends of 3 fishery development trajectories.

general, long-lived species and species with low recruitment rates will also have large unsustainable catches during development.

The catch pattern for trajectory 3 shows a gradual slow increase with random variation about it. Because the development was slow, there was never a period of large nonsustainable catches.

Abundance

Figure 4.20 shows trends in abundance during the three trajectories in Figure 4.17. In trajectory 1, abundance is driven down very low and then gradually recovers. In trajectory 2, there is only a minor overshoot beyond the optimum stock size. In trajectory 3, the abundance is driven down slowly but not as far as the biological optimum.

Catch rate

Figure 4.21 shows trends in catch rate for the Figure 4.17 trajectories. The catch rate drops very slowly in trajectory 1 until collapse occurs. It is

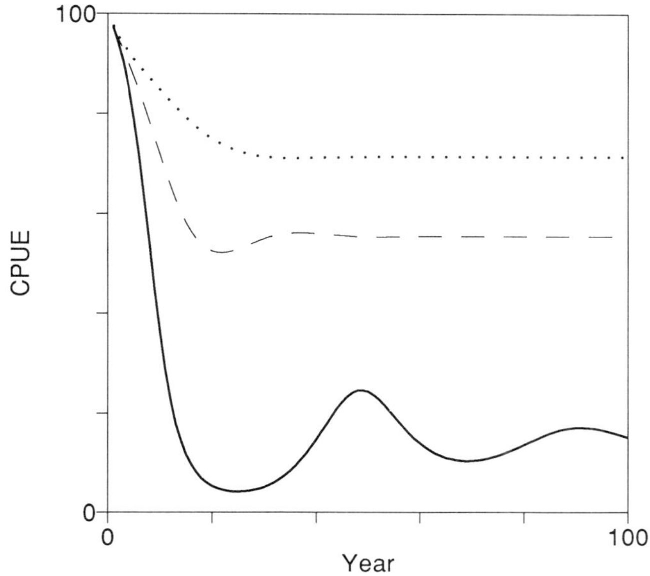

Figure 4.21. Trends in catch rate (CPUE) for the three fishery development trajectories in Figure 4.17.

the slow drop in catch rates that is normally associated with serious over-exploitation and collapse. In trajectories 2 and 3, the catch rate drops more or less proportionally to abundance.

Profitability

Profitability shows similar trends to catch rate, since in general an average vessel's profitability will be related to catch rate. Note, however, in Figure 4.22, that once profitability in all fisheries falls to near zero, the growth in fleet size comes to a stop.

4.7. Summary and Critique

Fleet dynamics is probably the most understudied subject in fisheries. Although it is clear that overinvestment is the single most serious fisheries management problem, we have been unable to find many published time-series data on how many vessels participated in particular fisheries. These data exist, and are hiding out in the grey literature, but as yet people do not consider this worth putting in the journals. There are literally a handful of studies of investment; there are thousands of papers by economists on how investment should take place, but hardly any that actually look at real data.

Catching power is probably the best studied of the topics considered here

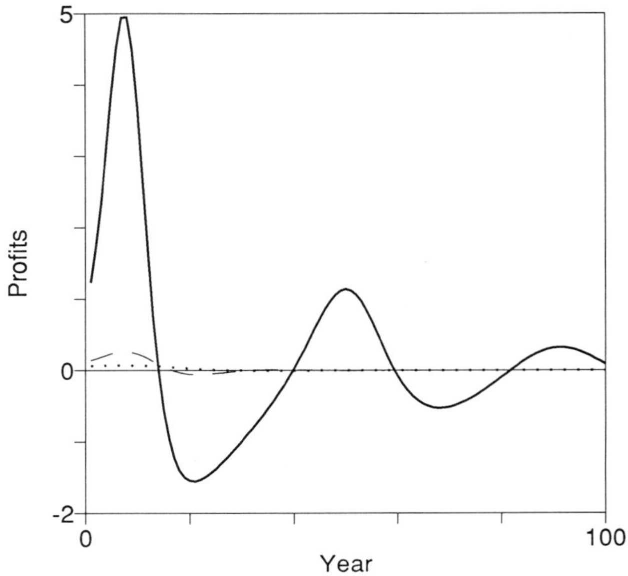

Figure 4.22. Profitability trends for the three fishery development trajectories shown in Figure 4.17.

but there is still a great deal of work to be done, particularly on how fishing gear and fishermen's skill interact to produce effective catching power. Since so many fisheries seem to show increasing catching power in the absence of major changes in equipment, investigations of fishermen's skill are very important.

Studies of fishermen's movement are just becoming popular. Gordon (1953) discussed how fishermen should behave, however, it is only in the last few years that people have seriously started to look at actual data to see where fishermen are fishing and what return rates they are getting. Many fisheries agencies have very rich data bases, often with daily records of most fishing vessels. Access to these data bases is often difficult because of confidentiality problems, but they can be accessed, and should be. Fisheries science will be far richer, and our understanding of how fisheries behave more advanced, if half of the energy that goes into biological studies of fish were devoted to behavioral studies of fishermen.

Bibliographic Notes

The references given below provide an entrace to most of the published papers on fleet dynamics with the exception of the large literature on standardization of fishing effort. Papers that take a very broad view, and examine the biological and economic development of specific fisheries are Clark and Lamberson (1982) for the whale fishery, Clark and Kirkwood (1979) for a prawn fishery, Bjorndal and Conrad (1987) for the North Sea herring Fishery, and Wilen (1976) for the pelagic fur seal fishery. There is a vast literature on standardization of catching power (Parrish and Keir 1959 for instance), but these old methods are superceded by the GLM methods used by Allen and Punsly (1984).

Analysis of, or models of spatial fisheries are the Allen and McGlade papers (1986,1987), McGlade and Allen (1984), Caddy (1975), Hilborn and Ledbetter (1979), Hilborn and Walters (1986), Hilborn and Kennedy (1991), MacCall (1990) and Kleiber and Edwards (1988).

Studies of discarding are Saila (1983), Pikitch (1987) and Pikitch et al. (1988).

For a mathematically more rigorous, and more economically oriented view, Colin Clark's (1985) book *Bioeconomic modelling and fisheries management* is an excellent study of many aspects of fleet dynamics.

Part III

Estimation of Parameters

5
Observing fish populations

5.1 Introduction

In the previous two sections of this book, we examined why we do stock assessment and how fisheries behave. We have studied alternative models of fish and fisherman dynamics, but we have not yet looked at how to determine which models are actually most consistent with the data, nor how to estimate the parameters of the models. This section, entitled "Estimation of parameters" is the traditional hard core of fisheries stock assessment: the nuts and bolts techniques used to answer the important question, "what happens if . . . ?"

However, before we jump into the actual mechanics of the techniques, we must first look at fisheries data: how are they collected and what do we need to know about them when we use them? There are great dangers in uncritically using data without understanding where they came from, how errors and variability arise, and what alternative types of data may be obtained.

There is an unfortunate division of effort in many fisheries agencies. Those who collect the data often do not have the technical skills to analyze them, and those who analyze them often have little field experience with the realities of collection. We, of course, belong to the group who use a lot of data we have not collected and do not fully understand. We hope that one of the contributions of this book will be to make the analytic techniques more accessible to biologists who do actually collect and know the data. As a general warning, analysts using data should make every effort to become familiar with the data-collection procedure. You should visit data-collection sites, discuss with field staff problems they see in the system, and generally become aware of the myriad of ways that the data appearing on your computer screen may be an imperfect representation of the actual fishery.

The purpose of this chapter is to provide a broad survey of how data are collected and what can actually be observed about fish populations. These topics form the first two sections of the chapter. The third section considers the specific problem of estimating abundance, and the fourth section reviews estimation of movement.

Abundance estimation is also a by-product of several of the standard analytic techniques, particularly catch-at-age analysis, virtual population analysis, and depletion methods. The choice of including abundance estimation as a section of this chapter rather than making it a chapter of its own is arbitrary. Equally arbitrary is our decision to include a section on estimating movement. We want to discuss it at some point, and this chapter seemed a reasonable place.

5.2. Modes of Observation

Biologists involved in stock assessment have a number of alternative sources of data. Most common are fishery-dependent data, such as catch and effort statistics, but information is becoming increasingly available from sources independent of the fishery, such as research surveys, tagging experiments, and tracking of individual fish.

Catch and effort data

Traditional stock assessment might also be called "analysis of commercial catch and effort data," since such data has formed the backbone of most fisheries stock assessments. Indeed it is almost impossible to perform stock assessment without knowing the history of commercial catch. We want to know how the stock responds to exploitation, and, unless we know the exploitation history (catches), we really cannot begin. Establishment of a catch data collection system is probably the top priority for a new fishery; most parameter estimation techniques discussed in subsequent chapters require catch data.

Commercial effort data are quite frequently used. They may be used very crudely by assuming that uncorrected effort is proportional to fishing mortality, or the data may be somehow massaged and scaled into estimates of effective effort which are hopefully better related to fishing mortality.

Commercial catch and effort data can be obtained in three ways: on-board observer programs, port sampling, and log books. Each of these approaches has merits and risks.

On-board observers

The most direct and reliable method for collection of commercial catch and effort data is to have trained observers on board fishing vessels to record data directly. Under almost all circumstances only a portion of the fishing effort will be sampled, and one must make some assumption about the behavior of vessels with observers on board relative to unobserved vessels. In a global fisheries perspective, observer programs are rather rare. These programs are expensive and are intrusive to fishermen. They are most commonly found in situations where foreign fishing vessels are required to have observers on board as part of access agreements. Care must be taken in the design of on-board data collection, particularly when observers are required to obtain age- or species-composition samples. It is very difficult to randomly sample catches during the fishing process.

Observers are particularly valuable when questions of by-catch and discarding are of concern, because normally by-catch and discarding do not show up in port sampling; it is usually not in the fishermen's interest to report such information in log books.

Landings records — port sampling

The most common method for collection of commercial catch and effort data is port sampling. This comes in two forms: (1) as part of the fish sale process, where there are records of landings and sales, and these records are then passed on to the management agency to compile the catch estimates from sales; or (2) paid observers actually meet incoming vessels and collect catch data from the vessels directly. This second method is often used in recreational fisheries under the name of *creel survey*, where a fraction of landings are sampled to estimate the catch per unit effort.

Log books

The third, and least direct method of catch and effort estimation involves the fishermen keeping log books for subsequent analysis by the management agency. Log books may be a requirement of fishing, and all fishermen may turn them in, or a fraction of fishermen may keep log books, usually on a volunteer basis. Log books, can, in principle, provide an enormously valuable data base. In many Australian trawl fisheries, for instance, fishermen maintain a log book of each trawl of the net, giving starting and finishing locations, as well as catch. This allows the management agency to spatially map fishing effort and catch and provides data at a level of detail that would be unobtainable without 100% observer coverage of all vessels.

Of course, log book systems assume accurate reporting by fishermen. Wise use of log book data requires a very good understanding of how fishermen fill out the log books, and the fishermen's attitudes toward the log books. Log book programs will generally not provide any age composition information about the catch, nor species composition unless all species are commercially valuable and sorted by the fishermen.

Research surveys

One of the major problems with using commercial catch and effort data to estimate stock distribution and abundance is that fishermen go where the fish are. Fishing effort is normally concentrated on the highest densities of fish and attempts to assess the range, or total abundance of fish from commercial catch and effort data can be expected to be biased.

Management agencies normally attempt to avoid the biases of commercial catch data by using research surveys, either by chartering commercial fishermen or using government research vessels. Indeed, in much of the world there is growing distrust of using fishery-dependent data and there is much more reliance on survey data. The major problem with surveys is cost. Operating vessels is so expensive that the number of samples is usually not nearly as large as would be desired by the scientific staff using the data.

Visual surveys — transects, flights

Many fish can be observed visually. This can involve swim transects and visual counts for slow moving fish such as sessile invertebrates, visual counts of marine mammals from chartered vessels or planes, or counts of tuna schools from aircraft. There are a variety of well-established statistical techniques for estimating abundance from visual transects. All these techniques hinge on the question of what proportion of individuals in the area surveyed are seen. This question involves the problem of observer reliability, but more bothersome are concerns about the visibility of the individual animals. Transect surveys for invertebrates always require some calibration of the proportion of individuals who are buried in crevasses or under sand and not visible. Similarly visual transects for marine mammals require assumptions about the proportion of individuals on the surface at any time.

Electronic survey — mobile or stationary

Most fish are difficult to count visually, and the use of electronic equipment, particularly hydroacoustic surveys, is growing (Johannesson and Mitson 1983). There are three key elements in an electronic survey: (1) the transect design must be set up with all of the associated problems of stratification; (2) there must be some calibration between the strength of signal received from the "targets" and the number of individuals; and (3) the species mix of the targets must be determined.

The transect design is a standard statistical problem (found in all surveys) and has no major properties unique to hydroacoustic or electronic surveys. More problematic is the calibration; the electronic transducers count reflected electronic impulses, not fish, and the problem of determining how many fish are actually there is one of the major problems in hydroacoustics. This can prove to be very difficult. A survey performed by an Australian fisheries agency of orange roughy (*Hoplostethus atlanticus*), which occur very deep (>500 fathoms), announced that there was an estimated stock of 5 millions tons. Since the price of orange roughy was over $1,000 per ton this led to an enormous speculation in licenses and great political pressure to raise the quota from its previous level of 20,000 tons.

In fact, the fishermen were unable to catch even the quota of 20,000 tons, and subsequent surveys determined that the 5,000,000-ton figure was false bottom and not fish at all. This may be an extreme example, but great care in determination of target strength is always warranted.

Target strength may be determined experimentally by sounding known targets or more commonly by sampling with fishing gear schools of fish that have been electrically sampled. Similarly, sampling of individuals with fishing nets is normally used to determine species composition.

Despite the problems of validation of target strength and species composition, hydroacoustics promises to grow in importance as a survey technique, and we believe that we will see growing demand for statistical techniques that can use hydroacoustic survey data in conjunction with traditional catch and effort data.

Fishing gear

Commercial fishing gear is often used on research cruises to estimate fish abundance, sample length and age, and so on. The major advantage of research survey use of fishing gear over commercial data is that the research survey can use a planned sampling design, rather than relying on fishermen's choices of where to fish. Additionally conditions of gear use (tow speed, soak time, etc.) can be better controlled. Use of research survey fishing is well established, particularly for trawl gear (Doubleday and Rivard 1981). The major problem with fish gear surveys is estimating the proportion of the fish captured by the gear. Trawl surveys normally take the area swept by the gear, and expand the catches by the proportion of the area surveyed. The bottom line is always "What proportion of fish in the area swept were captured?"

Tagging

Tagging of individual fish, usually by placing metal or plastic physical tags on the fish, is very commonly used in fisheries. Tagging programs can be enormous in size; every year nearly 30 million juvenile salmon are marked with small stainless steel tags in western North America alone. Tagging programs can also be very expensive; a program designed to mark yellowfin tuna (*Thunnus albacares*) in the western Pacific by the South Pacific Commission estimated that it would cost \$5,000 for every yellowfin tag that will be returned to the tagging agency.

Tagging studies are used to estimate abundance, survival, movement, and fishing mortality, and there are many books that deal with analysis of tagging data in great detail (Seber 1982, Burnham et al., 1987). There are a number of important considerations in tagging studies that we wish to review here. First consider program design: the key elements of tagging programs are tagging of fish and recovery of tags. Most fisheries tagging programs rely on commercial fishermen's catch and return of tagged fish. Most tagging programs concentrate most of their resources on tagging large numbers of fish, and tend to neglect the tag recovery end of the project; putting tags on fish is fun and somewhat glamorous, whereas groveling around on docks, trying to determine if fishermen are missing tags, or throwing them away, is dirty undesirable work. The design of any tagging study should consider the appropriate allocation between releasing more tags, understanding fishing mortality or effort distribution, and tag recovery effort.

If we examine how tagging data are structured and analyzed, we can understand better how to design tagging studies, and what can be inferred from them. Assume a simple discrete model of tag dynamics, where fish are released on one occasion, and then go through three annual stages of (1) fishing mortality, (2) natural mortality, and (3) tag shedding. As in the population dynamics models, we could also describe these dynamics with a differential equation assuming all three processes take place continuously. Let us define the following terms:

- R is the number of tags released
- A_t is the number of tagged fish alive at time t
- s is the annual survival rate from natural mortality
- α is the proportion of fish dying immediately after tagging ("instantaneous" tagging mortality)
- β is the tag reporting rate
- γ is the tag shedding rate
- C_t is the tags caught and returned at time t
- U_t is the proportion harvested at time t

Given these definitions, the following relationships hold:

$$\begin{aligned} A_1 &= R(1-\alpha) \\ C_t &= A_t U_t \beta \\ A_t &= A_{t-1}(1-U_{t-1})s(1-\gamma) \\ A_t &= R(1-\alpha)[(1-U)s(1-\gamma)]^{t-1} \end{aligned} \tag{5.2.1}$$

The last equation drops the t subscript for U indicating a constant fishing

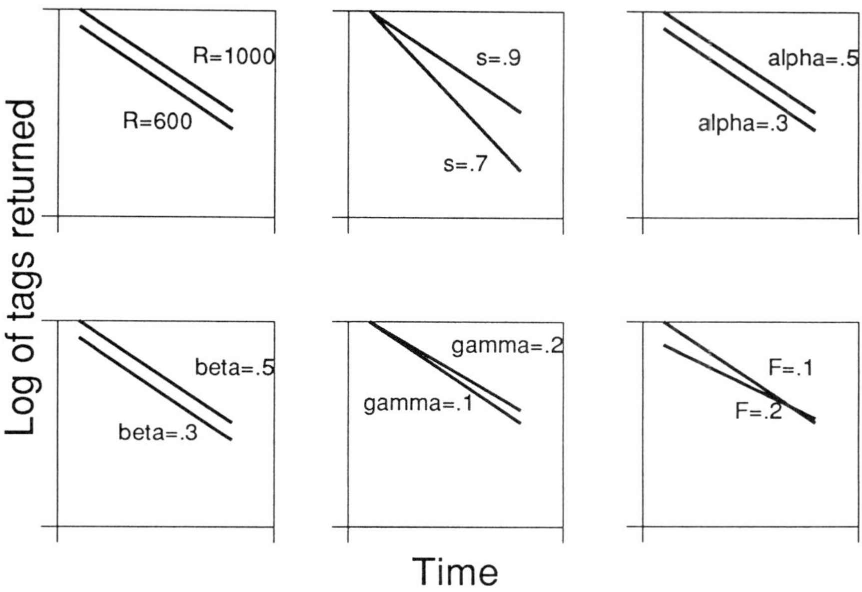

Figure 5.1. Expected tag return patterns for different parameters of the tagging model.

harvest rate [otherwise we would replace $(1 - U)$ with a product of $t - 1$ terms, outside the brackets]. The knowns in tagging studies are usually the releases R and the tag returns C_t. Ideally we would like to estimate the fishing mortality, survival, tag shedding, tagging mortality, and tag reporting rates. Given constant fishing mortality rates, the number of tags returned over time will be an exponential decay, generating a straight line when the logarithm of returns is plotted. If we examine how changes in parameters affect the expected return of tags (Figure 5.1) we see some distinct patterns emerge.

Note that three parameters R, β, and α all have the affect of raising or lowering the line, but do not affect its slope. Examination of Equation 5.2.1 shows that these three parameters always enter the C_t prediction together in exactly the same way. Since we normally know R, we can therefore estimate $(1 - \alpha)\,\beta$, but we cannot separate the two parameters. Figure 5.1 shows that s and γ also have exactly the same effect, that is to change the slope. This says we cannot tell from catches alone whether the fish are dying, or simply losing tags each year.

Finally, note that the fishing mortality rate changes both the slope and the intercept of the line. Fishing mortality rate is, in principle, estimable separately from the other two groups of parameters. In effect, a tagging study can let us determine three things, $(1 - \alpha)\ \beta$, s γ, and U. Designers of tagging studies normally try to resolve the uncertainty about these parameters in a number of ways. Double tagging experiments, in which some individuals are tagged with two tags, are used to try to determine tag shedding, and to a lesser extent tagging mortality (Wetherall 1982). Tagging studies also often hold individuals that have been tagged for some time to see if they stay alive. This is of questionable value since it is hard to assure that individuals that live in a tank, protected from predators, would not have been killed in the wild.

Probably the most important part of tagging studies that is often neglected is a systematic determination of tag return rates. There are two approaches. Paulik (1961) describes the statistics used to sample commercial catch to determine the frequency of tags, and then how to compare this percentage with the number of tags being returned from commercial fishermen. Hilborn (1988) describes the statistics that can be used to examine fish after fishermen have processed them to determine if tags are not being detected.

Tracking

Recent advances in miniaturization now permit radio or acoustic tags to be placed on many types of fish (Kanwisher et al. 1974, Young et al. 1976). This enables researchers to track individual fish continuously either by following the fish at a distance with a boat, or potentially by aircraft, satellite, or fixed-point detectors. A related technical development is passively interrogated tags which have an individual readable tag for each individual fish; these have been used primarily in monitoring movements of salmon downstream past fixed detectors. All of these tracking techniques permit much more detailed examination of short-term behavior (a few days or weeks) than has been available with other techniques. Tracking techniques have not yet proved terribly important beyond studies of basic behavior and ecology.

5.3. What Can Be Observed

In the previous section we have considered the technical means of making observations on fish populations; we now must consider what actually is observed and how various types of data are used in fisheries stock assessment.

Detection rates—catches, encounters, sightings

The most commonly used types of data can be grouped under the general heading of detection rates; the rate at which some sort of detector (net, acoustic array, visual observer) encounters fish. These detection rates generally have two uses. Catch is the result of detection of fish by fishing gear. Catch data have many uses, most importantly knowing what has been taken from a stock in order to assess how the stock responds to exploitation. The second uses of detection rates are usually in determination of abundance. Encounter rates form the basis for most forms of surveys, be they line transects, trawls, or whatever.

Sampled catch

Sampling of catch is one of the major ways of collecting data about fish populations. The sampling may take place on board commercial vessels, on research cruises, or from sampling at ports. Once you have fish in the hand, there are a number of pieces of information that may be obtained from them.

Age from hard parts

The most valuable information obtained from sampled catch, at least for temperate waters, is age. Age is normally obtained by examination of scales or otoliths for annual marks, and the accumulation of age distribution statistics provides the basis for a very large proportion of stock assessments in temperate waters. The techniques of catch-at-age analysis and virtual population analysis depend upon age distribution.

There are two major issues in aging: validation of the aging technique and determination of how many individuals to age. Validation of aging is terribly important, first because one must have some assurance that what are being called annual increments are truly annual increments, and secondly because errors in aging cause biases in the estimation techniques that use age data, and we need to know the extent of aging error when we use these techniques Beamish and McFarlane (1983).

It has been common not to assess, or at least not to report, systematic estimates of aging errors. However, the current trend has been to recognize that errors in aging are inevitable and not a sign of poor work; rather failure to assess aging errors is a sign of poor work.

Validation of aging is best done by blind aging of known-age individuals. However, known-age individuals are often quite difficult to obtain, and one often has to rely on repeated "blind" measurement of the same otolith or scale by different individuals. Of course, just because everyone agrees that a certain otolith has three rings does not mean it is a 3- year-old. In the absence of individuals of known age, validation will always underestimate the aging error.

Length and weight

Length and weight data from the catch is probably the most commonly collected type of fisheries data simply because it is the easiest. One can measure dozens of fish for length in the time it takes to collect a single otolith, and one can measure the lengths of hundreds of fish in the time it takes to collect and examine a single otolith or scale. It is little wonder that length and weight data are so common.

In conjunction with age data, length or weight data can be used to construct growth curves. In some circumstances, length data can be used as a surrogate for age data (see Chapter 13). In tropical waters especially, there is a great deal of interest in length-based stock assessments.

Sexual condition

Fish can usually be easily examined for sexual condition. This is often quite useful in determination of the stock structure, as well as general life history. The age/fecundity relationship is also used in some age-structured models. Many species of fish exhibit sex-specific growth, survival, and vulnerability to fishing gear.

Genetics and morphometrics

Stock structure can often be determined from physical differences between the fish. Such work was initially done on the basis of morphologic features of the fish but has been largely supplemented and superseded by electrophoresis and more recently DNA studies. Whatever technique is used, the purposes are normally the same, to determine what the breeding structure of the fish stock is, which serves, in turn, to help determine the appropriate scale for management.

Stomach contents

It is possible, in some fish, to tell what they have been eating from what is found in their stomachs. Stomach-contents analysis was initially used for basic biology and ecology work and had little relevance to fisheries management. But as concern about species interactions has become more widespread, techniques have been developed to use stomach-contents data in stock assessment. Probably the largest such program is the multispecies VPA (Pope and Knight 1982), which attempts to use stomach-contents data in conjunction with commercial catches to determine a community-wide history of abundance.

Location and movement

Tagging and tracking data provide individual location and movement data. This type of data sees two primary uses; first in stock identification and secondly in measurement of exchange between fisheries. The constant problem of determining what is the appropriate stock for setting regulations can be clarified by movement data from marking. If marked fish freely move between two areas, then it is more likely these are the same stock.

Movement data are also frequently used in jurisdictional disputes concerning migratory or mobile fish. The large-scale marking programs on Pacific salmon have been primarily motivated by the need to determine who catches the fish originating in different waters.

5.4. Estimating Abundance

Every stock assessment biologist would like to know the size of the fish stock he/she works with; in fact in many cases this may seem like the most important question. Three subsequent chapters are devoted to estimating abundance, catch-at-age analysis, virtual population analysis, and depletion estimates. All three of these techniques rely on a time history of exploitation from the stock to determine abundance. What if you want the answer now and do not have time to wait for a history of experience? The three most commonly used short-term methods are surveys, tagging, and analysis of commercial catch-per-unit-effort distributions.

Design of surveys for abundance indices

One of the central problems in fishery stock assessment is to obtain an abundance index that is proportional to stock size and will hence reflect stock trends and responses to changing management regimes. Many assessment procedures pretend that CPUE from commercial catch and effort statistics is adequate for this purpose, but this pretense can go badly wrong. In some cases, CPUE will remain high until the stock is badly depleted (due to fish aggregation and efficient search), whereas in other cases, it will fall very rapidly while the stock is hardly being touched (due to "mining" a few highly vulnerable concentrations of fish). As fisheries develop, there are usually substantial changes in the spatial distribution of both fish and fishermen, and these changes may not be reflected clearly in aggregate catch and effort statistics until a disastrous situation has developed.

Deliberate surveys aimed at measuring relative abundance patterns should be a central part of every fisheries management system where agency funding and/or cooperative arrangements with fishermen make it practical to expend at least some fishing effort in "representative" patterns over space and time, rather than in patterns that the fishermen find to be most profitable. There is much confusion about what it means to obtain a representative survey sample of relative abundance, so this section provides an overview of some key considerations in survey design. For further details about statistical considerations in survey design, we recommend Cochran (1977), Green (1979), and Doubleday and Rivard (1981). For a good comparative overview of survey design problems in terrestrial field ecology, see Krebs (1989).

What should a representative survey accomplish?

A well designed abundance survey should accomplish at least two objectives. First, it should provide an estimate of average fish density (per swept area for nets, or as fish encounters with gear per time for line and passive net gear) over the entire spatial range where the stock(s) of concern might be found. Second, it should permit mapping of the spatial distribution of density within the entire range, at least in terms of presence or absence of fish (i.e., it should provide measurements of the boundaries of the stock distribution, which may not coincide at all with administrative or economic boundaries associated with the fishing process). To meet the second objective, note that the survey will have to deliberately extend beyond the known boundaries of the fish distribution, or at least beyond those boundaries that the fishermen have found economical to work or that managers have set for conservation or political purposes. This point leads to perhaps the most important single rule for fisheries survey design: never use the current distribution of fishing activity as the key or sole criterion for deciding the spatial boundaries of the abundance survey area.

Additional or auxiliary objectives might be to provide information about locations of economically worthwhile concentrations of fish, seasonal patterns of movement, changes in distribution associated with life history development (e.g., movement offshore as fish grow), locations and sizes of spawning aggregations, and population composition (size, age, sex, genetic substock structure). However, it is important to keep in mind that while some of these objectives may be of interest to fishermen who want to find more fish, scientists who want to understand more about the biology of the stock, and managers trying to evaluate alternative regulatory schemes (such as movement of fishing boundaries), the central and ultimate goal should be to provide consistent information about how the stock is changing in its distribution and abundance over time. By losing sight of this goal, it is entirely too easy to develop survey systems that focus too narrowly on particular sampling objectives every year, and thus end up providing spotty or inconsistent information on trends over space and time.

Given that a central objective should be to provide indices of relative abundance over space and time, it hardly needs saying that the basic survey sampling device should be one that catches fish in any local area in proportion to their abundance in that area. Thus it should not be a net, trap, line, or electronic counting/sounding device that easily "saturates" (fills with fish, provides only presence/absence information) as the local abundance of fish increases. If the use of such a device is unavoidable, then survey development should be accompanied by careful "soak time" experiments to measure precisely how rapidly the device is saturated under a range of conditions in terms of local abundance of fish.

Survey stratification

Surveys are generally conducted over areas large enough to be spatially heterogeneous, such that it will be known *a priori* that some areas or habitat types are better than others for finding fish. Provided that the boundaries and sizes of such areas can be precisely located and mapped before the survey begins, each area should be treated as a separate sampling stratum. Separate sampling programs and abundance estimates are then developed for each stratum, and the overall abundance index is constructed as a weighted average or total over the strata with the estimate for each stratum receiving a weight proportional to its size and inversely proportional to uncertainty about its estimate.

Generally, stratification provides a gain in precision for the overall abundance index estimate whenever the variation among observations within each stratum is less than would be expected from a random sample taken over the whole survey area while ignoring stratum boundaries. Thus it is not necessary that the strata represent perfectly homogeneous or uniform areas of fish abundance; the key point is simply to choose the stratum boundaries so that each stratum is at least a little more homogeneous than the overall survey area. However, it is critical that you be able to locate the stratum boundaries precisely in advance of gathering samples, and that you stick to your boundary definitions when it comes time to analyze the data. Suppose you find on analyzing your data that some strata generally had low density, except in samples taken near the boundary with another, more productive stratum. You might be tempted to move the boundary a bit, so as to define two strata each with more homogeneous observations. You might get away with this change without biasing the estimates, especially if the samples were taken on a systematic grid within each stratum in the first place (see next subsection), but you will most likely bias downward your estimate of the variance of your overall survey estimate. That is, you will underestimate the uncertainty of your overall abundance index, by failing to treat the stratum sizes as random variables (which they are if you move them around in response to the data). Think of it this way: a moving stratum boundary is not a sampling stratum at all (in the sense of what was assumed in defining your textbook recipes for stratified sample estimates and their variances); rather, it is like a contour placed on a map after measuring a collection of elevations and has components of uncertainty related to the uncertainties in the measurements themselves.

While stratum boundaries should be treated as fixed in the analysis of any data set where the sampling locations and sampling rates were defined in relation to these boundaries, it is not necessary or wise to treat the boundaries as immutable from year to year. Most often the first few surveys reveal that the boundaries could have been better placed in the first place, and it would be silly not to take advantage of this information to improve the survey design over time. However, care must be taken, particularly in the selection of sample sites within each stratum, to insure that the overall abundance index (weighted mean over sample strata) remains comparable over time. If new strata are added, correspondingly less weight must be placed on the original strata from which the new ones were split. If the survey is extended into whole new areas, then either the older survey estimates have to be corrected by including density estimates in them for the new areas or by restricting the temporal trend comparisons to those strata that were sampled at all times. To avoid having to make nasty and subjective decisions about what to do in such cases, the best strategy is to make sure that the overall survey area is large enough in the first place that adding new areas is unlikely to be necessary.

Random versus systematic sampling within strata

Basic statistics texts, and references such as Krebs (1989), admonish that there is no way to guarantee an unbiased estimate of the mean abundance within any stratum except by choosing sampling sites within the stratum completely at random, with every possible site having equal probability of being included in the sample each year. If instead you choose (for convenience or to study some gradient) to take the samples along a transect within the stratum, you invite getting a nonrepresentative sample by having the transect lie parallel to (rather than across) some biological gradient so that the observations along it are representative of only one part of the gradient and are misleadingly similar to one another. If you choose to take the samples on a systematic grid, so as to cut across gradients in any direction that they may occur and to allow you to map density patterns, there is still a minor risk of having all the grid points lie in troughs or peaks of density created by some factor that causes periodic density variation with a period equal to the distance between grid points. If the grid points are too far apart, you may of course miss any density concentrations or schools entirely; however, this risk is the same or even greater for a random sample of the same size.

Cochran (1977) presents a very important theorem about the relative precision of systematic versus random sampling: the estimate of the mean from a systematic sample is more precise than the estimate from a random sample if and only if the variance among observations within the systematic sample is greater than the variance among observations from a random sample. That is, systematic sampling will give a more precise estimate of the mean density (though the estimate may be biased) only if the systematic transect or grid is oriented so as to cut across spatial gradients so that its observations are more variable than would be obtained by randomly choosing sample points while ignoring the gradients. If the systematic sample is constructed so that its observations are less variable than would be obtained by random sampling (as would be expected for example by taking trawl samples along a single depth contour rather than cutting across depth contours), then it will be less precise than a random sample and will also very likely provide a biased picture of mean density.

One disadvantage of systematic sampling is that the variance (precision) of the sample mean cannot be assessed from data on variation among observations within the sample (recall that the variance of the mean from simple random samples is given by s^2/n, where s^2 is the variance among observations and n is the sample size). The only safe way to determine the precision of the mean of a systematic sample is to gather repeated samples at randomly chosen transect or grid starting points, then look directly at variation among the means of these samples.

On balance, we feel that it is generally best to use systematic grid sampling for abundance surveys. There are almost always gradients in fish densities, so that the systematic sample will be more precise than random sampling. However, the sampling should never involve one or a few transects, since the risk of having these oriented along some unknown gradient is just too great. Other advantages of grid sampling are that (1) it is generally most efficient in terms of ship time and sample point location, (2) it permits the most precise possible mapping of spatial patterns of density and boundaries of distributions, (3) it minimizes the risk of missing density concentrations that are of roughly the same diameter as (or larger than) the distance between grid points, and (4) it facilitates comparisons of distribution and density patterns over time.

Given the management resources (time, money, manpower) to take only a limited number of abundance samples, a key question becomes whether to keep the total sampled area small (with a high density of grid points within it) or to instead spread it out so as to be sure that the grid will extend beyond the spatial boundaries of at least the major fish concentrations. We have noticed a tendency among fisheries scientists to opt for greater concentration of samples, rather than to have a lot of empty (and supposedly "wasted") samples near the boundaries of the grid. This is a very bad practice that eventually comes back to haunt the management agency in a form that some managers call "black-hole arguments." When faced with apparent stock declines and restrictive regulations, fishermen often argue that plenty of fish are still there but have moved into unusual areas (black holes usually offshore and deeper than are normally fished commercially or surveyed). Sometimes such black-hole arguments are quite valid: fish do move in response to environmental factors, and it is obviously risky to have an abundance survey system that provides and index of environmental changes rather than actual abundance changes. On balance, we feel that it is generally best to err on the side of low sample densities, in favor of making sure that the survey covers a large enough total area even if many of the survey samples routinely come up empty.

Multistage surveys

Fish abundance is likely to vary on several scales in space and time, and it is generally impossible to cover all scales adequately with a single survey grid and sampling device. This is particularly a problem with highly aggregated or schooling fish, where a survey grid using regular fishing gear (seines, trawls, etc.) would have to involve far too many grid points to be confident of hitting a significant and consistent number of aggregations.

In such cases, it may be best to conduct the survey in two or more stages, each involving one spatial scale and method of observation. The first stage will involve some very crude and fast method of covering large areas, so as to define general range limits and locations of aggregations. Sampling at this stage might involve aerial surveys, echo sounding, or other devices that can cover large areas without a high time/effort cost per sample point. The second stage will then use the distribution (preferably mapping) results of the first stage to define sampling strata and sample locations for devices that provide more precise estimates of local densities. Further stages might involve repeated samples at the second stage locations, transects to define aggregation boundaries, subsampling of catches at the survey locations, and so forth.

When even the first (crude) stage sampling has to involve devices such as echo sounders that can cover only a small part of the total survey area for each unit of fishing gear (vessel) involved in the survey, the key step in survey design will be to find some way to involve more gear in the survey. Options here include development of economic incentives for fishermen to participate in the survey (e.g., extra fishing time for cooperating vessels), adding survey requirements to fishing license arrangements, expansion of fisherman reporting requirements to include information such as echo sounding charts obtained while steaming to and from preferred fishing grounds, and development of "vessels of opportunity" programs, where sampling devices such as echo sounders are installed on non-fishing vessels that regularly pass through the survey area.

Tagging

Again there is an extensive literature on use of tagging data to estimate population abundance. Seber (1982) and Burnham et. al. (1987) are the standard references. Unfortunately, we can not be terribly optimistic about tagging studies to estimate abundance; such estimates are usually the least reliable output of tagging studies. Tagging studies always make some use of the marked-to-unmarked ratio to estimate total population abundance, and the sad fact is that the chances that marked and unmarked individuals behave the same is rather small. Study after study has shown serious bias in population estimation from tagging studies. We simply do not recommend tagging studies as a way of estimating abundance for most fish populations, unless it is practical to mark a very large percentage (25% or more) of the total population.

Commercial catch data

Basic observed patterns

It is simply irresistible to try to use catch per unit effort data to estimate or as an index of fish abundance. The question is, what type of relationship should we expect between CPUE and abundance?

Figure 5.2 shows three possible relationships between CPUE and abundance. On the first curve, the CPUE stays high as abundance drops. We call this type of distribution *hyperstability*. This type of relationship can be expected on a small spatial scale when handling times are large. It can be expected in almost any fishery where search is highly efficient, so that most effort concentrates on the areas where fish are most abundant, and the fish remain concentrated as abundance declines. Exploitation of marine mammals in spawning aggregations on beaches or bays and purse seining for easily discovered schools of clupeoids would be expected to result in hyperstability.

The second curve shows CPUE proportional to abundance, which arises when handling times are small and search is random. Some forms of net fishing and hook-and-line fishing would be expected to display this type of relationship within a single site, so long as significant gear saturation did not occur. We refer to this type of relationship as *proportional*.

The third curve we call *hyperdepletion* because the CPUE drops much faster than abundance. The stock appears to be depleted, yet abundance has not

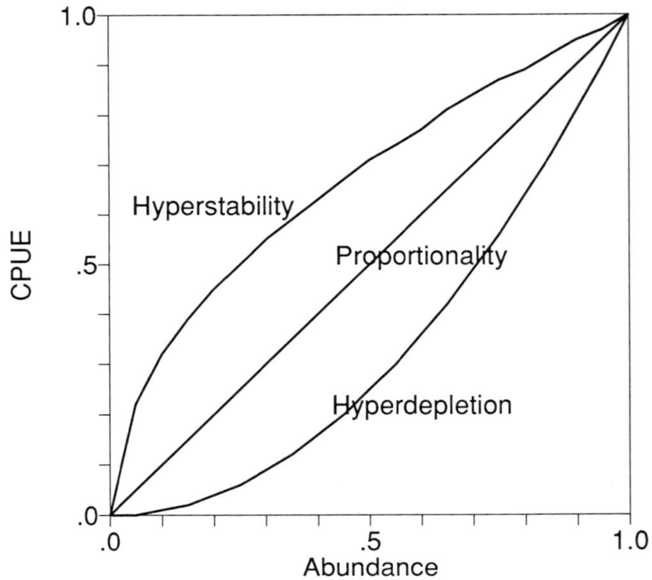

Figure 5.2. Three possible relationships between abundance and CPUE.

dropped as much as CPUE. One mechanism that can operate within a single spatial site to generate this type of relationship is differential behavior of fish in response to the gear, such that there is a small but highly vulnerable subset of fish which is depleted, leaving behind a much less vulnerable, but still abundant subset of the stock. An example would be cryptic invertebrates where much of the population may be hidden so as to remain nearly invulnerable to fishing.

In the following sections, we consider the types of behavior of fish and fishermen that are likely to lead to the different types of behavior shown above, and how to use CPUE as a measure of abundance. The observed CPUE typically results from thousands of individual units of fishing effort, whether these are tows of nets, minutes of soak time for pots, or hook hours for hook-and-line fisheries. The spatial pattern of abundance, the spatial pattern of fishing effort, and the relationship between abundance and capture success on an individual site all combine to produce the aggregate CPUE. In order to understand this aggregate, we must examine each of these components.

Basic models

The simplest assumption regarding the relationship between commercial catch and abundance is that the catch rate (CPUE) is directly proportional to abundance. In a discrete time model the assumption would be

$$C = NEq \tag{5.4.1}$$

where C is the catch, N is the stock abundance, E is the fishing effort, and q is the efficiency of the gear. This model can be written in continuous form as

$$\frac{dC}{dt} = NEq \tag{5.4.2}$$

This assumption forms the basis for most biomass dynamic models, as well as most catch-at-age analysis. It has been demonstrated to be wrong in almost every case where it has been possible to test—simply stated it is almost impossible for this relationship to be true, and those who use it are running severe risks.

In order for catch rate to be proportional to abundance, fishing effort must be distributed at random with respect to the fish. In most fisheries, any fisherman who fished at random would soon be out of business, and anyone who has ever designed an abundance survey and set out the fishing sites randomly (without stratification) knows you will not catch many fish.

No fishermen fish at random! Fishermen go where they believe the fish are, and almost all fishermen have enough idea about where fish can be found that the resulting effort is not random but concentrated on good fishing sites. The result is that the catch rate reflects the abundance in high density sites, not a random or systematic sample of sites. The first rule of using catch rate data to determine abundance is *you must spatially stratify*.

Spatial stratification

The first step in understanding commercial catch data is to draw maps of catch and effort. These will almost always show concentration of effort, and this should be where fish are most abundant. If by some miracle effort is uniformly distributed in space, it may be safe to assume that catch rate is proportional to abundance.

To analyze the catch rate data spatially, we must break the data into spatial strata. This may be done by visual mapping, or the strata may be related to the statistical breakdown of the catch statistics themselves. However, it is done, let us assume we have catch (C) and effort (E) data for a series of i spatial strata, and we also have the area (A) of each area. If we allow that for each area the catch rate is proportional to the abundance by the relationship

$$C_i = N_i E_i q_i \tag{5.4.3}$$

then

$$N_i = \frac{C_i}{E_i} \times \frac{1}{q_i} \tag{5.4.4}$$

and the total N over all areas is

$$N_{\text{total}} = \sum_i \left(\frac{C_i}{E_i} \frac{1}{q_i} \right) \tag{5.4.5}$$

If we assume that the actual efficiency of the gear q is the same in all areas, and that the q_is just reflected different size of areas, then we can use the area A_i and a single q as follows:

$$N_{\text{total}} = \frac{1}{q} \left(\sum_i A_i \frac{C_i}{E_i} \right) \tag{5.4.6}$$

If the q's are actually different between gears, then the q_i's can be put in the denominator of the right-hand side.

All of these calculations assume that we have the C's, A's and the E's. The first two quantities are reasonably easy(relative to other fisheries data), but the usual hang up is the E's. The question of "corrected" effort is discussed in the section on fleet dynamics; in practice we would recommend using the GLM method discussed there to generate a proper adjusted index of abundance to plug into the above equations in place of the quantity C_i/E_i.

A major problem in spatial stratification is what to do with strata where no effort occurs. An obvious way to treat these strata is to assume that there are no fish, but this will have an obvious biasing effect if there really are fish in these strata. If most of the fishing effort is concentrated in a few strata, the estimate of total abundance may be highly sensitive to the assumption made about numbers of fish in unsampled strata. Alternatives to assuming no fish in unsampled strata are to interpolate between sampled sites or to extrapolate trends. The best method will depend upon the details of the fishery, and we recommend trying several different approaches to test the sensitivity of the final estimate to the assumption made. Many methods of spatial stratification are being examined, with kriging (Clark 1979) a popular topic at conferences in the late 1980s.

Figure 5.3 shows spatial mapping of CPUE as is performed on a regular basis by the Inter-American Tropical Tuna Commission (IATTC). Note that the CPUE is high in a small proportion of the total area.

On-site dynamics

If we examine the relationship between catch rate and abundance in a small enough spatial area, we can perhaps hope to meet the assumption of

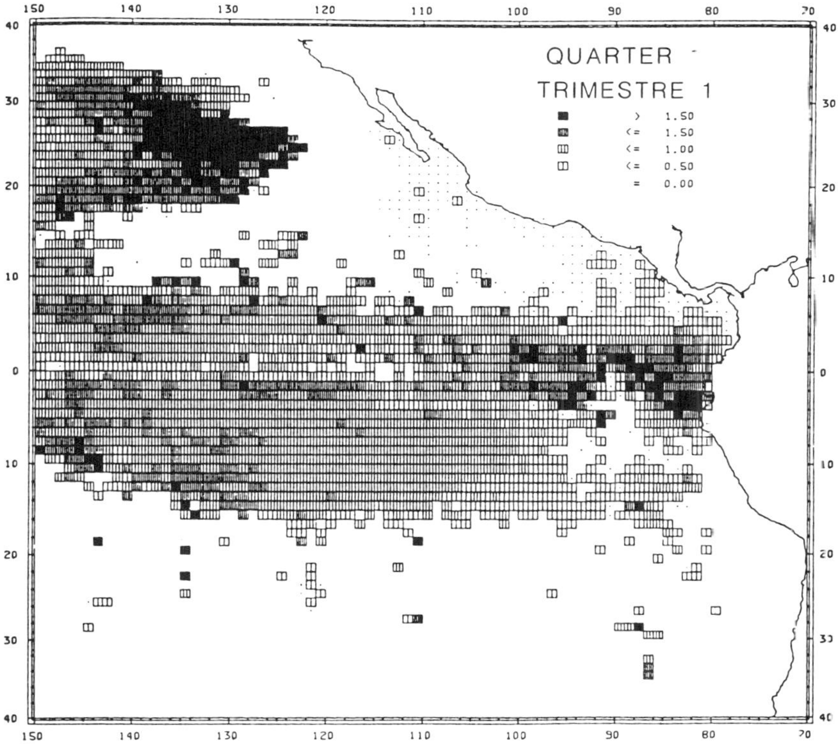

Figure 5.3. Spatial distribution of catch per unit effort for bigeye tuna in eastern tropical Pacific. From Miyabe and Bayliff 1990. Figure 12. Reprinted by permission of the Inter-American Tropical Tuna Commission.

random search by fishermen with respect to fish. Such an assumption is made when using spatially stratified CPUE as a measure of total abundance given above. Unfortunately, there are many aspects of fishermen's behavior that will cause CPUE to be not proportional to abundance even on a very small spatial scale. These behaviors include the following:

1. Efficiency of search. Traditional models assume random search which is required for the CPUE to be proportional to abundance. If fishermen are efficient at finding fish even on a small spatial scale, then the CPUE on that scale will not be proportional to abundance.
2. Handling time. For CPUE to be proportional to abundance, there can be no significant proportion of fishermen's time budgets allocated to handling time. For this reason, encounter rates per unit time searched, rather than per some cruder measure of effort, are sometimes used as an index of abun-

dance. Within any spatial scale for which fisheries data can be collected, spatial patchiness will exist. For example, on a one-acre reef, an abalone diver will concentrate on patches of abalone. Within any grid system, be it in one-degree squares or smaller, purse seiners will undoubtedly seek out fish schools.

3. Fishermen interaction. Fishermen rarely work independently of one another. Sharing of information by fishermen will lead to nonrandom search. Interference, such as limitation in the number of good fishing sites (Ledbetter 1986), will cause CPUE to not be proportional to abundance.
4. Depletion. Even on a small spatial scale we would expect CPUE to decline with cumulative effort due to short-term depletion. Thus on the smallest spatial scale, we should not expect CPUE to be proportional to abundance.

For all the above reasons, we can imagine that CPUE may not be directly related to abundance, even on as small a scale as data can realistically be collected. Some simple detailed studies of both fishermen's behavior with respect to handling time, cooperation or interference, and spatial search could quickly and easily provide some valuable information on how closely CPUE on a small spatial scale will be related to abundance.

Variable catchability. Another interesting variation on the handling time theme was considered by Cooke and Beddington (1985). They assumed that the catchability is not constant but rather is a random variable. This is almost certainly true in most fisheries; anyone who has done any hook-and-line fishing knows that there are periods when the fish are biting and periods when they are not. Tides, phases of the moon, seasonal movements of fish, diurnal cycles, and so on all contribute to time-varying catchability.

If there are any significant components of handling time in the mechanics of the fishing process, then in periods of higher catchability more time will be spent handling, and the gear will effectively undersample the periods of high catchability, and oversample periods of low catchability. If, for instance, catchability is high during the new moon, then more time will be spent during the new moon handling the gear, and proportionately less time will be spent searching. The net effect will be that the CPUE will not be an average of the periods of high and low catchability, but rather weighted more toward the periods of low catchability. This will cause a systematic biasing of the relationship between CPUE and abundance.

Cooke and Beddington show that hyperstability should be expected, and the degree of hyperstability will depend upon the extent of variability in catchability and the importance of handling time.

Choice of area to fish

Two models for the choice of fishing site are the assumption that fishermen's effort is randomly distributed over the population, and second that effort is allocated to the best fishing areas. This second model is probably the more realistic, but does assume (1) fishermen have perfect information about CPUE in all areas and (2) all areas have equal costs and desirability of fishing.

If these assumptions are correct, then the relationship between CPUE, effort, and abundance is dictated by the spatial pattern of fish abundance combined with modifications due to local behavioral effects described in the previous section. Ignoring these local effects, the basic predictions are that 1. CPUE will decline slowly with abundance if fish tend to occur in the same density, regardless of total abundance—this would be the case for most schooling fish as considered by Paloheimo and Dickie (1964) but does assume that fishing does not change the density (an assumption we question later) 2. if a few high-density concentrations exist, then CPUE may decline more quickly than abundance, whereas the high-density locations are fished out but a large stock occurring at lower densities remains.

If we relax the assumption that effort always goes first to the area with the highest CPUE, we find that the relationship between CPUE and abundance may be quite surprising. For example, consider a fishery in which there are two fishing areas, initially at the same abundance. Assume that one area is close to the home port and the other is either far away, dangerous, expensive to fish, or for some other reason undesirable. As the fishery develops, nearly all effort will initially be spent in the close area, until at some point some fishermen are willing to trade off the undesirable aspects of the other area for its high CPUE. If we assume that the CPUE in the less desirable area must be twice as high as CPUE in the desirable area before effort will switch, then all effort will concentrate in the desirable area until it is driven down to half its virgin size. At this point, total biomass will be 75% of the virgin stock, but CPUE will be 50%of its original level (over the whole fishery). It is then that effort will start shifting to the less desirable area, and the CPUE will actually increase even though total abundance may still be declining. Thus, even this rather trivial example illustrates how in the same fishery CPUE can decline faster than abundance and then actually increase with decreasing abundance. To our knowledge, no previous studies have considered the potential impact of differential costs or desirability. Hilborn and Ledbetter (1979) describe what appears to be differential desirability of fishing locations in the British Columbia salmon fleet. Hilborn and Walters (1987) present a method for simulation of these types of effects.

Another feature of spatial effort allocation occurs during development of fisheries. Often new fishing grounds are being discovered, and the total spatial area of the fishery increases. The effect will be similar to that of differential costs and desirability; overall CPUE can remain stable or even increase as the overall abundance declines. However, in this situation, spatially disaggregated CPUEs should show stock declines as each new fishing ground is fished out.

Determinants of effort

A third component of fishermen's behavior is the determination of total effort expended by fishermen. Total fishing effort per time is often related to CPUE (Hilborn and Ledbetter 1979, Millington 1984). If fishermen cannot meet costs, they will not do much fishing. Thus, if there is any spatial or temporal aggregation in the CPUE data, time or space strata with good CPUE will be overrepresented and strata with poor CPUE will be underrepresented.

As a further example, imagine a fishery in which fishermen must choose which area to fish, and the choice might be random with respect to the abundance of fish. However, those who chose the areas with poor CPUE would do some initial fishing, decide that CPUE is too low, and either quit fishing or go elsewhere. Those who chose sites with good CPUE would expend more effort. Thus, search could be totally random, yet CPUE aggregated over all areas would not reflect abundance; areas with high CPUE would again be overrepresented.

Fish distribution, biology, and behavior

This section examines alternative ways that fish may be distributed over space, and the implications of this for fishing success. First we distinguish between two large categories of fish, those that are essentially sedentary or closely tied to a spatial location, and those that redistribute themselves rapidly over space (which we call diffusive, following Clark 1982.)

Sedentary stocks. For the purposes of this discussion, sedentary stocks are those whose abundance in a particular place depends upon the growth and removal dynamics within the area rather than what goes on in adjacent areas. Sedentary marine invertebrates or highly territorial fish would be typical examples. In these stocks, the recruitment may actually be affected by the status of stocks elsewhere, but at the extreme each spatial location can be thought of as an independent stock.

In an unfished state, the distribution of abundance would then depend upon the distribution of "carrying capacities" among the locations or sites. If all spatial locations were similar, we would expect the stock distribution to be uniform. If some places were much better than others, densities would be much higher in them than in others. As a fishery develops on these discrete stocks, the dynamics of each stock would depend primarily on its fishing history.

If some areas are less desirable than others on some basis other than CPUE,

Table 5.1. Distribution of northern anchovy larvae in 1952 and 1965.

Larvae per 10 m^2	% area at this density	
	1952	1965
0	63	46
1–10	13	9
11–100	13	11
101–1,000	10	20
1,001–10,000	1	14
10,001–100,000	0	0

Data extracted from Figure 6–1 of MacCall (1984). Area sampled in the 2 years was roughly the same, although there were more stations in 1965. Total spawning abundance in 1952 was less than 10,000 tons, in 1965 it was 750,000 tons.

there will be a tendency for hyperdepletion. The less desirable areas will become a reserve of underutilized areas with higher CPUE.

Diffusive stocks For diffusive species, the abundance at any one location will depend on the abundance elsewhere. For instance, many species are fished in spawning concentrations. The spatial extent of the spawning and the density of fish in the concentrations will depend upon total abundance and fish behavior. Some species may move about so as to maintain a constant density and therefore expand the area of the aggregation as abundance increases; other species may maintain the same area and simply increase density. MacCall (1984) has considered such patterns in some detail and views it as a form of "habitat selection." Although spawning aggregations form a convenient example, the distribution at any time in the life history is obviously governed by the distribution of habitats and fish behavior in relation to habitat choice. Winters and Wheeler (1985) show how total abundance, area covered, and catchability are closely related in several stocks of herring in Eastern Canada.

We can distinguish between two major classes of diffusive behavior: 1. Behavior where the fish maintain a more or less constant density by adjusting the area covered— *constant density model*; 2. Behavior where a constant proportion of the stock goes to each area so that the density in each area is proportional to abundance— *proportional density model*

Under the constant density model, more areas will show high density (rather than each area increasing in density) as total abundance increases. Table 5.1 shows the distribution of anchovy larvae from MacCall (1984) and illustrates this point. The proportion of the area surveyed (essentially the same in the

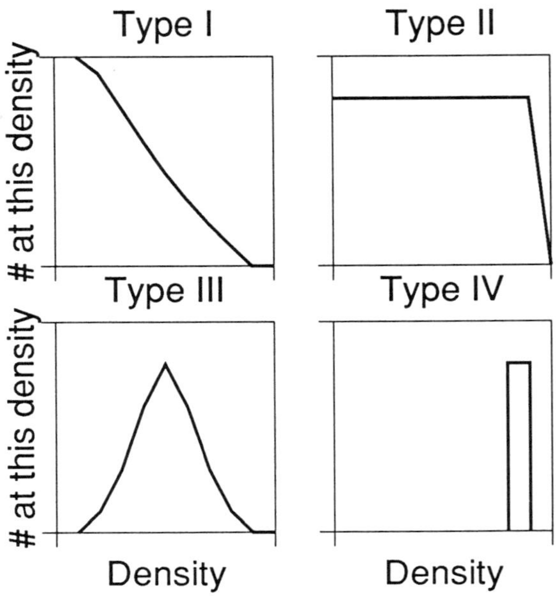

Figure 5.4. Four types of concentration profiles.

2 years) covered by different densities of larvae is shown for 1952 (low abundance) and 1965 (high abundance). The major change was that the area covered by the high-density concentrations increased in 1965. The spawning stock in 1952 was estimated to be 10,000 tons, in 1965 750,000 tons. If the density at each site had been simply 75 times higher in 1965, we would have expected to see numerous sites with densities over 10,000. MacCall uses the analogy of a bowl that gets filled: as the abundance becomes higher, the area covered by high concentrations expands much more rapidly than the density in each site.

Territoriality is an obvious mechanism that would lead to constant density diffusion patterns. Although a mechanistic analysis has not been done for many (perhaps any) species, distributions in which the area covered by high-density concentrations increases with abundance are thought to be characteristic of clupeoids and have been documented on some occasions (Saville and Bailey 1980).

Concentration profiles

Clark (1982) used the concept of concentration profiles to describe the distribution of fish at any point in time. Figure 5.4 shows the four types of concentration profiles Clark discussed. In type I profiles, there are few locations with high densities of fish and progressively more locations with decreasing densities. In type II profiles, the number of fish available is the same at all concentrations. Thus, thinking in terms of fish per unit area, there might be 1 ha with 10 tons/ha, 10 ha with 1 ton/ha, and 100 ha with 0.1 tons/ha. In type III profiles, there is some middle level of density that is the most common, with high- and low-density areas being less abundant. Finally, in type IV profiles, all fish occur at the same density.

The y axis in such graphs indicates the number of fish occurring at this density or total area covered by fish at this density. Clark used number of fish, but it is often easier to think of the area covered by different concentrations as in MacCall (1984). The two systems are reasonably easy to interchange, with the exception of type II, where what is a flat line when the y axis is number of fish would be a type I profile when the total area covered was the y axis.

Clark assumes that fishermen will go to the best area first and extract all of the fish in that area. He further assumes that the CPUE in that area will be proportional to its' initial density. This mechanism is a natural extension of the mining analogy. Once a ton of ore is removed, the concentration of the remaining ore remains unchanged. This is not generally true in fisheries.

Implicit in Clark's analysis is that CPUE is directly related to the initial abundance but remains unchanged as successive fishing effort is applied to the same spatial area. Thus there can be no local area effects that produce hyperstability, or the CPUE would not be proportional to initial density. For CPUE to stay constant as more units of effort are applied to each area, the density would have to stay constant. This may occur in some schooling fish, where, as soon as some of the stock is harvested (even within days or hours), the fish redistribute themselves over a smaller area to provide the same density. This is the only biological mechanism we can think of that would be consistent with Clark's assumption. Generally, at any spatial scale, the removal of part of the stock will reduce the density that the next units of fishing effort will encounter.

However, if we assume that the CPUE in a given site for a single small unit of effort will be proportional to density and that density decreases as the stock is removed, then we obtain quite different results. Figure 5.5 shows the four types of concentration profiles, followed by two other sets of figures. In the second panel we see how CPUE decreases with increasing effort, if in all cases each additional unit of effort is applied to the best area; this selection behavior will cause the areas with the highest initial densities to be fished down until their CPUE is equal to the CPUE in areas with lower initial densities. Effort will then alternate between (or be distributed among) these areas of equal CPUE until their densities are in turn reduced to a density comparable with even poorer sites. Therefore, the catch per effort decreases with additional effort in all profile types, but drops faster in profiles I and

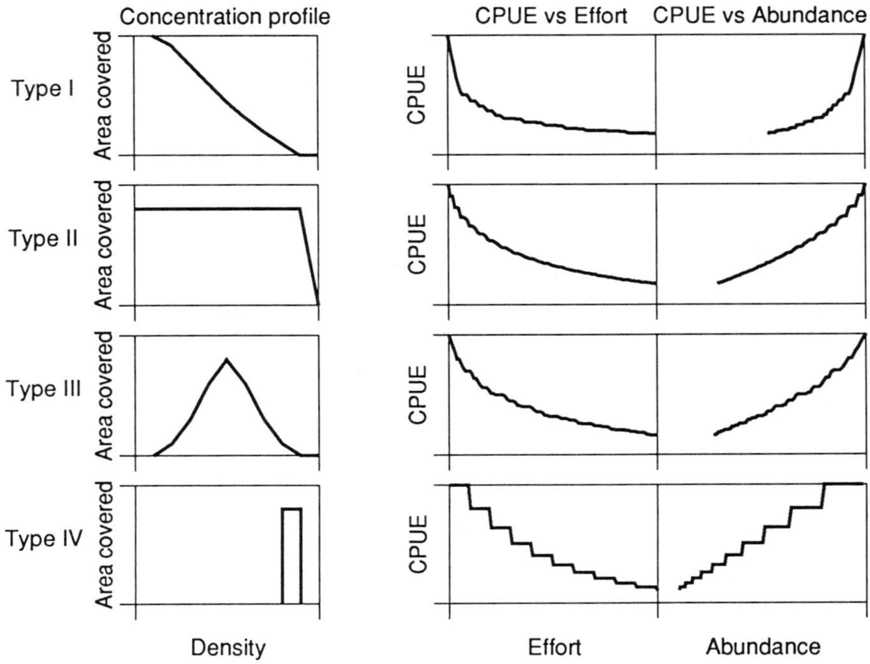

Figure 5.5. The relationship of different types of concentration profiles, CPUE and abundance.

III where there are a small number of areas with initially high density. Profile type IV with all areas starting at the same density shows the slowest reduction in CPUE as additional effort is applied.

The third panel shows the most important result, the relationship between CPUE and abundance. Profile I shows very strong hyperdepletion, the CPUE is very high while the few high-density areas are fished, but, as the effort increases and the lower density sites are exploited, the CPUE drops rapidly, although total abundance has decreased little. Hyperdepletion is also strong in profile III and present but weak in profile II. Only in profile IV do we see a linear relationship between CPUE and abundance.

In these examples we were assuming sedentary stocks with no recruitment; that is we were simply observing the relationships among CPUE, effort, and abundance as an initial biomass was exploited. The results would be obviously much different if the stocks reestablished their concentration profiles by diffusion or if there were local area effects, such as handling time. This is an area were some formal mathematical analysis (similar to that of Clark 1982), or simulations as described in Hilborn and Walters (1987) could perhaps be done.

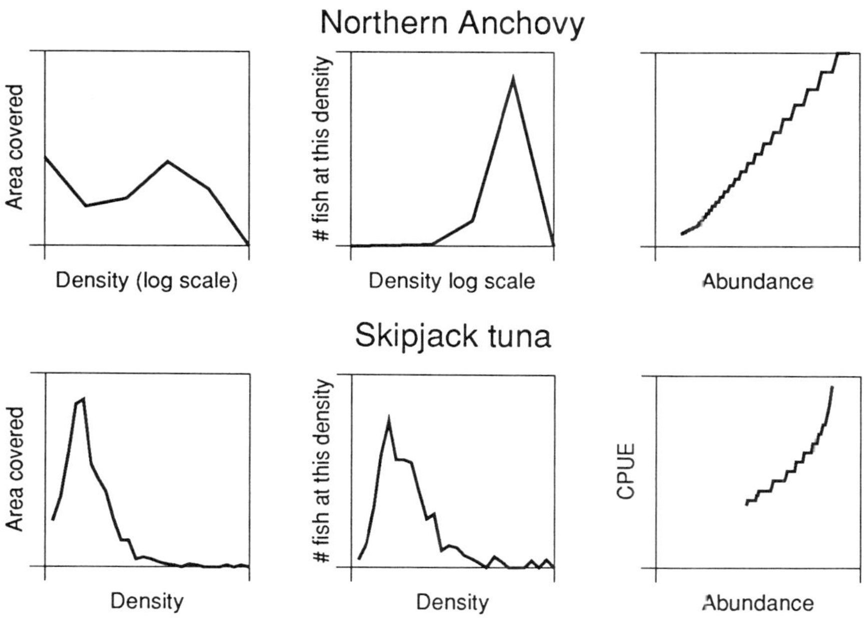

Figure 5.6. Concentration profile for northern Anchovy larvae in 1965 (top panel) and skipjack tuna in the western Pacific (bottom panel).

California Anchovy Figure 5.6 shows a concentration profile for northern Anchovy larvae in 1965, plotted giving both number of fish and total area covered versus density. Also shown is the resultant relationship between CPUE and abundance. Although it does not fall neatly into any of the classes, it is probably closest to a type IV profile, particularly when plotting number of fish versus density. Most of the animals are initially available in the high-density areas. Note that the x axis is logarithmic in this case.

Experience in fishing clupeoids shows that CPUE in a single site is not related to abundance; rather, the fish are easily found and handling time (or hold capacities) may limit CPUE. We would therefore expect the total relationship between CPUE and abundance to show hyperstability. This would result primarily from the local area dynamics of search and handling, not from the concentration profile. If the diffusive dynamics of the clupeoid caused them to reestablish the high-density concentrations then the combination of concentration profiles and diffusive behavior could cause hyperstability without the local area effects.

Western Pacific skipjack tuna. The bottom half of Figure 5.6 shows a concentration profile for skipjack tuna (South Pacific Commission—SPC—data base). The densities are measured on a 1° square basis, using the average pole-and-line CPUE (tons per day) for the year 1980. All areas covered by the SPC data base are included. In this case, the differences between the two possible y axes for the concentration profiles are small; it is a type III profile with a rather long tail. This results in hyperdepletion as seen in the third panel.

It is difficult to assess the probability of hyperdepletion in skipjack data, since most surface tuna fisheries catch skipjack and yellowfin together. Thus, fishermen are unlikely to allocate effort to areas of high skipjack CPUE first. Further, information is often quite imperfect, and the distances covered in this data base are so large that movement costs cannot be ignored.

Most longline CPUE data for yellowfin do seem to show hyperdepletion. Catches drop very rapidly as the fishery begins and then stabilize, although total catch often rises by an order of magnitude after the initial drop in CPUE has taken place. If a good historical data base on longline CPUE were available, investigation of the concentration profiles would be most rewarding.

Interannual dynamics. The previous discussion of concentration profiles considers only what happens within a particular year, assuming no recruitment or diffusion. The long-term dynamics of CPUE, effort, and abundance depend upon the diffusive behavior of the stock, recruitment, growth, and mortality, as well as fishing. In addition, it is quite possible that diffusion may act rapidly within a fishing season to reestablish the concentration profiles. The constant density type of diffusive behavior will obviously lead to hyperstability, particularly if the diffusive behavior acts within a season. The proportional density model of diffusive behavior will contribute to proportionality. Elsewhere (Hilborn and Walters 1987), we have described a simple simulation procedure to investigate the interannual factors simultaneously for specific fisheries and specify methods for simulating both the constant density and proportional density models. However, it is possible that some analytic results can be obtained for these problems, and we encourage the more mathematically inclined to pursue this problem.

Summary—estimating abundance from commercial CPUE

Hyperstability is one of the best and worst features of a fishery. For the fishermen, it means not suffering decreases in CPUE as abundance changes. For the manager, it offers the terror of a stock declining without any change in CPUE to tell him trouble is afoot. Some of the major fishery collapses in the world have been ascribed to hyperstability. The numerous clupeoid collapses (Peruvian anchoveta and North Sea herring among the more spectacular) and many marine mammal collapses such as the Alaska fur seal (*Callorhinus ursinus*) and the sea otter (*Enhydra lutris*) on the west coast of North America, are among those often attributed to it (Clark 1982). Clark cites these as likely cases where a type IV concentration profile was maintained over time as total stock declined.

Are type IV concentration profiles responsible for fisheries collapses? We question this conclusion and offer a simpler and actually more traditional explanation. The hyperstability in these cases was more likely due to the fact that most of the stock in these species occurred at densities higher than the economic threshold for profitable operation. Finding the clupeoid schools or mammal breeding sites is easy. The stocks were fished out because they could be found and easily captured, not because they all tended to be found at the same density. The densities (at least for marine mammals) certainly did change significantly as exploitation continued. The problem was that even at low densities the fishery was economically profitable.

Concentration profiles have a role to play in that these species do not exist in large numbers at densities low enough to be unprofitable for harvesting. There is no tail of the concentration profile below an economic minimum density. Severe overexploitation resulted from the very low economic threshold for profitable fishing. If search is efficient enough or price is high enough, even a type I or type II concentration profile can be exploited to near extinction. The type III or type IV concentration profiles were contributing, but were not the only, causes of the overexploitation. It is not the distribution itself, but rather the proportion of the stock that is found above the minimum economic density, that makes a fishery prone to overexploitation.

Clark also cites returning salmon which form highly vulnerable schools as a case of type IV concentration profiles. The large spawning runs that are highly vulnerable to fishing nets would seem an obvious example of a behavior that produces this type of concentration profile at one point in the life history. However, the densities, and the CPUE that fishermen can obtain can vary by 10 or 100 times from a good year to a bad year. It is very unrealistic to think of returning salmon as always occurring at the same density. Again, the reason that salmon are so vulnerable to overexploitation is that even when the density is quite low, a few vessels could operate profitably. Thus the bionomic equilibrium is low due to price.

Our basic desire is to know the relationship between CPUE and abundance. This relationship will result from the interaction of the following:

1. Operational characteristics of fishing gear in a single area, including search efficiency, handling time, and variability in catchability
2. Determinants of effort by fishermen, especially the relationship between CPUE and total effort
3. Determinants of area fished by fishermen, including quality of information, desirability of different areas, and different CPUE
4. Concentration profiles
5. The dynamics of the movement or diffusive behavior of the stock.

Table 5.2. The contribution of various components of fishermen's and fish behavior and biology to hyperstability, proportionality and hyperdepletion.

Fishermen's Behavior	
Local area effects	
nonrandom search	hyperstability
handling time	hyperstability
Interaction between fishermen	
information sharing	hyperstability
interference	hyperstability
Effort response to abundance	hyperstability
Choice of area to fish	
effort to best CPUE area	hyperdepletion except for type IV concentration profile
attributes other than CPUE	tends to cancel hyperdepletion or hyperstability
Fish Behavior	
Sedentary stocks	depends on concentration profile
Diffusive stocks	
constant density model	hyperstability
proportional density model	proportionality
Concentration profiles	
type I	strong hyperdepletion
type II	hyperdepletion
type III	hyperdepletion
type IV	proportionality

Table 5.2 summarizes how the above factors will contribute to hyperstability or hyperdepletion. In general, the operational characteristics of fishing will tend to cause hyperstability, whereas the distributional characteristics will tend to cause hyperdepletion. It is therefore extremely difficult to make *a priori* estimates of what to expect. Hyperstability appears to be far more common, being well documented in many species.

Hyperdepletion appears to have occurred in the South Australian rock lobster (*Jasus novaehollandiae*) fishery. The fishery began in the 1930s and involved a few thousand tons per year, but with very high catch rates; in 1946, the CPUE was 16 kg per pot lift (Lewis 1981, 1983). After World War II, the fishery picked up and by 1949 over 1,000 tons were harvested each year. The CPUE had declined to 10 kg per pot lift. By 1957, the harvest was almost 2,000 tons per year . . . and CPUE was down to 4 kg per pot lift. If CPUE was proportional to abundance, this would indicate that the stock was depleted to one fourth of its original abundance. However, throughout the 1960s, 1970s, and 1980s, the yield remained at 2,000 tons per year, and CPUE stayed steady at first, then declined as effort increased in the 1970s and 1980s. It appears that the fishery initially developed in a few areas of underwater limestone bluffs that provided very high densities of lobsters and very high catch rates. However, when the fishery expanded, it spread beyond these limestone bluffs to a much larger area that supported lobsters at lower densities. The drop from 16 kg per pot lift to 4 kg per pot lift represented a shift in the fishery from the few limestone concentrations to the much larger (and lower density) area of the rest of South Australia. During the development phase of this fishery, abundance appears to have dropped much more slowly than CPUE.

A second example of hyperdepletion also comes from South Australia and was related to us by Neil Carrik. The fishery for prawns (*Penaeus latisulcatus*) in the Gulf of St. Vincent involves approximately 20 prawn trawlers. When the fishery opens, there are a few locations where prawns are in very high concentrations, and these are initially targeted in the first few hours of the fishery. Catch rates are very high, and these concentrations are fished down quite rapidly. The fishery then spreads out to fish most of the prawns, which are found at much lower densities. Again the CPUE initially declines very rapidly, although the high-density concentrations represent only a small fraction of the total prawn population available for capture.

We can offer some direct advice to biologists who wish to use information about CPUE to make inference about changing abundance. Paloheimo and Dickie (1964) advised spatial mapping of CPUE; this should be done whenever possible to determine what the concentration profiles look like, and how the spatial distribution of the stock and fishing effort are changing. Data collection systems should always attempt to measure the spatial distribution of catch and effort.

The relationship between CPUE and abundance within a single site must also be examined. We suggest two approaches. First operational studies of the fishing gear can provide great insight into likely relationships between CPUE and abundance. Time budgets and the operational characteristics of the search process will be highly informative. When time spent handling becomes larger than time spent searching, hyperstability should always be suspected.

Even more useful are direct measurement of abundance, either by surveys or by depletion experiments. In a depletion experiment, the CPUE can be measured as the fish in a small area are rapidly depleted. This should provide excellent information on the relationship between CPUE and abundance over a wide range of abundances on a single site. Depletion experiments are probably one of the most valuable yet underutilized techniques available to fisheries scientists.

There is a need for much more work on how the diffusion dynamics of the stock, both within a year and between years, interacts with the concentration profile, fishermen's behavior, and local area dynamics. It would be most useful to have spatial mappings of densities (or CPUE) for a number of stocks, and to see how these have changed over time. Few people have examined how fishermen's effort or choice of areas to fish change in response to CPUE, yet these types of data are often sitting on file in many fisheries agencies. Agencies that do not collect spatial location of the catch should do so if at all possible and need to carefully consider the appropriate spatial scale of the data collection system.

5.6 Estimating movement

Earlier in this chapter, we discussed tagging studies and suggested that although tagging studies have often proved to be unreliable for estimating abundance, they have a great deal of potential for estimating fish movement. Most analyses of movement patterns from tag return data have been graphical; arrows are drawn connecting release and recovery sites. While this type of analysis does have some uses for very specific qualitative questions, quantitative measures of movement are often desired. In this section, we describe a simple computational approach for the analysis of movement pattern from tagging data. The approach is very general, and can serve as a template for much more complex movement analysis and/or tag analysis on a single area.

For simplicity, assume we have two areas, and the dynamics of the two populations can be written as

$$
\begin{aligned}
N_{t+1,1} &= N_{t,1} - N_{t,1}(1 - s_{t,1}) - N_{t,1}U_{t,1} + R_{t,1} - N_{t,1}p_{12} + N_{t,2}p_{21} \\
N_{t+1,2} &= N_{t,2} - N_{t,2}(1 - s_{t,2}) - N_{t,2}U_{t,2} + R_{t,2} - N_{t,2}p_{21} + N_{t,1}p_{12}
\end{aligned}
\tag{5.5.1}
$$

where N is the number of fish alive, s is a survival rate, U is a harvest rate, and p is a probability of moving from one area to the other. The subscripts denote time and area. This model was first proposed (in continuous form) by Beverton and Holt (1957) and used for analysis of movement patterns by Sibert (1984). Even though this is a simple discrete time two area model it is trivially extended to a continuous model, or one with any number of areas.

If we now consider the dynamics of a group of tags released in each area at some time, then the number of tags still present in each area will be

$$
\begin{aligned}
N_{t+1,1,1} &= N_{t,1,1} - N_{t,1,1}(1 - s_{t,1}) - N_{t,1,1}U_{t,1} - N_{t,1,1}p_{12} + N_{t,1,2}p_{21} \\
N_{t+1,1,2} &= N_{t,1,2} - N_{t,1,2}(1 - s_{t,2}) - N_{t,1,2}U_{t,2} - N_{t,1,2}p_{21} + N_{t,1,2}p_{12} \\
N_{t+1,2,2} &= N_{t,2,2} - N_{t,2,2}(1 - s_{t,2}) - N_{t,2,2}U_{t,2} - N_{t,2,2}p_{21} + N_{t,2,1}p_{12} \\
N_{t+1,2,1} &= N_{t,2,1} - N_{t,2,1}(1 - s_{t,1}) - N_{t,2,1}U_{t,1} - N_{t,2,1}p_{12} + N_{t,2,2}p_{21}
\end{aligned}
\tag{5.5.2}
$$

Now the subscripts for $N_{t,i,j}$ denote number of tagged fish at time *t* released in area *i*, currently present in area *j*. Note that we ignore tag shedding and tagging mortality, although these can be added quite easily (though they are usually not estimable). If we consider that the number of tags recovered is the number present in an area, times the fishing mortality rate, times the tag return rate, then

$$X_{t,i,j} = N_{t,i,j} U_{t,j} \beta \tag{5.5.3}$$

where X is the number of tags recovered and β is the tag return rate. If we know how many tags we released in each area, and when they were released, and the U's, s's, p_{12}, p_{21} and β are either known or are parameters to be estimated, then we can find the value of these parameters that give us the best fit between the observed recoveries, and those predicted by simulation using the above equations.

Obviously we cannot possibly estimate all of these parameters, particularly we would never be able to estimate time varying U's, s's and β from a single release in each area. In fact, we doubt that any tagging study would let you estimate them. They either need to be estimated in separate experiments, or we have to assume that they are constant over some broad stretches of time, and perhaps space. For instance, Hilborn (1990) used this type of model to examine tuna movements among 7 areas. He assumed that the s's were different by area, but constant over time, that the β's were all 1, and that the U's were either proportional to fishing effort in each area, or that the U was constant over time in each area. The need for this type of auxiliary information or assumption is found in all tagging studies.

However, once you have decided what parameters you really are going to try to estimate, and for movement studies the p's are the key ones, you must then specify some minimization criterion to decide which combination of parameters provides the best fit between the observed recoveries and the predicted ones. Sibert (1984) used simple least squares; Hilborn (1990) used both a poisson and multinomial likelihood function.

The implementation details of using a poisson and multinomial likelihood are given in the chapter on parameter estimation. The main purpose of giving this model here is to show how simply a quantitative movement model can be formulated.

5.7 Summary and critique

A very high proportion of fishery stock assessments have been based either on analysis of catch and effort data using biomass dynamics models, catch data using VPA, or simple yield-per-recruit analysis using growth and vulnerability information. The more recent trends in stock assessment are to

(1) consider in detail how the structure of the data collection system affects the reliability of the traditional methods and (2) extend the range of information using surveys, environmental variables, and so on. Both of these are very positive trends; we regard it as especially valuable to try to use totally different methods for determining stock trends. Therefore, comparing abundance surveys to abundance trends indicated by catch data is now commonly done, often with great discrepancies found between the two approaches.

There are no firm rules about what methods will or won't work for you. Probably the most important rule is *know your data*. It is a very long way from the concept of an experimental design to the final stages of analysis. The second general rule when considering abundance is *spatial mapping*. If you don't do it, you have only yourself to blame when things go wrong!

Bibliographic Notes

One of the best general readings for the problems of measuring fish statistics, particularly in low-budget situations, is Caddy and Bazigos (1985). Cochran (1977) is the standard and most useful reference on sampling; Krebs (1989) provides much of the same material from an ecological perspective, as does Green (1979). Doubleday and Rivard (1981) is the best summary of trawl techniques, and Johannesson and Mitson (1983) is a good summary of acoustic methods. Seber (1982) and Burnham et al. (1987) are the two best references on abundance estimation, particularly for mark recapture. Clark (1979) provides an overview of methods for spatial mapping. Altukhov and Salmenkova (1981) is an example of how genetic studies have been used to define stock identity. Beamish and McFarlane (1983) reviewed the need for age validation.

6

Relating Models to Data: Parameter Estimation

6.1. Introduction

In previous chapters, we have considered a number of models of how fish and fishermen behave and discussed the kinds of data that can be obtained. In a sense, this represents two aspects of fisheries research; theoreticians sit in offices making simple abstract models about how fisheries may or should work, and field staff actually go out on fishing boats, count fish, and collect data. It is often the case that these two groups do not communicate. Many theoreticians are not terribly interested in data, and many field staff are interested only in what they can observe, not abstractions of reality.

Stock assessment brings these two approaches together. We cannot make predictions of how fisheries will respond to alternative management actions without models, and these models must, of necessity, be simplified abstractions of reality. To be useful, the models must mimic the real world to at least some degree of precision. The business of finding a useful model involves two steps: (1) selection of a model structure that appears most appropriate to the problem at hand, and (2) selection of appropriate parameters for this model. The distinction between model choice and parameter estimation is a fuzzy one. Most individual models can be considered as special cases of a more general model with parameters to determine which special case is operating. Therefore, for most of this chapter we will consider model choice synonymous with parameter estimation.

In our discussion of fisheries models in earlier chapters, we often used the term parameter in reference to survival rates, growth rates, catchability coefficients, and so on. More precisely, a parameter is defined as a quantitative property (of a system) that is assumed to remain constant over some defined time span of historical data and future prediction. Typical quantitative properties of fish populations that we usually treat as parameters include natural survival rates, growth rates, and fecundity at age.

A very large proportion of stock assessment work is parameter estimation, and a general understanding of the basic principles and techniques involved is essential for anyone working in stock assessment. There are three essential requirements for parameter estimation:

1. A formal model with parameters to be estimated
2. Data from a population to use to estimate the parameters
3. A criterion to judge the goodness of fit to the data of any particular combination of model and parameter estimates.

Parameter estimation involves finding parameter values that provide the best fit between the model and the available data, according to the criterion. Be careful that you understand the limits of any such activity: finding parameter estimates that fit the data well does not imply in any way that the model will make correct predictions, or that the best-fit estimates are correct. Many estimates may give equally good fits, yet make wildly different predictions; worse, good fits may be obtained with estimates that are grossly incorrect, due to peculiarities in the historical data and/or model structure.

There are two widely used criteria for goodness of fit. The most commonly used is least squares, in which the parameter estimates are those that minimize the sum of squared differences between the predicted observations from the model and parameters, and the observed data. The sum of squares can be written as

$$SS = \Sigma\,(\text{observed} - \text{predicted})^2 \tag{6.1.1}$$

When the observed values involve a collection of different types of data measured in different units and with different variances, each element in the above sum must be divided by an estimate of the relative variance of that observed value; this is called a *weighted least squares* criterion. Most statistical methods rely on the sum of squares criterion.

An alternative to least squares is maximum likelihood. Maximum likelihood chooses the values of the parameters that maximize the probability that the actual observations would have occurred if the parameters were true. (More precisely: define the model so that it specifies probabilities for observations as a function of parameter values; take the maximum-likelihood parameter estimates to be those values for which the probability of the actual observations is highest.) For many estimation problems, particularly those that are some variant of linear regression, the least squares and maximum-likelihood estimates will be the same.

In this chapter, we provide a review of basic techniques associated with parameter estimation. In the early sections, we will assume almost no background in mathematics and statistics, so many readers will find these sections rather straightforward. However, the later sections will require better training in mathematics, and some knowledge of linear algebra will be most helpful. We provide a review of regular linear models as applied to fisheries problems, and then move on to maximum-likelihood and Bayesian estimation. Along the way we will touch upon recursive formulae which are very helpful for rapid computation of time-series regressions, and introduce the general linear model. We discuss bias problems that are particularly troublesome when fitting models to time-series data. Finally, we deal with the bootstrap method of calculating confidence intervals.

Our goal in this chapter is quite ambitious; we cover multiple regression, generalized linear models, maximum likelihood, Bayesian estimation, and nonlinear search techniques. There are complete text books and regular university courses devoted solely to these topics, and to treat them in a few pages is rather presumptuous. However, we expect that most readers will have a good basic understanding of statistics, certainly of multiple regression and analysis of variance. The sections on multiple regression and generalized linear models should not broach any totally new concepts. Given a basic understanding of statistics and probability, the concepts of nonlinear search, Bayes theorem, and maximum likelihood are rather simple to grasp at an elementary level. Indeed, they are so intuitive that many readers may not need to use the cited reference books. However, anyone desiring a more than superficial understanding of these techniques will need to examine the references.

In the earlier sections of this chapter, we provide complete examples, which in some cases overlap with material in the following chapters on specific techniques. For a number of more advanced methods, such as generalized linear models, the details of the examples are given in the chapters that use the methods, rather than repeating the material in this chapter as well. For many of the methods, this chapter may be sufficient for students who might occasionally need to refer to the formulas. Our general hope is that we provide an overview of the techniques, an appreciation of the general approach, and an understanding of how to tell if your parameter estimation is likely to be reliable.

6.2. Graphic Grid Search Methods

Suppose that you have any arbitrary procedure or model for generating a collection of predictions given a set of parameter values and that there are observed values for comparison to some of the predictions. The procedure may just be a simple equation or may be a complex computer algorithm. The only necessary feature of the procedure is that it associate (or generate) a set of predictions for each value assigned to its parameters. All of the models discussed in Chapter 4 generate such predictions.

When the number of parameters is small (one or two are most convenient), we can simply graph the goodness-of-fit criterion against the parameter val-

Table 6.1. Fraser River chum stock-recruitment data.

Brood	Spawners	Recruits
1961	164	236
1962	180	468
1963	325	1293
1964	185	579
1965	430	925
1966	212	325
1967	822	1933
1968	390	1434
1969	303	534
1970	356	361
1971	579	1239
1972	453	653
1973	565	1210
1974	235	386

Data from P. Starr (personal communication).

ues. That is, generate the predictions for a range of parameter values, compute the goodness-of-fit criterion for each parameter value, and then plot goodness of fit versus parameter value. This becomes cumbersome with more than two parameters. In the next few sections, there are numerous illustrations where goodness of fit is graphed against the parameters (see Figures 6.2 and 6.3 for instance). In general, it is extremely good practice to *always plot goodness of fit against parameters* when the number of parameters is small enough.

An extension of the graphic method is simple grid search, where you search over a range of possible parameter values (even in three or more dimensions) to find the best parameter values. This remains practical for three and four parameters, but gets rather difficult beyond that. Most parameter estimation is mathematically more sophisticated than graphic and grid searches, simply because some attempt is made to use particular characteristics of the model structure to find the best estimates analytically or with a minimum number of search steps.

6.3. Linear Least Squares: One Parameter

We begin with the simplest problem in estimation, that of a single parameter using least squares. Table 6.1 gives data on spawning stocks and subsequent recruitments for chum salmon (*Oncorhynchus keta*) on the Fraser River in British Columbia. Figure 6.1 shows a plot of recruits versus spawners for these data. This plot is suggestive of recruitment overfishing; recruitment appears to be roughly proportional to spawners, so more spawners

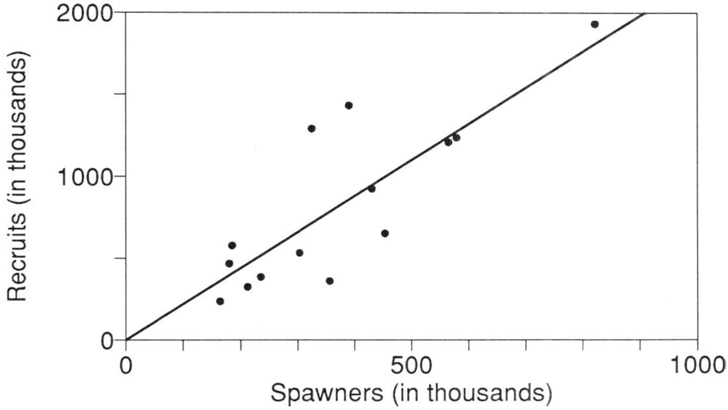

Figure 6.1. Spawner recruit data for Fraser River chum salmon.

should result in more recruits available to harvest. Suppose we assume a simple linear stock-recruitment relationship:

$$\text{recruits} = b \times \text{spawners}$$

or

$$R = bS \tag{6.3.1}$$

where b is recruits per spawner. To estimate b we can do a simple regression constrained to pass through the origin, with the slope (b) equal to the recruits per spawner.

In the context of least squares, our model can be written

$$\hat{R}_i = bS_i \tag{6.3.2}$$

where $\hat{R}_i$ is the predicted recruits in the ith brood year, S_i is the observed spawners in the ith brood year, and b is a parameter to be estimated. We wish to find a value of b that minimizes the sum of squares

$$\Sigma\,(R_i - \hat{R}_i)^2 \tag{6.3.3}$$

If we write the model as

$$R_i = bS_i + \epsilon \tag{6.3.4}$$

where R_i is the actual number of recruits in brood year i, then ϵ is the unexplained prediction error (residual). We can see that

$$\epsilon = R_i - \hat{R}_i \tag{6.3.5}$$

and the least squares parameter estimates are those which will minimize $\Sigma\epsilon^2$.

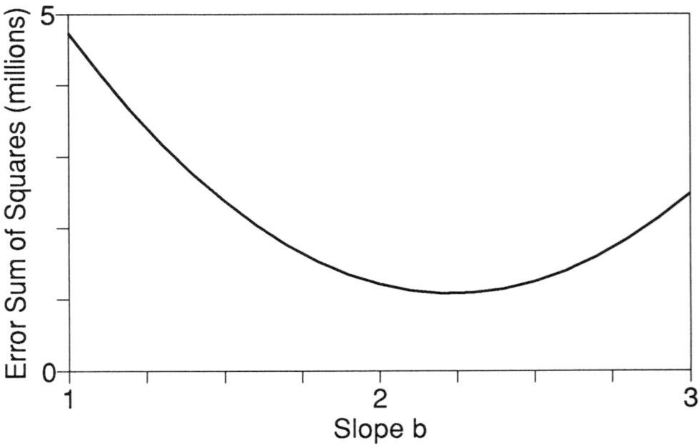

Figure 6.2. Sum of squares plotted against slope of the stock recruitment line for Fraser River chum salmon.

Figure 6.2 shows the relationship between $\Sigma\epsilon^2$ and b. Note that there is a minimum value at approximately $b = 2.2$. This value of b is the slope that is represented by the thin line in Figure 6.1. We found the estimate of $b = 2.2$ by trying different values of b between 1 and 3 in steps of 0.1. This is an example of simple graphical search.

For relationships of parameters and data that are linear or generally of the form

$$Y = bX + \epsilon \tag{6.3.6}$$

the value of b that minimizes the sum of squared residuals is

$$\hat{b} = \frac{\Sigma XY}{\Sigma X^2} \tag{6.3.7}$$

where $\hat{b}$ is an estimate of the true population parameter b. Equation 6.3.7 is merely an easy way to find the value of b that fits best by the sum of squares criterion. Although it is quite easy to minimize the $\Sigma\epsilon^2$ by searching as we did in Figure 6.2, it is much more difficult to do so in estimation problems where we are trying to estimate three or more parameters. In catch-at-age analysis, for instance, we might attempt to estimate 20 or 30 parameters, and a direct search in 20 or 30 dimensions is impossible.

Estimation methods really consist of two parts: (1) computational methods to find the parameter estimates and (2) methods to assess confidence in the parameter estimates. We obtain an estimate of $\hat{b} = 2.27$ from the data in Table 6.1. However, inspection of Figure 6.2 shows that almost the same sum of squares could be obtained from a value of $b = 2.0$ or $b = 2.4$. Thus if someone asserted that the real value of b was 2.0, we should not be too strong in our conviction that it was 2.27. It is, of course, possible that the real (long-term average) value is 3.0 or 1.0, and we were unlucky to be working with a very unrepresentative sample for some reason.

To calculate our confidence about b, as opposed to our estimate $\hat{b}$, we can make use of information provided by the data about the parameter σ_ϵ^2, the variance of residuals. If ϵ has a high variance, then there is a lot of variability in the data not explainable by the relationship between R and S, and we must admit a higher risk that the data are not representative of long term average behavior for the population. The estimate of σ_ϵ^2 is generally called s_ϵ^2 and is calculated as

$$s_\epsilon^2 = \frac{\Sigma\, \epsilon^2}{n - 1} \tag{6.3.8}$$

where n is the number of data points. From Table 6.2 $s_\epsilon^2 = 83271$ and the standard deviation s_ϵ is 288. Table 6.2 shows all the calculations for the data in Table 6.1.

Uncertainty about the estimated slope $\hat{b}$ can be summarized in terms of the variance of the estimate, $\sigma_{\hat{b}}^2$. This variance is estimated by $(s_{\hat{b}}^2)$, which is calculated as

$$s_{\hat{b}}^2 = \frac{s_\epsilon^2}{\Sigma\, X^2} \tag{6.3.9}$$

which is $83271/2390219 = 0.0348$. $s_{\hat{b}}$ is the standard deviation of $\hat{b}$, and is 0.19 in this case.

Confidence limits for the real slope b are given as

$$\hat{b} - s_{\hat{b}} t_{n-1} < b < \hat{b} + s_{\hat{b}} t_{n-1} \tag{6.3.10}$$

where t_{n-1} is the t distribution with $n - 1$ degrees of freedom. $t \approx 2$ for large n and 95% confidence, thus $1.86 < b < 2.61$ with 95% confidence.

6.4. Linear Least Squares: Two Parameters

Just as there was a general form for the estimation of one parameter with the relationship $Y = bX + \epsilon$, there is a comparable form when we have two parameters b_0 and b_1 to estimate in the linear model:

$$Y = b_0 + b_1 X + \epsilon, \tag{6.4.1}$$

which may look more familiar in the form

Table 6.2. Calculations for Fraser River chum data.

XY	X^2	$\hat{Y}$	ϵ	ϵ^2
38,704	26,896	367	−131	17,056
84,240	32,400	402	66	4,308
420,225	105,625	726	567	320,933
107,115	34,225	414	165	27,377
397,750	184,900	961	−36	1,311
68,900	44,944	474	−149	22,170
1,588,926	675,684	1,837	96	9,128
559,260	152,100	872	562	316,082
161,802	91,809	677	−143	20,538
128,516	126,736	796	−435	189,039
717,381	335,241	1,294	−55	3,055
295,809	205,209	1,013	−360	129,324
683,650	319,225	1,263	−53	2,806
90,710	55,225	525	−139	19,407

$$\Sigma X = 51{,}99 \qquad \Sigma Y = 11{,}576 \qquad \Sigma XY = 5{,}342{,}988 \qquad \Sigma X^2 = 2{,}390{,}219 \qquad \Sigma \epsilon^2 = 1{,}082{,}533$$

$$s_e^2 = 83{,}271.77 \qquad s_\epsilon = 288.57 \qquad \hat{b} = 2.24 \qquad s_{\hat{b}}^2 = 0.03484 \qquad s_{\hat{b}} = 0.19$$

$$\hat{b} - s_{\hat{b}} t_{n-1} = 1.86 \qquad \hat{b} + s_{\hat{b}} t_{n-1} = 2.61$$

$$y = a + bx$$

or

$$y = mx + b \tag{6.4.2}$$

and is, of course, the normal linear regression equation for a straight line with a slope of b_1 and an intercept of b_0. We want to find the values of b_0 and b_1 that minimize the sum of squares $\Sigma(Y - \hat{Y})^2$.

For example, if we wish to estimate the parameters of the Ricker stock-recruitment function

$$R = Se^{a-bS}, \tag{6.4.3}$$

the estimation equation is normally obtained by log transforming the model to

$$\log\left(\frac{R}{S}\right) = a - bS, \tag{6.4.4}$$

which is of the same form as Equation 6.4.1 with $\log\left(\frac{R}{S}\right) = Y$, $a = b_0$, $b = b_1$, and $X = -S$.

Figure 6.3 shows the relationship between the sum of squares and the

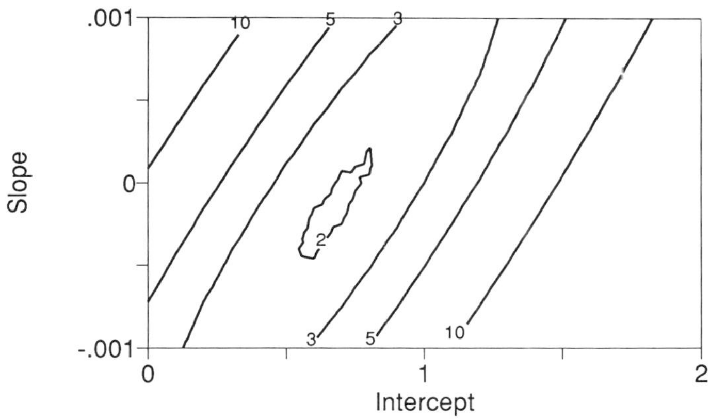

Figure 6.3. Sum of squares for two parameter model of Fraser River chum salmon.

parameter estimates for the Fraser River chum data fitted to the two-parameter Ricker model. Plotted on the x axis is the value of b_0 and plotted on the y axis is the value of b_1. The minimum value of the sum of squares (1.9) is reached at $b_0 = 0.7$ and $b_1 = 0.0002$. The resolution of the contours drawn in Figure 6.3 is insufficient to tell exactly where the minimum is, but we can see a well defined bottom of the bowl.

There is a diagonal valley in Figure 6.3, which shows that a low sum of squares can be achieved by many combinations of the parameters. A combination of 0.8 and -0.002 yields approximately the same sum of squares as parameter values 0.5 and 0.0005. This means that there is a "correlation" between the estimates of b_0 and b_1, and we cannot distinguish very well between combinations of these parameters. This problem of correlation is discussed in detail later.

The graphical method of Figure 6.3 is fine for visual inspection to see which parameter combinations give good fits, but, in the linear case, a formula can be used to find the best parameter estimates. The standard formulae for two variable linear least squares regression are

$$\begin{aligned} \hat{b}_1 &= \frac{\Sigma\, xy}{\Sigma\, x^2} \\ \hat{b}_0 &= \bar{Y} - \hat{b}_1\bar{X} \\ \epsilon_i &= Y_i - (\hat{b}_0 + \hat{b}_1 X_i) \\ s_\epsilon^2 &= \frac{\Sigma\, \epsilon^2}{n - 2} \end{aligned} \tag{6.4.5}$$

where

$$\begin{aligned} x &= X - \bar{X} \\ y &= Y - \bar{Y}. \end{aligned}$$

Be careful about formulae like these for point estimation; they carry no information about what combinations of estimates might fit the data just as well. For the linear model, the length, width, and orientation of valleys like the one in Figure 6.3 can be found from the parameter variances (length and width) and correlation (orientation); these formulae are given in Equations 6.6.10, 6.6.11, and 6.7.1.

6.5. Linear Least Squares: Three Parameters

As you may have suspected by now, least squares can be used to estimate any number of parameters in linear models. Before proceeding to the general case, we will look at the three-parameter case separately because it is often used in biomass dynamics models. The general form is

$$Y = b_0 + b_1X_1 + b_2X_2 + \epsilon \tag{6.5.1}$$

where Y is the dependent variable, X_1 is one independent variable, X_2 is another independent variable, and b_0, b_1, and b_2 are parameters.

The formulae for estimates of b_0, b_1, and b_2 are

$$\begin{aligned}
\hat{b}_1 &= \frac{\Sigma\, x_2^2 \,\Sigma\, x_1y - \Sigma\, x_1x_2 \,\Sigma\, x_2y}{\Sigma\, x_1^2 \,\Sigma\, x_2^2 - (\Sigma\, x_1 \,\Sigma x_2)^2} \\
\hat{b}_2 &= \frac{\Sigma\, x_1^2 \,\Sigma\, x_2y - \Sigma\, x_1x_2 \,\Sigma\, x_1y}{\Sigma\, x_1^2 \;\Sigma\, x_2^2 - (\Sigma\, x_1x_2)^2} \\
\hat{b}_0 &= \bar{Y} - \hat{b}_1\bar{X}_1 - \hat{b}_2\bar{X}_2 \\
\epsilon_i &= Y_i - (\hat{b}_0 + \hat{b}_1X_{i,1} + \hat{b}_2X_{i,2}) \\
s_\epsilon^2 &= \frac{\Sigma\, \epsilon^2}{n - 3}
\end{aligned} \tag{6.5.2}$$

The computational formulae are

$$\begin{aligned}
\Sigma\, x_1^2 &= \Sigma\, X_1^2 - \frac{\Sigma\, X_1 \,\Sigma\, X_1}{n} \\
\Sigma\, x_2^2 &= \Sigma\, X_2^2 - \frac{\Sigma\, X_2 \,\Sigma\, X_2}{n} \\
\Sigma\, x_1y &= \Sigma\, X_1Y - \frac{\Sigma\, X_1 \,\Sigma\, Y}{n} \\
\Sigma\, x_2y &= \Sigma\, X_2\mathrm{Y} - \frac{\Sigma\, X_2 \,\Sigma\, Y}{n} \\
\Sigma\, x_1x_2 &= \Sigma X_1X_2 - \frac{\Sigma X_1 \,\Sigma X_2}{n}
\end{aligned} \tag{6.5.3}$$

6.6. Linear Least Squares: Any Number of Parameters

This section deals with general least squares estimation of any number of parameters in linear models. We use matrix notation because it is both pow-

Table 6.3. Hecate St. rock sole data set up for regression estimates of the Schaefer model parameters. $\Delta U = (U_{t+1}/U_t) - 1$.

Year	Landings	ΔU	=	$b_1 \times 1$	+	$b_2 \times$ cpue	+	$b_3 \times$	Effort
1954	835	−0.27	=	$b_1 \times 1$	+	$b_2 \times 1.01$	+	$b_3 \times$	826.73
1955	1324	−0.15	=	$b_1 \times 1$	+	$b_2 \times 0.74$	−	$b_3 \times$	1,789.19
1956	512	−0.17	=	$b_1 \times 1$	+	$b_2 \times 0.63$	−	$b_3 \times$	812.70
1957	211	1.04	=	$b_1 \times 1$	+	$b_2 \times 0.52$	−	$b_3 \times$	405.77
1958	797	−0.22	=	$b_1 \times 1$	+	$b_2 \times 1.06$	−	$b_3 \times$	751.89
1959	109	0.78	=	$b_1 \times 1$	+	$b_2 \times 0.83$	−	$b_3 \times$	131.33
1960	626	0.05	=	$b_1 \times 1$	+	$b_2 \times 1.48$	−	$b_3 \times$	422.97
1961	660	−0.24	=	$b_1 \times 1$	+	$b_2 \times 1.55$	−	$b_3 \times$	425.81
1962	503	−0.14	=	$b_1 \times 1$	+	$b_2 \times 1.18$	+	$b_3 \times$	426.27
1963	754	−0.25	=	$b_1 \times 1$	+	$b_2 \times 1.02$	+	$b_3 \times$	739.22
1964	484	1.00	=	$b_1 \times 1$	+	$b_2 \times 0.77$	+	$b_3 \times$	628.57
1965	318	−0.46	=	$b_1 \times 1$	+	$b_2 \times 1.54$	+	$b_3 \times$	206.49
1966	1468	0.02	=	$b_1 \times 1$	+	$b_2 \times 0.83$	+	$b_3 \times$	1,768.67
1967	1045	−0.47	=	$b_1 \times 1$	+	$b_2 \times 0.85$	+	$b_3 \times$	1,229.41
1968	1386	0.73	=	$b_1 \times 1$	+	$b_2 \times 0.45$	+	$b_3 \times$	3,080.00
1969	1230	−0.44	=	$b_1 \times 1$	+	$b_2 \times 0.78$	+	$b_3 \times$	1,576.92
1970	455	−0.11	=	$b_1 \times 1$	+	$b_2 \times 0.44$	+	$b_3 \times$	1,034.09
1971	895	−0.28	=	$b_1 \times 1$	+	$b_2 \times 0.39$	+	$b_3 \times$	2,294.87
1972	230	0.64	=	$b_1 \times 1$	+	$b_2 \times 0.28$	+	$b_3 \times$	821.43
1973	180	−0.30	=	$b_1 \times 1$	+	$b_2 \times 0.46$	+	$b_3 \times$	391.30
1974	220	1.00	=	$b_1 \times 1$	+	$b_2 \times 0.32$	+	$b_3 \times$	687.50
1975	562	−0.14	=	$b_1 \times 1$	+	$b_2 \times 0.64$	+	$b_3 \times$	878.13
1976	892	−0.35	=	$b_1 \times 1$	+	$b_2 \times 0.55$	+	$b_3 \times$	1,621.82
1977	427	0.22	=	$b_1 \times 1$	+	$b_2 \times 0.36$	+	$b_3 \times$	1,186.11
1978	394	−0.30	=	$b_1 \times 1$	+	$b_2 \times 0.44$	+	$b_3 \times$	895.41
1979	361	−0.23	=	$b_1 \times 1$	+	$b_2 \times 0.31$	+	$b_3 \times$	1,164.52

From Stocker (1981).

erful and concise. Readers unfamiliar with matrix algebra will have some rough going in this section, but should make the effort required to understand it because it is an invaluable tool for fisheries stock assessment. We recommend Searle (1966) for a good initiation to matrix algebra for biologists. The later section on nonlinear estimation problems uses similar notation, and readers will find the nonlinear section more straightforward if this section is mastered thoroughly.

The linear relationship between m parameters and a data set of n points consisting of a dependent variable Y and m independent variables can be written as

$$Y_i = b_1 X_{i,1} + b_2 X_{i,2} + \ldots + b_m X_{i,m} + \epsilon_i \qquad (6.6.1)$$

The relationship can be made to have an intercept (nonzero Y for zero values of all X's) by setting $X_{i,1} = 1$ for all i.

Table 6.3 shows how the North Hecate Strait rock sole (*Lepidopsetta bilineata*) data would look in this format when set up to estimate the discrete

Table 6.4. Hecate Strait rock sole data in matrix form.

$$
\begin{pmatrix} -0.27 \\ -0.15 \\ -0.17 \\ 1.04 \\ -0.22 \\ 0.78 \\ 0.05 \\ -0.24 \\ -0.14 \\ -0.25 \\ 1.00 \\ -0.46 \\ 0.02 \\ -0.47 \\ 0.73 \\ -0.44 \\ -0.11 \\ -0.28 \\ 0.64 \\ -0.30 \\ 1.00 \\ -0.14 \\ -0.35 \\ 0.22 \\ -0.30 \\ -0.23 \end{pmatrix}
=
\begin{pmatrix}
1 & 1.01 & 826.73 \\
1 & 0.74 & 1789.19 \\
1 & 0.63 & 812.70 \\
1 & 0.52 & 405.77 \\
1 & 1.06 & 751.89 \\
1 & 0.83 & 131.33 \\
1 & 1.48 & 422.97 \\
1 & 1.55 & 425.81 \\
1 & 1.18 & 426.27 \\
1 & 1.02 & 739.22 \\
1 & 0.77 & 628.57 \\
1 & 1.54 & 206.49 \\
1 & 0.83 & 1768.67 \\
1 & 0.85 & 1229.41 \\
1 & 0.45 & 3080.00 \\
1 & 0.78 & 1576.92 \\
1 & 0.44 & 1034.09 \\
1 & 0.39 & 2294.87 \\
1 & 0.28 & 821.43 \\
1 & 0.46 & 391.30 \\
1 & 0.32 & 687.50 \\
1 & 0.64 & 878.13 \\
1 & 0.55 & 1621.82 \\
1 & 0.36 & 1186.11 \\
1 & 0.44 & 895.41 \\
1 & 0.31 & 1164.52
\end{pmatrix}
\begin{pmatrix} b_1 \\ b_2 \\ b_3 \end{pmatrix}
+
\begin{pmatrix} \epsilon_1 \\ \vdots \\ \epsilon_{26} \end{pmatrix}
$$

Schaefer model parameters. Note that $X_1 = 1$ for all observations (so Equations 6.4.1 and 6.5.1 are special cases of 6.6.1 with $X_0 = 1$ for all observations). The discrete Schaefer model can be rewritten as a linear regression model:

$$\frac{U_{t+1}}{U_t} - 1 = r - \frac{r}{kq} U_t - qE_t \tag{6.6.2}$$

where U is CPUE. Equation 6.6.2 converts to the general linear form of Equation 6.6.1 where $b_1 = r$, $b_2 = -r/kq$, $b_3 = -q$, $Y = (U_{t+1}/U_t) - 1$, $X_1 = 1$, $X_2 = U_t$, and $X_3 = E_t$.

Note from Table 6.4 that we have a vector of Y's, one for each observation, a vector of b's (m of them), a vector of ϵ's (a residual for each observation) and a matrix of X's. Equation 6.5.1 can be rewritten in matrix notation as

$$\mathbf{Y} = \mathbf{Xb} + \epsilon \tag{6.6.3}$$

Table 6.4 shows how the North Hecate rock sole data would look in matrix form.

Equation 6.6.3 is a general form for any number of parameters and observations and has the very nice property of having a simple solution for the least squares parameter estimates that minimize $\Sigma\ \epsilon^2$:

$$\hat{\mathbf{b}} = (\mathbf{X}^T\mathbf{X})^{-1}\mathbf{X}^T\mathbf{Y} \tag{6.6.4}$$

We can make the computation of Equation 6.6.4 a little more understandable if we assume we have only one parameter to estimate and reanalyze the Fraser River chum data from Table 6.1. In matrix-vector notation the one parameter model is just

$$\mathbf{R} = \mathbf{S}b + \epsilon \tag{6.6.5}$$

We have one parameter to estimate, a vector of Ys (14 of them) and a vector (or a 1 by 14 matrix) of Xs. Equation 6.6.4 can be rewritten in this case as

$$\hat{b} = \left[(X_1 \quad X_2 \quad \ldots \quad X_{14}) \begin{pmatrix} X_1 \\ X_2 \\ \vdots \\ X_{14} \end{pmatrix} \right]^{-1} (X_1 \quad X_2 \quad \ldots \quad X_{14}) \begin{pmatrix} Y_1 \\ Y_2 \\ \vdots \\ Y_{14} \end{pmatrix} \tag{6.6.6}$$

If you remember your matrix multiplication, Equation 6.6.6 is equal to

$$\hat{b} = (\Sigma\ X^2)^{-1}\ \Sigma\ XY \tag{6.6.7a}$$

or

$$\hat{b} = \frac{\Sigma\ XY}{\Sigma\ X^2} \tag{6.6.7b}$$

which is the easily remembered equation for slope in a one-parameter regression.

Returning to the general case of Equation 6.6.5, the matrix $(\mathbf{X}^T\mathbf{X})$ is a square matrix with the number of rows and columns equal to the number of parameters and can be written as

$$\mathbf{X}^T\mathbf{X} = \begin{pmatrix} \Sigma\ X_{i1}^2 & \Sigma\ X_{i1}X_{i2} & \ldots & \Sigma\ X_{i1}X_{im} \\ \Sigma\ X_{i2}X_{i1} & \Sigma\ X_{i2}^2 & \ldots & \Sigma\ X_{i2}X_{im} \\ \vdots & \vdots & \ddots & \vdots \\ \Sigma\ X_{im}X_{i1} & \Sigma\ X_{i,m}X_{i2} & \ldots & \Sigma\ X_{im}^2 \end{pmatrix}. \tag{6.6.8}$$

The estimate for residual error variance is still the sum of squared residuals divided by the number of data points less the number of parameters estimated:

$$s_\epsilon^2 = \frac{\sum_{i=1}^{n} [Y_i - (\hat{b}_1 X_{i1} + \hat{b}_2 X_{i2} + \ldots + \hat{b}_m X_{im})]^2}{n - m} \tag{6.6.9}$$

Another useful statistic is the variance-covariance matrix (**A**) for $\hat{b}$ which is calculated as

$$\mathbf{A} = s_\epsilon^2 (\mathbf{X}^T\mathbf{X})^{-1} \tag{6.6.10}$$

which will have the form

$$\mathbf{A} = \begin{pmatrix} s_{b_1}^2 & s_{b_1 b_2} & \cdots & s_{b_1 b_m} \\ s_{b_2 b_1} & s_{b_2}^2 & \cdots & s_{b_2 b_m} \\ \vdots & \vdots & \ddots & \vdots \\ s_{b_m b_1} & s_{b_m b_2} & \cdots & s_{b_m}^2 \end{pmatrix} \tag{6.6.11}$$

The $s_{b_i}^2$ determine confidence limits for the individual b's, whereas the $s_{b_i b_j}$ define joint confidence limits. When a covariance element ($s_{b_i b_j}$) is large, the joint confidence region for b_i and b_j is an elongated ellipse.

6.7. Parameter Correlation and Confounding

A final very useful statistic for regression is the parameter correlation matrix **C** which says how correlated the parameters estimates are:

$$C_{ij} = \frac{A_{ij}}{(A_{ii} A_{jj})^{1/2}} \tag{6.7.1}$$

If the parameters are uncorrelated, the C's will be zero, but more commonly there will be correlation between parameters. If you look back at Figure 6.3, you will see that the sum of squares surface has a valley that runs from the lower left to upper right. This says that the parameters b_0 and b_1 are positively correlated.

Very strong parameter correlation is a pervasive problem in fisheries statistics. In catch-at-age analysis, we are almost always struggling to separate the natural mortality rate from fishing mortality. The two usually have very high (0.99 is not unusual) negative correlations. Parameter correlations arise from the fact that the independent variables are correlated with one another. When the independent variables are correlated it is not logically possible to tell with certainty which independent variable caused a change in the dependent variable. This is called *confounding* of effects.

Table 6.5 shows the variance-covariance matrix **A** and the parameter correlation matrix **C** for the Ricker model (Equation 6.4.4) fitted to the Fraser

Table 6.5. Regression statistics for Fraser River chum data fit to the Ricker model.

Parameter	Biological Meaning	Estimate	Standard Deviation
b_0	Ricker a	0.66	0.24
b_1	Ricker b	0.0017	0.0059

Variance-Covariance Matrix **A**

$$\begin{pmatrix} 0.0596 & -1.2 \times 10^{-4} \\ -1.2 \times 10^{-4} & 3.49 \times 10^{-7} \end{pmatrix}$$

Parameter Correlation Matrix **C**

$$\begin{pmatrix} 1 & -0.90 \\ -0.90 & 1 \end{pmatrix}$$

River chum data. Also shown are the standard deviations and the values of the parameters. The ratio of the standard deviation to the parameter estimate for the Ricker a value is 0.36, whereas the ratio for the Ricker b value is 3.4. Obviously the confidence about the a value is much higher than about the b value. Referring again to Figure 6.3, it should be clear why. The bottom of the sum of squares valley spans a reasonably small range of Ricker a values but a large range of b values. There is a strong negative correlation between the parameters (−0.90), which is represented in Figure 6.3 by the valley being long and narrow.

Table 6.6 shows the regression statistics for the North Hecate rock sole data. If you remember, these data showed quite good contrast in abundance and effort. As can be seen from Table 6.4, the parameter estimates are all reasonably good; in no case is the standard deviation greater than the parameter estimate. The variance-covariance matrix is always difficult to interpret, and it is easier to look at the correlation matrix **C**. The worst correlation is between the first two parameters, r and $(-r/kq)$. If we remember that (r/kq) is the rate at which growth of the population slows with increasing density, then the strong negative correlation says that the stock could have a lower intrinsic rate of increase (r) and less density dependent reduction in growth. Effectively this means that r could be smaller and k larger or r larger and k smaller. The negative correlation between the first and third parameters says that r could be smaller and $-q$ larger. The positive correlation of 0.42 between the second two parameters says that if $(-r/kq)$ is larger, then $-q$ can also be larger.

6.8. General Linear Models

The most powerful and general application of linear models is the general linear model (GLM), which is finding wider and wider use in fisheries sta-

Table 6.6. Regression statistics for North Hecate Strait rock sole data fit to a discrete logistic model.

Parameter	Biological Meaning	Estimate	Standard Deviation
b_1	r	0.63	0.31
b_2	$-r/kq$	0.55	0.27
b_3	$-q$	−0.00018	0.0015

Variance-Covariance Matrix **A**

$$\begin{pmatrix} 0.099 & -0.073 & -3.6 \times 10^{-5} \\ -0.073 & 0.074 & 1.7 \times 10^{-5} \\ -3.6 \times 10^{-5} & 1.7 \times 10^{-5} & 2.3 \times 10^{-8} \end{pmatrix}$$

Parameter Correlation Matrix **C**

$$\begin{pmatrix} 1 & -0.85 & -0.75 \\ -0.85 & 1 & 0.42 \\ -0.75 & 0.42 & 1 \end{pmatrix}$$

tistics. The basic GLM can be stated just as we did in equation 6.6.1 and 6.6.3:

$$Y_j = p_1 X_{i,1} + p_2 X_{i,2} + \ldots + p_n X_{i,n} + \epsilon_j \tag{6.8.1}$$

where Y_i is an observation, the p's are parameters to be estimated, the X's are independent variables, and ϵ is the error. To an inexperienced observer this might look an awful lot like the regular multiple regression discussed earlier, but the difference is that the X's may be not only continuous variables, but discrete (0,1) treatment effects also. A linear model where all of the $X_{i,j}$ take only 1 or 0 values (turning on or off the effects of p_j in each observation i) is called an *experimental design model*.

In this book we include two examples of how GLM are used. In catch-at-age analysis (Chapter 11) we use a GLM to estimate initial cohort sizes, natural mortality, and fishing mortality, where some of the X's are treatment factors (0,1) representing the recruitments of successive cohorts and other X's are years of natural mortality and cumulative fishing mortality experienced by the cohort.

The second example is the estimation of standardized catch rates (Chapter 4) where we use a GLM to estimate time trends in CPUE, correcting for differences in catching power of the vessels and spatial variation in vulnerability. Each of these two examples requires a detailed understanding of the particular problem, and we prefer to have the examples in the text chapters rather than in this estimation chapter.

GLM's also include other forms of linear models such as contingency tables and probit analysis, but these are not frequently used in analyzing fisheries statistics.

6.9. Recursive Least Squares

We often wish to follow or predict the evolution of parameter estimates over time as more data are collected. We could do this by using Equation 6.6.4 and recalculating as each point is collected. A computationally simpler and faster alternative is recursive estimation, which does not require reprocessing all the data each time a new data point is collected.

The recursive method relies on having at any time a vector of the current parameter estimates $\hat{\mathbf{b}}_t$, and a variance-covariance matrix of the parameter estimates $\mathbf{P}_t$. As each new data point $(Y, X_1, \ldots, X_m)$ is collected, $\hat{\boldsymbol{b}}$ and $\mathbf{P}$ are updated by the following equation:

$$\begin{aligned}\mathbf{P}_t &= \mathbf{P}_{t-1} - \mathbf{P}_{t-1}\mathbf{X}_t(\sigma_\epsilon^2 + \mathbf{X}_t^T\mathbf{P}_{t-1}\mathbf{X}_t)^{-1}\mathbf{X}_t^T\mathbf{P}_{t-1} \\ \hat{\mathbf{b}}_t &= \hat{\mathbf{b}}_{t-1} - \frac{1}{\sigma_\epsilon^2}\mathbf{P}_t(\mathbf{X}_t\mathbf{X}_t^T\mathbf{b}_{t-1} - \mathbf{X}_tY_t)\end{aligned} \qquad (6.9.1)$$

where $\mathbf{X}_t$ is a vector of the independent variables of the estimation equation at time t, Y_t is the observation at time t, and σ_ϵ^2 is the residual error variance. Close scrutiny of Equation 6.9.1 will reveal some properties that are obvious and some that are surprising.

If we first look at the equation for updating $\mathbf{P}$, we see that $\mathbf{P}$ must decrease (or remain the same) as more data are collected, independent of the actually observed Ys. Even though this may initially seem surprising, think back and look at Section 6.5, where we found that the variance-covariance matrix in a regular regression does not depend upon the Ys, but only on the $\mathbf{X}$s. Also note that σ_ϵ^2 appears in Equation 6.9.1, so that if the data start diverging from our prediction, this will inflate our estimate of σ_ϵ^2 and thus increase our estimates of the variance-covariance elements. Since σ_ϵ^2 is unknown, we use the deviations between our predictions and observations as an estimate of σ_ϵ^2.

6.10. Nonlinear Estimation

All the problems presented so far have involved linear estimation, that is, they can all be reduced to an equation with the parameters appearing on the right-hand side in an additive form with each parameter multiplying an independent variable that does not itself depend on the parameters. Linear parameter estimation problems are used for the Ricker stock-recruitment model, both the discrete and Schnute's form of the Schaefer model, and Palohei-

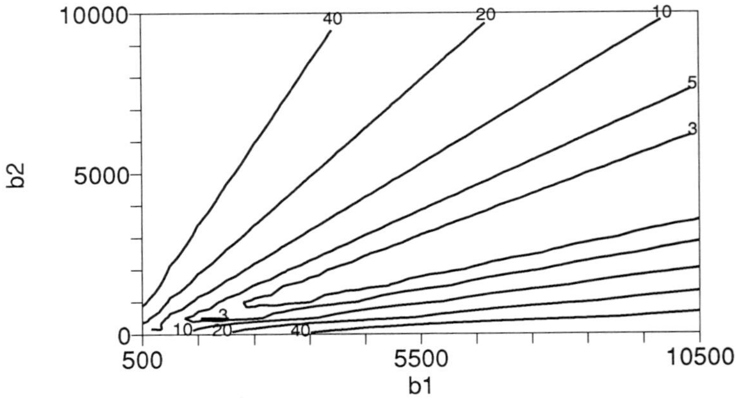

Figure 6.4. The sum of squares for different parameter combinations of the Beverton-Holt model using Fraser River chum salmon data.

mo's method for catch-at-age analysis. Nonlinear techniques are used for Beverton-Holt's stock-recruitment curve, the Deriso delay difference model, and Doubleday's catch-at-age analysis method.

The Beverton-Holt stock-recruitment model can be written as

$$R = \frac{b_1 S}{b_2 + S} \tag{6.10.1}$$

Note that b_1 and b_2 do not separate into a nice linear form. In the chapter on stock and recruitment, we show how Equation 6.10.1 can be transformed into a linear form, but we are unhappy with the error assumptions involved in that transformation. We would prefer to write the Beverton-Holt equation as

$$\log\left(\frac{R}{S}\right) = \log\left(\frac{b_1}{b_2 + S}\right) + \epsilon \tag{6.10.2}$$

where ϵ is a normally distributed residual. This gives the lognormal distribution of errors about the average stock-recruitment relationship that we find desirable in stock-recruitment analysis.

Parameter estimation for Equation 6.10.2 is a typical nonlinear problem. We would like to find the values of b_1 and b_2 that minimize the sum of squares between the predicted log (R/S) and the observed log (R/S). We can do this graphically just as we did in Figure 6.2. Figure 6.4 shows the sum of squares for the Beverton-Holt parameters for Fraser River chum salmon. We can find the minimum, but in this case, there is an incredible amount of parameter correlation, and a wide range of parameters give equally low sum of squares.

For one or two parameters, a graphical method can be very useful, but for larger problems (Deriso's or Doubleday's models) you must have a more efficient numerical method of finding the best fitting parameter estimates. Unfortunately, there is no direct solution for nonlinear systems as there is for linear ones. The reason for this is that in linear systems the derivative of the Y with respect to the parameter estimates is independent of the parameter estimates. The derivatives are in fact the X's (see Equation 6.10.5 below). This means that the sum of squares is quadratic (a parabola) in each parameter. If you look back at Figure 6.2 you can see the quadratic shape. The bottom of the parabola can be found quite easily; Equation 6.6.4 is the easy solution for the general case.

Nonlinear problems do not have a quadratic sum of squares surface. Notice how in Figure 6.4 the contours of equal fit are not just simple ellipses. This means that we must use an iterative numerical search scheme to find the minimum. There are many different methods used in such searches; the basic reference is Bard (1974).

The easiest method to understand is called Newton's method, and has the advantage of being closely related to the least squares method. The key to Newton's method is what is called the "Jacobian matrix" **J**, the elements of which are defined by

$$J_{ij} = \frac{\partial \hat{Y}_i}{\partial b_j} \tag{6.10.3}$$

In other words, **J** is the sensitivity of the predicted observations to the parameter estimates. **J** is analogous to the **X** matrix of independent variables in linear regression because for a system of linear equations where

$$\hat{Y}_i = \hat{b}_1 X_{i1} + \hat{b}_2 X_{i2} + \ldots + \hat{b}_m X_{im} \tag{6.10.4}$$

the sensitivities of $\hat{Y}_i$ to b_j for linear regressions are just

$$\frac{\partial \hat{Y}_i}{\partial b_j} = X_{ij} \tag{6.10.5}$$

For nonlinear systems, the derivatives needed for Equation 6.10.3 may sometimes be obtained analytically, but more often are easiest to obtain numerically. In a nonlinear model, the derivative estimates will depend on the parameter values as well as the independent variables. The derivatives for the Beverton-Holt equation (6.10.2) are

$$\frac{\partial \hat{Y}_i}{\partial b_1} = \frac{1}{b_1}$$
$$\frac{\partial \hat{Y}_i}{\partial b_2} = -\frac{1}{b_2 + S_i} \tag{6.10.6}$$

Therefore the Jacobian matrix (**J**) for the Beverton-Holt stock recruitment model would be

$$\begin{pmatrix} \frac{1}{b_1} & -\frac{1}{b_2 + S_1} \\ \frac{1}{b_1} & -\frac{1}{b_2 + S_2} \\ \vdots & \vdots \\ \frac{1}{b_1} & -\frac{1}{b_2 + S_n} \end{pmatrix} \tag{6.10.7}$$

where S_i is the spawning stock in year i. To estimate the parameter vector b, we need a starting estimate. We can rarely just set all the b's equal to zero and start from there. More often, we need to start with "reasonable" guesses and much of the art (as opposed to the science) of nonlinear estimation is figuring out good starting guesses. In some cases, you can get starting estimates from linear approximations to the model; in other cases, you simply have to try different starting values until you find ones that lead to convergence. When the sum of squares surface is complex, with false minima, you may need to try several starting b's to make sure they converge to the right place. Once you have guessed at an initial **b**, Newton's method is to iteratively update it by the following equation:

$$\hat{\mathbf{b}}_{\text{new}} = \hat{\mathbf{b}}_{\text{old}} + \lambda(\mathbf{J}^T\mathbf{J})^{-1}\mathbf{J}^T\mathbf{v} \tag{6.10.8}$$

where $\mathbf{b}_{\text{new}}$ is the next estimate of the true **b**, λ is a parameter usually set to between .1 and 1.0, **J** is the Jacobian matrix and **v** is the vector of residuals produced by $\mathbf{b}_{\text{old}}$. If $\hat{\mathbf{b}}_{\text{old}}$ are all zeros and $\lambda = 1.0$, then equation 6.10.8 is identical to Equation 6.6.4, since **J** is equivalent to **X**, and $\mathbf{v} = \mathbf{Y}$ if $\hat{\mathbf{b}}_{\text{old}} = 0$.

Newton's method assumes that the sums of squares surface is roughly quadratic, and when this assumption is acceptable the method iteratively approaches the global minimum. The parameter λ is usually set less than 1.0 to help avoid overshooting the minimum. There are many alternative nonlinear search procedures, and Newton's method is actually one of the most primitive. One of the most widely used methods is called Marquardt's algorithm. Most computer installations have different algorithms available.

The book *Numerical Recipes* by Press et al (1986) gives computer code for several nonlinear search algorithms, and the computer disk accompanying this book also contains a nonlinear search program.

There are two major problems with nonlinear estimation. First, it can be slow. When more than two or three parameters are being estimated, it can take dozens or even hundreds of evaluations of the model. In extreme cases (20 parameters, 200 observations), an estimation may take several hours of computer time on a minicomputer. The second problem is false minima. Under some circumstances, there may be local minima where the estimation can become stuck because any small movement in any direction leads to a poorer fit. We have personally not encountered this problem very often in production models, stock and recruitment, or fitting of tagging models, but it is always a danger. You should always start the estimation with different initial guesses to see if you obtain convergence at the same values.

6.11. Maximum Likelihood

Maximum likelihood is an alternative criterion for estimating parameters, and, although it often appears to be quite difficult, the basic concepts are straightforward and can be used by anyone with an elementary understanding of probability. In general, maximum-likelihood estimators are to be preferred over least squares because maximum likelihood is based on careful consideration of how random "errors" arise and are distributed, whereas least-squares estimators for most nonlinear problems are essentially *ad hoc*. Most serious work in fisheries parameter estimation is now being done by maximum-likelihood estimation. The basic idea behind maximum likelihood is to find the values of the parameters for which the observed data are most likely. Consider the following simple example.

A tagging experiment was performed where 100 fish in a lake were tagged. Make the usual assumptions: a completely closed population with no natural or tagging mortality, no immigration or emigration, and tagged fish are as likely to be recaptured as untagged fish. We wish to estimate the size of the population in the lake (X), and to do so we again capture 100 fish the following day. Of these 100 fish, 20 are tagged from the previous day. We now wish to estimate the value of X from these observations.

To find the maximum likelihood estimate of X, we begin by hypothesizing a probability distribution for the data in relation to the parameters. For this mark-recapture experiment, the appropriate distribution will be the binomial (when all assumptions are met and we assume that X is large enough that we can consider the experiment to be sampling with replacement) , where the probability of observing m marked individuals from a sample of n is given by the following equation:

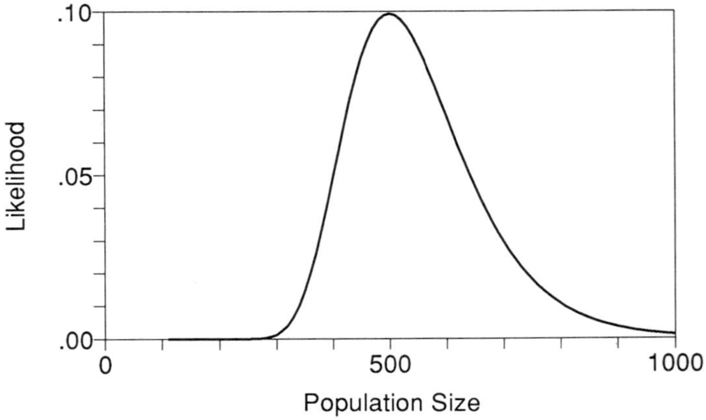

Figure 6.5. Likelihood versus population size.

$$\Pr\{m|n,p\} = \left[\frac{n!}{m!(n-m)!}\right]p^m(1-p)^{(n-m)} \qquad (6.11.1)$$

here p is the proportion of the fish in the population that are marked, which is $p = 100/X$ in this case, m is 20, and n is 100. The maximum likelihood estimate of X is derived by searching over X to find the value for which $\Pr\{m|n,p\}$ is maximum.

We can write the likelihood of the data (m) given X as

$$L\{\text{data}|X\} = \left[\frac{n!}{m!(n-m)!}\right]\left(\frac{100}{X}\right)^m\left(1-\frac{100}{X}\right)^{(n-m)} \qquad (6.11.2)$$

Figure 6.5 shows this likelihood as a function of the population size; the maximum likelihood estimate is $X = 500$.

When dealing with one- or two-parameter models, graphical plots of the likelihood function are quite useful, just as they are when dealing with least-squares estimates. Also, in the case of simple likelihood functions like that given in Equation 6.11.2, the maximum-likelihood estimate can be found by simple calculus. However, when finding maximum likelihood estimates for several parameters, nonlinear search routines, such as described in Section 6.10, are usually needed. Often, one must minimize -1 times the likelihood, as most nonlinear routines are function minimizers rather than maximizers.

Another computational trick that is often employed is to maximize the logarithm of the likelihood, because one often deals with very small likelihoods and when the gradients (Jacobian matrix) are calculated, severe errors due to machine accuracy may result if logarithms are not used.

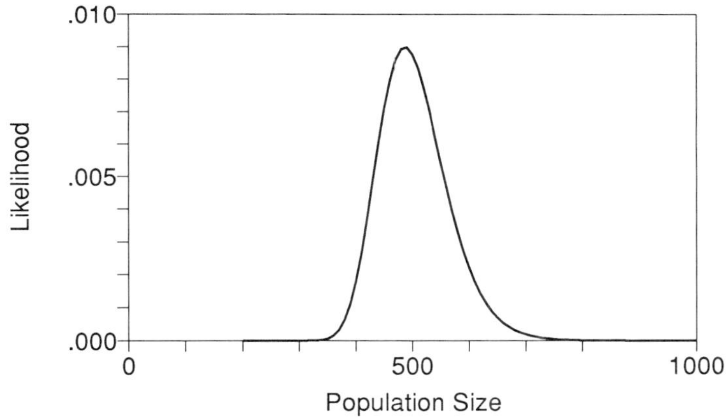

Figure 6.6. Likelihood of different population sizes with two recapture periods.

Multiple observations

Likelihoods can just as easily be calculated for multiple observations when the observations are independent of one another. In this case, the likelihood of a set of observations is simply the product of the likelihoods of the individual observations. More formally stated

$$\mathrm{L}\{O_1 \ldots O_n\} = \mathrm{L}\{O_1\} \times \mathrm{L}\{O_2\} \times \ldots \times \mathrm{L}\{O_n\} \qquad (6.11.3)$$

For example, continuing the previous tagging experiment, we marked the 80 unmarked fish we captured on the second day, and on the following day recaptured 80 fish of which 30 had tags. The likelihood of any population size is now the likelihood of observing 20 marks out of 100 recaptures given that 100 fish were marked, times the likelihood of observing 30 marks out of 80 recaptures given that 180 fish were marked. Figure 6.6 shows the likelihood for each possible population size for this example.

Likelihoods with the normal probability distribution

The tagging example given above relies on the binomial probability distribution for formulating the likelihoods. Other probability distributions can easily be applied to maximum-likelihood estimation. One of the more common probability functions is the Gaussian or normal distribution. The likelihood for the normal distribution is the familiar bell-shaped curve, which is described by the following equation:

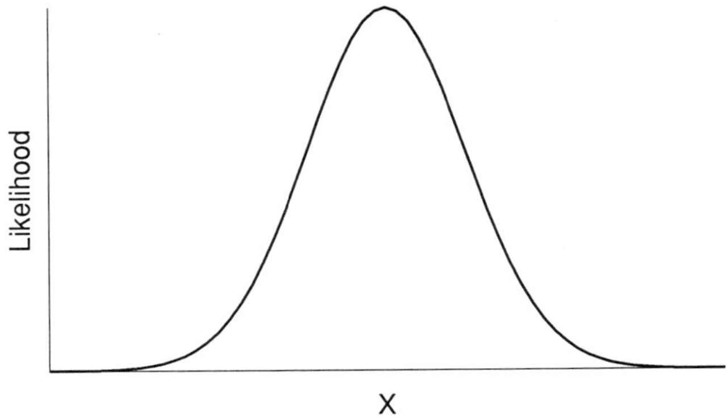

Figure 6.7. A normal likelihood function.

$$L\{X|\mu,\sigma\} = \frac{1}{\sigma\sqrt{2\pi}} exp(-(X - \mu)^2/2\sigma^2) \tag{6.11.4}$$

You are undoubtedly familiar with the bell-shaped curve shown in Figure 6.7, but people are often a bit vague about what the *y* axis really is. It is the likelihood: the relative probability of an observation of each specific *X* value. The normal likelihood function, when combined with a little computing power, lets us escape from most of the burden of statistical tables and formulae.

Consider again the problem of estimating the slope of the recruits versus spawner relationship shown in Figure 6.2 from a maximum-likelihood perspective. Our basic model will be

$$R_i = bS_i + \epsilon_i \tag{6.11.5}$$

where ϵ_i is normally distributed with a mean of zero and a standard deviation of σ. Using likelihood methods we have two parameters to estimate, b and σ. We wish to see how the likelihood varies for different values of these parameters. The likelihood of any individual observation can now be written

$$L\{R_i|b,\sigma,S_i\} = \frac{1}{\sigma\sqrt{2\pi}} exp(-(R_i - bS_i)^2/2\sigma^2) \tag{6.11.6}$$

and by taking the product of the likelihoods for all observations, we can calculate the likelihood for any pair of values of b and σ.

Figure 6.8 shows the likelihood contours for the data give in Table 6.1. Not surprisingly, we obtain the same estimate for b that we did using least squares because for linear problems, least-squares estimates are identical to maximum likelihood estimates.

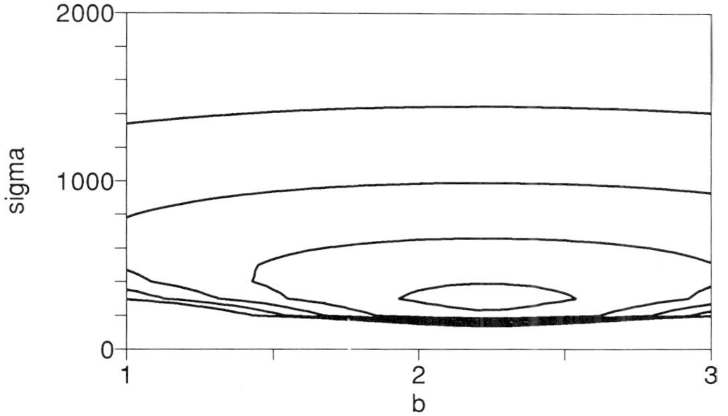

Figure 6.8. Likelihood contours for the Fraser River chum data given different values of b and σ.

Likelihoods with the Poisson probability distribution

Another probability distribution that is often used in stock assessment for maximum likelihood estimation is the Poisson distribution. The Poisson distribution is most commonly used with rare occurrences, such as the recapture of tagged fish. For example, imagine we tag 400 individuals at time zero, and the tags are recovered in a commercial fishery over the next seven, equally spaced time periods. Assume that we recapture 118, 81, 62, 38, 26, 17, and 12 tags at times 1 to 7 respectively, and therefore 46 tags are not recaptured. This type of tagging experiment is known as a Schnable census, and is discussed extensively in Seber (1982) and Burnham et al. (1987).

If we ignore natural mortality but include a constant rate of fishing mortality, we can write the relationship for the number of tags recovered (C_t) as

$$N_t = N_0 e^{-Ft}$$
$$C_t = N_t(1 - e^{-F}) \qquad (6.11.7)$$

where N_t is the number of tagged fish alive at time t, N_0 is the number of fish tagged at time 0, and F is the fishing mortality rate.

Assuming we wish only to estimate the fishing mortality rate (we ignore the information on proportion of captured fish tagged which would tell us about the population size), we need a likelihood to tell us the probability of capturing X tagged fish, given that N are alive and a specific value of F. The expected number of tags captured at time t given N_t and F is $N_t(1 -$

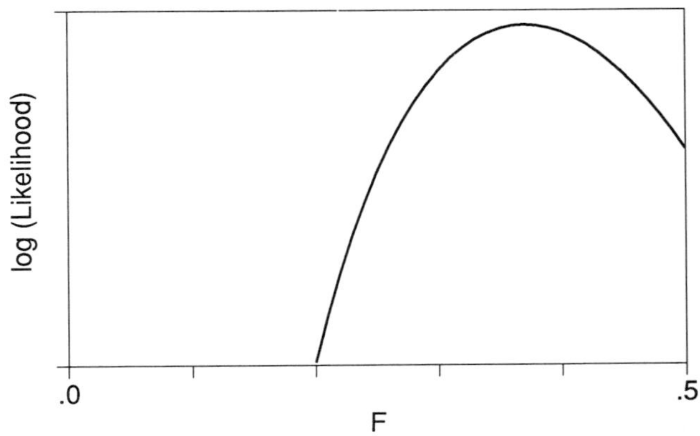

Figure 6.9. Log likelihood of the data for different F's.

e^{-F}), which we call μ. The probability of capturing X tags, given μ, is given by the Poisson distribution as follows:

$$P(X) = \frac{\mu^X}{e^{\mu} X!} \tag{6.11.8}$$

Thus if $(1 - e^{-F})$ is 0.01 and we have 1,000 tags alive at time 1, we would expect 10 recoveries, and the probability of observing 8 recoveries would be

$$P(8) = \frac{10^8}{e^{10} 8!} \tag{6.11.9}$$

Figure 6.9 shows the likelihood of all the data (product of the seven likelihoods for all recovery periods) plotted as functions of F.

Likelihoods with the multinomial distribution

The multinomial distribution is commonly used in fisheries statistics particularly for the analysis of tagging data. It is applicable where there are a known number of trials and each trial may fall into one of a known number of classes. For example, for the tagging experiment described in the previous section, there were 400 trials and 8 possible classes: recaptured at time 1, time 2, . . ., time 7, and not recaptured.

If we allow only fishing mortality and no natural mortality, then for any F, the probability of a tag being captured in the ith time interval p_i is

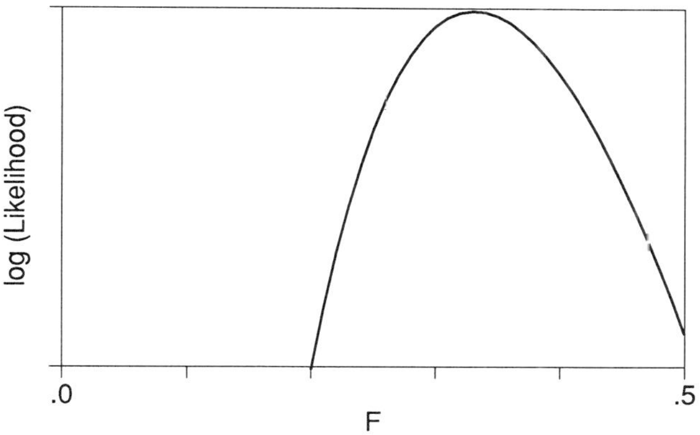

Figure 6.10. Log likelihoods of the data given different F values calculated using the multinomial distribution.

$$p_i = \left[\prod_{t=1}^{t=i-1} (1 - e^{-F}) \right] e^{-F} \tag{6.11.10}$$

This simply says that the probability of being alive and being caught at time i depends on the probability of not being caught before, times the probability of being caught at time i. The last "class" is individuals that were never caught, and the probability of being never caught is $(1 - \Sigma p_i)$.

The likelihood of the observed recoveries at time i (n_i) given the values of the p's is

$$\frac{N!}{\Pi_i\, n_i!(N - \Sigma_i\, n_i)!} \left(\prod_i p_i^{n_i}\right) \left(1 - \sum_i p_i\right)^{(N - \Sigma_i n_i)} \tag{6.11.11}$$

where N is the number of tags released (number of trials), n_i is the number of tags recovered at time i, and the $N - \Sigma_i n_i$ terms are those tags not recaptured. We can then search over values of F and plot the logarithm of the likelihood. This is shown in Figure 6.10.

Confidence intervals with maximum likelihood

In any estimation problem, we want some idea of the confidence bounds of our parameter estimates, as well as the single point estimate. When using the normal distribution, this is easily accomplished simply by integrating over the likelihood function to find the appropriate confidence intervals. Assuming we have chosen sufficiently small discrete steps on the parameter axis, we normalize all likelihoods so that they sum to 1.0. Then if we want the 90% confidence interval, we simply find the value of the parameter that has 5% of the total likelihood below it as the lower bound, and the value of the parameter that has 5% of the total likelihood above it as the upper bound.

In general one cannot integrate over the normalized likelihood function to find the confidence intervals (Hudson 1971). The practice we recommend in that case is to use a Bayesian approach, as discussed in the next section, which does enable simple computation of confidence intervals.

6.12. Using Bayes Theorem

Bayesian estimation is very much like maximum likelihood, except that (1) it assumes there is a prior assessment of the relative probabilities of alternative hypotheses or parameter values and (2) instead of using the parameter value with the highest posterior probability as the estimate, a "risk function" is employed to determine which parameter to use. In this section we will consider only the first part of Bayesian estimation, we will consider the risk analysis components later in Chapter 17. There is a rather long tradition of debate among statisticians about the role of Bayesian estimators, which has centered primarily on the legitimacy of prior probabilities. This debate has cooled down considerably with Bayesian statistics now playing a significant role in the statistical world.

For fisheries stock assessment, the problem is even more clear cut. Few, if any, stock assessments begin with absolutely no idea about the relative credibility of different parameter values. Biological limits, knowledge from other fisheries, and prior experience with the stock in question are almost always present. For instance, when estimating growth curves, we know that fish can only grow so fast, and that growth over time cannot be negative for a species. Therefore, we have a prior assessment of the probability of different hypotheses. Similarly, when doing a production model analysis for a species, we always know something about the relevant parameters. Past experience elsewhere in the world with the same species will tell us about possible values the intrinsic growth rate might take. Knowledge of the fishing gear will tell us about the potential catchability (almost sure to be less than 1.0 and greater than 0.0), and physical limits will to some extent bound the possible k values.

Given these arguments, Bayesian statistics are usually going to be most appropriate for fisheries stock assessment problems. Few papers using Bayesian methods have appeared in the fisheries literature, although the number is growing and we suspect that within a few years most stock assessment papers will use them.

Bayesian statistics rely on Bayes' theorem, which defines an important property of conditional probability:

$$\text{Pr}\{\text{hypothesis}|\text{data}\} = \frac{\text{Pr}\{\text{data}|\text{hypothesis}\} \times \text{Pr}\{\text{hypothesis}\}}{\Sigma\ \text{Pr}\{\text{data}|\text{hypothesis}\} \times \text{Pr}\{\text{hypothesis}\}} \qquad (6.12.1)$$

This equation breaks into three parts. Pr{hypothesis|data} is the posterior probability (expressed as an absolute probability between 0.0 and 1.0) of the hypothesis given the data (and prior information). Pr{hypothesis} is the prior probability of the hypothesis before the data are considered. Pr{data|hypothesis} is simply the likelihood of the data if the hypothesis is true, exactly as it was in the previous section. The denominator is exactly the same as the numerator except summed over all possible hypotheses; it is the total probability of getting the data, over all hypotheses admitted as possibly having produced the data.

To use Bayes' theorem to generate a posterior distribution we therefore need three things:

1. A list of all possible hypotheses
2. A prior probability for each hypothesis, normalized so that the sum of all prior probabilities is equal to 1.0
3. A likelihood function to calculate the probability of the data if the hypothesis is true.

From our example in the previous section, we already have items 1 and 3. In order to calculate the likelihoods, we had to search over a range of possible parameter values. We normally bound and discretize the possible parameter values. Item 3, the likelihood, is exactly what we were obtaining in the examples of the previous section.

The only missing item is therefore the prior probability distribution. There are no constraints on how this is derived. One may have quite good estimates of the parameters and their likelihoods from prior experience with the same species or even the same stock. Alternatively, one may have only vague constraints on parameters. In this second case, one often adopts the so called "uniform prior," in which all hypotheses considered are assigned equal probabilities.

As an example, reconsider the simple single mark-recapture experiment in the previous section (Figure 6.5). We can specify our hypotheses as all possible population sizes between 110 and 1,000 in steps of 10, assign each hypothesis equal prior probability (1/number of hypotheses), and use the same binomial likelihood we used earlier. Figure 6.11 shows the Bayesian posterior probabilities using these conditions. Comparison with Figure 6.5 shows almost no difference.

This is a quite common result. Except in cases where there are either very little data, or the prior probability distribution is very constraining, the dif-

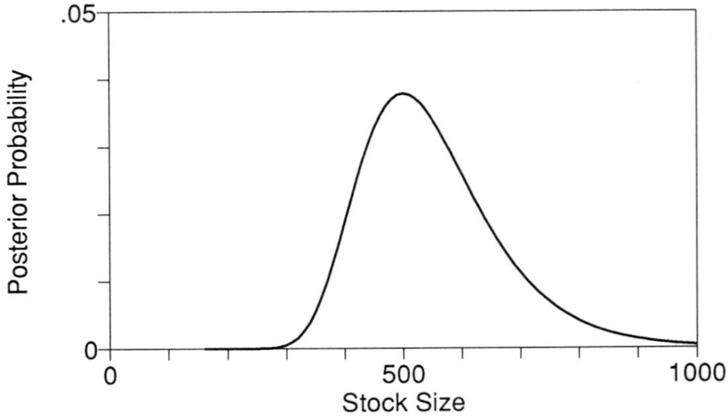

Figure 6.11. Bayesian estimates of population sizes.

ference between the maximum-likelihood estimate and the Bayesian estimate is small.

Bayesian confidence intervals

Confidence intervals from the Bayes' posterior distributions are simple. Since the sum of all posterior probabilities is equal to 1.0, the 90% confidence bounds are simply all hypotheses excluding the least likely 10% of admitted hypotheses. For a one-dimensional problem, we simply chop 5% off each tail of the posterior distribution. To find the 90% confidence interval, we simply look up the stock sizes that represent 5% and 95% of the cumulative probability from Figure 6.11.

6.13. Fitting Time Series

Many types of fisheries data are time series. Catch-effort series used to fit production models, stock-recruitment, catch-at-age, tag recoveries, and so on, are all examples of time series. What distinguishes time series from "standard" data is that in time series data there is no functional independence of one observation from previous observations. For instance, when relating length to weight for 100 sample fish, we can think of each fish as an independent observation of some true underlying relationship. With time series data, however, the individual predictions are not independent. The predicted biomass, catch-at-age, or CPUE at any time depends upon previous values. A key consequence of this dependence is that variations that arise from the dynamics of the system (rather than measurement) will impact future states (stock sizes, CPUES, etc.) of the system. For example, one good recruitment year will impact subsequent stock sizes for several years.

If we assume that the only variation in the system comes from *process error*, then we can write the system dynamics as follows:

$$X_{t+1} = f(X_t, p_t, u_t, w_t) \tag{6.13.1}$$

where X is the state of the system, p is a vector of parameters, u is a set of controls, and w is some random process error that reflects the stochastic dynamics of the system. In many systems, we may also want to add explicit time lags to the model if the state does not contain all the information needed to contain the lags inherent in the system. For example, if our model contains the full age structure, we may not need any explicit time lags, but if it is a simple production model, we might wish to have the state next year depend in some way on the state two or more years before.

We can write the discrete Schaefer model in a stochastic form as follows:

$$B_{t+1} = B_t - C_t + rB_t\left(1 - \frac{B_t}{k}\right) + w_t \tag{6.13.2}$$

where B is the biomass, C is the catch, r and k are the usual parameters, and w is a normally distributed random variable with mean zero and standard deviation σ_w. This equation simply says that there is a randomly distributed error on the production of the stock. This is called process error because it involves stochasticity in the dynamic process of the model. This is not necessarily the best error assumption but it is the simplest for this illustration. It is a slightly different assumption about process error than we made in Section 6.6 when fitting the same model using a linear model.

If we could observe B_t perfectly and specified a likelihood relationship between the predicted and observed biomass, we could quite simply find the values of r, k, and σ_w that fit the data best. For the above model this would be

$$\hat{B}_{t+1} = B_t - C_t + rB_t\left(1 - \frac{B_t}{k}\right)$$

$$d = B_{t+1} - \hat{B}_{t+1}$$

$$L\{d|r,k,\sigma_w\} = \frac{1}{\sigma_w\sqrt{2\pi}}\, exp(-(d)^2/2\sigma_w^2). \tag{6.13.3}$$

It would then be a simple matter to multiply all of the likelihoods for the individual data points together to obtain the total likelihood of the data. When we assume that only process error is of concern, then each observation (value of B_{t+1}) is treated as an independent observation, and we effectively ignore the time series nature of the data.

The alternative assumption to process error is called *observation error* and assumes that we cannot observe the state of the system perfectly. This is undoubtedly true for most systems. Observation error can be written as

$$X_t^{\text{obs}} = g(X_t, v_t) \tag{6.13.4}$$

where X_t^{obs} is the observed value of X at time t, g is an observation error function, and v is the parameter(s) of the observation error process. For example, continuing our simple production model example we might specify the following model:

$$B_t^{\text{obs}} = B_t + v_t \tag{6.13.5}$$

where v is a normally distributed random variable with a mean zero and standard deviation σ_v.

Fitting models with an observation-error assumption is quite different than with the process-error assumption. We cannot predict the B_{t+1}^{obs} from B_t^{obs} because B_{t+1}^{obs} depends on B_t, not its observed value, and we do not know B_t; we can only observe B_t with error. One method commonly used to the fit model under the observation-error assumption is to pretend that there is no process error, and then make a prediction of B_{t+1} for all t's given the observed C's, the parameters r and k, and an initial starting biomass B_0. We can quite simply use Equation 6.13.2 to predict the B's. Once we have the B's that must be true if the C's, r, k, and B_0 are true, we then compare the predicted B's to the observed B's using either maximum likelihood or, quite commonly, simple least squares. This approach is illustrated in detail in Section 8.4 of the chapter on biomass dynamics models.

The approach to fitting time series with observation error is quite simple and elegant; you simply make predictions about the deterministic trajectory of the system from the parameters and the initial starting conditions and then compare the observed values to the deterministic predictions. This approach has been used for production models by Ludwig and Walters (1989) and Punt (1988) for fitting delay-difference models (Chapter 9), and for analysis of tagging data by Hilborn (1990).

There is currently quite a bit of research focused on the merits of assuming process error or observation error. Although no results are yet totally conclusive, work by Ludwig and Walters (1989) indicates that when both types of errors are present (as they almost always are), it is better to assume observation error. It is more difficult to allow both observation and process error simultaneously. In fact, it is essentially impossible to allow for both, without an estimate of the ratio between the variances of the two processes. Ludwig et al. (1988) have had some success fitting production models assuming equal process and observation error, and Punt (1988) found that the Ludwig et al. method commonly outperformed other methods.

6.14. Using a Collection of Heterogeneous Data

Parameter estimation, as we have discussed in the previous sections, involves finding the parameter values of a specific model that best fit a set of data. For every example we have used so far, the observations consist of a single observable quantity: recruitment, catch per unit effort, catch-at-age, and so on. The models we have discussed consider only a single type of data and "narrow" model.

The real world of fisheries statistics is much more heterogeneous; most management programs collect lots of different types of data, and many of these types of data are useful for making predictions, or inference, about the same quantities. For instance, we often want to estimate stock abundance. There are a lot of ways to do this; VPA and catch-at-age analysis produce estimates of abundance, and we may have surveys, analysis of CPUE trends, and sometimes a tagging study. We may have two or even three estimates of abundance, each from a different model and different data. We may have separate estimates of total mortality, fishing mortality, or natural mortality from tagging studies, VPA, catch-at-age analysis or some other technique. If we are using production models, we have estimates of r, k, and q from the production model, but we may have other, more diverse sources of information. We may have tagging estimates of population size, a few estimates of fishing mortality rates from tagging studies, occasional surveys of biomass, estimates of q from tagging studies, or perhaps estimates of r from knowledge of similar stocks in similar environments.

Most fisheries stock assessments have used these different types of data in an independent way. Each type of data usually fits nicely into a traditional estimation approach, which produces estimates of parameters. You therefore end up with more than one estimate of the same parameter. If these estimates are close enough, the analyst heaves a sigh of relief and has even more confidence in the estimates. More commonly, the estimates are not too close, and the analyst either ignores the one he has less confidence in, takes the average of the two, or presents both to decision makers and admits uncertainty. This dilemma is often seen where VPA and survey estimates of abundance are both available. Figure 6.12 shows population abundance estimated by these two methods for Bering Sea pollock (*Theragra chalcogramma*).

One of the most important trends in stock assessment is the development of methods to deal with such heterogeneous data in a more rigorous, statistically consistent manner. Such methods have commonly used the term auxiliary data to describe the data that are outside the normal estimation framework, but we prefer the term heterogeneous data because it does not assign a secondary status to the auxiliary data, but recognizes that none of the data and information have primacy.

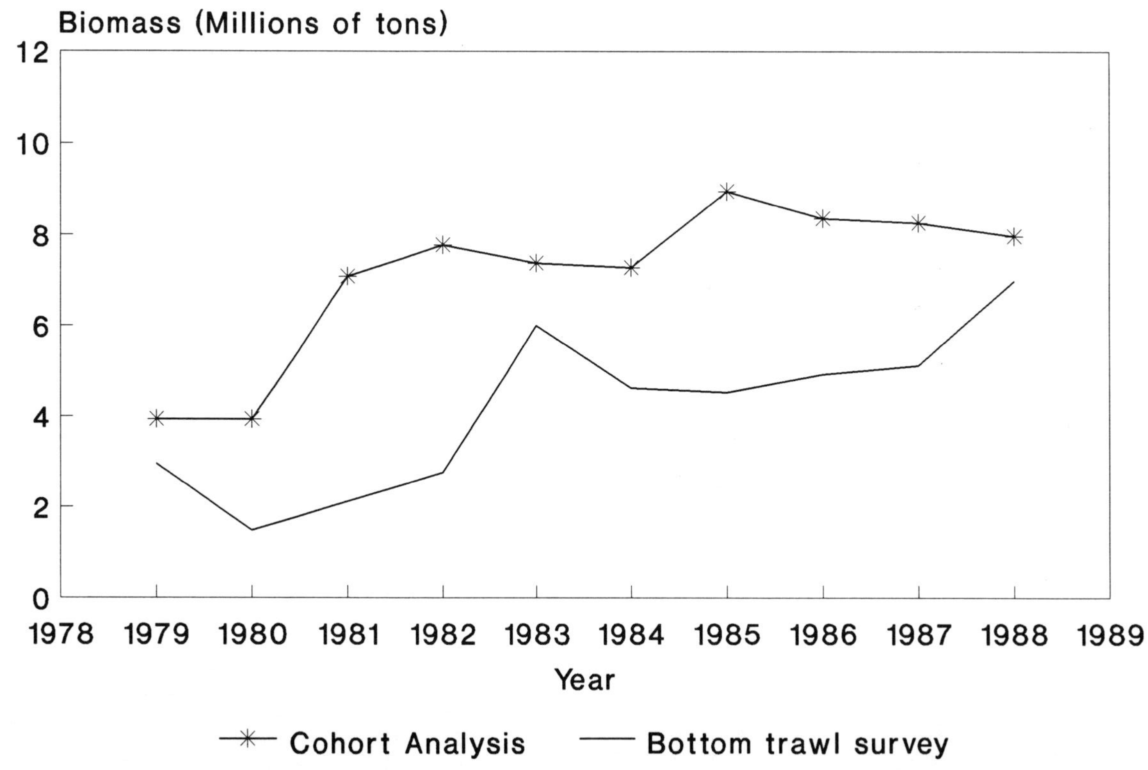

Figure 6.12. Estimates of abundance of Bering Sea pollock using VPA and from surveys. Data from National Marine Fisheries Service.

Bayesian approaches

Bayes' theorem (Section 6.12) provides one of the most appealing statistical frameworks for analyzing heterogeneous data. Bayes' theorem allows us to incorporate both prior information and heterogeneous data collected later. For example, imagine if we were using a production model, but had prior information about r from other stocks of the same species, and had "auxiliary" information about biomass from a survey. We could formulate the Bayesian estimation with a prior distribution of r based on what we know of the other stocks, and allow k and q very uninformative uniform priors. We could then calculate the likelihood of each observation of CPUE, given the parameter estimates and the dynamics of the production model, and the separate likelihood of the observed survey results. To our knowledge, no one has treated production models in a completely Bayesian framework, although it could easily be done.

Fried and Hilborn (1988) used Bayesian approaches to fit heterogeneous data for in-season estimation of run size in a Pacific salmon fishery. The problem was to estimate what the total return of sockeye salmon (*Oncorhynchus nerka*) will be at the end of the season for the largest salmon fishery in the world at Bristol Bay, Alaska. They considered three types of information: (1) a preseason forecast of total run based on spawning stock size, oceanic conditions, and catches of juveniles, (2) in-season estimates from a gillnet "test" fishery approximately 100 miles in front of the commercial fishery, and (3) the "inshore run," or total returns to the commercial fishery and the spawning grounds.

With respect to any fishing season, the preseason forecast is static (it is made several months before the fishery begins) and does not change during the year. The reliability of the test fishery and the inshore run do change as the season progresses. In fact, the inshore run, defined as the total catch plus escapement to the river to date, is 100% accurate at the end of the season.

Figure 6.13 shows the historical reliability of the preseason forecast, and the reliability of the test fishery and the inshore run for two specific dates. From such graphs, we can calculate the likelihood of a total run, given the preseason forecast, test fishing results on a particular date, and inshore run on a certain date.

Figure 6.14 shows the changing standard deviation of each of these heterogeneous pieces of data as the season evolves. The preseason forecast does not change over the season, test fishing results improve for a week or two, and then actually get less reliable, and the in-shore run data improve continuously throughout the season.

Fried and Hilborn's (1988) approach was to calculate the predicted total run in the following way. The Bayesian prior distribution of total run was

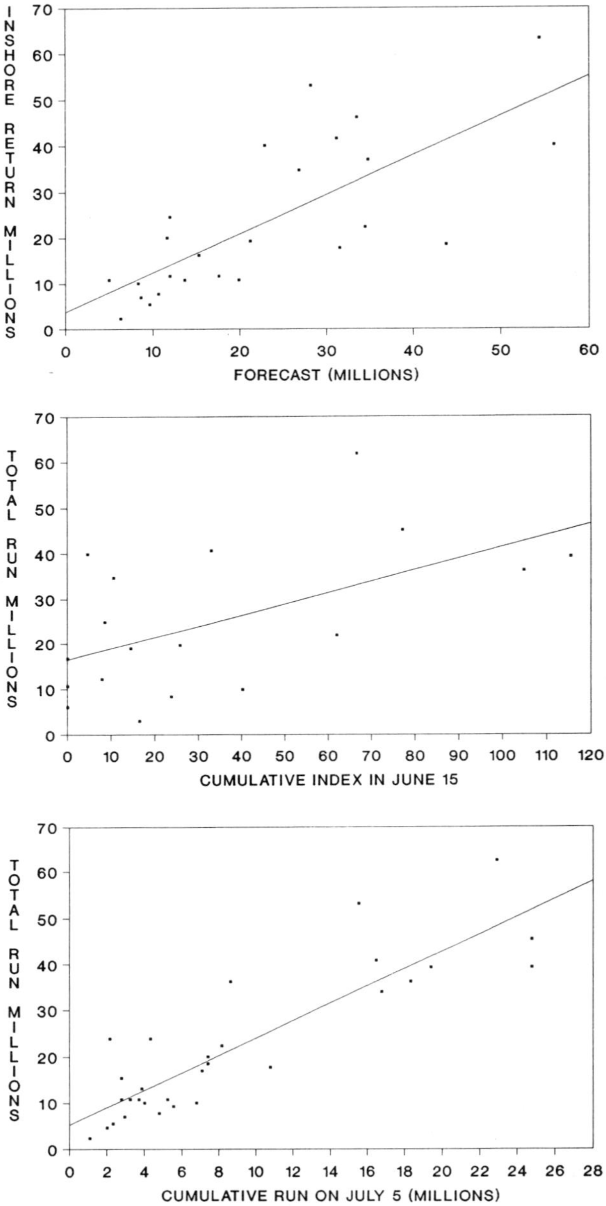

Figure 6.13. The reliability of different predictors of the sockeye salmon return to Bristol Bay, Alaska.

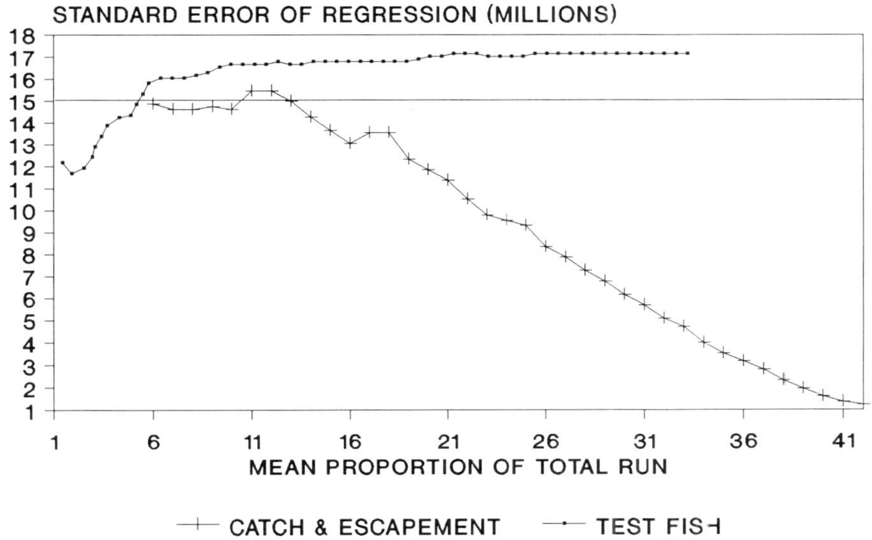

Figure 6.14. The standard deviation of different estimates of total return as the fishing season progresses.

calculated from the preseason forecast. Bayes' theorem was then used to update this estimate using the test-fishing data. The likelihood of the total return given the observed test-fishing index to date was plugged into Bayes' formula. Emerging from this calculation is a posterior distribution for total run. Bayes' theorem was then used to revise this estimate by using the likelihood of the total return given the observed inshore run to date. The posterior that emerged from this calculation is the best inshore run prediction for the date in question. Figure 6.15 shows the evolution of total run estimates for the 1985 season.

Formal likelihood methods

Likelihood methods also provide a convenient framework for combining heterogeneous predictions. Assume we have estimates of stock size available from both catch-at-age analysis and from a survey, and these estimates are derived by a maximum likelihood method: $\mathrm{L}\{c|B\}$ is the likelihood of the data given biomass B and model of the catch-at-age analysis, and $\mathrm{L}\{s|B\}$ is the likelihood of the survey data given the biomass. We can then find a maximum likelihood estimate of B by finding the value of B that maximizes the expression

$$\mathrm{L}\{c|B\} \times \mathrm{L}\{s|B\} \tag{6.14.1}$$

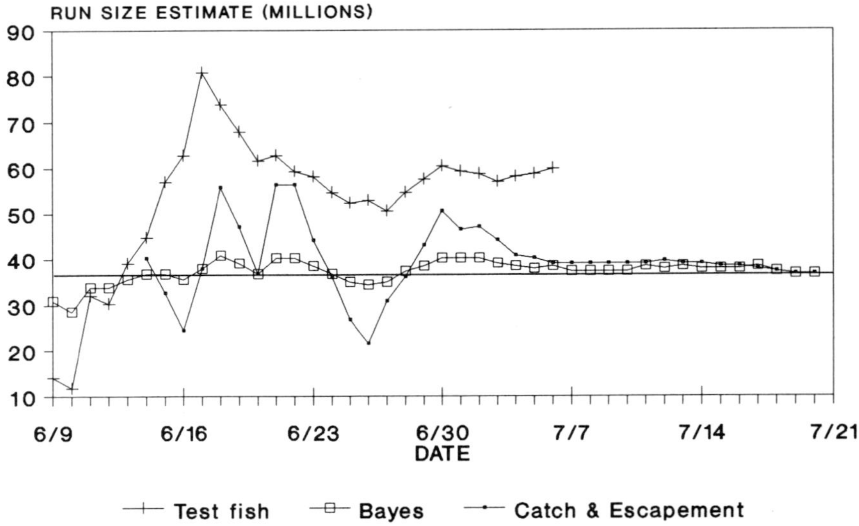

Figure 6.15. The evolution of the different run size predictions for 1985.

If we had ignored the preseason forecast in the example of in-season run size prediction for Bristol Bay sockeye salmon, and skipped the analysis of the prior distribution, then we could have estimated total run size using only the test fishing and inshore return data in a simple maximum-likelihood analysis.

Inverse-variance weighting

Another variation on this theme is the rule of inverse variances, which says that if we have n independent estimates of a quantity Q and there is an estimated variance of each estimate $s^2_{Q_i}$, then the best estimate of Q is the weighted mean of the independent estimates, using the inverse of the variances as the weight.

$$\bar{Q} = \frac{\dfrac{1}{s^2_{Q_1}} Q_1 + \dfrac{1}{s^2_{Q_2}} Q_2 + \ldots + \dfrac{1}{s^2_{Q_n}} Q_n}{\dfrac{1}{s^2_{Q_1}} + \dfrac{1}{s^2_{Q_2}} + \ldots + \dfrac{1}{s^2_{Q_n}}}. \tag{6.14.2}$$

If the estimates of Q are not independent, then there are extensions of Equation 6.14.2 that allow for covariances. Maximum-likelihood formulations lead to the inverse variance weighting method when models are linear and observations are normally distributed.

Each of the three methods mentioned above, Bayesian estimation, maximum likelihood, and inverse-variance weighting require a formal analysis of either likelihoods or variances. If the analysis of heterogeneous data can be done in this formal context, it is not difficult at all to combine the data in a statistically consistent way.

Ad hoc sum of squares

We often have trouble specifying formal statistical models for our various bits (and bytes) of information. A great deal of fisheries analysis is still done using sum of squares, and despite the mathematical elegance of maximum likelihood, it is not in many cases clear that formal maximum-likelihood methods are greatly superior to simple sum of squares.

Most stock assessments that have combined different types of data have done so by simply providing weights to the different elements. For example, in the chapter on biomass dynamic models (Chapter 8), we provide an example of adding survey estimates of abundance to the normal time-series fitting of a production model. This approach can be written generally as

$$SS_{\text{Total}} = SS_{\text{Model}} + wSS_{\text{Heterogeneous data}} \tag{6.14.3}$$

If we assign the heterogeneous data no weight ($w = 0$), then our fit will just be the best fit to the model. However, if we make w very large, then the fitting procedure will do everything possible to get a good fit to the heterogeneous data. If our heterogeneous data consisted of just a single biomass estimate from a survey, and we made w very large, then the sum-of-squares minimization would force the biomass estimate from the model to fit the survey biomass estimate. In theory, we could determine w from our statistical confidence in the heterogeneous data, compared to our confidence in the model. In practice, we usually explore the effect of different weights.

Such ad hoc methods were first used extensively in fisheries estimation by Fournier and Archibald (1982). Another example is Deriso et al. (1985).

6.15. Calculating Confidence Bounds by Bootstrapping

One of the most useful methods for calculation of confidence bounds is called the *bootstrap* and was developed by Efron (1981,1982,1987). The basic idea is very straightforward and is only remarkable in that it actually works. Imagine that we have a set of observed data X_i^{obs} for observations $i = 1 \ldots n$. Using some model, we fit the model and get our best predictions X_i^{pre}. The model may be a very simple one, fitting length to weight, or a complex time-series model tracking stock abundance over time.

In the simplest bootstrap, called the *nonparametric bootstrap* (Efron 1982, 1985), the residuals from the model fit are calculated as

$$r_i = X_i^{\text{obs}} - X_i^{\text{pre}} \tag{6.15.1}$$

We then generate a new data set by sampling from the residuals, with replacement, and adding a "new" residual to each predicted data point.

$$X_i^{\text{new}} = X_i^{\text{pre}} + r_j \tag{6.15.2}$$

Note that the subscript j is drawn from the values $1 \ldots n$ randomly with replacement. We then use the X_i^{new} values to fit the model again, and obtain new parameter estimates for the model. We repeat this process a few hundred times: generate n new data points and then estimate the parameters of the model. We then have a few hundred estimates of the parameters, which we can use to look at the parameter variances and covariances directly.

Another variation on the bootstrap, called the parametric bootstrap, fits a probability distribution to the residuals and then randomly samples from this distribution when generating the X_i^{new} values.

6.16. Bias Problems in Parameter Estimation

It is an unfortunate fact that when we use the parameter estimation techniques described above, we can almost always expect that the parameter estimates will be biased. Bias means that the average value of the parameter estimate, if the experiment (data collection and fishing policy) were repeated a large number of times, would not be the true parameter. Biases can be a trivial annoyance if they are small; a 5 or 10% error in a parameter estimate is often of no great concern. However, biases can be highly dangerous when they are large, and even sometimes when they are reasonably small. Consistent overexploitation by 20% would result in a serious loss of yield from a fishery.

In many well-studied parameter estimation problems in stock assessment the biases may be large, not 10 or 20% but 200 or 400%. Thus one is foolhardy to do any fisheries parameter estimation without some analysis of the likely biases in the methods. There are two sources of bias common to most fisheries parameter estimation. One is the so called "errors in variables" problem where the independent variables of the estimation are not measured without error as is assumed in regular regression. This problem is discussed in some detail in the chapter on stock and recruitment. The second source of bias is called a time-series bias and arises from the time series nature of most fisheries data.

Errors in variables

One of the basic assumptions of all linear regression methods is that the independent variables are measured without error. In the context of stock-assessment techniques, this would mean that stock size, fishing effort, and catch are measured perfectly. This is seldom, if ever the case; in fact these quantities are normally measured with considerable error. Ricker (1973a) recognized this problem and suggested the use of what is called functional regression, essentially estimating the slope of a relationship by taking the average slope of Y on X and X on Y. Walters and Ludwig (1981) and Ludwig and Walters (1981) explored this problem in more detail, particularly in the context of stock-recruitment analysis and recognized it as the classic problem called errors in variables (Kendall and Stuart 1977).

When the independent variables are measured with error, the parameter estimates are biased. We can therefore expect that most fisheries parameter estimates will be biased. Ricker's functional regression approach is not widely applied because (1) it is only applicable for single-variable regressions which excludes most common stock-assessment techniques and (2) it is an ad hoc method with no well-understood ability to correct for bias. Ludwig and Walters (1981) developed some methods for attempting to correct for the errors in variables bias, but these are highly complex calculations that do not work particularly well.

Most people now accept errors in variables as an intrinsic problem in estimation for fisheries, and rather than trying to correct the estimates, they try to understand how large the bias is likely to be. This is usually done by Monte-Carlo experiments, as discussed in the next section and as outlined more generally in Section 6.17 on the operating model concept.

Time-series bias

The second source of bias arises from the time-series nature of most fisheries data, where the observations are not truly independent of the previous states of the fishery and random process errors. This topic is also discussed in the specific context of stock and recruitment. Both of these biases will affect all fisheries parameter estimation, but they are almost totally unexplored except in the stock and recruitment context. There is some analytic work that may be relevant to these problems, but so far it has not been applied. The current state of the art is to do simulations assuming your parameters are correct and see what types of bias arise.

Principle: Always simulate the experiment you are analyzing and determine the biases from simulated data.

It should be routine practice in fisheries stock assessment to take the results from an estimation and resimulate the data to see what estimates are produced. Such a simulation experiment can be performed as follows, using the Fraser River chum salmon stock-recruitment data of Table 6.1. For the

Table 6.7. Parameter estimates for the Ricker parameters for various levels of observation error.

σ_o	a	b	σ_p
0.0	0.678	−1173	0.15
0.1	0.683	−1267	0.15
0.2	0.685	−1410	0.15
0.3	0.688	−1513	0.15

The median of 100 Monte-Carlo trials is given. Real parameters were $a = 0.66$, $b = -3787$, and $\sigma_p = 0.16$.

Ricker stock-recruitment model (of the form given in Equation 6.16.1 below) we estimate the parameters $a = 0.66$, $b = -3{,}787$, and $\sigma_p = 0.16$. Note that the negative value of b simply means that over the range of observations the recruits per spawner actually increase. We will assume that the data points are a single spawning stock producing a recruitment in the next year. In fact, chum salmon return at 3–5 years of age and the recruitment for any brood year occurred over a series of years. Even though we may therefore calculate the harvest rate by examining the relationship between recruitment in one brood year and spawning stock in the next, this does not represent the actual harvest rates that occurred in each year. However, the harvest rates calculated in this way do reflect the general pattern of variation in harvests. We thus define the model as follows:

$$
\begin{aligned}
R_i &= S_i\, exp(a(1 - S_i/b) + \epsilon\sigma_p) \\
S_{i+1} &= R_i(1.0 - h_i) \\
\hat{\mathbf{S}}_{i+1} &= \mathbf{S}_{i+1} e^{\epsilon\sigma_o}
\end{aligned}
\qquad (6.16.1)
$$

where σ_p is the standard deviation of the process error estimated as 0.16, ϵ is a normally distributed random variable with a mean of 0.0 and a standard deviation of 1.0, h_i is the harvest rate in the ith year, $\hat{S}_{i+1}$ is the estimated spawning stock size in year $i + 1$, and σ_o is the standard deviation of the observation error.

We had 14 years of data, therefore we will use the above model to generate simulations of 14 years, using the known values of a, b, and σ_p, and then estimate these parameters from the simulated data. Each such simulation is called a Monte-Carlo trial, and it is always a good idea to do at least 100 such trials when testing for bias. Table 6.7 shows the results for different levels of observation error.

In this example, there seems to be very little problem with either time series or errors in variables bias. The values of a and σ_p are very close to the true values. Although the value of b does seem biased, we must remember from our analysis of these data that *b* is extremely uncertain, and we should not expect to see this parameter reliably repeated by simulation.

The main reason there are such small biases is that the harvest rates varied drastically or at least appeared to vary drastically given the way we compute the harvest rates from the data. This in turn means that the stock size can range over a reasonable set of values. If we were to assume that harvest rates had been constant, the biases would be much more severe.

6.17. The Operating Model Concept

The ultimate aim of any parameter estimation should be to assess the relative levels of confidence that should be placed on different parameter estimates. Specifically we want to know three things: (1) how biased the estimator is likely to be, (2) how much uncertainty there is about the parameter estimate (the variance of the estimate), and (3) how confounded the different parameter estimates are when more than one parameter is estimated.

In the days before widespread availability of computers, these questions were left to a very small group of statisticians who could, for certain specific models, provide some answers to these questions in the form of arcane statistical formulae. The fisheries scientist could, at best, obtain the parameter estimates, and perhaps some confidence bounds that were valid only under particular (and often obscure) assumptions. Estimates of bias were very difficult to obtain, as were calculations of parameter confounding for all but the most simple of linear models. All of this has now changed, and every fisheries scientist should be able to assess bias, confidence, and confounding. Even more significantly, whereas in the past the statisticians could only estimate, for example, bias *if the model were correctly specified*, we can now make these determinations while recognizing that the model may be incorrectly specified.

For example, we can now quite simply ask how biased the estimates of the stock-recruitment relationship will be if we fit a Ricker curve to data generated by a Beverton-Holt model. This is called model misspecification, because the model used to estimate the parameters is not the model that really generated the data. Only the most naïve stock assessment biologist would actually believe he ever correctly specifies a model. The world of fish and fishermen is much more complex than any estimation model could ever be. At best we can hope that the estimation models we use will capture the important aspects of the aggregate behavior of many fish and fishermen interacting in ways that we hope are complex enough to result in statistically predictable patterns. Thus model misspecification is unavoidable, and the question is how do we assess the possible implications of this misspecification on the reliability of parameter estimation.

The approach that is now quite widely in use has come to be known as the operating model approach (Linhart and Zucchini 1986). This approach has four steps.

1. Write down a model of how you believe the real system behaves. This model should be as realistic as your conceptual understanding permits, and you should not restrict the factors included in it to just those for which you have hard data. It should include the biological dynamics, as well as how the data are collected. Both process and observation errors should be included. Age structure, spatial variability, individual variation, and so on, can all be part of this model, which is called the operating model.
2. Develop an estimation model and procedure for the system, using maximum-likelihood or Bayesian estimation, but possibly a more ad hoc least-squares method. Normally the estimation model will be much simpler than the operating model. For instance, the operating model might include age-structure while the estimation model might be a simple production model.
3. Determine what information is available from other sources, and incorporate this into the operating model and into the estimation model using Bayes' theorem or other methods.
4. Generate a large number of data sets from the operating model, and then use the estimation model to find the parameter estimates for each data set. From these simulated data sets and estimated parameters, you can assess the bias, confidence, and parameter confounding of the parameter estimates.

This approach has been used primarily with production models (Uhler 1979, Hilborn 1979, Mohn 1980, Ludwig and Walters 1985) and has recently seen widespread and systematic use in catch-at-age analysis (Kimura 1989, Fournier and Warburton 1989). This approach was used quite intensively by International Council for the Exploration of the Sea (ICES) in a workshop held in Iceland, where a large number of teams with different methods for analyzing catch-at-age data were given different data sets that had been generated from simulation models. Only the workshop organizers knew what the model generating the data was. The International Whaling Commission IWC has been particularly good about using it for systematic evaluation of whale population estimation and management strategies (IWC 1988). In its most common use, the operating model and estimation model are either identical or only differ in how process and observation error occur. Hilborn (1979) used a full age-structured operating model and a Schaefer-type production model, as did Ludwig and Walters (1985). Fournier and Warburton (1989), using an operating model approach, show that a composite estimation scheme using both age-structured and production models may outperform either estimation model when used by itself.

One of the impediments to development of scientific fisheries management has been the lack of true controls and replicates. We can do the stock assessments, but we usually cannot show that some alternative assessment or explanation would equally well explain the observed pattern (lack of control) or that the assessment is robust to random errors (lack of replication). The operating model approach allows a laboratory testing of estimation techniques, and should be an integral part of any stock assessment biologist's toolkit.

The simplest implementation of the operating model approach is to first fit the real field data to obtain parameter estimates. Then, using the estimation model as the operating model (no intrinsic model misspecification), using the same perturbation history that was found in the field data, and using realistic observation and process error estimates, see how biased the parameter estimates will be. For simpler models such as stock and recruitment or production models this is a simple exercise accomplished in a few hours. It should always be done! Once you have done this simple exercise, you will generally find it easy to undertake a more comprehensive analysis of possible model misspecification, by making your operating model progressively more realistic and redoing the estimation tests.

6.18. Summary and Critique

The estimation methods discussed in this chapter are an integral part of almost any stock assessment. Whereas linear methods have been most widely used because of their computational ease, nonlinear estimation methods are an essential part of the tools one should have available. Familiarity with maximum-likelihood and Bayesian methods is also becoming a necessity for anyone wishing to understand the current literature and develop new methods.

Bayesian methods, with their explicit use of information derived from other sources, are where we expect to see the most significant developments in the coming years. They provide the ability to incorporate into a single analytic framework a set of diverse sources of information about the stock.

More important than the individual statistical approaches is the operating model approach, which allows an evaluation of bias and sensitivity to model misspecification. Bias and model misspecification are an unavoidable aspect of parameter estimation—you simply cannot pretend they are not there.

Bibliographic Notes

This chapter touches briefly on a wide range of topics, most of which have entire textbooks devoted to them. We hope to have raised the important issues, and summarized some useful formulae, but most readers will need to look into the standard reference books.

For information on linear regression, see Draper and Smith (1966). McCullagh and Nelder (1983) is the standard book on generalized linear models. Young (1974) has done the most work on recursive least squares. Bard (1974) is an excellent reference for nonlinear estimation, as is Daniel and Wood (1980). See Edwards (1972) for likelihood estimation. de la Mare (1986) describes the method for fitting a time series of abundance data that is now recommended.

There are no standard references on fitting of time series to a collection of heterogeneous predictions; the journal articles cited in the text are the best summaries. Box and Jenkins (1970) is a standard reference on time-series methods, but their approach has little relevance to the type of time-series fitting now commonly used in fisheries. Efron (1981, 1982, and 1985) describes bootstrap methods. For an especially good discussion on maximum-likelihood methods as applied to mark-recapture models, see Burnham et al. (1987). Berger (1985) is a good reference book on Bayesian analysis.

7
Stock and Recruitment

7.1. Introduction

The most important and generally most difficult problem in biological assessment of fisheries is the relationship between stock and recruitment. If there were no repeatable relationship between spawning stock and resultant recruitment, then managers would only need to worry about optimization of yield per recruit. There would be no need to worry about complex and expensive regulations to prevent overfishing, except in terms of the relatively simple matter of preventing fish from being caught at sizes so small as to be wasteful in terms of potential production from body growth. The history of fisheries management would be much different. The classic stories of fisheries management, from the history of blue whales (*Sibbaldus musculus*) to California sardines (*Sardinops sagax*) to Peruvian anchoveta (*Engraulis ringens*) are all stories of stocks fished so hard that recruitment was greatly reduced and the stocks collapsed. Cushing (1971) identified four stocks which he felt had collapsed due to overfishing: the Hokkaido-Sakhalin herring (*Clupea harengus*), the Norwegian herring (*Clupea harengus*), the Japanese sardine (*Sardinops melanosticta*), and the California sardine (*Sardinops sagax*).

Unfortunately, the risk of recruitment overfishing has traditionally not been widely recognized in fisheries. Gulland (1983:70), in discussing the importance of stock and recruitment, states "more commonly the number of recruits is effectively independent of the adult stock size over most of the observed range of stock size." This suggests that stock and recruitment may commonly be ignored, and many fisheries biologists assume until there is very strong evidence to the contrary that recruitment will be independent of stock size.

Fisheries managers simply cannot ignore the fact that if you fish hard enough on any stock, you will reduce recruitment. While recruitment may be largely independent of stock size as a fishery develops, experience has shown that most fisheries will reach the point where recruitment begins to drop due to overfishing. The problem for biologists is to try to understand the relationship between stock and recruitment at least well enough to know how much the stock can be reduced before recruitment starts to drop.

As techniques for measuring spawning stock size and recruitment have improved and as stocks have been exploited harder, a significant relationship between spawning stock size and recruitment has emerged for a wide variety of species. Any fishery manager who acts as if recruitment will remain constant as a fishery increases is foolish. The reduction in recruitment with increasing exploitation of many fish stocks is now clear, with well documented cases of nearly all marine mammals, clupeoids, salmon, and many invertebrates being recruitment overfished. Tuna and some flatfishes are the major groups of fish where good relationships between stock and recruitment have failed to appear, and this is almost certainly due to the fact that they are not fished particularly hard. The bluefin tuna (*Thunnus maccoyii*), which are fished very hard, are now thought to be recruitment overfished.

Why stock and recruitment?

Stock-recruitment analysis normally consists of looking at the empirical relationship between the spawning stock size, and the subsequent recruitment of the year class produced by that spawning. This involves aggregating many different life history stages that can, in fact, often be studied separately. Why aggregate over much of the life history? There are two reasons: (1) it is usually the spawning stock, not other life stages, that can be controlled through management, and (2) it is important to have direct measures of the consequences of this control. For management measures such as hatchery enhancement, it may be interesting or useful to study the life stages leading to recruitment in more detail, but it is important to recognize that for population management the results of such studies must eventually be synthesized into overall predictions of recruitment responses to changing stock size. Any model for such synthesis must ultimately be tested by comparing its predictions to the observed overall relationship between stock and recruitment, so in the end there is no way to escape analysis of the overall relationship. Detailed study of life stages may help to discern why the overall recruitment relationship has some particular shape or quantitative pattern, but for management purposes the key is to know what the pattern is, whether or not we can explain why the pattern has arisen.

7.2. Observed Relationships

Before we delve into the theory and models associated with stock-recruitment relationships, we will look at several cases where spawning stock and resultant recruitment have been measured over a number of years with major changes in spawning stock size.

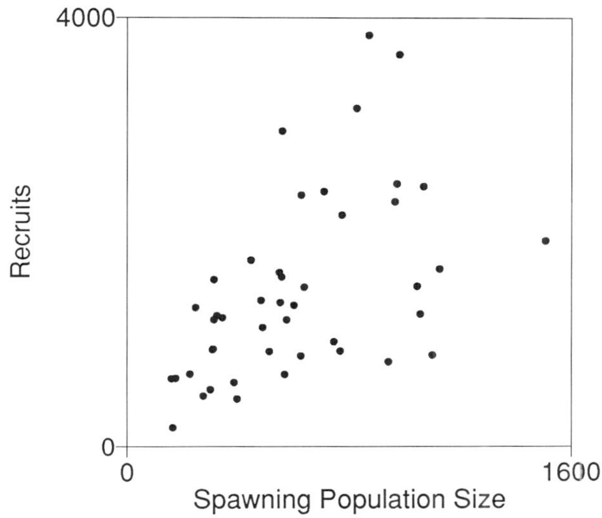

Figure 7.1. Stock-recruitment data for Skeena River sockeye salmon. Numbers in thousands. Data from Shepard and Withler (1958).

Skeena River Sockeye Salmon

Figure 7.1 shows the relationship between the number of adult sockeye salmon (*Oncorhynchus nerka*) spawning and the resultant production of adult sockeye in the Skeena River in British Columbia, Canada (Shepard and Withler 1958). This example has been widely cited as a demonstration of patterns of variation that become apparent when many years of data are available. Pacific salmon have played a prominent role in studies of stock and recruitment, because the spawning stock can be measured in fresh water, and aging techniques have been developed long enough to determine the resulting production of adults from a particular spawning stock size. The data in Figure 7.1 display several features that are common to most stock-recruitment data sets:

1. There is a trend for larger spawning stocks to produce larger recruitments.
2. There is a tendency for the total recruitment to stop increasing above some spawning stock size and possibly start to decrease.
3. The data are highly scattered and the trend may be difficult to discern.
4. There is a tendency for variability about the curve to be higher at larger spawning stock sizes.

It is clear from the general trend in Figure 7.1 that larger spawning stocks

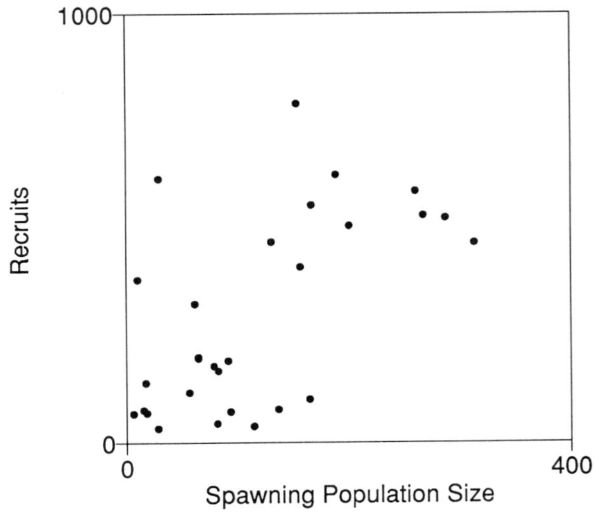

Figure 7.2. Stock-recruitment data for Icelandic summer spawning herring. Data from Jakobsson (1980).

have on average produced larger recruitments, but there is some indication that recruitment may not increase when escapements are over approximately 1 million.

Icelandic summer spawning herring

Figure 7.2 shows stock-recruitment data for Icelandic summer spawning herring (*Clupea harengus*) from Jakobsson (1980). In this case spawning stock and recruitment are estimated from cohort analysis. These data show the same general trends as in the Skeena River sockeye data. Clearly, low spawning stock sizes are much more likely to produce low recruitments than large spawning stock sizes, but we can see that small spawning stocks may produce good recruitments. With the exception of three data points, the data fall into two rather neat categories, a low spawning stock-low recruitment group and a high spawning stock-high recruitment group. Here it is less clear that the variability increases with increasing spawning stock size.

Exmouth Gulf Tiger Prawns

Figure 7.3 shows the relationship between the spring spawning stock and the autumn recruitment for tiger prawns (*Penaeus esculentus*) in the Exmouth Gulf in Western Australia (Penn and Caputi 1985). In this case, the

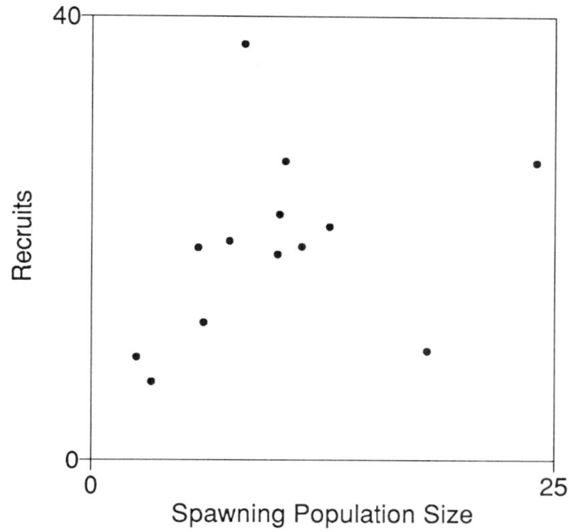

Figure 7.3. Stock-recruitment data for Exmouth Gulf tiger prawns. Data from Penn and Caputi (1985).

data represent abundance indices based on commercial catch and effort data. Again, we see a clear trend toward larger recruitment with larger spawning stocks.

There are two major outliers but the underlying trend is clear. These two outliers are apparently related to changes in rainfall during January and February.

The previous examples indicate decreasing recruitment at low stock sizes. Another common pattern, especially with demersal fishes, such as cods and the Pacific halibut (Figure 7.4), is for recruitment to apparently increase or stay constant as spawning stock declines, over the range of data available. As we shall see later, this pattern can be due to two very different mechanisms: (1) exponentially decreasing juvenile survival with increasing spawning stock size (the so-called dome-shaped recruitment curve), or (2) recruitment on average independent of spawning stock but with the stock size "tracking" long-term cycles in recruitment variation due to some environmental factor.

General observations

These four examples were not selected at random. They represent three quite distinct types of fish (or invertebrates) and were also chosen because the underlying stock-recruitment relationship is reasonably clear. Many other data sets show stock-recruitment relationships that are much less clear. One

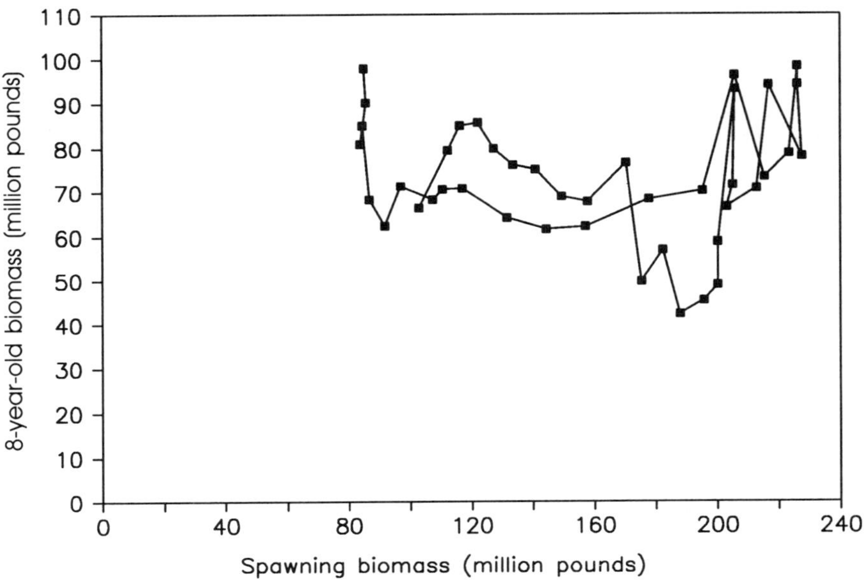

Figure 7.4. Stock-recruitment data for Pacific halibut (*Hippoglossus stenolepis*). Data from International Pacific Halibut Commission.

should never expect to find nice clean curves when stock-recruitment data are assembled, and the examples above are a bit unusual in even showing trends that are immediately evident on visual inspection of the data. The more common situation where no pattern is visually evident may be due to several causes, most likely either because stock and recruitment are not being measured accurately or because stock and recruitment simply are not closely related. Both of these possibilities will be discussed later.

7.3. Biological Processes

In the preceding section, we examined some relationships between stock and recruitment. In this section, we consider the underlying biological processes that will affect stock and recruitment.

Density-independent mortality

The simplest assumption regarding stock and recruitment is that a certain number of eggs are produced per unit of the spawning stock, and these eggs have some variable probability of surviving to the age at which recruitment is measured, where the probability of survival is unrelated to the spawning stock size or number of eggs produced (i.e., there is no significant intraspe-

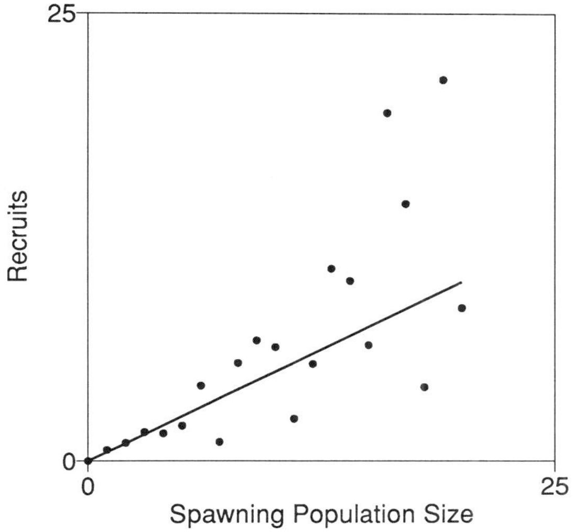

Figure 7.5. Stock-recruitment average relationship and sample data points for density-independent survival.

cific competition or cannibalism). This is called the density-independent model. It simply says that the agents of mortality affecting young fish, including predation, lack of food, currents, and so forth, act independently of how many eggs (or spawners) there are.

The solid line in Figure 7.5 shows what the average stock-recruitment relationship looks like for the density-independent model; recruitment is directly proportional to spawning stock size. In reality, we would not expect the data to be on the line, but rather to be like the points shown in Figure 7.5. Due to random chance there would be good years and bad years providing scatter around the average line. Many ecology students (and some texts) confuse density independence with high variability; the density independent model simply asserts that probability of survival is independent of stock size, whether or not this probability varies widely due to other factors.

Biologically, density-independence is a simple and reasonable assumption, but one that must have limits. No population can reproduce with the same average probability of success as the stock size increases indefinitely. Eventually shortage of food or habitat must reduce the survival from egg to recruitment or perhaps the number of eggs per unit of spawning stock. However, over a reasonably broad range of stock sizes the stock-recruitment curve may look much like Figure 7.5. The Skeena River sockeye, Icelandic herring, and Exmouth prawn data sets presented in the previous section appear to fit the density independent model for the left half of the data. Many marine mammals such as seals and whales are also thought to follow this model over a very broad range of breeding stock sizes.

There is an important point to understand before we leave the density-independent model. Suppose you see a set of data where total recruitment appears to vary around some average level, totally independent of spawning stock or parental egg deposition. Should you claim that recruitment rate in this stock is density independent? The answer is a very definite *no!* If recruitment is independent of egg deposition, then the average probability of survival (recruitment/egg deposition) must on average be sharply declining as egg deposition increases; this decline implies density dependence in the recruitment process (or bad data).

Compensation

Eventually every population must be limited by the resources available. Such limitation could affect growth or adult survival rather than survival up to the age of recruitment. It is well documented that growth rates often change at higher stock densities. We almost always expect to see some reduction in recruits-per-spawner as spawning stock sizes increase. Such a relationship is called compensatory, meaning that the stock-recruitment curve rises less steeply at higher stock sizes or even becomes flat (recruitment independent of spawners).

Let us use a theoretical example of Pacific salmon to look at how various biological mechanisms could lead to a compensatory stock-recruitment relationship. We will look specifically at the relationship between the number of spawning females and how many juvenile fish survive to the smolt stage. Spawning stock is therefore females, and recruits are smolts.

Model I — unlimited habitat

Assume that each female lays 4,000 eggs, and, if the eggs are undisturbed, 0.1% of them will survive to be recruits. Further assume that if one female digs up the gravel that contains the eggs of a previous female, the entire first set of eggs will die. The top graph in Figure 7.6 shows what the spawner-recruit relationship would look like if there were no limit in habitat and females always avoided digging up each other's eggs. The stock-recruitment curve would be a straight line and recruits per spawner would be constant. Even though this seems biologically unrealistic, it could well be true for spawners of a severely overexploited stock that was declining toward extinction.

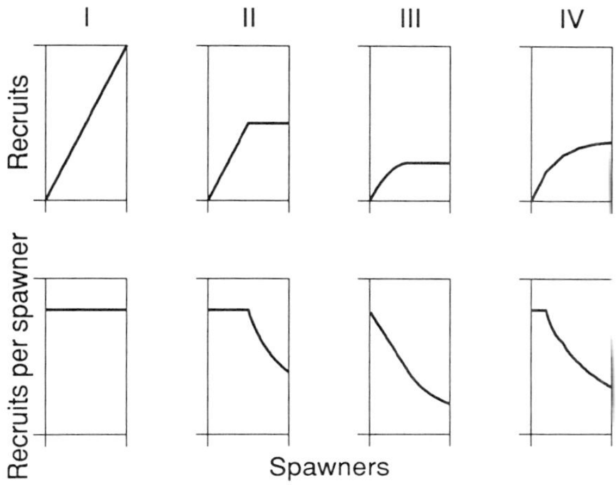

Figure 7.6. Stock-recruitment curves for alternative models of compensatory mortality.

Model II — strict territoriality

Assume that females are highly territorial and each one defends her territory perfectly. Assume that every possible territory is equally good in terms of its habitat quality (as indicated by probability of egg survival). Once all territories are full, subsequent females will be unable to spawn. In this case, we would see recruitment proportional to spawners until the habitat is full, and no increase in total recruitment thereafter. The recruits-per-spawner would be constant until all territories were full and then would begin to decline as the unsuccessful spawners added to the total spawning stock but total recruitment did not increase.

Model III — random egg deposition

At the opposite extreme from perfect territoriality, we have random egg deposition. As more and more females enter to spawn, they are more and more likely to dig up previously deposited eggs. When there are very many spawners, each newly arrived female will dig up eggs that were previously deposited so that no increase in total recruitment is possible. The recruits-per-spawner decrease continuously, because even the second female has a chance of digging up some of the eggs of the first one.

Model IV — gradations in habitat quality

Imagine that there are gradations in habitat quality, for example, good, medium, and bad. Let us assume that in good habitat 0.2% of the eggs survive, in medium habitat 0.1% and in bad habitat 0.05% survive. Further, let us assume that the females can find the best habitat and exhibit perfect territoriality, so that the first females will fill the best habitat, later females the medium habitat, even later females the poor habitat. If there are enough spawners so that all habitats are filled, the last females will not spawn.

This produces a stock-recruitment curve with an initial slope of 4:1, a subsequent slope of 2:1, a later slope of 1:1, finally followed by a flat curve. The recruits-per-spawner gradually decreases although the slope of the curve decreases in steps. Notice that the stock-recruitment curves for models III and IV are almost identical.

Other models

Many other models are possible, including compressible territories and continuous gradations in habitat. Both of these mechanisms will cause the stock-recruitment curve to be smoother without the sharp shifts in slope we saw in model IV. Both of these mechanisms will also produce stock-recruitment curves that approach a maximum recruitment at high spawning stock sizes.

Another type of model is one that causes the total recruitment to decrease at large spawning stock sizes. The usual hypotheses associated with this model are cannibalism and disease transmission in which a very large spawning stock results in extremely poor survival of the eggs. Oxygen limitation, where very heavy egg deposition leads to low oxygen for all eggs, could be another possibility. These types of mechanisms are called *overcompensation*.

Predation can cause density-dependent decreases in egg or juvenile survival rate (i.e., compensation), but only if the predators have very special behavioral characteristics such as "switching on" as juvenile density increases (type III functional response) or numerical aggregation where and when juveniles are abundant. See the section below on depensation.

Cushing (1971, 1973) and Harris (1975) distinguish between stock-dependent mortality and density-dependent mortality. The four compensatory mechanisms pictured in Figure 7.6 are density dependent, with the survival of eggs depending upon the density of eggs at any point in time (so as egg or juvenile numbers decline over time during the life of a cohort, there is continuous and rapid adjustment in survival rates of the remaining individuals). Stock dependent survival means the survival of eggs depends upon the initial or parental stock size, which can lead to the phenomenon of overcompensation.

Models for other life histories

Many fish do not have a spawning life history similar to Pacific salmon where the potential mechanisms of territoriality and gradations in habitat quality can be easily imagined. For species with pelagic eggs and larvae, for example, competition for food, stimulation of predator populations, and intraspecific competition might lead to compensation. We have a long way to go before we understand the dynamics of many fishes' life histories, yet when we look at stock and recruitment data we almost always see evidence of decreasing recruitment per spawner at larger spawning stock sizes.

On theoretical grounds, the best null hypothesis in the analysis of stock-recruitment data might be density-independence in average survival. Instead it is most common to take as a null hypothesis that recruitment is independent of stock, which can only happen if there are strong compensatory mechanisms (see density-independence section above). It is not difficult to imagine mechanistic explanations for constant recruitment, by using terms like "ecological bottlenecks" or "limiting factors," but such explanations are often very complex when you try to get away from the simplistic terminology and analogies and really try to think through what must be happening to the fish in terms of survival rates. Constant recruitment is the pattern that was often observed in many early analyses of stock-recruitment data, but as yet has very little theoretical basis. There is considerable room for further work on this question.

Depensation

Compensation is a decrease in recruits-per-spawner as spawning stock increases. Depensation is said to occur when there is an increase in recruits-per-spawner as spawning stock increases. Two mechanisms have been postulated for such an effect: (1) predation, in which the number of prey eaten by predators is reasonably constant so that, as the total egg production goes up, the percentage eaten by predators decreases, and (2) inability to find mates (or reduced fertilization success) at low densities (also known as the *Allee effect*). Nearly all discussion of depensation in the fisheries literature concentrates on predation as a depensatory mechanism, although Allee effects have been hypothesized for sessile marine invertebrates such as scallops and abalone that broadcast eggs and sperm into the water at spawning.

Imagine a spawning stock that will produce 2,000 eggs per spawner with an average density-independent survival to recruitment of 1%. If we have a predator population that will consume 2,000 potential recruits and can always find and eat them if they are available, the first 100 spawners will produce a total of 2,000 potential recruits but all of these will be eaten by predators. The 101st spawner will produce 20 surviving recruits, and each additional spawner will produce 20 until at very high spawner stocks the total production per spawner will approach 20 even allowing for the total loss of recruits from the first 100 spawners.

Alternative models of predation, in which the number of prey eaten de-

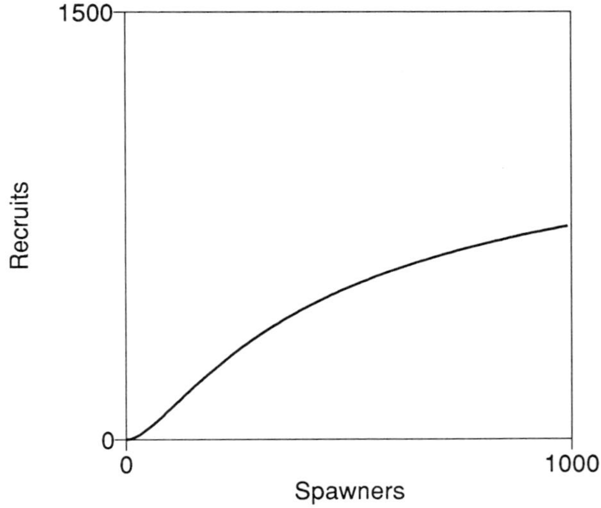

Figure 7.7. Stock-recruitment relationship with depensatory mortality.

pends to some extent on the number of prey available, can yield a stock-recruitment curve as shown in Figure 7.7, where the initial slope of the stock-recruitment curve is low, but then rises to show a normal compensatory pattern at higher spawning stock sizes.

Stock structure

The discussion thus far has assumed that we have a single stock where all individuals are equal. Many, if not most, fisheries operate on stocks that are in some sense divided, in both space and time. The act of fishing will change the stock structure and the apparent relationship between stock and recruitment.

Imagine a stock that is spatially subdivided, abalone on different reefs, herring in different inlets, or Pacific salmon in different parts of a river. In an unfished state, any habitat that can produce 1 adult recruit per spawner will have fish in it, and we would expect that most places will be populated. However, as soon as a fishery begins, the marginal habitats that can barely support a population in the unfished state will be fished out, and only habitats that are more productive will maintain populations. As harvest rates increase to higher levels, a smaller proportion of the habitats will support populations. Ricker (1973c) and Hilborn (1985b) discuss this and point out that the concept of stock-recruitment curves being fixed in time is probably wrong for most fish stocks simply because few fish stocks are not spatially

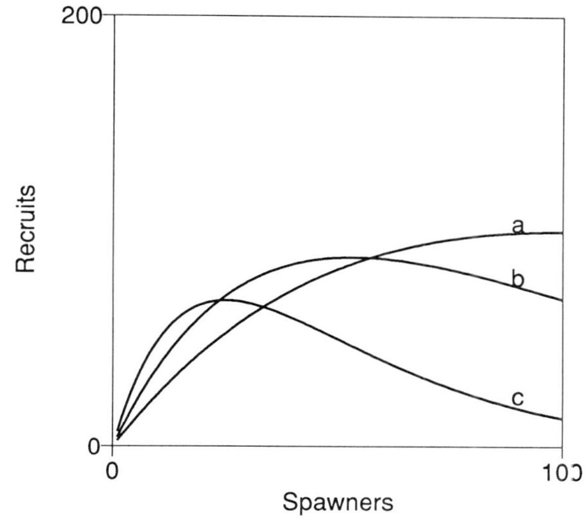

Figure 7.8. Stock-recruitment relationships for a stock that consists of spatially separated substocks when all substocks are present (a) and when less productive substocks have been eliminated by overexploitation (b and c).

subdivided. Figure 7.8 shows what the stock-recruitment curve would look like for a stock in which all of its subunits are still present, and when the less productive subunits have been eliminated by exploitation.

7.4. Measurement of Stock and Recruitment

We have been considering stock and recruitment as if its definition were obvious. This is rarely the case. As late as 1971, it was widely held that most fish stocks showed rather weak relationships between stock and recruitment (Cushing 1973, Gulland 1983). However, as better techniques became available for measuring spawning stock size and resultant recruitment, and as more stocks were fished very hard and generated a good contrast in spawning stock size, clear spawner-recruit relationships emerged.

Measuring spawning stock

The ideal measurement of spawning stock is the number of eggs. This is often estimated by multiplying the number of individuals at age times the average fecundity of individuals of that age and summing over ages. Such data are frequently available through VPA (Pope 1972) and biological sampling for fecundity. The following is a list of alternative measures of stock size, in descending order of reliability.

1. Number of females alive at each age times fecundity by age
2. Number of individuals alive by age times average fecundity by age
3. Total biomass of individuals at or above age of first reproduction
4. An index of abundance of the population in the year the eggs are deposited

Before the wide application of VPA, it was common to use the fourth method for measuring spawning stock, and CPUE was often used as the index of spawning stock size. Because of the imprecision of using CPUE as a measure of eggs deposited, considerable errors were introduced into the measurement of the x axis in the stock-recruitment curve. As we will see in more detail later, errors in assessing spawning stock size tend to make the relationship between stock and recruitment look as if recruitment were independent of spawning stock size. When performing stock-recruitment analysis, one should make every effort to understand how reliable the measure of spawning stock size is. If this measure is not particularly reliable, one should not expect to see a clear stock-recruitment relationship, even if one really exists.

Measuring recruitment

Recruitment can be defined as the number of individuals still alive at any specified time after the egg stage. In species such as Pacific salmon, in which the number alive can be measured at several points in the life history, one can talk about stock-recruitment curves up to the fry stage, the smolt stage, or the adult stage. However, in most fisheries we are limited as to where we can measure the number of individuals still alive, and this is usually at the point of recruitment to the fishery. The number of recruits at this stage is generally measured by VPA or other methods of catch-at-age analysis.

There are stocks where juvenile surveys are conducted or where fish younger than captured in the commercial fishery can be measured, such as for many sessile invertebrates. When measuring recruitment, we are torn between two conflicting desires. On the one hand we would like to measure recruitment close to the egg stage in order to see as clearly as possible the relationship between eggs deposited and juvenile fish production. On the other hand, we want to understand recruitment to the fishery and the relationship between residual stock after fishing and number of recruits to the fishery later on. We want all the natural variation that occurs between the egg stage and recruitment to the fishery to be included in our analysis of stock and recruitment.

7.5. Models and Properties

Analysis of stock-recruitment data is most often performed by fitting various curves to the data and referring to the curve as the *stock-recruitment relationship*.

Warning: Do not mistake average curves for the true relationship.

This is a dangerous practice, primarily because it tends to make one think of the relationship between spawning stock and recruitment as predictable rather than as the stochastic phenomenon it really is. There are several potential approaches to describing stock-recruitment curves, including:

1. *Rough and ready*. The Icelandic herring stock-recruitment data from Figure 7.2 could be described as follows. Stocks between 0 and 150,000 tons normally produce recruitments of between 30 and 200 with occasional production as high as 600. Spawning stocks over 200,000 tons produce average recruitments of approximately 500 with an occasional very high production. This type of description provides managers with the guideline of maintaining the stock at or about 200,000 tons and could be perfectly useful.
2. *Stock-recruitment curves with variances*. Fitting stock-recruitment curves is not intrinsically a bad thing, so long as the amount of variance associated with the relationship is reported and used in any subsequent analysis. That is, recognize clearly that the curve attempts to describe only how mean recruitment rate varies with stock size, and also look at the pattern of variation around the mean. We shall, in fact, devote most of our attention in this section to this approach.
3. *Tabular or Markov models*. In this approach, we divide the potential spawning stocks and recruitments into finite intervals and calculate the probability of any spawning stock interval producing a recruitment in any of the possible recruitment intervals. This is really just a more formal version of the rough-and-ready method.

Basic principles

Ricker's principles

Ricker (1975) discussed desirable properties of stock-recruitment curves (average relationships) and listed four basic properties that he felt should generally be applicable:

1. A stock-recruitment curve should pass through the origin; when there is no parental stock there is no recruitment
2. The curve should not fall to the abscissa at higher levels of stock, so that there is no point at which reproduction is completely eliminated at high densities
3. The rate of recruitment (recruits-per-spawner) should decrease continuously with increases in parental stock
4. Recruitment must exceed parental stock over some part of the range of possible parental stocks

Properties 1–3 are found in almost all stock-recruitment curves used, including those we will discuss in some detail below. Property 1, while superficially reasonable, is really only applicable when there is no immigration to the stock from outside. When recruits may come in as pelagic larvae or at any other life history stage between spawning and recruitment, it is quite possible that a zero spawning stock could still result in some recruitment. Ideally, one deals with data from a unit stock where immigration is not possible, but in reality this is often not true. Property 2, as Ricker noted, is not logically necessary, or even biologically required, but does accord with observations. Property 3 is reasonable when only compensatory biological mechanisms are present. Any depensatory mechanisms (predation, Allee effects) will violate this property, because recruits-per-spawner may increase at low levels of spawning stock. Similarly, some compensatory mechanisms, such as perfect territoriality, will not reduce recruits-per-spawner at all over certain ranges of the spawning stock size. Generally, it is risky to assume that Property 3 holds, unless the data clearly favor it.

Property 4 is really only applicable to semelparous species (those who spawn only once), such as Pacific salmon, since it is only in those species that stock and recruitment are measured in the same units and all spawners die so that one can speak of replacement on a fish-per- fish basis. For long-lived species, the more general property is that recruitment must be high enough over some range of stock sizes to more than replace annual natural mortality losses (mortality rate times stock size) for those stock sizes.

Other principles

Although we will primarily consider the Ricker (1954) and Beverton-Holt (1957) stock-recruitment curves, both of which meet requirements 1–3 stated above, we believe there are two more general principles associated with spawner-recruit relationships. These are continuity and stationarity.

Continuity. Continuity means that the average spawner-recruit curve shows no sharp jumps and does not change rapidly over a very small range of stock sizes. Considering variability in survival rates due to factors other than stock size, we should not expect that Mother Nature can tell the difference between 100,000 and 101,000 fish, and it is pointless to expect any sharp jumps in average behavior due to a small change in number of spawners. More precisely: if you could do an experiment involving holding the spawning stock size constant for many years at each of two similar spawning stock size "treatment" levels (say 10% apart), you should not expect the mean recruitment rate to be much different between the two treatments; applying this argument to every possible pair of treatment levels results in the conclusion that mean recruitment should vary smoothly with stock size. Almost all stock-recruitment curves meet this assumption.

Stationarity. A more important assumption is that the average stock-recruitment relationship is constant over time. When estimating a stock-recruitment relationship from data, one must assume that no underlying change occurred while the data were collected.

More precisely, again, imagine that you could do an experiment involving holding the spawning stock constant at two levels, in the following sequence: gather many years data at the first level, switch to the second for many years, then switch back to the first level for a final run of many years. The recruitment relationship is said to be stationary if the distribution of recruitment rates at the first spawning stock level is the same before and after the switch to the second level. Why is such a property important? If it does hold, we can expect historical data and experience to provide a reliable basis for making predictions about the consequences of policy changes that will affect spawning stock, at least within the range of historical experience. The assumption of stationarity is implicit in any analysis that purports to estimate parameters from historical data for use in prediction. In many cases, we know that stationarity is a bad property to assume. For example, changes in spatial structure of spawning stocks (noted above) will lead to persistent changes in the stock-recruitment relation. Walters et al. (1985) documented that the stock-recruitment relationship for herring in British Columbia depends upon cod (*Gadus macrocephalus*) abundance, which is not likely to remain stable over long periods. However, unless dealing with a multispecies model, or a model that has explicitly changing external environmental variables, you will be forced to assume that there is a stationary stock-recruitment relationship. Walters (1987) discusses the potential sources of nonstationarity in stock-recruitment relationships and what can be done about them. We will deal with these questions in a later section.

Beverton and Holt's model

Form and properties

The Beverton-Holt spawner-recruit curve (Beverton and Holt 1957) relates recruitment to spawning stock, using the equation

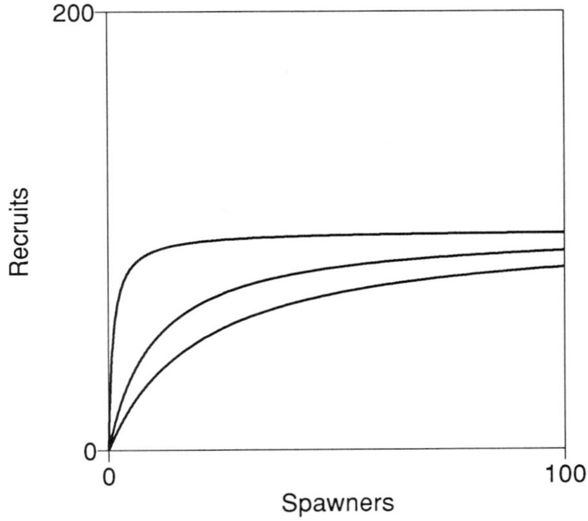

Figure 7.9. Beverton-Holt stock-recruitment curves.

$$R = \frac{aS}{b + S} \tag{7.5.1}$$

where R is the recruitment, S is the spawning stock, a is the maximum number of recruits produced, and b is the spawning stock needed to produce (on average) recruitment equal to $a/2$. The initial slope (maximum recruits/spawner) is a/b. This curve is shown in Figure 7.9. Its basic property is that recruitment constantly increases toward an asymptote as spawning stock increases. Remember, of course, that this is the average relationship. The Beverton-Holt relationship can be written in several ways, each of which has the same shape. One alternative is

$$R = \frac{a'S}{1 + b'S} \tag{7.5.2}$$

where the maximum number of recruits is now a'/b', and a' is the maximum R/S as S approaches 0.0. Another useful alternative is

$$R = \frac{a'S}{1 + (a'/b'')S} \tag{7.5.3}$$

where a' is again the maximum R/S at low stock size and b'' is the maximum number of recruits when S is very large.

Which of these three possible forms we use depends on how we wish to relate the parameters to measurable quantities. We will quite commonly have some idea of the maximum average recruitment, and thus a form that has this as a parameter is desirable. If we can also (or instead) measure the initial slope, which is essentially eggs per adult times survival rate from egg to recruitment at optimal conditions (low density), then Equation 7.5.2 or 7.5.3 may be the preferred form. If, however, we only have stock-recruitment data as in Figures 7.1–7.3, it may be easier to think of a spawning stock size that produces half of the maximum recruitment, in which case Equation 7.5.1 may be the preferred form.

Biological assumptions

The Beverton-Holt stock-recruitment curve is based on the assumption that juvenile competition results in a mortality rate that is linearly dependent upon the number of fish alive in the cohort at any time. That is

$$\frac{dN}{dt} = -(q + pN)N \qquad (7.5.4)$$

where N is the number alive in the cohort at time t, q is a density-independent mortality rate and pN is a mortality rate component that is proportional to the density of the cohort at time t. Thus when the cohort is larger, the individuals disappear faster. If we assume that each spawner produces a certain number of eggs that begin the cohort (N at time 0 equals fecundity times spawners), and then solve Equation 7.5.4 for the number alive at some specific later time of recruitment, we arrive at Equation 7.5.3 (see Beverton-Holt 1957, p. 48 for derivation). The a and b parameters of the equation represent combinations of fecundity, q, and p parameters.

Equation 7.5.4 can represent a wide variety of biological phenomena, including competition for food or space, and would be a reasonable description of the random egg deposition model discussed earlier. However, when dealing with stock-recruitment curves, we are averaging so many biological processes that it is better to think about the curve as a general statistical description and not try to determine the proper stock-recruitment curve from general principles.

Ricker's model

Form and properties

Ricker (1954) proposed a frequently used model of stock and recruitment that can be written

$$R = Se^{a(1-S/b)} \qquad (7.5.5)$$

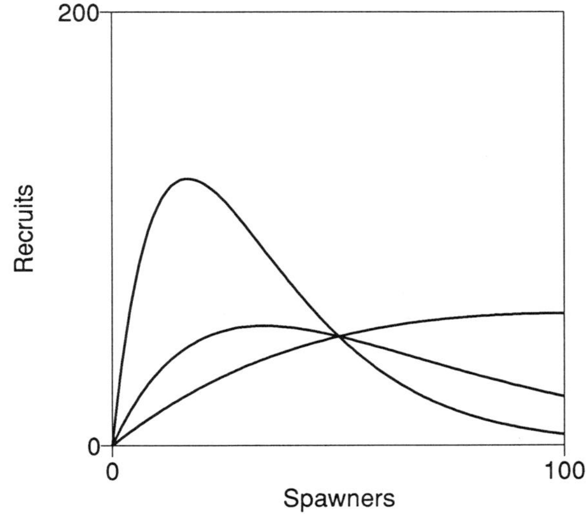

Figure 7.10. Ricker curves with different a values.

where R is the recruitment, S is the spawning stock size, e^a is the initial slope of the curve, and b is the value of S at which $R = S$. Ricker curves with different a values are shown in Figure 7.10.

Unlike the Beverton-Holt curve, the Ricker curve shows declining recruitment at higher stock sizes. The form of Equation 7.5.5 is often used for species such as Pacific salmon where recruitment is measured as adults, the point at which $R = S$ has a biological meaning. When dealing with other species where spawning stock may be measured in tons and recruitment as individuals recruited to the fishery at a specific age, Equation 7.5.6 is a preferred form of the Ricker model:

$$R = aSe^{-bS} \tag{7.5.6}$$

In this form, a is the recruits-per-spawner at low stock sizes, and b describes how quickly the recruits-per-spawner drop as S increases.

Biological assumptions

The biological assumption behind Ricker's model is that the mortality rate of the eggs and juveniles is proportional to the initial cohort size, that is, mortality is stock-dependent rather than density dependent. In the Beverton-Holt model, it was proportional to the cohort size at each time in the life history. To derive the Ricker curve we could write the rate of change for cohort size as

$$\frac{dN}{dt} = -(q + pS)N \tag{7.5.7}$$

where N is the cohort size at any time prior to recruitment, S is the initial spawning stock size (proportional to initial cohort size), $q + pS$ is the instantaneous mortality rate for the cohort, and q is the density-independent mortality rate. Solving for the cohort size at some specific time t we obtain

$$N_t = N_0 e^{-pSt} e^{-qt} \tag{7.5.8}$$

where N_0 is the initial cohort size. We also know that $N_0 = Sf$, where f is the number of eggs per spawner. Thus the recruits at time t (N_t in Equation 7.5.8) as a function of S is

$$R_t = aSe^{-bS} \tag{7.5.9}$$

where t and p have been combined into a new parameter b, and a is the survival from density-independent mortality times the fecundity.

Four commonly discussed mechanisms that can lead to a Ricker-shaped recruitment curve are cannibalism of the juveniles by the adults, disease transmission, damage by adults of one another's spawning sites, and density-dependent growth coupled with size-dependent predation. In cannibalism, the mortality rate of recruits would depend upon how many adults there were; thus the pS term in Equation 7.5.7 is clear, with p representing an instantaneous cannibalism rate per adult. In disease transmission, it is assumed that the initial cohort size (proportional to S) would determine how many of the individuals were infected with the disease. For the site damage mechanism, e^{-bS} is thought of as representing the probability of any site not being damaged if the spawning is distributed at random and each spawner may damage a relative area b.

Other models

The Beverton-Holt and Ricker models capture most of the needed behaviors for average stock-recruitment curves, and the available data seldom justify more complex or alternative shapes of models. However, there are others in the literature that should be mentioned.

Deriso model

Deriso (1980) proposed a stock-recruitment model that is interesting primarily in that the Beverton-Holt and Ricker models are specific cases of it. Its properties are discussed in Schnute (1985) in a useful manner. The model is

$$R = aS\,(1 - bcS)^{1/c} \tag{7.5.10}$$

where c is a third parameter that determines the general form of the model as follows:

$$\begin{aligned} c &= -\infty & R &= aS \\ c &= -1 & R &= aS/(1 + bS) \\ c &= 0 & R &= aSe^{-bS} \\ c &= 1 & R &= aS(1 - bS) \end{aligned} \tag{7.5.11}$$

The first form is density-independent recruitment, the second case is the familiar Beverton-Holt form, which could easily be set to produce constant recruitment, the third case is the Ricker form, and the last case has recruitment per spawner declining linearly with increasing spawning stock. The last case would violate Ricker's second principle, because recruitment will be negative with large S.

Depensatory models

The simplest way to approach depensatory models is to think of the stock-recruitment curve as producing potential recruits which are then subjected to predation. We can model the predation process using a standard Holling (1959) type II functional response

$$E = \frac{cR}{e + R} \tag{7.5.12}$$

where E is the number of prey eaten per predator, R is the number of prey available (recruits), c is the maximum number of prey that an individual predator can eat, and e is the prey density at which a predator will eat $c/2$ prey. If we have M predators, the total number of recruits surviving the predation process (N) will be

$$N = R - \frac{cMR}{e + R} \tag{7.5.13}$$

Peterman (1977, 1980) and Peterman and Gatto (1978) discuss various examples of depensatory actions ranging from native Indian food fisheries to predatory trout and provide evidence that depensation is not uncommon in fisheries. The Icelandic herring data shown earlier (Figure 7.2) could easily be imagined to fit such a relationship, although other mechanisms are certainly possible.

Another way to represent depensatory effects is to replace the aS or $a'S$ terms in the Ricker or Beverton-Holt recruitment models with the term aS^m

Table 7.1. Probability of recruitment table for Skeena River sockeye salmon.

Recruitment	Spawning Stock							
From:	0–	200–	400–	600–	800–	1000–	1200–	1400–
To:	200	400	600	800	1000	1200	1400	1600
3600–4000	0	0	0	0	0.333	0	0	0
3200–3600	0	0	0	0	0	0	0	0
2800–3200	0	0	0.091	0	0.167	0	0	0
2400–2800	0	0	0	0	0.167	0.200	0	0
2000–2400	0	0	0	0.429	0.167	0	0	0
1600–2000	0	0	0.182	0	0	0.200	0	1.000
1200–1600	0	0.333	0.364	0.143	0	0.400	0	0
800–1200	0	0.250	0.273	0.429	0	0.200	0	0
400–800	0.667	0.417	0.091	0	0.167	0	0	0
0–400	0.333	0	0	0	0	0	0	0
number of points	3	12	11	7	6	5	0	1
avg spawners	164	308	532	695	924	1077	0	1506
avg recruitment	487	917	1433	1585	2697	1535	0	1921
surplus yield	322	609	901	889	1773	458	0	415

where the power parameter m is greater than 1.0. In principle, m can be estimated from stock-recruitment data and should be much larger than 1.0 (e.g., 2.0 or larger) if depensatory effects are strong.

Tabular and Markovian approaches

A totally different approach to the description of stock-recruitment relationships is to not bother fitting any average curve, and instead simply describe the data tabularly by breaking the range of potential stocks and recruitments into intervals and computing the proportion of the times that a spawning stock within any specific interval produces a recruitment within each recruitment interval. This method has been used by Getz and Swartzman (1981) and Overholtz et al. (1986). Table 7.1 shows such an approach for the Skeena River sockeye data of Figure 7.1.

This type of approach is also called a Markov model in mathematical terms and the table is called a Markov transition probability matrix, but it is simpler to just think of it as a table. This approach is highly recommended when many years data are available, because it can accommodate any possible form of stock-recruitment curve and explicitly incorporates the type of variation seen in the data.

The major problem with the tabular approach is that it requires a lot of data. You should not even try such an approach without 30–50 data points. Even if we only use five intervals for spawning stock size and recruitment, we are attempting to estimate 25 parameters. This is quite a jump from the three parameters of the Beverton-Holt and Ricker models. Do not fool yourself by thinking that you have avoided a parametric approach by using a tabular representation of recruitment probabilities based directly on the data; when you attempt to make predictions about future recruitment patterns and variability using the tabular approach, all those tabled parameter estimates will come back to haunt you.

Structure of random variation around recruitment curves

The tabular approach explicitly incorporates random variation, but the average stock-recruitment curves, such as Beverton-Holt and Ricker, need an additional term to describe how the observed recruitment will vary around the average. There are two approaches to analyzing variation around recruitment curves. One is theoretical, where you propose models of how survival should vary and then see what this means in terms of the distribution of recruits for any specific spawning stock size. The second approach is to simply look at the distribution of variability around fitted curves.

Theory

If we think of the stock-recruitment process as a series of individual life history stages and the total survival from egg to recruit as the product of these survivals, we can write

$$s = s_1 s_2 s_3 \ldots s_n \qquad (7.5.14)$$

where s is the total survival over life history stages 1 to n, and s_i is the survival rate in life history stage i. If we take logarithms of both sides, we obtain

$$\log(s) = \Sigma \log(s_i) \qquad (7.5.15)$$

If we assume that the survival rate in each life history stage is an independent random variable, and that no single or few stages with peculiar patterns of variation dominate the sum, then we can use an important idea from basic statistics to predict the distribution of the sum (i.e., of overall survival rate). The Central Limit Theorem states that the sum of any long series of independent, identically distributed random variables [log (s_i) in this case] will have a distribution that approaches the normal distribution as the number of values summed increases. In practice, not many values (e.g., 10–20) need be added before the sum closely approaches a normal distribution. This means that the overall survival rate should be lognormally distributed, as in Equation 7.5.16

$$R = \bar{R}e^{w} \qquad (7.5.16)$$

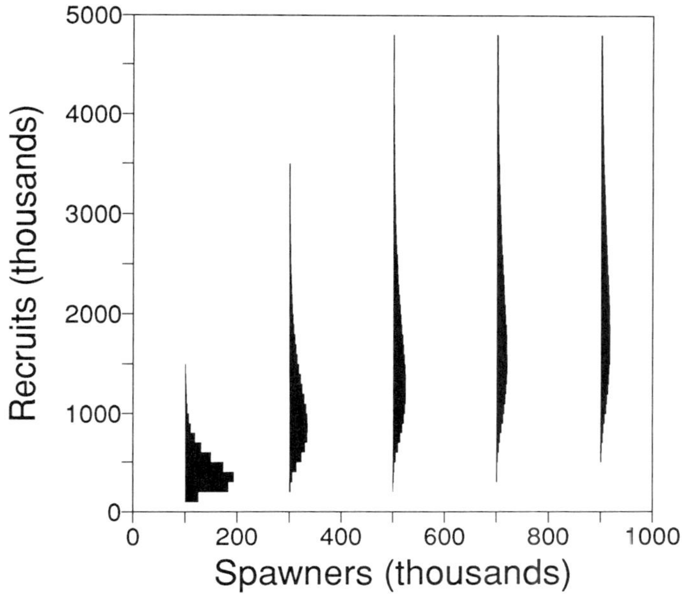

Figure 7.11. Random variation around a spawner-recruit curve expected under the assumption of independent survival through sequential life history stages.

where w is normally distributed with a mean of zero and a standard deviation of σ_w. A lognormal distribution will occasionally show very large recruitment; it has a long tail toward the upward end. Another feature of this model is that the amount of variation will be proportional to the average recruitment, so we expect to see lower variability at small recruitments and higher variability at large recruitments. These features are captured in Figure 7.11.

Observed patterns

Peterman (1981) examined a number of data sets to see if the observed pattern of variability conformed to a lognormal pattern. The Skeena River data in Figure 7.1 are typical of the pattern observed, and Peterman found that he could not reject the hypothesis that the distributions were lognormal. Of the data presented in Figures 7.1–7.3, the Icelandic herring data seem not to show a lognormal pattern, since there appears to be as much variability at small spawning stock sizes as at large ones. However, the tendency to generate a few large recruitments is evident in the herring data set as well.

In summary, it appears that lognormal distributions are found in many stock-recruitment data sets and that there are theoretical justifications for this assumption. We therefore recommend that a lognormal distribution be the starting assumption for stock-recruitment work, although particular data sets may show different patterns and other models may be preferred for specific species.

7.6. Estimation of Stock-Recruitment Parameters

There are two approaches to the estimation of stock-recruitment relationships; first are the statistical methods discussed in the section below and second are simple graphical and eyeball methods. Eyeball methods are dangerous and should never be used. You cannot trust your eye and hand to accurately represent average behavior in a data set, especially when there is considerable random variation. This problem is vividly illustrated by Figure 7.12 redrawn from Murphy (1983), which presents stock and recruitment data for the greater lizardfish (*Sardia tumbil*).

This eyeball fit shows a very steep recruitment curve whose slope increases initially, and an apparently very narrow region of optimal spawning stock.

Warning: Do not fit stock-recruitment curves by eye.

If we were to believe this recruitment curve, we would say that average recruitment could almost double as spawning stock went from roughly 1 to about 1.3 on the x axis and then decline sharply by the time it reached 2. While such a pattern is biologically conceivable, it would violate the principle of continuity mentioned above; we find it difficult to believe that nature can detect and respond to such small changes in the systematic and repeatable fashion shown.

The lower lines are the best-fit average recruitment curves of the Beverton-Holt and Ricker models using the methods discussed below. Obviously one's view of the stock-recruitment relationship in Figure 7.12 is quite different when using models that have been properly fit.

Another example where fitting by eye was used with some costly economic consequences is shown in Figure 7.13. Stock-recruitment plots for Rivers Inlet sockeye salmon were used during the late 1970s as one justification for closing the fishery 1979–1983 (and maintaining reduced fishing from 1984 to 1988) to rebuild spawning stocks toward a target of 1.0 million spawners. In eyeballing the data, biologists inevitably placed considerable weight on the two high recruitments seen at larger stock sizes. The fitted Ricker curve for 1948–1973 data (shown on the plot) fell far below these points, since the fitting procedure assumed the two high points to be log-

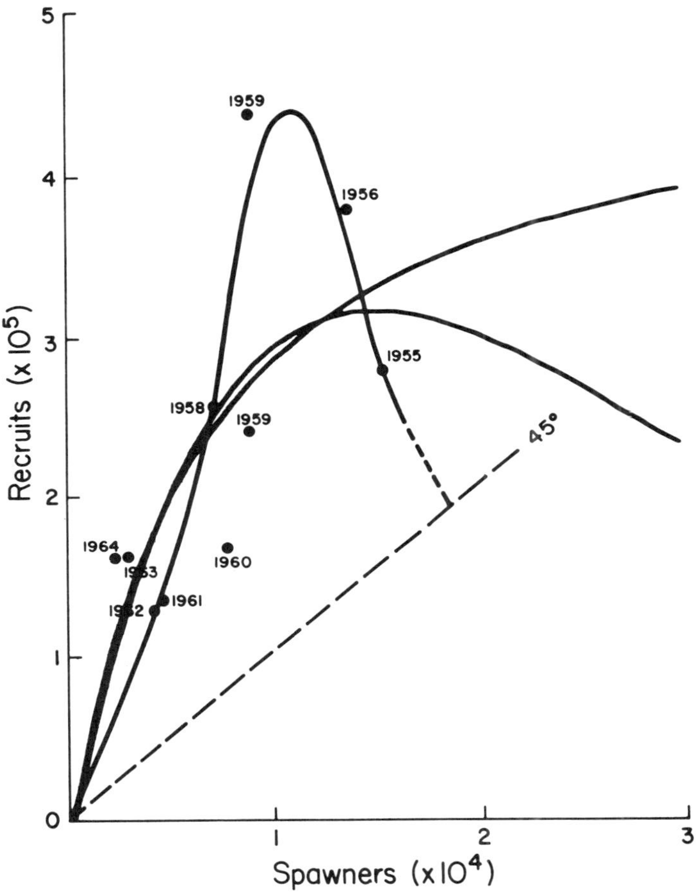

Figure 7.12. Stock-recruitment data for the greater lizardfish, redrawn from Murphy (1982). The highest line was fitted by eye and published in the paper by Murphy. The two lower lines are the Beverton-Holt and Ricker curves fitted by statistical methods.

normal "outliers" (see previous section), but no one wanted to trust the fit. Look at the recruitments for the 1979–1983 spawning years: would you have forecast better with an eyeball fit than with the Ricker curve? It is conceivable that in the fullness of time we will find that the optimum escapement for Rivers Inlet sockeye is indeed 1.0 million rather than the 0.4 million indicated by the Ricker curve, but it is beginning to look suspiciously like the fishermen have given up some 3.0 million sockeye from their catch since

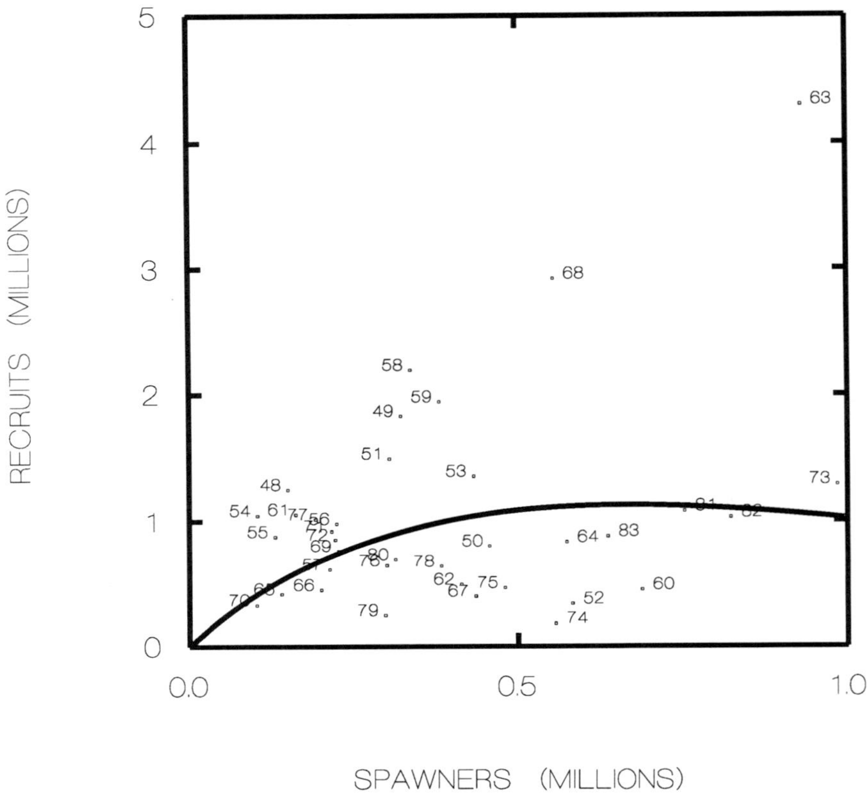

Figure 7.13. Stock-recruitment data for the Rivers Inlet sockeye salmon (1949–1982). Data from Canadian Department of Fisheries and Oceans.

1979 for nothing. At \$25 per fish, the eyeball fitting procedure may have contributed to a loss of some \$75 million from this fishery.

Ricker model

The normal method for fitting the Ricker model to a set of stock-recruitment data is to transform Equation 7.5.5 into the following form:

$$\log\left(\frac{R}{S}\right) = a - \frac{a}{b}S \tag{7.6.1}$$

and then treat this as a linear regression

$$y = b_0 + b_1 X + w \tag{7.6.2}$$

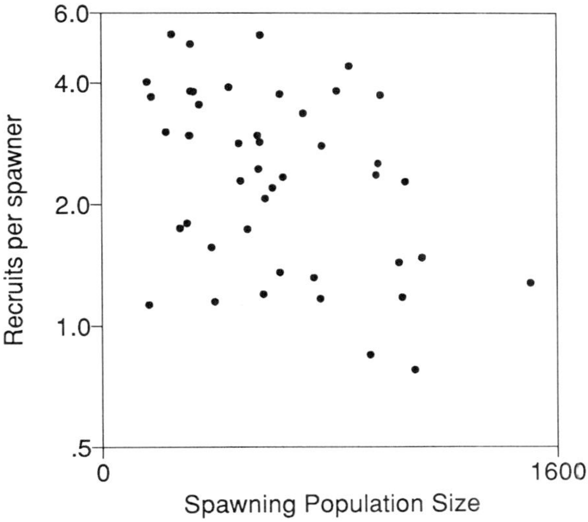

Figure 7.14. log (R/S) plotted against S for Skeena River sockeye salmon data.

where $y = \log\left(\frac{R}{S}\right)$ is the dependent variable, $b_0 = a$ is the intercept, $b_1 = -\frac{a}{b}$ is the slope, and w is the residual. We can view Equation 7.6.1 graphically by plotting the logarithm of R/S against the spawning stock size S. This is shown for the Skeena River sockeye data in Figure 7.14. The intercept in this relationship is the a parameter, where e^a is the initial slope of the stock-recruitment curve. One nice feature of Equation 7.6.1 is that it implicitly assumes lognormal errors. Thus the model that is really being fitted is

$$R = Se^{a(1-S/b)}e^{w} \tag{7.6.3}$$

The regression estimates two parameters, the slope and intercept, which are easily converted to the a and b parameters. We also obtain a third parameter σ, which is the standard deviation of the residuals. Because the expected value of e^w is not equal to zero when w is normally distributed with a mean of zero, the average stock-recruitment curve will not be the one described by the parameters a and b. Hilborn (1985c) has shown that since the expected value of e^w is $e^{\sigma^2/2}$, the average stock-recruitment curve will be a Ricker curve with the parameters a' and b' defined as

$$\begin{aligned} a' &= a + \frac{\sigma^2}{2} \\ b' &= \frac{a'}{a} b \end{aligned} \tag{7.6.4}$$

where *a* and *b* are the Ricker parameters estimated from the linear regression.

Beverton and Holt's model

The Beverton-Holt model can also be transformed into a linear regression for estimating parameters: Equation 7.5.1 becomes

$$\frac{S}{R} = \frac{b}{a} + \frac{1}{a} S \tag{7.6.5}$$

This form, although computationally convenient, does not conform to our assumption about lognormal errors; the form the errors take is actually quite difficult to interpret. The preferred method of fitting the Beverton-Holt model is to use Equation 7.6.6 :

$$\log (R) = \log \left(\frac{aS}{b + S}\right) + w \tag{7.6.6}$$

where *a* and *b* must now be estimated using nonlinear regression. The variance of *w* is simply the variance of the residuals (σ^2). Given the widespread availability of microcomputers to all fisheries biologists even in the poorest countries, anyone who has enough data to be fitting stock-recruitment relationships should do it properly and use Equation 7.6.6 unless there is a good reason to assume that lognormal errors are incorrect. One could of course try several methods and examine the residuals for normality.

As with the Ricker model, the parameters *a* and *b* of the Beverton-Holt model estimated from Equation 7.6.6 will not represent the average number of recruits for any level of spawning stock. The average will be

$$R = \frac{aS}{b + S} e^{\sigma^2/2} \tag{7.6.7}$$

which is a Beverton-Holt curve with the *a* parameter redefined as

$$a' = ae^{\sigma^2/2} \tag{7.6.8}$$

Management parameters from Ricker and Beverton-Holt models

When dealing with semelparous species (those that breed only once) stock-recruitment curves can be used to describe the entire life history of the fish stock, and we can calculate such parameters as optimum stock size for MSY, optimum harvest rate for MSY, and the stock size which produces maximum recruitment. Ricker (1975) presented a table of these formulae, and they are reproduced here in Table 7.2 with some minor modifications and editing.

Warning: This is only applicable for semelparous species, so unless you are working with such a species, do not try to use these relationships.

Before using these relationships remember that they only apply when the adults reproduce only once and the recruitment is measured after all non-fishing mortality. Pacific salmon are the only commonly managed fish species that nicely conform to these assumptions. A typical misapplication of these relationships can be found in the lizardfish graph (Figure 5.4) where the $y = x$ line is drawn with the implicit assumption that when $y = x$ the stock will be at equilibrium. If a stock size of 2 produces an adult population cohort size of 1, the stock may still be able to grow if many of the adults survive another year and reproduce again.

The one substantial change in Table 7.2 from Ricker's table is the calculation of optimum stock size and harvest rate for the Ricker model. Ricker gave these as complex transcendental functions of the parameters; however, Hilborn (1985) showed that the optimal stock size for MSY can be approximated by

$$S_{\text{MSY}} = b(0.5 - 0.07a) \tag{7.6.9}$$

and that the optimum harvest rate for MSY is close to

$$u_{\text{MSY}} = 0.5a - 0.07a^2 \tag{7.6.10}$$

Both of these relationships hold for the region $(0 < a < 3)$ which generally applies for fish species that meet the assumption of adults reproducing only once.

One interesting feature of the relationships in Table 7.2 is that in both cases there are two parameters to the models, one of the parameters is a slope parameter related to the number of recruits-per-spawner when compensatory effects are small (small stock size), and the other is a scaling parameter that describes how the recruits-per-spawner decrease with increasing stock size.

We call the slope parameter productivity, and the optimum harvest rate depends only on it. The second parameter we call capacity, and it determines how large the stock would be if it were not fished. The optimum stock size depends on both productivity and capacity. Productivity is related to density-independent survival rates (it will in fact be the eggs per spawner times the survival rate through all life history stages when there is no density-dependent effect). The capacity parameter will be determined by the nature of compensatory mortality. Although no systematic surveys have been per-

Table 7.2. Biologically important attributes of the Ricker and Beverton-Holt models.

$R =$	aSe^{-bS}	$Se^{a[1-(S/b)]}$	$\frac{aS}{b+S}$	$\frac{aS}{1+\frac{a}{b}S}$
Slope at origin	a	e^a	$\frac{a}{b}$	a
$S_{\max}$	$\frac{1}{b}$	$\frac{b}{a}$	∞	∞
$R_{\max}$	$\frac{a}{b}e^{-1}$	$\frac{be^{a-1}}{a}$	a	b
$S_{\text{MSY}} = S^*$	$\frac{\log a}{b}(0.5\text{–}0.07 \log a)$	$b(0.5\text{–}0.07a)$	$a\sqrt{\frac{b}{a}} - b$	$b\sqrt{\frac{1}{a}} - \frac{b}{a}$
MSY + S^*	$aS^*e^{-bS^*}$	$S^*e^{a[1-(S^*/b)]}$	$\frac{aS^*}{b+S^*}$	$\frac{aS^*}{1+(a/b)S^*}$
u_{MSY}	$0.5 \log a - 0.07(\log a)^2$	$0.5a - 0.07a^2$	$1 - \sqrt{\frac{b}{a}}$	$1 - \sqrt{\frac{1}{a}}$

formed, we suspect that the productivities will typically be similar within a species over much of its range (although presumably higher in better habitats than in poorer ones), whereas the capacity parameter will depend on the size of the area and should be quite variable from stock to stock of the same species.

Uncertainty and confidence bounds

Equations 7.6.1 and 7.6.7 provide us with parameter estimates for the average recruitment curves and the variances. We would also like to know how much confidence we have about the parameter estimates. There are several ways of estimating the level of confidence: (1) variations on traditional linear regression confidence bounds, (2) use of nonlinear regression approximations of confidence bounds, (3) numerical methods such as jackknife and bootstrap statistics, and (4) simplification to one-parameter models and use of simple linear statistics. None of these methods is well established or well understood. The purpose of this section is simply to review the alternative approaches and discuss what is currently known about their strengths and weaknesses.

Linear statistical estimates

Simple linear statistics provide estimates of the standard deviation of the intercept (σ_{b_0}) and the standard deviation of the slope (σ_{b_1}). If we are using the Equation 7.5.5 form of the Ricker model, we obtain clean confidence bounds on the parameter $a = b_0$, but rather messy bounds on b, since we estimate the slope of the regression (a/b) rather than b itself. Equation 7.6.11 gives an alternative form of the Ricker model and its estimation that provides clean confidence bounds for both parameters.

$$R = Se^{a-bS}e^{w}$$

$$\log\left(\frac{R}{S}\right) = a - bS + w \qquad (7.6.11)$$

The estimation of the optimum harvest rate using Equation 7.6.11 is also straightforward since it involves only a, whereas the estimate of S_{MSY} is very messy because both a and b are involved. It is additionally complicated because we should be using a' and b', which depend on the variance and, of course, our confidence in the estimate of the variance. For these reasons no one to our knowledge has attempted to discuss confidence limits for the Ricker model using simple linear statistics. Many authors do use the confidence bounds of the regression from standard statistics, but these do not provide confidence bounds on the management parameters.

The Beverton-Holt model does not present any neat linear form in which the parameters emerge directly from the estimation. Therefore it is considerably more difficult than the Ricker model for treating as a problem in linear confidence intervals.

Nonlinear approximations

The problem of no linear form producing direct parameter estimates of the Beverton-Holt model can be solved by nonlinear estimation using Equation 7.6.6. One can make approximations of the confidence intervals on the parameters using a number of standard computer packages. Three problems present themselves. First, these approximations are only approximations in the region of the best estimate of the parameters and are unlikely to be reliable for the highly variable data one obtains in stock-recruitment analysis. Although the 10 or 20% confidence limits might be well approximated, the chances of coming anywhere close to the 80 or 95% confidence limits are very small.

The second problem with nonlinear approximations is that they do not provide an estimate of the confidence in the variance estimate, which is required for a realistic appraisal of the confidence in the average yield curve.

Third, it is difficult and potentially misleading to provide estimates of confidence for the management parameters, which are computed as functions of the regression parameters and therefore involve components of uncertainty measured by the variances and covariances of these parameters.

Numerical methods: jack-knife and bootstrap

Probably the most promising techniques for producing reliable confidence limits are the so called jack-knife and bootstrap methods that rely on the intensive use of computers. Jack-knife is the simpler of the two. The basic idea is that if we have a data set of, for instance, 20 data points, we can generate 20 different data sets that consist of 19 points, by successively excluding each data point from the set. We then reestimate the parameters for each of the 20 new data sets. This provides us with 20 estimates of the parameters of the model, as well as the management parameters S_{MSY} and u_{MSY}.

The jack-knife method is quite simple, requiring only n replications of the estimation method for n data points. It is established that jack-knife is reasonably reliable for linear statistics (Efron 1979, 1981, 1982), but how well it works for derived parameters such as S_{MSY} and u_{MSY} is unknown and deserves more systematic investigation.

The bootstrap method is a computationally more intensive variation on jack-knife. From a data set of n points, we generate a large number (100–1,000) of new data sets of n points by randomly selecting (with replacement) data points from the original data. Some points might occur several times in the generated data sets, some not at all. Once a generated data set is prepared, parameter estimation is done and estimates of all the parameters stored away. We repeat this until we have generated 100–1,000 data sets, and estimated the parameters 100–1,000 times. We then take the standard deviation of the parameters obtained and treat these as estimates of the true standard deviation. Alternatively, and unlike jack-knife, we can use the actual frequency distribution of generated parameter estimates to examine the expected confidence bounds. This allows for asymmetry that using a standard deviation does not.

Both the bootstrap and jack-knife estimates are being used in production models (Butterworth and Andrew 1984, Punt 1988) and catch-at-age analysis (Deriso et al. 1985), but neither has yet been used in published stock-recruitment assessments.

Model Simplification Don Ludwig (unpublished) has proposed a simplified method for both fitting the Ricker model and analyzing the confidence bounds associated with the parameter estimates. His method begins with the observation that when one examines plots of log (R/S) versus S, the average x and average y values are usually well defined and can be treated as known quantities without error. He then writes the estimation equation for the Ricker model as

$$y_i = \bar{y} + b_1(x_i - \bar{x}) + w \qquad (7.6.12)$$

Now the only parameters to estimate are the slope b_1 and σ_w. This provides exactly the same answer for b_1 as one obtains from the standard two parameter regression. However, by assuming that all uncertainty is in the slope, the variance of the slope can be estimated as

$$\sigma_{b_1}^2 = \frac{\sigma_w^2}{\Sigma (x_i - \bar{x})^2} \qquad (7.6.13)$$

that is, the residual variance divided by the sum of the squared deviations from $\bar{x}$. We can then compute the confidence bounds on b_1 as

$$\hat{b}_1 - t_{n-2}\sigma_{b_1} < b_1 < \hat{b}_1 + t_{n-2}\sigma_{b_1} \qquad (7.6.14)$$

where t is the t statistic with $n - 2$ degrees of freedom. A warning is needed here: these confidence limits for b_1 will be too optimistic unless $\bar{x}$ and $\bar{y}$ are truly well determined, which is generally not true when there are less than 20–30 years data. The Ricker parameters a and b can be estimated from the $\hat{b}_1$ above by

$$a = \bar{y} - \hat{b}_1\bar{x} \qquad (7.6.15)$$

$$b = \bar{x} - \frac{\bar{y}}{\hat{b}_1}$$

Examples of fitting the Ricker and Beverton-Holt models

In this section we examine problems and results in fitting Ricker and Beverton-Holt stock-recruitment curves to the three data sets shown in Figures 7.1–7.3.

Curves

Figure 7.15 shows the stock-recruitment graphs from Figures 7.1–7.3 with the best fitting Ricker and Beverton-Holt curves. For the Skeena River sockeye data the two stock-recruitment curves are almost identical. The $y = x$ line is drawn on this graph to indicate the average sustainable yield, which is the height of the stock-recruitment curve above the $y = x$ line. We can see that over a very broad range of stock sizes the apparent sustainable yield is roughly constant. The fitted curve also indicates that thereshould be no immediate concern about decreased production at higher stock sizes; the curve does not bend over very steeply.

The Icelandic herring data are much more interesting and reflect some major limitations of using standard stock-recruitment curves. The Beverton-Holt curve bends over quite rapidly, whereas the Ricker curve continues to

(a)

(b)

(c)

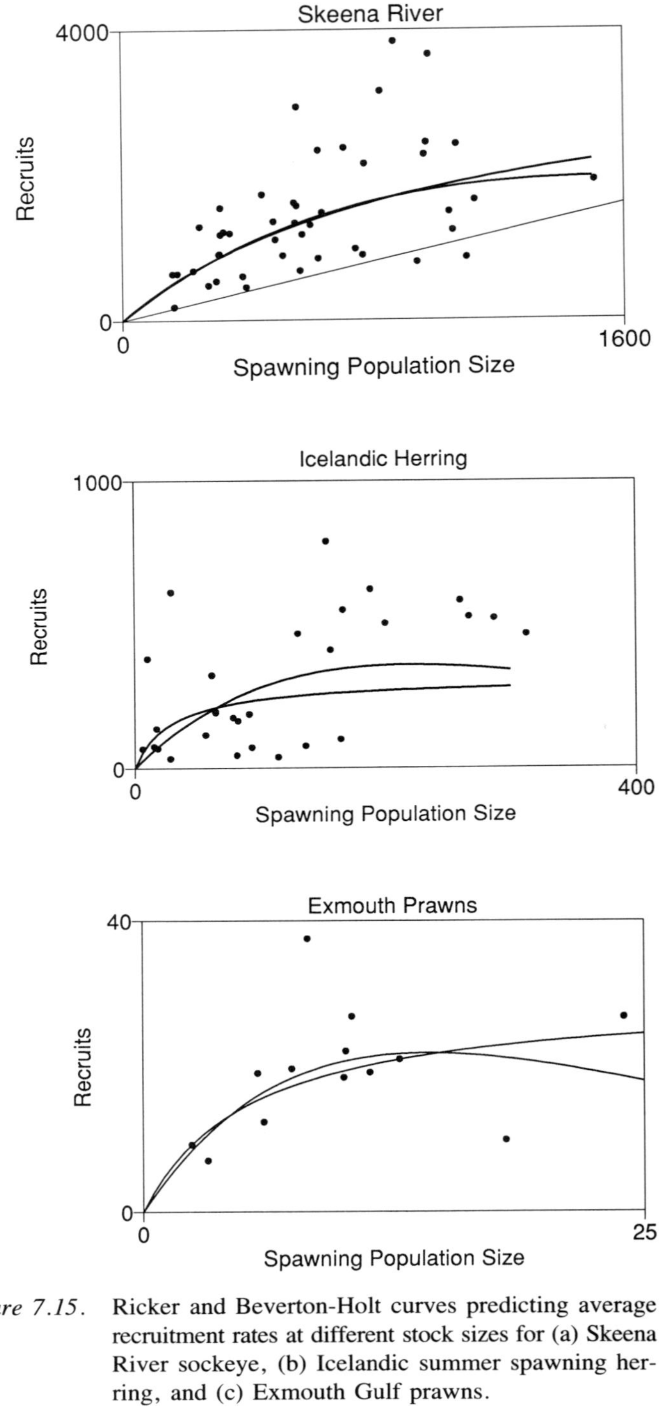

Figure 7.15. Ricker and Beverton-Holt curves predicting average recruitment rates at different stock sizes for (a) Skeena River sockeye, (b) Icelandic summer spawning herring, and (c) Exmouth Gulf prawns.

climb out to a spawning stock size of approximately 170. Neither curve appears to fit the data particularly well, and all the data points for high stock sizes are above the curves. An analysis of residuals would almost certainly conclude that there is a pattern not represented by the curve.

There are two major reasons for this apparently bad fit. First, the curves are fit to the logarithm of recruitment, so the curve will fit the lower recruitment points more closely (on an arithmetic graph) than the higher recruitment points. Secondly, in the stock size range 100–160 there are several very low recruitments and a few high ones. If the curves were to fit the high points more closely, large residuals (magnified by the logarithmic transform) would be generated.

We can say with some confidence, therefore, that these data do not conform to a Ricker or Beverton-Holt curve with logarithmic error assumptions. It seems quite likely that some systematic change was occurring, a possibility discussed by Walters (1987).

The Exmouth Gulf tiger prawn data fit either curve quite well except for the two major outliers noted in section 7.22. The Ricker curve bends over a bit above a spawning stock of 13, whereas the Beverton-Holt curve continues to increase at a reasonable rate over the entire range of the data. Any measure of goodness of fit will be dominated by the two outliers.

Sums of Squares

Table 7.3 shows sum of squares tables for the Skeena sockeye data for each model. The common null hypothesis is that the logarithm of recruits is constant, and the F tests compare the sums of squared residuals after fitting the models to the sum of squares assuming constant recruitment. Note that this is not the default null hypothesis when fitting the Ricker curve using a regression package: the null hypothesis for a regression fit is that the log(R/S) is constant. Also shown are sums of squares and F tests for improvement due to model fitting, for the null hypothesis that recruits-per-spawner is constant, which we suggested earlier may be a better null hypothesis than constant recruitment. These tests enable us to ask if the models do any better than just assuming constant recruitment or a density-independent spawner-recruit relationship. An alternative way to determine if recruits-per-spawner is constant would be to determine whether the Ricker b or Beverton-Holt b' parameters differ significantly from zero.

We can see that the Ricker and Beverton-Holt models provide very similar fits, something we observed graphically in the previous section. The sum of squares attributable to the model divided by the total sum of squares is the proportion of total recruitment variation explained (r^2), and we can see that for the constant recruitment hypothesis, we have explained much of the variability and the F statistic is quite large. Neither model does much better

Table 7.3. Sum-of-squares tables for Ricker and Beverton-Holt models fitted to the Skeena River sockeye salmon data. The notation is d.f. = degrees of freedom, s.s. = sum of squares, m.s. = mean square, and F = value of F statistic.

source	d.f.	s.s.	m.s.	F
Ricker model for constant recruitment				
model	1	6.879	3.439	14.963
residual	42	9.654	.230	
total	43	16.533		
Ricker model for proportional recruitment				
model	1	1.802	.901	3.920
residual	42	9.654	.230	
total	43	11.456		
Beverton-Holt model for constant recruitment				
model	1	6.786	3.393	14.622
residual	42	9.747	.232	
total	43	16.533		
Beverton-Holt model for proportional recruitment				
model	1	1.709	.855	3.683
residual	42	9.747	.232	
total	43	11.456		

than the proportional recruitment hypothesis; again something we noted graphically before. All F values are significant at the 0.05 level, but we can reject only the constant recruitment hypothesis at the 0.01 level.

Parameter estimates

Table 7.4 shows the parameter estimates and confidence bounds for the Skeena River sockeye data using the model $R = Se^{a(1-S/b)}$. The Skeena River sockeye data is the only data set for which we can estimate optimal harvest rate and optimal stock size, because it is the only semelparous stock under consideration. σ_a is given for both the linear estimation method and bootstrapping in the Ricker model. Note that both estimates are similar. This is true as well for σ_b in the Ricker model. Linear estimates of standard deviations for the Beverton-Holt model are not given.

The standard deviation σ_w is essentially the same for each model. The proportion of variation in stock size explained by the model under the constant recruitment null hypothesis is 41%, whereas that explained under the proportional recruitment null hypothesis is only 16%. The optimum harvest rate, standard deviations from bootstrapping, and 50% confidence intervals

Table 7.4. Parameter estimates and associated statistics for Skeena River sockeye data.

Parameter	Ricker	Beverton-Holt
a	1.23	3.66
σ_a (linear est)	0.154	—
σ_a (bootstrap)	0.162	1.00
50% c.i.	1.09–1.30	3.32–4.29
b	1932	3671
σ_b (linear est)	768	—
σ_b (bootstrap)	533	1092
50% c.i.	1717–2432	2906–4369
σ_w	0.22	0.23
correlation	−0.89	—
r^2 constant recruitment	0.41	0.41
r^2 proportional recruitment	0.16	0.14
optimum harvest rate	0.55	0.48
s.d. (bootstrap)	0.0541	0.062
50% c.i.	0.50–0.57	0.45–0.52
optimum stock size	851	915
s.d. (bootstrap)	246	271
50% c.i.	769–1099	725–1089

are given, and are roughly similar for both models. The same is true for the optimum stock size. Note, however, that the uncertainty about optimum stock size is much larger than about optimum harvest rate. When one considers that the 50% confidence intervals go from roughly 600,000 to 1,000,000 fish, one recognizes how uncertain we often are about optimum escapement. This is with 45 data points! These data are obviously not too relevant to current Skeena River management, because much has changed since 1909, but the example should serve as a warning to those who hope to understand the optimum stock size with a high degree of confidence, just by waiting until ample years of data become available.

Another useful statistic from regressions is the correlation between parameters. The a and b estimates are almost always highly correlated; in this case the value is -0.89. This is discussed in more detail in the next section.

The Icelandic summer herring data show several interesting features in Table 7.5. First, note how much uncertainty there is about the a parameter. No standard deviations are given for the Beverton-Holt parameters because some of the bootstrap estimates produced extremely large parameters that bias the standard deviation, but do not affect the calculation of confidence intervals. The standard deviation σ_w is much higher than we saw in the Skeena River data. The Ricker model fails to explain significantly more variation

Table 7.5. Parameter estimates and associated statistics for Icelandic Summer herring.

Parameter	Ricker	Beverton-Holt
a	4.44	10.83
σ_a (bootstrap)	1.21	6.29
50% c.i.	3.77–5.48	8.17–16.6
b	0.0046	309
σ_b (bootstrap)	0.00195	84
50% c.i.	0.0031–0.0056	272–385
σ_w	0.974	0.85
r^2 constant recruitment	0.00	0.13
r^2 proportional recruitment	0.15	0.26

than the constant recruitment hypothesis, whereas the Beverton-Holt model explains at best 26%. We see numerically that these models just do not fit the data particularly well.

The Exmouth tiger prawn data show no real surprises in Table 7.6, with a very small σ_w considering the two outliers, and a reasonably high proportion of the variation explained. The Ricker model seems to fit slightly better based on the r^2.

Parameter correlation

A common phenomenon when fitting stock-recruitment curves (and most fisheries statistics for that matter) is that parameters are often highly correlated. We saw for the Skeena River sockeye data that the correlation was −0.89. These data can be explained almost equally well by many different stock-recruitment curves, but the family of curves that fit the data well represents alternatives with large a and low b values, or low a and large b values. The simplest way to think about these alternatives is to realize that the data only determine the mean recruitment at the mean stock size very well, and the alternatives represent the family of curves that pass through this mean but with different slopes. If one simply plots the sum of squares versus a and b values, parameter correlation shows up as a diagonal valley.

Figure 7.16 shows another way to visualize parameter correlation. We have plotted the 100 bootstrap estimates of a and b for the Ricker model $R = Se^{a-bS}$ fit to the Exmouth Gulf prawn data. Most parameter estimates are clustered near the mean estimates, but there are long tails. Note that using this version of the Ricker model the parameters are positively correlated.

Table 7.6. Parameter estimates and associated statistics for Exmouth Gulf Tiger Prawns.

Parameter	Ricker	Beverton-Holt
a	4.185	5.63
σ_a (bootstrap)	0.80	2.44
50% c.i.	3.67–4.63	4.23–7.51
b	0.71	29.55
σ_b (bootstrap)	0.16	9.5
50% c.i.	0.060–0.079	25.5–38.2
σ_w	0.14	0.15
r^2 constant recruitment	0.43	0.38
r^2 proportional recruitment	0.58	0.53

Analysis of residuals

An additional step that should be taken in the analysis of stock-recruitment data is to carefully examine the residuals. Patterns in the residuals may indicate either an incorrect model fit to the data or an external variable providing systematic change. Figure 7.17 shows one way of examining the residuals by plotting the autocorrelation coefficients, in this case for the Skeena River sockeye data. There is a strong (significant at the 0.05 level) negative autocorrelation at lag 3. This means that if there was a point above the line at year t, there was usually a point below the line at year $t + 3$. This type of autocorrelation is a good indication of the need to look beyond simple stock-recruitment relationships in understanding the dynamics of the stock. Simply take the residuals $\hat{w}_t$ from the model fit, and calculate the lag k autocorrelation $r(k)$ as

$$r(k) = \left[\frac{T - 1}{T - 1 - k}\right] \frac{\sum_{t=1}^{T-k} \hat{w}_t \hat{w}_{t+k}}{\sum_{t=1}^{T} \hat{w}_t^2} \tag{7.6.16}$$

Some workers prefer to use the "Yule-Walker" estimates, which are as above except omitting the correction for degrees of freedom $(T - 1)/(T - 1 - k)$; the Yule-Walker estimates will necessarily decrease for high k where there are fewer observations to use in the estimates.

It is worth injecting a warning at this point. When there is a strong autocorrelation pattern in the natural factors that give rise to the stock-recruitment residuals, the fitting procedures can give badly biased stock-recruit-

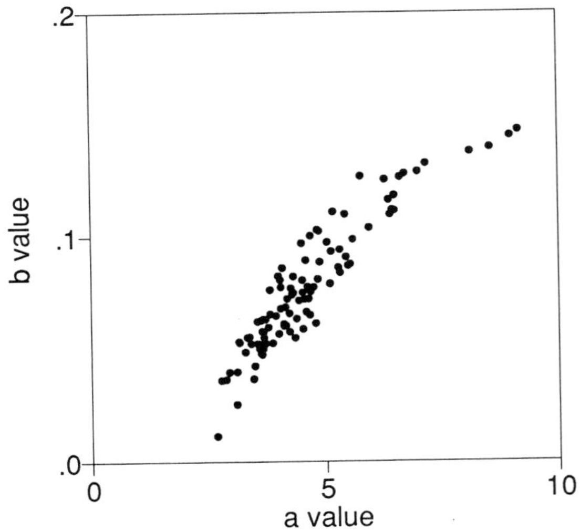

Figure 7.16. Correlations between a and b parameters for bootstrap estimates of Ricker model fit to Exmouth Gulf prawn data.

ment parameter estimates and will also lead to underestimates of the strength of these autocorrelations (see the section below on biases). This problem is especially serious for long-lived species, where the spawning stock will accumulate with a time delay following any run of years of relatively good environmental conditions for recruitment; if runs of bad years tend to follow runs of good years, then the bad years will coincide with high spawning stock accumulations. This will make the fitted recruitment curve bend down at high stock sizes, and this fit will ascribe variation (wrongly) to the curve rather than to the autocorrelation (runs) pattern in the environmental effects.

7.7. Extensions of Basic Models

Up to this point, we have been primarily dealing with two-parameter stock and recruitment curves that have rather simple and, frankly, boring behavior. These models are generally sufficient for the analysis of fisheries stock-recruitment data simply because the data do not often let us see more detail amidst all the variability. The depensation model discussed earlier provides a hint of what kinds of interesting things can happen when we start to consider different relationships at various life history stages. We will consider two models that treat several sequential life history stages.

The second potential area for extension of simple stock-recruitment models is in adding the effects of other species or environmental changes. The basic

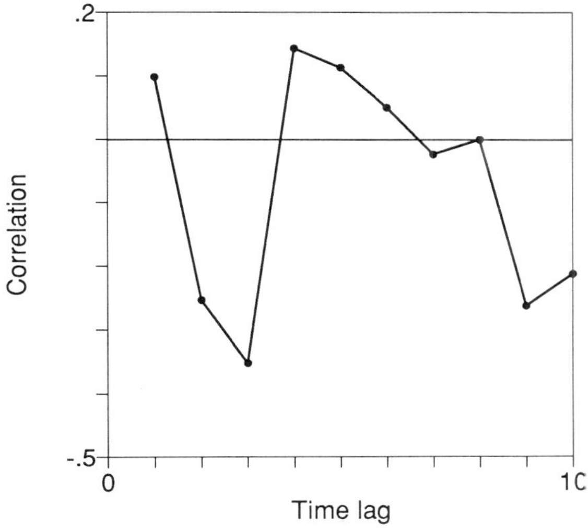

Figure 7.17. Autocorrelation coefficients of residuals of Ricker model fit to Skeena River sockeye data. Only the lag 3 coefficient is significant.

models consider all effects other than those caused by the abundance of the stock as unexplained random variation. As we learn more about the biology of the stock-recruitment process, we may be able to determine the cause of some of the external variability, and this may provide us with both a different perspective on the stock's internal dynamics, and perhaps a different prescription for management in the face of external changes.

Multistage models

Multistage models simply connect a series of stock-recruitment curves over various life-history stages into one aggregate relationship.

Paulik model

Paulik (1973) presented an interesting graphic method for showing how several stock-recruitment models can be combined over several life-history stages. Figure 7.18 shows an example from Paulik's paper with three stock-recruitment curves, from egg to fry, fry to smolts, and smolts to adults. The most interesting aspect of this approach is that reasonably straightforward curves at different life history stages can produce a complex curve for the aggregate stock-recruitment relationship.

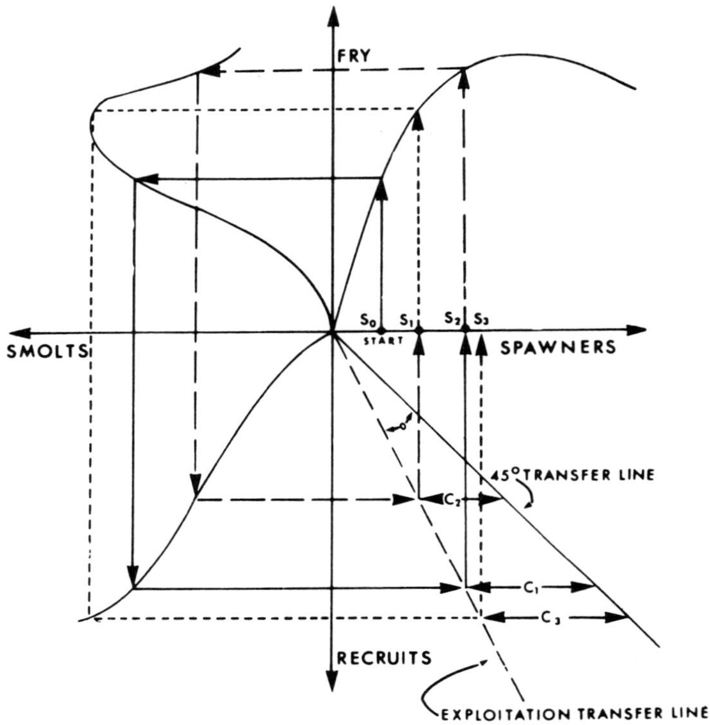

Figure 7.18. The Paulik (1973) 3-stage life-history model for Pacific salmon. The ordinate of one life stage becomes the abscissa of the next. It may help to rotate the page 90° clockwise when reading from one life stage to the next. From Paulik (1973). Reprinted by permission of ICES.

Beverton-Holt multistage model

Beverton and Holt (1957) showed how several Beverton-Holt curves can be combined over various life-history stages to produce a single stock-recruitment model, that is, in fact, also a Beverton-Holt curve. We can write the individual life-history stage models as

$$R_{i+1} = \frac{p_i S_i}{1 + (p_i/c_i)S_i} \tag{7.7.1}$$

where R_{i+1} is the number of recruits at the $(i + 1)$th life-history stage, p_i is the productivity for the stock-recruitment curve between stage i and $i + 1$, c_i is the capacity for that curve, and S_i is the number of fish starting that stage.

The total life-history stock-recruitment relationship up to stage i becomes

$$R_i = \frac{P_i}{1 + (P_i/C_i)S_1} \tag{7.7.2}$$

where R_i is the number of recruits at the end of the entire life-history stage i, S_1 is the spawners at the beginning of the first life-history stage, and P_i and C_i are aggregate productivities and capacities calculated as

$$\begin{aligned} P_i &= \prod_{j=1}^{i} p_j \\ C_i &= \frac{P_j}{\sum_{j=1}^{i} P_j/c_j} \end{aligned} \tag{7.7.3}$$

Moussalli and Hilborn (1986) show how one can use this model to calculate the impacts of changing productivity or capacity at any life stage on the optimal harvest rate or stock size. Any increase in productivity (survival) will increase the optimum harvest rate, but may decrease or increase the optimum stock size depending upon capacity limitations at various life-history stages.

Adding environmental factors

One will often have reason to believe that some environmental factor such as temperature, upwelling, or rainfall may affect the stock-recruitment relationship. For instance, Penn and Caputi (1985), in their study of Exmouth Gulf prawns, noticed that the two major outliers in their stock-recruitment analysis coincided with very unusual years for cyclones, the high outlier had very low rainfall in January and high rainfall in February, and the very low outlier (1970–1971) had very high January rainfall and very low February rainfall. They felt that these rainfall patterns had important effects on juvenile prawn survival in inshore coastal or estuarine nursery areas. One can add such environmental variables to a Ricker stock-recruitment curve as follows:

$$R = S\, exp[a - bS + c(E - \bar{E}) + w] \tag{7.7.4}$$

where E is some environmental factor such as rainfall, c is a coefficient expressing the magnitude of its effect, and w is the unexplained residual. We prefer to use the deviation from average E, which makes the coefficient c easier to interpret.

Stocker et al. (1985) fit a Ricker stock-recruitment curve to herring data from the Strait of Georgia with a number of environmental factors including sea surface temperature and discharge of the Fraser River. They also tried a number of other environmental variables but these two were significant. Any number of environmental terms can be added onto the basic stock-recruitment curve using

$$R = S\ exp[a - bS + \Sigma\ c_i(E_i - \bar{E}_i) + w] \tag{7.7.5}$$

where the subscript i now refers to environmental variable i.

Environmental variables can be added to the Beverton-Holt or any other stock-recruitment relationship in a similar way, although it will generally involve non-linear estimation. The Beverton-Holt form is illustrated by

$$R = \frac{aS}{b + S}\ exp[\Sigma\ c_i(E_i - \bar{E}) + w] \tag{7.7.6}$$

Why is this dangerous?

As tempting as it is to add environmental variables to stock-recruitment data, this is a potentially dangerous practice. Usually a biologist has hundreds of possible environmental time series which might be applied to the stock-recruitment data. Once a stock-recruitment analysis has been performed the biologist will be aware of which data points are outliers to the general trend, and will be highly sensitive to any environmental factors that appear to coincide with the major outliers.

Warning: Be very, very cautious in fitting environmental variables, as it is almost impossible to make sure the apparent correlation is not spurious.

Thus, it is very possible that over a few years of normal work, anyone will find some environmental factor that both correlates well with the residuals in the stock-recruitment data and has some potentially understandable mechanism to explain the interaction.

One should be particularly aware of this danger when the environmental factor is associated with a few of the data points, as in the Exmouth Gulf prawn data. Nearly all of the explanatory power of the environmental variable is associated with two data points. This is not to say that the January and February rainfall is not causal in this situation, but the chance of finding an environmental variable that correlates with the two big outliers but is not causal is quite high.

There is another good reason not to get carried away about adding environmental variables to a stock-recruitment model: the more variables you add, the better will the model fit but the worse will be each of its parameter estimates. It is very easy to end up with a model that fits the historical data precisely, but has utterly no predictive ability. The general term for this problem is *overfitting*.

Adding species interaction

One can include species interaction, either as predation or competition, in the same way that other environmental variables are added into stock-recruitment curves. Walters et al. (1985) considered the impact of cod upon stock and recruitment for herring in British Columbia. They found that poor herring recruitment per spawner could be well explained by years of high cod abundance. Consider the following modification of the Ricker model:

$$R = S\ exp[a - bS - cC + w] \tag{7.7.7}$$

where R is the recruits of herring, S is the spawning stock of herring, and C is the number of cod over age 2. The regression coefficient c now becomes an estimate of the mortality rate of potential recruits to the herring population, per individual cod. Walters et al. were able to compare this estimate of mortality rate per cod to estimates based on cod feeding rates and diet composition data from stomach contents and determined that the apparent impact shown by stock-recruitment analysis was quite consistent with the feeding data.

7.8. Where Stock and Recruitment Analysis Goes Wrong

In the above sections we have shown how to go about estimating stock-recruitment relationships from data. In this section we will explain why it often goes wrong and gives terribly misleading answers. Unfortunately, the types of misleading answers produced by stock and recruitment analysis are almost always the same; the answers mistakenly lead you to believe that recruitment will not decline very much with spawning stock. We think that bad stock-recruitment analyses have been a significant factor leading to overexploitation and stock collapse for some major fisheries.

Errors in measurement of stock and recruitment

Probably the most serious problem in stock and recruitment is obtaining reliable measures of spawning stock size and subsequent recruitment. Early methods, which relied on catch rates as indices of abundance, often produced apparent "shotgun" patterns in which the data looked totally random. Biologically this means that recruitment is not affected by spawning stock.
However, it was not at all recognized how sensitive the stock-recruitment

analysis is to errors in measurement of spawning stock size until Walters and Ludwig (1981) performed some simulations with explicit errors in measurement of spawning stock size. If one assumes that the spawning stock size is measured with some error v, normally distributed with a mean of zero and a standard deviation of σ_v, and the spawning stock size used in the analysis is given by

$$\hat{S} = S\ exp[v - \sigma_v^2/2], \tag{7.8.1}$$

then reasonably small measurement errors ($\sigma_v = 0.2$) can transform a true linear stock-recruitment curve into one which appears to show very little relationship between spawning stock size and recruitment. This is shown in Figure 7.19.

Notice in Figure 7.19 that errors in spawning stock measurement do not just make the pattern fuzzier. These errors have the much more pathological effect of spreading the observations along the x axis of the graph so as to make the slope of the *x-y* relationship appear smaller, or in the extreme y may appear to be independent of *x*. In stock-recruitment analysis, having *y* appear to be independent of *x* leads to overestimates of the slope of the recruitment curve for low spawning stocks, and to the general appearance that increase in spawning stocks will not cause any improvement in recruitment. The amount of bias caused by errors in measuring spawning stock size depends both upon the magnitude of the errors and the amount of real variation in spawning stock. Consider the following mental experiment. If real spawning stock size is constant and recruitment is variable, one would obtain a stock-recruitment relationship with a distribution of recruitments at one stock size. If we then allow for errors in measurement of spawning stock size (although it is in fact constant), we would see apparently the same average recruitment but spread over a range of spawning stock sizes. Thus recruitment would appear to be independent of stock size.

If, however, we have greatly different true spawning stock sizes, the actual stock-recruitment curve may begin to emerge from the biases introduced by errors in measuring spawning stock size. Although Walters and Ludwig do provide several equations relating the expected bias to parameters of the data set, in practice it is difficult to make a priori estimates of how biased the analysis will be. The best general advice is that one should first make a realistic appraisal of the reliability of the measurements of spawning stock. Then one should do many simulations with different amounts of variation in the true spawning stock size to see how the apparent variation in spawning stock size will relate to the true one, and how biased the estimates will be. Our experience is that when spawning stock sizes fluctuate no more than 2–4 times from lowest to highest, biases can be quite severe, but if small stocks are at least 1/10th of the largest spawning stocks, the biases will probably be of little concern.

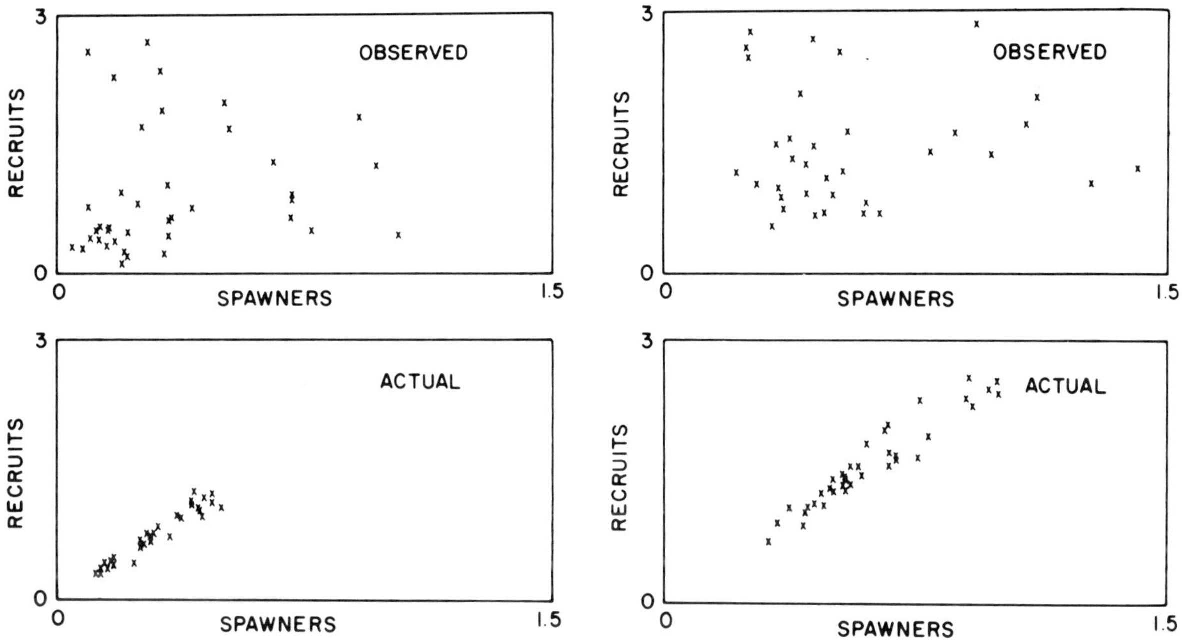

Figure 7.19. The real relationship between spawning stock and recruitment, and the apparent relationship when spawning stock size is measured with error. From Walters and Ludwig. (1981). Can. J. Fish. Aquat. Sci. 38: Figure 1.

The biases themselves are almost always the same; they make recruitment look to be less affected by spawning stock size than it really is. This means that optimum harvest rate will be overestimated and optimum stock size underestimated. The biases will be most severe with severely overexploited stocks where large variation in stock size is not possible. If the stock was already overexploited when reliable data collection began, there is a very high chance that stock-recruitment analysis will tell you that the stock is near optimum stock size or harvest rate when, in fact, it is severely overexploited.

Time-series bias of parameters

Hilborn and Starr (1984) simulated estimation of Ricker stock-recruitment parameters in the absence of errors in spawning counts and found there was still a bias. The bias was introduced because the independent variable of the regression (spawning stock) is not independent of the process errors around the mean recruitment relationship: positive process errors lead to higher spawning stocks later in time, and negative errors to lower stocks. This is called *time-series bias* and it will occur in any stock-recruitment data set where the spawning stock has been allowed to vary with recruitment rate rather than being deliberately controlled at experimental levels without regard to variability in recruitment (almost all stocks are allowed to vary in the way that will cause bias). The problem is that in order for the stock-recruitment parameter estimates to be unbiased, the spawning stock sizes must be controlled variables, independent of past data (Walters 1985, Caputi 1988). In fisheries, large recruitments are likely to produce large spawning stocks later, thus there will be correlation between the variability in the stock-recruitment process and the spawning stocks.

A good way to think of the time-series bias is to imagine a naive scientist who tries to experimentally determine the effects of different spawning stock levels on recruitment. He sets up an experiment with several different levels of spawning stock, planning to measure the average recruitment at each level. But every time he is working with a low spawning level and happens to get a good recruitment, he switches to a higher spawning stock treatment level. Then, when he is working with a higher spawning stock level, he does just the opposite and chooses to switch to a lower level whenever he gets a poor recruitment. What would he see from this experiment? His estimated average recruitment rates would be biased upward for low spawning stocks, and biased downward for high spawning stocks. If he fits a recruitment curve to his experimental results, the curve will be too steep for low stock sizes (just like the errors in variables effect of the previous section, he will think the stock is more productive than it actually is), and will bend over too soon. Most fisheries are operated as though a naive experimenter were at the helm:

good recruitments at low stock sizes are allowed to result in stock increase, and poor recruitments at high stocks are allowed to result in stock decrease.

This bias will occur in the absence of errors in measurement of spawning stock, but shares many of the same attributes. It will be worst when there is little contrast in spawning stock size, particularly if the stock is overexploited and varies within a very narrow range. Instead of being proportional on average to spawning stock size, recruitment will appear to be independent of spawning stock, and thus the optimum stock size will be underestimated. The best solution is to provide more variability in spawning stock size. Harvest policies that rapidly change spawning stock size will minimize the bias, whereas a fixed harvest rate or fixed effort policies that maximize the correlation between recruitment and subsequent spawning will maximize it. Again, simulation studies are probably the best method to appreciate how severe this bias will be for any data set.

The time-series bias can be grossly exaggerated in the presence of autocorrelated environmental effects. An extreme example is shown in Figure 7.20, which shows results of simulating a halibut-like population with a Beverton-Holt stock-recruitment relationship, but with the deviations around this relationship following a sine curve with a 20-year period (Walters and Collie 1989). Notice that although the data were generated using a Beverton-Holt relationship with recruitment rate largely independent of stock size, the sampled data end up fitting a Ricker curve much better than the correct curve. The Ricker fit causes us to underestimate both the magnitude of environmental effects and their autocorrelation structure (periodicity). Further discussions of this issue can be found in Walters and Collie (1989), and Armstrong and Shelton (1988).

For the Ricker stock-recruitment model, Walters (1989) has shown that the bias in the b parameter estimate is given approximately by the formula

$$\text{Bias} = \begin{pmatrix}\text{error}\\ \text{variance}\end{pmatrix}\begin{pmatrix}\text{mean}\\ \text{spawners}\end{pmatrix}(0.5 + r)\,\frac{(n/k) - 1}{SSX} \tag{7.8.2}$$

where error variance is the regression error variance, mean spawners is the regression mean for x, r is the assumed or estimated autocorrelation at lag 1 in residuals from the stock-recruitment curve, n is the number of stock-recruitment data points, k is the number of years from birth to recruitment (number of recruitment cycles in the data), and SSX is the regression sums of squares deviations of x from $\bar{x}$. Although this correction helps us to see what factors will affect the bias, there is a problem with using it in practice: where do you get r? If you fit the data and then estimate r from the residuals, you will grossly underestimate it, as shown for the extreme example in Figure 7.20. A safe starting point is to assume r = 0.3–0.5, then apply the

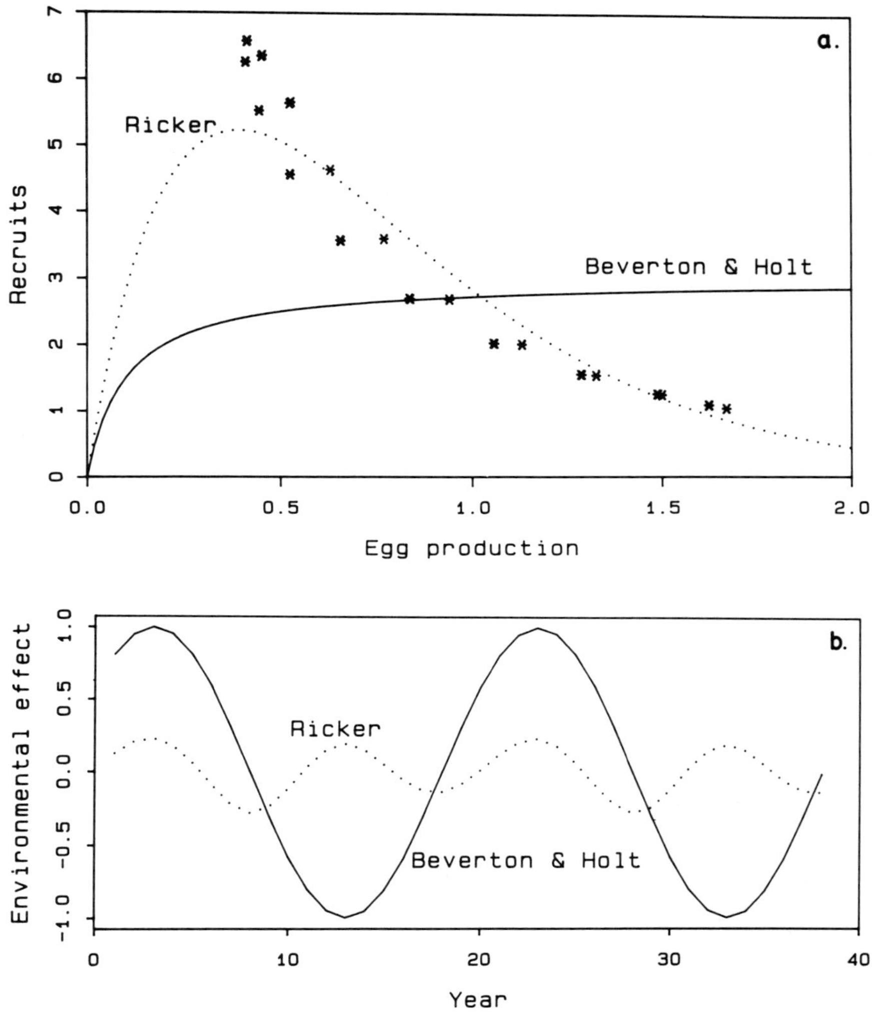

Figure 7.20. Simulated stock-recruitment data when recruitment is affected by cyclic environmental factor. From Walters and Collie. (1989). Can. J. Fish. Aquat. Sci. 45: Figure 1.

bias correction to give starting parameter values for further Monte-Carlo checking.

Nonstationarity

Stock-recruitment analysis normally assumes that the average stock-recruitment curve does not change over time. Although the world would be easier to manage if this were true, often the average stock-recruitment relationship will indeed change. In this section, we consider some potential causes of such change and what to do about it. For a more detailed discussion of non-stationarity, see Walters (1987).

Stock structure

One traditional assumption of most fisheries analysis is the unit stock. We assume that the stock is not spatially or genetically subdivided and that the age distribution of the stock is a complete description of the current state of the stock.

Most fisheries take place in a spatially diverse world, and the unit stocks are often spatially or temporally separated in spawning. What is called the stock for management purposes will often consist of many distinct and relatively isolated (in terms of immigration and emigration) subpopulations, each with its own stock-recruitment curve. Ricker (1973c) and Hilborn (1985b) have considered the implications of such aggregation for stock-recruitment analysis. In the virgin condition, all substocks are present, and the aggregate stock-recruitment curve represents the entire range of substock potential. However, after a period of severe exploitation, less productive stocks have been fished down, and make a smaller contribution to the total stock. Then the remaining stock appears to be more productive (higher recruits-per-spawner), but of a smaller total size.

The dangerous aspect of this problem is that even if a stock is overexploited, analysis of stock and recruitment data will often indicate that it is not. The data will indicate that the stock is quite productive but not particularly large. This is exactly the same bias we find due to errors in spawning stock size measurement as well as the time series bias.In principle the problem of nonstationarity due to spatial structure can be dealt with by modelling each structural unit separately. But in practice, there is seldom enough information to do this precisely enough to insure against making even bigger errors than by looking at the aggregate stock as a unit. A particular difficulty is with processes of dispersal and recolonization of depleted spatial areas; these processes may be critical to predictions from disaggregated models yet involve data that are very slow and difficult to collect (and are almost never available in practice).

External variables

Fish live in an ecosystem and are connected to physical and biological elements of the system in an intimate fashion. We should therefore not be surprised that stock and recruitment dynamics can be affected by physical and biological changes in the environment. Earlier we saw how one can

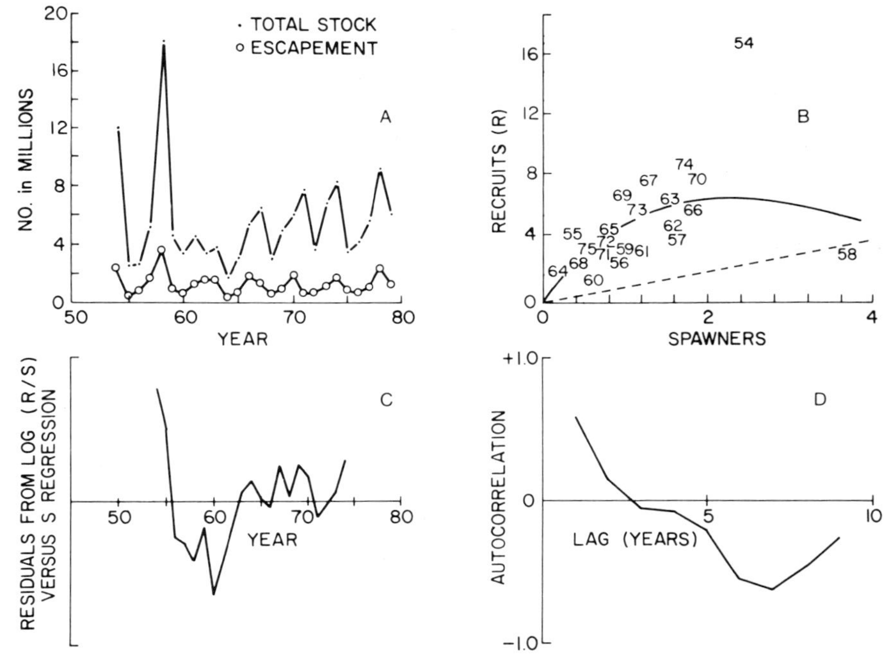

Figure 7.21. Analysis of Fraser River sockeye salmon data for possible nonstationarity. From Walters. (1987). Can. J. Fish. Aquat. Sci. 44:(supp. 2), Figure 2.

incorporate external variables, either environmental or other species, into a stock-recruitment relationship. We should expect, however, that we will often have undetected external variables affecting the stock-recruitment relationship. We would like to know how to detect a systematic change in the stock-recruitment relationship due to such variables, and what to do about it.

Walters (1987) suggests the following three alternatives for analysis of nonstationarity in time series of stock and recruitment: (1) covariance analysis of the data in time blocks, (2) representation of the parameters as polynomials in time, and (3) time series analysis of model residuals. Figure 7.21 shows an analysis of residuals for Fraser River sockeye salmon from Walters' paper. Note the reasonably strong serial correlation in residuals, particularly the sequence of negative residuals in the late 1950s.

Lack of contrast

The most serious problem in most stock and recruitment data sets is a lack of contrast in spawning stock levels. To understand how recruitment will respond over a range of spawning stock sizes, the stock must have been observed over a broad range of spawning stock sizes, with enough observations near the extremes of the range to overcome the time-series bias effects discussed above. Variation in spawning stock size is also the best antidote to errors in variables biases, and by far the most informative action that can be taken under the general heading of management experimentation.

While we may learn the most by pushing a stock over a broad range of spawning stocks, there are serious costs associated with such experimentation. Large spawning stocks cannot usually be achieved without significant reductions in catch and increases in associated economic and social costs. Small spawning stock sizes cannot be reached without the danger of serious overexploitation and long recovery times. A major challenge in fisheries stock assessment and management is balancing the need for information from contrasting stock sizes with the economic and social desirability of stability.

7.9. Summary and Critique

In the preceding sections, we have seen that a number of tools are available for the analysis of stock and recruitment, but there are many pitfalls awaiting the unwary biologist who wants to fit a curve and get an answer. The following steps are recommended for any analysis of stock-recruitment data.

1. Assess the reliability of the measures of spawning stock (and recruitment). The reliabilities should be measured as variances.
2. Analyze the stock-recruitment relationship by as many methods as possible, including these: (a) Beverton-Holt model, (b) Ricker model, and (c) Markov table.
3. Analyze the residuals from the Beverton-Holt and Ricker models to assess possible nonstationarity.
4. Simulate the estimation process, using estimated stock-recruitment parameters as "true" parameters and incorporating observation and process errors. A few hundred such simulations should provide some idea of how biased the parameter estimates may be.

In addition to the concerns about error and bias in fitting stock-recruitment relationships, the most important message from this chapter is that *you must think about stock-recruitment relationships rather than average curves*. For all but the most simpleminded (and usually dangerous) policy analysis the variability about the average stock-recruitment relationship is more important than the average recruitment levels.

Analysis of stock-recruitment data provides an enormous number of traps for the unwary—good luck.

Bibliographic Notes

The emphasis in this chapter has been on how to estimate and describe the stock-recruitment relationship from data; we have given little emphasis to the biology of the recruitment process. Shepherd and Cushing (1980) describe a mechanism that could account for observed patterns in larval survival, while a volume edited by Parish (1973) and a paper (Cushing 1973) provide good reviews of the biological dynamics of recruitment. Cushing (1988) provides a more recent review of stock-recruitment. Ricker (1954) and Beverton and Holt (1957) also provide considerable discussion of the biology of recruitment.

The most thorough discussion of fitting stock-recruitment relationships remains that in Ricker (1975), although so much has changed in our understanding of the statistical biases involved that his procedures must be used with great caution. Prior to the 1980s almost all fitting of stock-recruitment relationships was done by linear transformations of the Ricker and Beverton-Holt curves. No consideration was given to error structure or biases. Walters and Ludwig (1981) first considered the importance of errors-in-variables bias, and Hilborn and Starr (1984) and Walters (1985) described the bias due to the time series nature of the data.

Little if any consideration has been given to uncertainty in the stock-recruitment relationship and this, in turn, precludes any systematic examination of alternative management policies in a logical fashion. Walters and Hilborn (1976) were the first to consider the implications of management policy choices for uncertainty in the spawner-recruit relationship, and raised the question of whether harvest rates should be deliberately varied in order to provide better information. Gulland (1973) provides an alternative view on the importance of stock and recruitment in decision making.

Many people have proposed "alternative" stock-recruitment curves which have different shapes and properties. We see this as a rather counterproductive trend, in that there remains an unfortunate emphasis on the average stock-recruitment curve, rather than examination of the actual data and in particular the nature of variation. The use of tabular summaries of the spawner-recruit relationship (Markov tables) is preferred because they explicitly recognize the importance of variability. Getz and Swartzman (1981) used such a tabular approach.

8
Biomass Dynamic Models

8.1. Introduction

Biomass dynamic models are the simplest stock assessment models that are commonly used. Stock-recruitment models are in most cases used only as a part of a complete age-structured assessment. Only for semelparous species, and in practice only for Pacific salmon, can stock-recruitment models be used by themselves to evaluate alternative fishing plans. We presented stock-recruitment before biomass dynamics as a more logical introduction to the problem of parameter estimation, since most stock-recruitment models involve fitting only two parameters, whereas most biomass dynamic models involve fitting at least three parameters.

Biomass dynamic models are commonly called production models, or even surplus production models. We prefer to use the term biomass dynamic models because one can consider surplus production, or net production in the context of a full age-structured model: the unique characteristic of the Schaefer model and its relatives is that they describe the dynamics of the stock in terms of biomass, rather than numbers at age. Thus the term biomass dynamic models is much more appropriate.

The use of biomass dynamic models has a somewhat checkered history; they have formed the major assessment tool for many fisheries, particularly in the assessments made by several tuna agencies (the International Convention for the Conservation of Atlantic Tunas—ICCAT, — and the Inter-American Tropical Tuna Commission—IATTC), as well as agencies dealing with regular finfish, such as the International Commission for South East Atlantic Fisheries (ICSEAF). Despite their rather wide use, biomass dynamic models have been looked down upon as poor cousins of age-structured analysis; indeed, in most circumstances biologists would prefer to use the age-structured tools of VPA and catch-at-age analysis than rely on biomass dynamic models.

The reputation of biomass dynamic models has improved slightly in the recent past. One reason for their poor standing was their frequent failure to make sense of data sets: the estimated optimum effort or MSY was often absurd. Hilborn (1979) showed that this was generally not due to a model failure, such as lack of age structure and time delays, but was instead data failure, due to poor contrast between fishing effort and stock abundance. Since these are the same data failures that plague age-structured analysis, one cannot begrudge the simplicity of biomass dynamic models on this account. The second revisionist attitude comes from the work of Ludwig and Walters (1985, 1989), who showed that biomass dynamic models may provide better estimates of management parameters than age-structured approaches, even when important parameters such as growth and vulnerability are known.

It is quite difficult to age many fishes, particularly tropical ones, and age-structured analysis is often not practical in these fisheries. Furthermore, in many fisheries, again tropical ones especially, the catch consists of many species, and the catch data are difficult if not impossible to collect by species. Management regulations are also difficult to make species specific. In these circumstances, treating the entire catch as a biomass dynamics pool may be more appropriate than trying to look at single species dynamics.

Finally, in cases where both full age-structured and biomass dynamic models have been applied, the answers have often come out the same (Quinn et al. 1985) In such cases, one can question the need for the more detailed analysis.

Thus even though biomass dynamic models remain something of a second class citizen in the hierarchy of fisheries models, they are still an important and occasionally indispensable tool. Every fishery biologist should be familiar with their powers and limitations.

8.2. The Concept of Surplus Production

The changes in a population's biomass from one time to the next can be simply written as

$$\begin{aligned}\text{next biomass} = \text{last biomass} &+ \text{recruitment} + \text{growth} \\ &- \text{catch} - \text{natural mortality}\end{aligned} \tag{8.2.1}$$

if we ignore immigration and emigration. The two sources of increase in the population are recruitment of new individuals (birth) and gain in weight by individuals already present in the population. Catch and natural mortality constitute the two sources of loss. In the absence of fishing and by combining recruitment and growth into a single term called production, we can rewrite Equation 8.2.1 as

$$\text{next biomass} = \text{last biomass} + \text{production} - \text{natural mortality} \tag{8.2.2}$$

If production is greater than natural mortality, the population will be growing; if it is less than natural mortality, the population will be declining. The term *surplus production* is generally used to represent the difference between production and natural mortality. Surplus production represents the amount the population biomass will increase in the absence of fishing, or the amount of catch that can be taken while maintaining the biomass at a constant size.

Fishing plans can be couched in terms of surplus production. The International Pacific Halibut Commission (IPHC) allowed the biomass of Pacific halibut (*Hippoglossus stenolepis*) to increase through most of the 1980s. The stock was rebuilding from low abundance. The fishing plan adopted by the IPHC took 75% of the surplus production as catch and let the other 25% add to the rebuilding of the stock.

Figure 8.1 shows a simple representation of the expected relationship between stock biomass and surplus production. Most views of fishery population dynamics hold that at low stock sizes there must be low surplus production because there are few individuals in the population to grow and reproduce, and at large population sizes the surplus production must again decline to zero because of slower growth, higher mortality rates, and limitations on recruitment.

Surplus production can be used with any level of detail in representation of population dynamics. You can discuss the surplus production for a highly disaggregated spatial and age-specific model, or for a simple model where only population biomass in one spatial location is considered. This simple type of model, which we refer to as a *biomass dynamic* model, can be written as

$$\text{new biomass} = \text{old biomass} + \text{surplus production} - \text{catch} \quad (8.2.3)$$

We can then describe surplus production as some function of biomass (as a curve from Figure 8.1, for instance) and have a complete model of fish population dynamics.

This is an extremely simple view of population dynamics. All the complexities of age structure, spatial structure, and so on, are ignored and the population is described by a single number—population biomass. It is, however, a frequently used tool in fisheries, particularly in cases where catch-at-age data are difficult to obtain. Biomass dynamic models are the most commonly used stock-assessment model in tuna and most tropical fisheries and are widely applied in many temperate fisheries.

Using biomass dynamic models in formulating fisheries management plans depends greatly on the nature of the available data. When the fish biomass can be directly estimated, the graphical relationship between biomass and surplus production can be directly fit, whereas when the biomass cannot be directly measured and only an index of abundance is available, the esti-

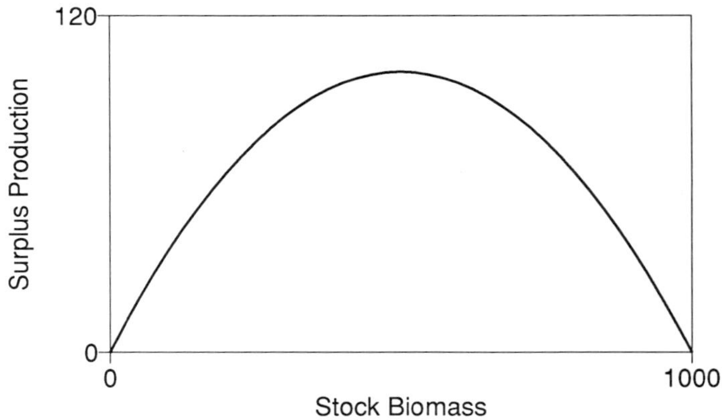

Figure 8.1. Expected relationship between surplus production and biomass for the simple Schaefer biomass dynamic model.

mation procedures become more complex and highly model dependent. Because biomass is very rarely measured directly, almost all applications of biomass dynamic models have used an index of abundance.

8.3. Estimation with Known Abundance

If we have a method to estimate the biomass, we can transform our previous equations to yield the following relationship:

$$\text{surplus production} = \text{new biomass} - \text{old biomass} + \text{catch} \quad (8.3.1)$$

which provides a way to graphically analyze how surplus production is related to biomass.

Figure 8.2 shows the relationship between estimated abundance of Pacific halibut and the estimated surplus production for the stock. No clear pattern emerges from these data; low abundances seem on average to have produced similar surplus production to that produced at high abundances.

Analysis of data of this form is almost exactly analogous to the analysis of stock-recruitment data. In fact, while the more conventional biomass dynamic models of Schaefer (1954) and Pella and Tomlinson (1969) are most commonly used for this type of analysis, Ludwig and Hilborn (1983) used the Ricker model to fit surplus production versus biomass data with some success.

Since it is so rare to have direct estimates of biomass, this approach is much more commonly applied when both (1) an estimate of some index of abundance over time, and (2) an independent estimate of the relationship between true abundance and the index are available. The relationship from

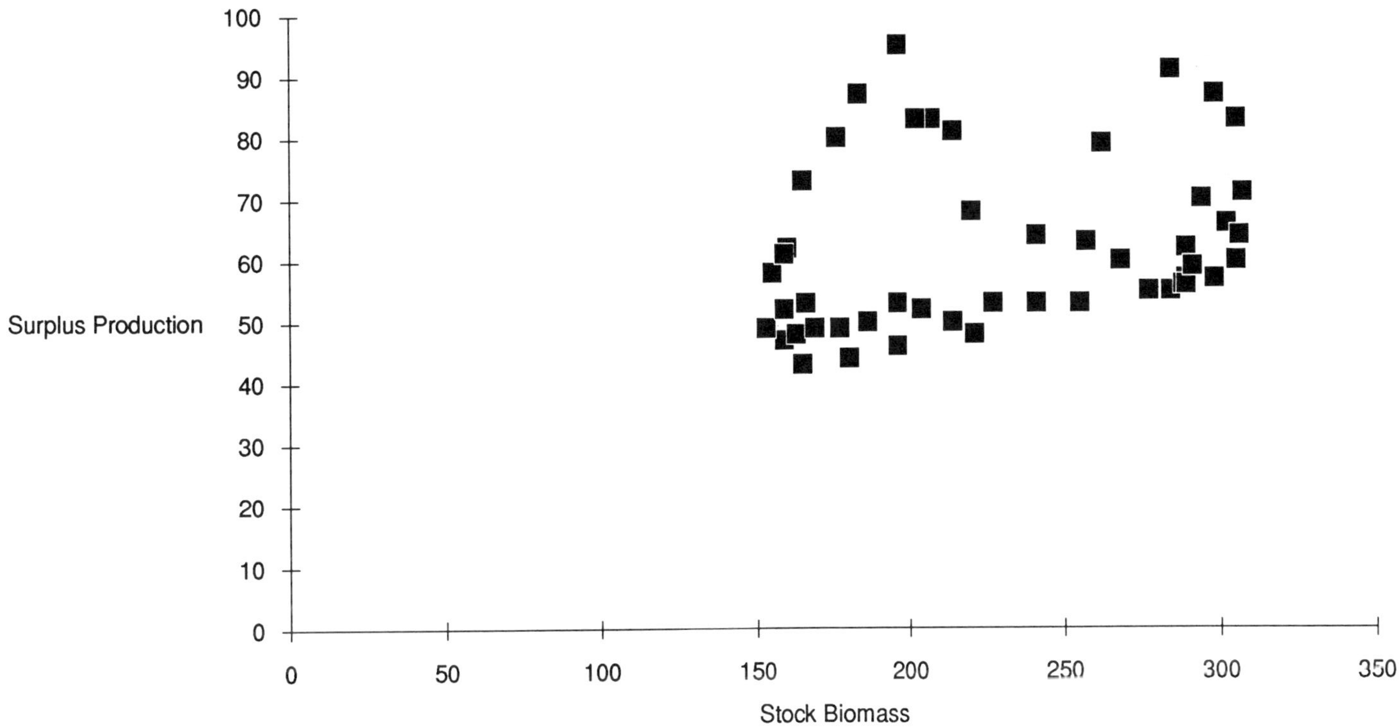

Figure 8.2. Estimated surplus production plotted against apparent biomass for Pacific halibut 1935–1983. (Data from IPHC).

(2) is then used to reconstruct a biomass time series from the index data (1). The most common example of this type of approach would be when a time series of CPUE is available and an estimate of the catchability is obtained from tagging studies. Then biomass over time is reconstructed as (biomass) = (CPUE)/(catchability). This approach was used by Schaefer (1954) in his original paper. The data shown in Figure 8.2 are in fact derived from an estimated catchability.

8.4. Estimation with an Index of Abundance

When no direct estimate of total biomass is available, we must not only specify a relationship between stock biomass and surplus production, but also a parameter (or parameters) that relate the index of abundance to actual abundance. There are three basic approaches that have been used in such models, and these are discussed in the next three sections.

We begin first by a consideration of alternative biomass dynamic models and their properties: how such models will behave when exploited and the relationship between the parameters and quantities of interest to managers such as potential yield, optimum harvest rates, unfished stock size, and so on.

We then consider approaches to estimating parameters for these models from real data. This is where the most important part of stock assessment using biomass dynamic models is found and where some of the greatest blunders in fisheries stock assessment have happened. After the basic estimation procedures are discussed, we consider what is known about the biases and robustness of the estimation procedures and consider some alternative approaches to the most preferred estimation approach, time-series fitting.

The ability to determine how the stock responds to exploitation (the model parameters) depends critically on the perturbation history of the stock and this is explored in some detail. Finally, one of the most powerful additions to biomass dynamics modelling is considered, that is, how to include information other than catch and index of abundance data in a biomass dynamic model estimation.

Alternative models and their properties

The Schaefer model

The first widely used biomass dynamic model was formulated by Schaefer (1954) based on earlier work by Graham (1935). Schaefer's model is normally written

Table 8.1. Summary of management parameters of interest for the Schaefer model.

Maximum surplus production (also MSY)	$rk/4$
Stock size for MSY	$k/2$
Rate of exploitation at MSY	$r/2$
Effort required to achieve MSY	$r/2q$
Maximum rate of exploitation	r
Effort at maximum rate of exploitation	r/q

$$\frac{dB}{dt} = rB\left(1 - \frac{B}{k}\right) - C \tag{8.4.1}$$

where B is the biomass of the stock, r is an intrinsic rate of population growth, k is a parameter which corresponds to the unfished equilibrium stock size, and C is the catch measured as a rate (e.g. tons per time).

It is assumed that the catch rate C is proportional to the stock size and to fishing effort

$$C = qEB \tag{8.4.2}$$

where B is the stock biomass, E is fishing effort, and q is a parameter that describes the effectiveness of each unit of fishing effort. The above model for C implies that the catch per unit of fishing effort (U) is an index proportional to stock abundance, or

$$U = C/E = qB \tag{8.4.3}$$

The properties of the Schaefer model are reasonably straightforward—after all, this model is only the simple logistic growth model of first year ecology, with a term for the catch added on along with the presumption that catch is proportional to effort and stock size. The management parameters of interest are summarized in Table 8.1.

Its main feature is the symmetric relationship between surplus production and biomass; surplus production is zero at a biomass of zero at the low end and at a biomass of k at the high end. The surplus production is maximized at a biomass of $k/2$. If the stock is unfished and starts out at some level greater than zero, the biomass will grow (or decay) toward a biomass of k. If the initial biomass is very near zero and there is no fishing, the initial growth rate will be roughly rB.

Pella and Tomlinson model

Pella and Tomlinson (1969) considered an extension of the Schaefer model:

$$\frac{dB}{dt} = rB - \frac{r}{k}B^m - C \tag{8.4.4}$$

which is identical to Equation 8.4.1 except for the addition of the parameter m. If $m = 2$, the two equations are identical. The philosophical reason for adding this parameter is that the normal Schaefer model dictates that the surplus production curve be perfectly symmetric in relation to stock size, whereas the addition of the parameter m allows the production relationship to be skewed to the left, $m < 2$, or to the right, $m > 2$. We have seen in Chapter 3 that either of these is quite possible and the addition of the third parameter would seem to be desirable.

In practice this model has been used quite extensively, primarily because of the popularity of the computer program GENPROD (Fox 1975), which provides a method of estimating the parameters using the equilibrium assumptions discussed below. Although there are theoretical grounds for including the extra parameter, it is unfortunate that there are few if any data sets on real fish populations for which one can realistically estimate the asymmetry of the production relationship. Our view is that the Pella and Tomlinson extension of the Schaefer model is a nice theoretical construct but should be used only with great care in fisheries stock assessment. One can fix the value of m based on theoretical considerations and thus reduce the model back to three unknown parameters. The very substantial contribution of the Pella and Tomlinson paper is in the estimation method they used. As discussed below, they developed a method that remains the best available technique, but unfortunately it is rarely used.

Difference models

Walters and Hilborn (1976) used a simple difference equation of the Schaefer model

$$B_{t+1} = B_t + rB_t\left(1 - \frac{B_t}{k}\right) - C_t \qquad (8.4.5)$$

where B_t is the biomass at time t, r and k have the same meaning as in the differential equation version of the Schaefer model, and C_t is the catch during time t and is defined as

$$C_t = qB_tE_t \qquad (8.4.6)$$

where q has the same meaning as earlier, and E_t is the fishing effort over year t.

This model is essentially identical to the differential equation. So long as the rates (r and qE) are low, the two models will behave quite similarly. If, however, the quantity qE goes over 1 for this model, predicted catch can exceed the population. This model also has the potential for what is known mathematically as "chaotic" behavior for high r values, but this is of no real interest in fisheries stock assessment. The relationship between the parameters of the model and quantities of interest to managers are identical to the regular Schaefer model.

Estimation procedures

Biomass dynamic models are rather straightforward; the estimation of their parameters is not. There are three approaches used in estimating the parameters of the biomass dynamic models when only an index of abundance (CPUE, etc.) is available: (1) the assumption of equilibrium conditions, (2) transformation of the equations into linear form, and (3) time-series fitting.

Equilibrium methods

At equilibrium, the relationship between CPUE and effort is linear under the assumptions of the Schaefer model. Gulland (1961) suggested using this relationship to assess the optimum effort and CPUE. However, instead of using effort on the x axis, he suggested using an average effort over the last n years where n is the number of age classes being fished. The logic behind this is that the CPUE one observes depends upon how hard the stock has been fished in the recent past. This implicitly assumes that the CPUE does not depend on how large the stock size has been. Garrod (1969) suggested using a curvilinear relationship rather than a linear one.

No matter what type of logic one uses for such plots, the implicit assumption is that the stock is at equilibrium, and we know this is rarely if ever true.

Warning: Never use equilibrium methods.

One of the more popular variations on the equilibrium assumption has been the method of Fox (1975). Fox provided a computer program (GENPROD) that has seen wide distribution and publication in fisheries books. We believe the equilibrium methods are so risky to use that we will not describe them in this book. The basic problem is that equilibrium methods usually overestimate surplus production (and optimum fishing effort) whenever they are applied to data gathered during a stock decline (e.g., during fishery development). They assume that every catch observed was in fact sustainable, which of course cannot be true if the stock was actually declining.

The one type of equilibrium analysis that appears to be useful on occasion involves spatial contrast in fishing effort. Munro (1980) has plotted yield from small-scale artisanal fisheries versus fishing effort, comparing yield per unit reef area for different villages (Figure 8.3). The assumption again is that each reef unit is at equilibrium with respect to its effort. One can then just look on a graph to see what happens to yield as effort increases. No formal estimation is required, and but we must always be wary of the equilibrium assumption.

Regression methods

The second basic approach involves transforming the equations into a linear form and then fitting by linear regression. These approaches recognize the dynamics, but often make rather odd assumptions about the error structure. Regression methods are, however, computationally much quicker than the better time-series methods presented later, and are useful for illustrative purposes.

Schnute's method Schnute (1977) showed that the Schaefer model could be transformed into the following dynamic equation:

$$\log\left(\frac{U_{t+1}}{U_t}\right) = r - \frac{r}{kq}\left(\frac{U_{t+1} + U_t}{2}\right) - q\frac{E_{t+1} + E_t}{2} \tag{8.4.7}$$

by integrating the Schaefer model over time steps of 1 year. This is a form of linear regression that can be used to estimate the parameters of the Schaefer model without making an equilibrium assumption. The principles and mechanics of this estimation are analogous to the regression estimation method using the difference equation model of Walters and Hilborn discussed below. The dependent variable of the regression for Schnute's method is an index of relative population change, whereas the independent variables represent average relative abundance and fishing intensity over each year.

Walters and Hilborn The difference equation model of Walters and Hilborn (1976) is analogous to the estimation procedure derived by Schnute for the Schaefer model. Using the following two basic relationships

$$B_{t+1} = B_t + rB_t\left(1 - \frac{B_t}{k}\right) - qE_tB_t \tag{8.4.8a}$$

and

$$B_t = \frac{U_t}{q} \tag{8.4.8b}$$

and substituting the observation relationship for B_t from U_t into the first equation, we obtain

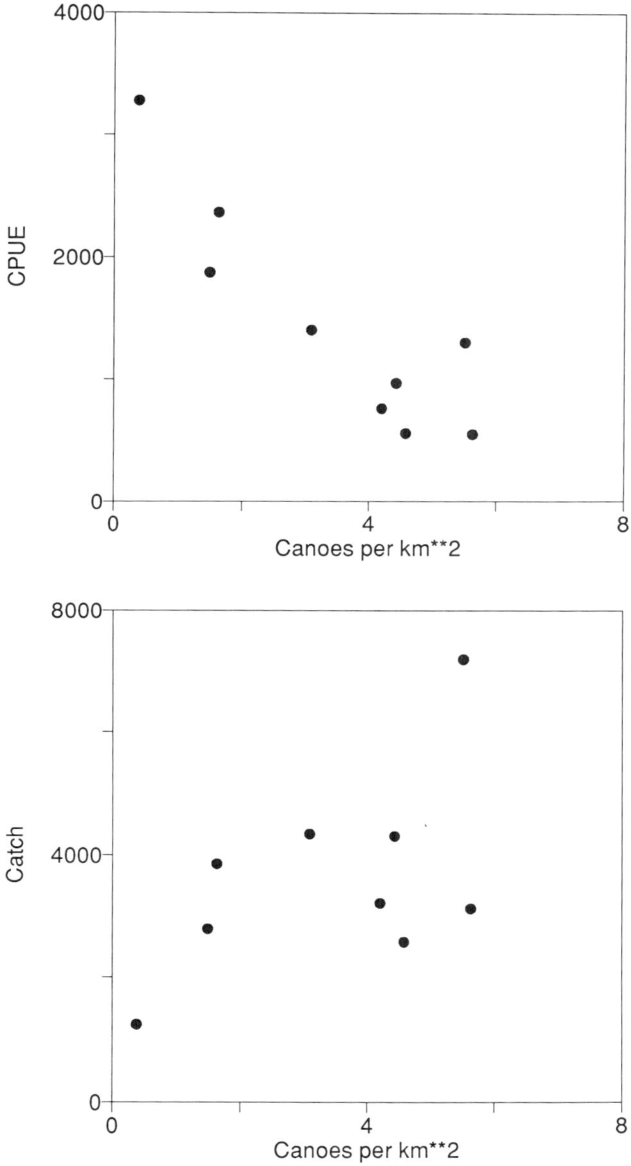

Figure 8.3. The relationship between fishing pressure measured as canoes per unit of reef area and yield from the reef. Data from Munro and Thompson (1983).

$$\frac{U_{t+1}}{q} = \frac{U_t}{q} + \frac{rU_t}{q}\left(1 - \frac{U_t}{kq}\right) - U_tE_t \tag{8.4.9}$$

By rearranging, we obtain

$$\frac{U_{t+1}}{U_t} - 1 = r - \frac{r}{kq}U_t - qE_t \tag{8.4.10}$$

This apparently complex algebraic statement can be translated into

$$\begin{pmatrix}\text{rate}\\ \text{of change}\\ \text{of biomass}\end{pmatrix} = \begin{pmatrix}\text{intrinsic}\\ \text{growth}\\ \text{rate}\end{pmatrix} - \begin{pmatrix}\text{density dependent}\\ \text{reduction in}\\ \text{growth rate}\end{pmatrix} - \begin{pmatrix}\text{exploitation}\\ \text{rate}\end{pmatrix} \tag{8.4.11}$$

which is another way of saying that, in the absence of harvesting and at low densities, the population will grow at the intrinsic rate of growth. Without harvesting the population size index U will increase until the density dependent reduction term is exactly equal to r. Finally, it is easily seen that if the third term (exploitation rate) is held greater than the first term, the population will go toward zero. This estimation equation has the computational advantage that it conforms to a standard multiple linear regression of the form

$$Y = b_0 + b_1X_1 + b_2X_2 \tag{8.4.12}$$

where Y is the dependent variable, X_1 and X_2 are the independent variables U_t and E_t, and the b's are the parameters of the regression, r, $-r/kq$, and $-q$.

This format may look attractive and is certainly computationally easy; however experience has shown that unless there has been a very informative perturbation history, this estimation is unlikely to provide reliable parameter estimates. It can also be badly biased (Uhler 1979); the bias may be reduced by using U_{t+1} as the regression dependent variable, with U_t, U_t^2, and E_tU_t as the independent variables (and no b_0 intercept term).

When people first started experimenting with regression methods for estimating Schaefer model parameters, they often obtained negative parameter estimates; r or q were estimated as less than zero, which is biologically impossible. It was felt that this indicated model failure, that the assumptions of the model were just too simple, and that by not explicitly incorporating lags to recruitment, and so on, these simple biomass dynamic models were failing to capture some important aspects of the data.

Indeed, one of the attractions of the equilibrium assumptions frequently employed is that they almost never give you impossible answers. A dynamic estimation method (Schnute or Walters and Hilborn) may estimate the MSY as negative, whereas fitting the same data by the Fox method will provide a positive MSY. When a dynamic method gives you an absurd answer you should conclude that there is little useable information in the data rather than fall back on a method that will give you a believable but likely wrong answer.

This reflects a fundamental contrast in views of stock assessment; to many stock assessment biologists, obtaining any answer, as long as it is biologically plausible, is preferred to being unable to estimate the parameters of interest. An answer, even if wrong, is considered better than no answer.

However, it is now recognized that most apparent model failures are in fact *data failures*. The information available about the model from the data is insufficient to provide any information about the stock dynamics and the apparent best explanation of the data is that r or q is negative. A typical example, discussed in detail later, is the classic "One Way Trip," in which effort increases every year, and CPUE decreases each year. This very uninformative type of time series often provides negative parameter estimates. Often the best fit to the data is obtained by assuming that k is small and that the stock started out above its unfished equilibrium, and the decline in CPUE over time had nothing to do with exploitation. In these cases q is often estimated as negative.

Observation error/time-series fitting

What are now considered to be the best methods for estimating production model parameters are called observation error/time-series fitting methods, which were first proposed and used by Pella and Tomlinson (1969). These methods are unrelated to the traditional time-series analysis methods of Box and Jenkins (1976), which have not proven terribly useful in analysis of fisheries stock dynamics. The basic idea in a time-series fitting is to take an initial estimate of the stock size at the beginning of the data series available, then use the model to predict the whole time-series. The parameter values are then adjusted to provide the best fit of the predicted-to-observed time-series of relative abundance or catch data. In the most general case, this involves estimating the normal parameters of the Schaefer model, r, k, and q, plus one additional parameter, the starting biomass B_0.

Nonlinear parameter estimation techniques are necessary to find the best fit of the predicted biomass, given the observed catches. The normal fitting criterion is minimization of the squared deviations between observed and predicted CPUE.

Using the discrete form of the Schaefer model to remove the need for too much algebra and calculus, the equations can be written as follows

$$\hat{B}_t = \hat{B}_{t-1} + r\hat{B}_{t-1}\left(1 - \frac{\hat{B}_{t-1}}{k}\right) - C_{t-1}$$
$$\hat{U}_t = \hat{q}\hat{B}_t$$
$$\epsilon_t = (\hat{U}_t - \hat{U}_t)^2 \tag{8.4.13}$$

where the C's and U's are observed data and B_0 is a parameter to be estimated. Figure 8.4 shows a time-series fit of the Hecate Strait rock sole (*Lepidopsetta bilineata*) data, to the continuous form of the Schaefer model. The top bars show the observed catch in each year. Note that in the early years of the data the catches were small and the stock built up. The catches then increased and the stock declined. Catches were then reduced, but the stock failed to increase.

An alternative approach is to predict the catches, given the known efforts and the predicted biomasses as follows:

$$\hat{C}_t = \hat{B}_t q E_t \tag{8.4.14}$$

One then minimizes the difference between observed catches and predicted catches. One can also minimize the deviations of the logarithms of predicted catches or indices of abundance, or try other transformations.

Additional options in time-series fitting have to do with how the starting population size B_0 is estimated. In practice, there is usually very severe parameter confounding between the normal three parameters of the Schaefer model, r, k, and q. The additional parameter B_0 makes matters even worse. In practice, therefore, we normally do not attempt to estimate B_0 as a separate parameter, but we assume that $B_1 = C_1/(E_1 q)$. This says in effect that for a given estimate of q we have an estimate of the starting population size. One could use a running average of the first few C/E's if they show high variability, and the assumption that the first data point is representative seems dubious. For situations where data are available right from the start of the fishery, it may be reasonable to assume that $B_0 = k$, which eliminates B_0 from the estimation.

There has been no systematic survey of the relative performance of various alternative forms of time-series fitting. We can make no generalizations about what should be minimized or how you should deal with B_0. Any stock-assessment biologist dealing with a fishery that is worth millions (or even hundreds of thousands of dollars) should take the time and care to try a rather wide range of time-series fitting options, using both his own observed data and simulated data.

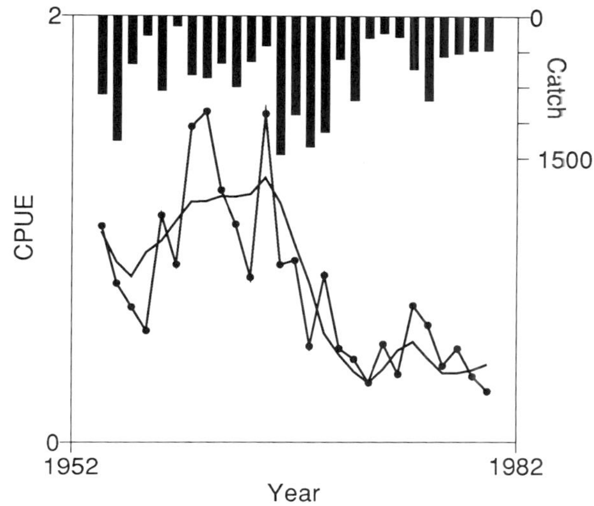

Figure 8.4. Observed and predicted CPUE for Hecate Strait rock sole.

The importance of perturbation histories

Principle: You cannot understand how a stock will respond to exploitation unless the stock has been exploited

---

In the analysis of stock and recruitment, this means that unless we have contrast in spawning stock size we cannot understand the stock-recruitment relationship. In the context of parameter estimation for biomass dynamic models, this means that you must have historical variation in stock size and fishing pressure to estimate the parameters of the model with any reliability. If we look back at Equation 8.4.10 and think graphically about its structure, we get Figure 8.5, which shows the underlying assumption of the Schaefer model.

Figure 8.5 illustrates that, under the Schaefer model, if population size is low (low C/E) and there is no fishing effort, we expect the population to grow at an annual rate of r. If the population size increases up to k ($C/E = kq$), we would expect the population growth rate to go to zero even in the absence of fishing. Similarly, even at a very small population size, if the fishing mortality rate qE approaches r, the population growth rate will go to zero.

The regression-fitting procedures attempt to fit the best plane to the observed data points. To fit a plane to the data, we must have contrast in the data. Specifically we must have a minimum of three data points at somewhat different regions of the graph. Ideally, these would include a data point at

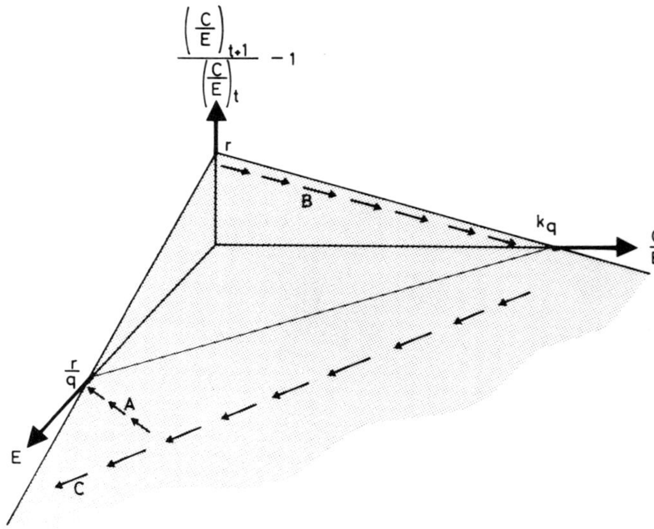

Figure 8.5. The assumed relationship between rate of change of the C/E, effort, and stock size (as indexed by the C/E) for the Walters and Hilborn estimation of the Schaefer model. From Hilborn (1979). J. Fish. Res. Board Can. 36: Figure 12.

low stock sizes with low fishing effort (to provide information about r), a data point at high stock sizes with low fishing effort (to help estimate kq), and a data point at high fishing effort to estimate q. Because the stock sizes (and therefore C/E) are serially correlated, especially during fishery development, it is almost impossible to obtain three such widely divergent and informative combinations of situations.

The one way trip

Once we begin to think of data sets in terms of the contrast they provide, we have a convenient framework for classification of data series. The most common type of data available consists of a fishery with continuously increasing fishing effort and declining catch per effort. This type of time-series we call the one-way-trip and it is, unfortunately, both the most common and the most difficult to interpret.

Figure 8.6 shows the C/E versus E for the Atlantic Ocean yellowfin tuna (*Thunnus albacares*) population. If we think back to Figure 8.5, we can see that such a time-series of data cannot possibly provide good information about the whole plane to be estimated, since any single line of data in the

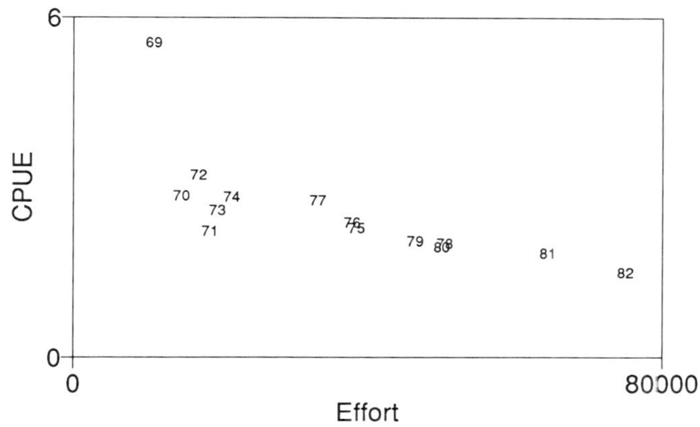

Figure 8.6. C/E versus E for Atlantic Ocean yellowfin tuna. Data from Fonteneau (1984).

C/E versus E space can be fit by an infinite number of planes. If we attempt to fit these data using the Schnute method, we get predictably unreliable results, as shown in Table 8.2.

The one-way-trip provides such biases that the estimates shown in Table 8.2 are totally unreliable. The most useful information is that the standard deviation for each parameter is roughly as large as the parameter estimate, and this tells us not to take the values seriously. Figure 8.7 fit shows three separate time-series fits to the data sequence for the Atlantic yellowfin data. These three lines have almost identical sums of squares, and yet represent widely different parameter values as shown in Table 8.3.

Note that the k value is doubled from fit 1 to 2 to 3, yet the sum of squares remains unchanged. Also note that the k value is roughly 10 times higher in fit 1 than obtained by the Schnute method shown in Table 8.2. We obviously have almost no information about k from this data set. That means we can only estimate MSY (equal to $rk/4$) with any precision if it turns out that the product rk is well determined in spite of k being poorly determined. Only if the stock has remained very near equilibrium throughout the stock history will rk be well determined, an assumption we cannot check without having an independent estimate of q.

The problem with the one-way-trip is that there are always a large number of possible explanations for what has been observed. The usual explanation is that the stock began somewhere below k and has been declining as the exploitation rate increased with increasing E's. The decrease in stock would have been moderated to some extent by production generated by the r term. Unfortunately, it is almost impossible to tell from one-way-trip time series if the stock began just below k or very far below it. In fact, usually there

Table 8.2. Parameter estimates and confidence bounds of the Schaefer model from the Atlantic yellowfin tuna data.

Parameter	Median Estimate	Standard Deviation
MSY	93,992	99,710
Optimum harvest rate	0.622	0.586
r	1.244	1.171
k	257,980	1,262,298
q	0.0000095	0.0000093

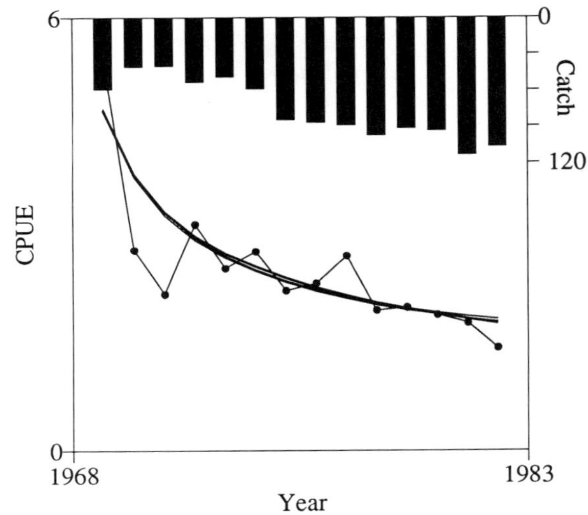

Figure 8.7. Three time-series fits to the Atlantic yellowfin data.

is almost no information about the k value at all. This is shown quite dramatically by the fits given in the previous paragraph.

A variation on this possible explanation of the one-way-trip is that $r = 0$ and what we have seen is simply a depletion experiment. This will almost always also fit the observed trend in the data very well, at least until there has been a long period of stability at an intermediate level of abundance. Another common result is that the parameter estimates suggest that the population size began above k, and the decreasing trend in abundance was simply natural decay back toward k with no effect of fishing.

Given these alternative explanations, there is little information about management parameters such as optimum exploitation rate or MSY from such a time-series of data.

Table 8.3. Parameter estimates from three fits to Atlantic yellowfin data by time-series methods.

Fit Number	r	k	q	MSY	Sum of squares
1	0.18	2,103,419	9.8×10^{-7}	98,175	3.82
2	0.15	4,000,000	4.5×10^{-7}	148,303	3.83
3	0.13	8,000,000	2.1×10^{-7}	261,365	3.83

Up and down the isocline

A somewhat more informative time-series of data can result when there is a period of increasing effort followed by a period when the effort is decreased gradually. This is best illustrated by the data series of Pacific halibut shown in Figure 8.8.

Regulation of the fishery began in the 1920s when the fishery was felt to be overfished and effort was reduced, which led to increasing CPUE. Effort was again increased and CPUE decreased until effort was once again reduced and CPUE began to increase. In principle, this does not provide the contrast between CPUE and E needed to identify all three of the parameters of the Schaefer model, because there is a strong correlation between CPUE and E.

We call this type of time-series up-and-down the isocline, because the changes in fishing effort have been gradual enough to keep CPUE and E correlated. We can believe with some confidence that the equilibrium relationship between E and CPUE is bracketed between the ascending and descending portions of Figure 8.8. However, we should not believe that the stock was always at equilibrium—in fact this is biologically impossible—only that the equilibrium line is between the two paths of ascent and descent.

When we attempt to estimate the parameters of the model using Schnute's method, we obtain the results shown in Table 8.4. This analysis shows reasonable confidence in the parameters that reflects the fact that the stock has seen a considerable range in exploitation. While E and CPUE are correlated, the correlation is not so bad as to prevent us from obtaining apparently good parameter estimates.

If the exploitation history had moved the stock very slowly up and down the isocline, so that CPUE and E were perfectly correlated, we would have no information about any of the individual parameters of the Schaefer model, but we would know the location of the isocline perfectly. This would allow us to estimate two of the management parameters quite well (the optimum effort and optimum CPUE) and by taking the product we could get the MSY. We can get the parameters because the intersections of the isocline with the E and CPUE axes coincide with 2 E_{opt} and 2 U_{opt} (see Figure 8.5). Thus we could be in the interesting position of having a good idea of how hard to

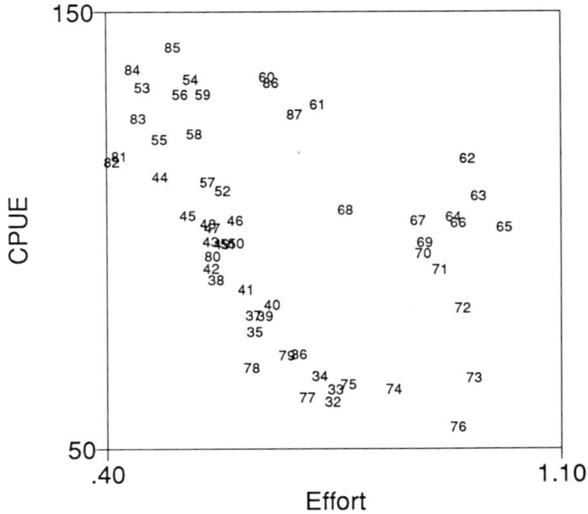

Figure 8.8. Time series of CPUE versus E for Pacific halibut.

fish and what sustainable yield we might expect, although we could not distinguish whether we had a large unproductive stock or a small productive one.

Good contrast

Ideally, we will have strong contrast between CPUE and effort. Figure 8.9 shows one such data set for Hecate Strait rock sole (Stocker 1981). The effort and CPUE have varied over roughly a fourfold range, and there have been relatively rapid changes back and forth between low and high effort levels.

Table 8.5 shows the parameters as estimated by the Schnute method. It is interesting, despite the better contrast in the data, that the standard deviations of the estimates are considerably wider than in the halibut data. There are two reasons for this. There are many more years data in the halibut case, and the halibut appear to be "better behaved" (that is, the data are intrinsically less variable). The rock sole data also illustrate that despite a rather good amount of contrast, we may remain quite uncertain about the potential yield from a fishery.

Comparison of perturbation histories by Monte Carlo simulation

In the previous three sections we have attempted to compare the amount of information that can be extracted from various types of perturbation his-

Table 8.4. Parameters of the Schaefer model for the Pacific halibut data.

Parameter	Median Estimate	Standard Deviation
MSY	77	29
Optimum h.r.	0.157	0.043
r	0.31	0.086
k	1,006	672
q	0.287	

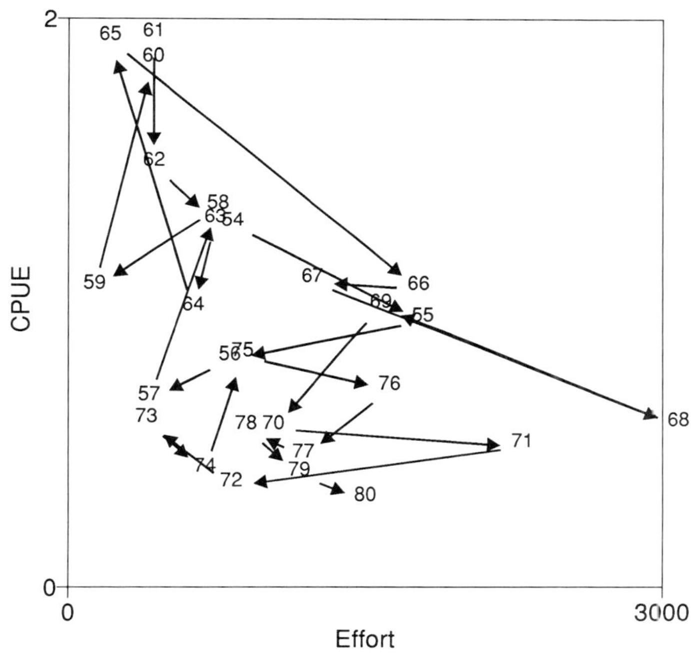

Figure 8.9. The history of the Hecate Strait rock sole fishery.

tories. The comparison is made somewhat difficult because of differences in the length of the data series, the degree to which the actual perturbation history conforms exactly to what we want to illustrate, and of course to the intrinsic unpredictability of nature. To better illustrate the relationship between perturbation histories and information obtainable from a biomass dynamic model, we simulate a stock under various perturbation histories. Our method is to take a known set of Schaefer parameters $r = 0.4$, $k = 1000$, and $q = 0.01$ and simulate 100 time series of data under each perturbation history. For each of the 100 sets of data, we then plot the parameters es-

Table 8.5. Parameters of the Schaefer model for the North Hecate Strait rock sole data.

Parameter	Median Estimate	Standard Deviation
MSY	482	1,529
Optimum h.r.	0.145	0.229
r	0.29	0.46
k	3,594	30,243
q	0.00034	0.00028

timated by the Schnute method. This illustrates parameter correlation and the general level of certainty.

Good contrast. Figure 8.10 shows a simulated data set with good contrast. The stock is reduced gradually by increasing effort from an initial virgin state. After the stock is reduced to roughly half its virgin size, effort is greatly reduced and the stock is allowed to recover. The recovery phase provides a good indication of the r value, whereas the depletion phase provides an estimate of q. Once r and q are identified, the k value is partially known, although it still shows the most uncertainty.

The one-way trip. Figure 8.11 shows a similar picture for a one-way trip. Effort increases continuously, and CPUE decreases more or less constantly. The CPUE is reduced to roughly a quarter its starting conditions, q is reasonably well defined, r is pretty well defined, and k is very poorly defined. Our knowledge of MSY and the optimum effort is only slightly worse than the good-contrast case described above.

Up and Down the Isocline. Figure 8.12 illustrates a time series where effort is increased gradually, and then decreased gradually. There is a very high correlation between effort and CPUE, so in principle there must be some serious parameter confounding. Both q and r are reasonably uncertain. However, note that the management parameters of sustainable catch, optimum CPUE, and optimum effort are rather well defined.

From the up-and-down time series, we know the equilibrium isocline must lie within the data points; that is below the descending points and above the ascending ones. Therefore we know the isocline reasonably well. The management parameters are all functions of the equilibrium isocline, so although we do not know the r, k, and q values, we do know the optimum CPUE, the optimum effort, and the MSY.

Recovery only. Figure 8.13 shows a time series of data where the stock begins at very low abundance, and is allowed to recover under low fishing pressure. After the stock recovers, effort is increased slightly, but not enough to reduce the stock abundance. From such a time series, only r can be estimated, and, as a result, the optimum harvest rate is known as well. All of the management parameters including k and q, are almost totally uncertain.

Table 8.6 shows the mean parameter estimates and their coefficient of

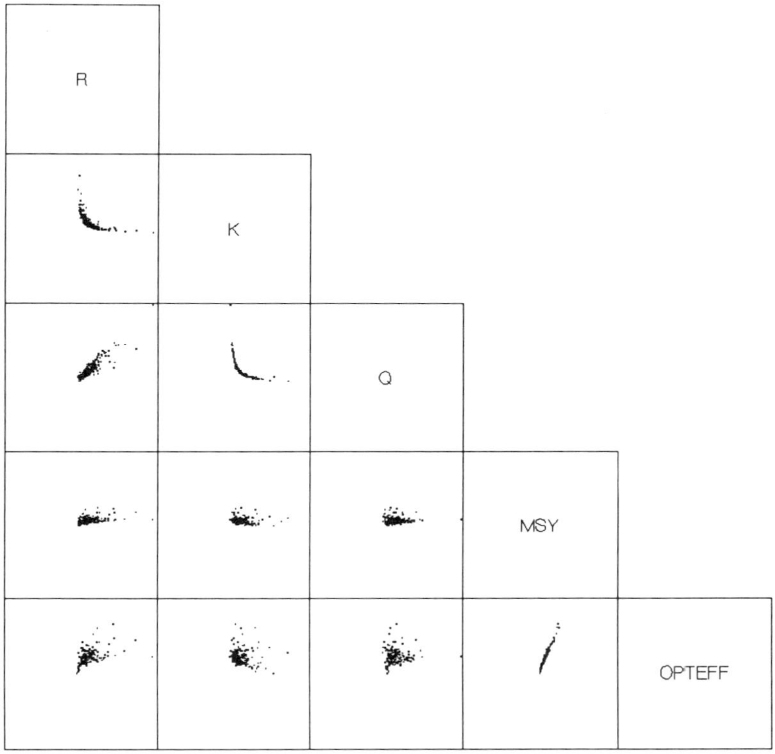

Figure 8.10. Monte Carlo simulations of the good contrast time series. Each point is one Monte-Carlo simulation. This diagram allows one to look at each parameter plotted against every other parameter. Each parameter is scaled from a minimum of −8 times its true value to +8 times its true value.

variation. In general, r and q can be reasonably well estimated, but only the good contrast series provides a reliable estimate of k. The good-contrast series provides the best coefficients of variation, except for the q value in the up-and-down series and the r value in the growth series.

Measures of goodness of fit

One question that always arises in fitting biomass dynamic models is how well the model actually fits the data. Traditional measures of goodness of fit have been derived from the regression statistics used to estimate the parameters.

For the equilibrium-fitting methods, the proportion of variance explained

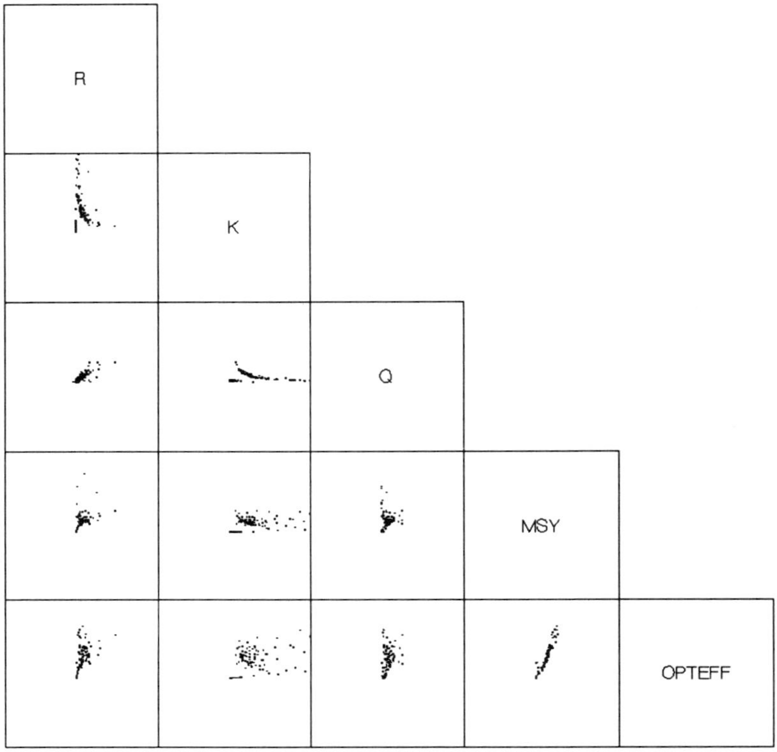

Figure 8.11. Monte Carlo simulations of the one-way-trip time series.

by the regression (regression r^2) has normally been given. It is almost always a very high value, and $r^2 = 0.95$ or higher are not at all uncommon. This is both deceptive and dangerous. Because most equilibrium methods are fit to one-way-trips, and graphically the fit is either in the C versus E or the CPUE versus E space, it is quite easy to obtain an excellent fit to the data. However, this in no way implies any certainty about our understanding of the underlying production potential of the stock. *This is one of the major reasons why equilibrium methods are so dangerous and should be avoided.*

Schnute (1977) proposed using the proportion of the variance in (U_{t+1}/U_t) that is explained as a measure of goodness of fit. This is essentially the traditional regression r^2 but Schnute is demanding that the model explain year-to-year changes in the relative growth of CPUE, which is much more demanding than just explaining a one way trend in CPUE or C. This performance statistic is almost always very low; often less than 10% of the variability is explained. For time series fitting, it is usually quite obvious how weii the model fits the data, and we could certainly calculate some statistics

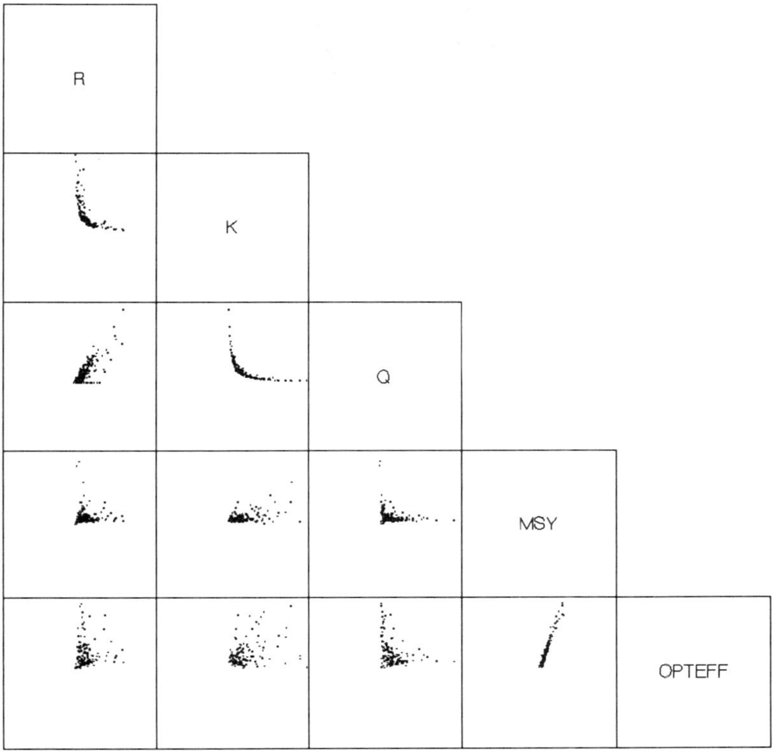

Figure 8.12. Monte Carlo simulations of the up-and-down the isocline time series.

about the percentage of variability in CPUE that is explained, although this may not be very meaningful.

Warning: The important question in model fitting is not how well the model fits the data, but how many other biological alternatives would fit the data equally well.

This is why we must examine the uncertainty in model parameters, rather than concentrate on a simple-minded goodness of fit measure.

Another simple test that we often use is a plot of CPUE versus E, with the best-fit isocline drawn. If we trace the time sequence of the points in this graph, any point up and to the right of the isocline should be followed by a year with lower CPUE, and similarly any point below and to the left of the isocline should be followed by higher CPUE the next year. If the data behave roughly according to the Schaefer model, most of the data points should fit this criterion. Figure 8.14, the Hecate Strait rocksole data, shows the points that fail to meet this criterion with an circle.

Table 8.6. Parameter estimates and their coefficient of variation from 100 Monte-Carlo trials for four perturbation histories

Perturbation	r	k	q
One-way trip	0.13 (1.5)	290,000 (4.5)	0.003 (1.3)
Growth	0.39 (1.6)	1,200,000 (5.1)	0.013 (2.2)
Up and down	0.39 (1.2)	38,000 (0.2)	0.009 (0.1)
Good contrast	0.47 (0.05)	1,142 (0.65)	0.013 (0.62)

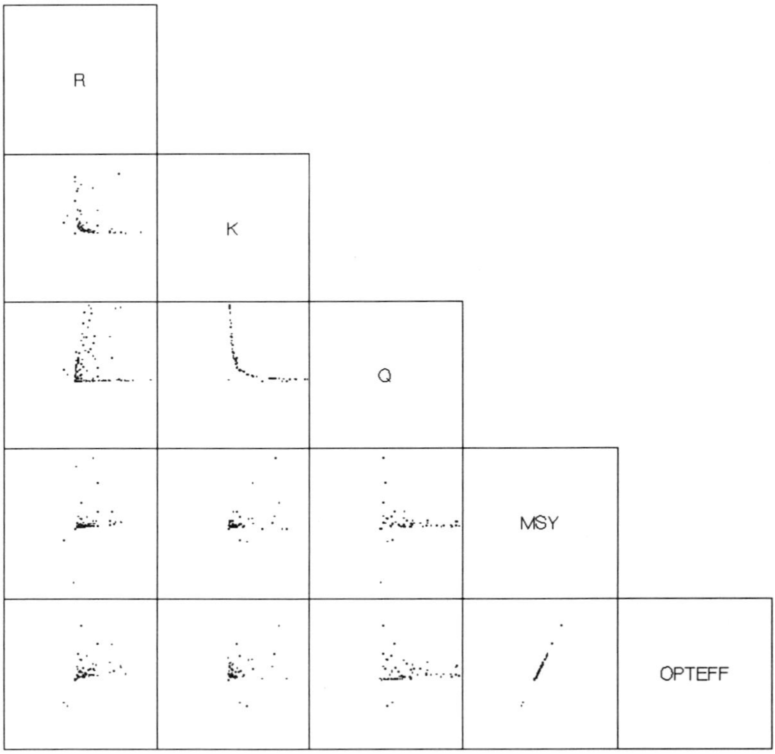

Figure 8.13. Monte Carlo simulations of the recovery time series.

We could then use the "correct" and "false" predictions in a sign test for a nonparametric measure of goodness of fit. However, this would imply a form of statistical rigor that seems unwarranted.

Biases and robustness

When the various methods for estimating the parameters of production models were first proposed, there was little thought given to the potential

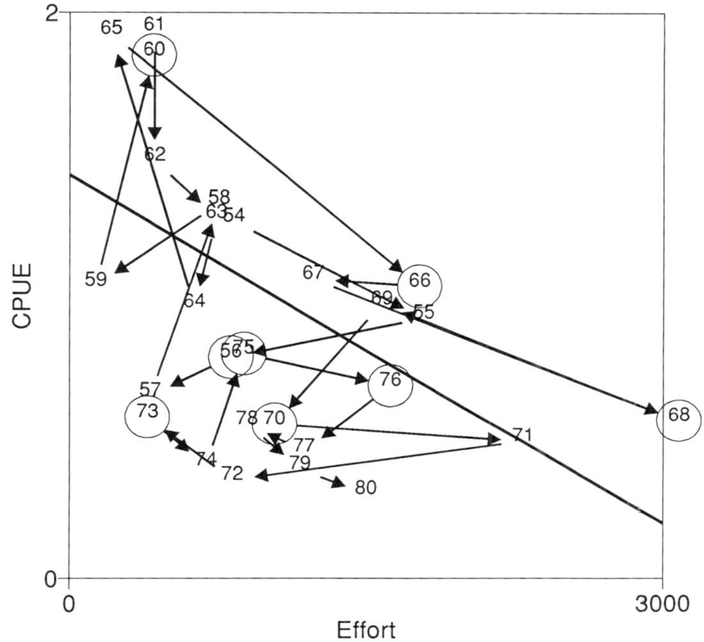

Figure 8.14. Hecate Strait rock sole data highlighting (in circles) points that fail to follow the basic prediction of the Schaefer model.

biases. Since most of the estimators were derived from some variation of traditional regression models, we somewhat foolishly believed these estimators might be reasonably behaved. However,as the estimators began to be used, particularly against known data generated by simulation, serious bias problems became obvious. Uhler (1979), Hilborn (1979), Mohn (1980), Roff and Fairbairn (1980) and Roff (1983) all explored variations on production models for biases and all found very serious problems. It is not at all uncommon for parameters to be biased by a factor of two.

Unfortunately, there has been little success in obtaining analytic results, and most results come from Monte Carlo trials of specific situations. Therefore, it should be considered mandatory for anyone using production models (or any stock assessment technique) to generate simulated data using the parameters that emerge from their analysis to examine bias. You cannot blindly hope that the parameters that emerge from your analysis are unbiased or even close until you have done some simulations.

Adding auxiliary information

With rare exceptions, time series of indices of abundance (CPUE, etc.) and catch data will prove unsatisfactory in obtaining reliable estimates of the productive potential of a fish stock. Both a highly informative perturbation history and a well behaved stock will be required, and the chances of having either are remote. The sad fact is that most fisheries time series are, by themselves, uninformative.

The use of auxiliary information, or combining information available in time series of catch and abundance index with other information, is proving promising. Fournier and Archibald (1982) and Deriso et al. (1985) pioneered the idea of using auxiliary information in catch-at-age analysis. It has equal potential in production models (this chapter) and their extensions (Chapter 9).

The general statistical concepts in combining different types of information are discussed in the Chapter 6. We will assume that the reader is familiar with that chapter and here we will show how auxiliary information in conjunction with traditional data can be used in a biomass dynamic model. In addition to the time series of catch and index of abundance data, the following types of information are often available to stock-assessment biologists:

1. We might know that the stock was unfished at or near the beginning of the time series. This means that in effect the initial index of abundance U_1 is equal to kq.
2. We might have biomass survey information from acoustic surveys, trawl surveys, and so on, available for 1 or more years.
3. We might have *a priori* estimates of some of the parameters r, k, or q. For instance q might have been estimated from a tagging study or r from other similar populations, or we might know enough about the basic biology of the population or species to have an idea of what value r should have. Similarly k might be estimated on an area- or habitat-available basis: there might be other stocks of the same or similar species where k and total area are both known.

Incorporation of various forms of auxiliary information into biomass dynamic models is quite straightforward and involves adding an additional goodness of fit criterion onto the goodness of fit for the time series alone. If our time-series procedure was to minimize the sum-of-squares deviations between predicted index of abundance $\hat{U}_t$ and observed index of abundance U_t for n years of data, then our criterion would be

$$\text{minimize} \sum_{t=1}^{n} (U_t - \hat{U}_t)^2 \tag{8.4.15}$$

We now consider m individual pieces of auxiliary information, and let X_i be the auxiliary information itself (an estimated biomass for instance), $\hat{X}_i$ be the predicted value of the information given the parameters of the model,

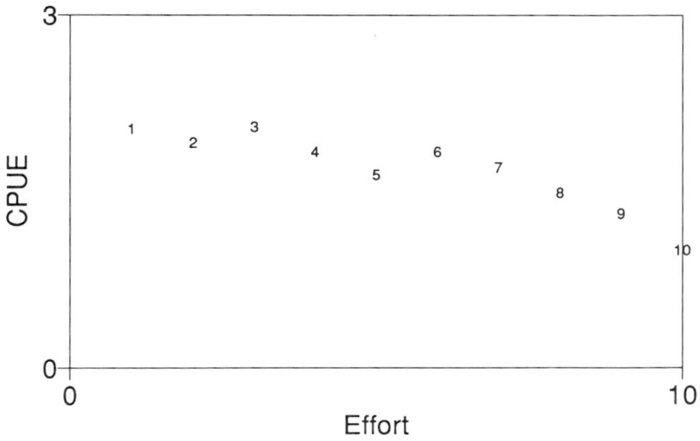

Figure 8.15. A one-way trip of 10 data points.

and w_i be the weight given to the deviation between the predicted and observed value of the piece of information. The goodness-of-fit criterion now becomes

$$\text{minimize} \sum_{t=1}^{n} (U_t - \hat{U}_t)^2 + \sum_{i=1}^{m} w_i(X_i - \hat{X}_i)^2 \tag{8.4.16}$$

In most circumstances we would have only a single piece of auxiliary information, so normally $m = 1$. If we assign w_i a very high weight (as large or larger than the sum of squares obtained from the time series fit alone), then we are effectively constraining the estimation procedure to fit the auxiliary information exactly. As we assign w_i lower and lower weights, we give the estimation procedure more and more flexibility to deviate from the auxiliary information. The weights w_i can be viewed as the ratios of the variance of the time-series observations to the variances of the auxiliary observations; assigning low weight is then interpreted as assuming that an auxiliary observation was measured with low accuracy (high variance).

For illustration, we have simulated a brief, highly uninformative one-way trip shown in Figure 8.15. For this simulation $r = 0.2$, $k = 100$, $q = 0.02$, $\sigma_p^2 = 0.2$ (process error), and there is no observation error.

Table 8.7 shows the fits obtained for the time series fit alone and the time-series fit combined with auxiliary information about r, k, q, kq, and known biomass at time 1. In each case the weight assigned to the auxiliary information was large enough to constrain the estimation to fit the auxiliary information exactly.

We can see that without auxiliary information the value of q is reasonably well known, because, as discussed earlier, a one-way trip provides information about q, as it is essentially an open depletion experiment. However,

Table 8.7. Parameter estimates obtained by time-series fitting from an uninformative one-way-trip with and without auxiliary information. Real values are $r = 0.2$, $K = 100$, $q = 0.02$, MSY = 5, optimal effort = 5.

Auxiliary Information	r	k	q	MSY	Optimal Effort
none	0.05	1,205,191	0.025	15,623.8	1.06
known k	0.18	100	0.026	4.4	3.34
known r	0.20	97	0.026	4.8	3.76
known q	0.06	2,117,869	0.020	33,715.0	1.58
known kq	0.31	80	0.025	6.1	6.11
known B_0	0.05	3,289,195	0.024	42,638.8	1.05

the best time series fit says that the stock was grossly overexploited when it began and has simply been driven to even lower numbers. The suggested management policy is to let the stock recover, which the parameter estimates predict would provide enormous payoffs, because the MSY is estimated a few orders of magnitude too high. Providing auxiliary information only in the form of an independent estimate of q or biomass does not help because q is reasonably well known without auxiliary information, and, because q is known, the biomass is also well known. This is somewhat discouraging since biomass surveys are often the easiest type of data to obtain, and this suggests that knowing the biomass alone will not tell you whether the stock was unexploited or initially highly overexploited. However, given q you can at least plot apparent surplus production versus stock size (see earlier section) to get some idea of why the overall relationship is poorly determined.

Interestingly enough, the other three forms of auxiliary information are quite helpful, and with them we obtain good estimates of all the parameters. Referring back to Figure 8.5, note that the example in Figure 8.15 illustrates how the one-way trip defines one line that the plane of Figure 8.5 must pass through, and that this line fixes the q value. Supplying any other point in the space is sufficient to define the plane quite well.

The addition of auxiliary information to model estimation is computationally simple and we believe it provides one of the most important areas for further work in quantitative stock assessment.

Extensions

Polovina's extension

Polovina (1989) used biomass dynamic models to examine the Hawaii slipper lobster fishery from three isolated areas. In each area, he had only 5 years of data, so that attempting to estimate the three parameters of the

Schaefer model would have been rather optimistic. Aggregated over all three areas, he would have attempted to estimate 9 parameters from 15 observations. Polovina used the Schnute estimation method, which means he really had only 4 observations for the estimation (since Schnute's method uses $(U_t + U_{t+1})/2$. His extension of the basic model was to assume that the biology of the lobster was the same in each area, so that the r value could be assumed to be identical in each area. He was not willing to assume that the unfished stock size in each area k was the same, since this would presumably be related to the amount of available habitat, nor did he assume that the catchability q was the same, since this could be quite dependent on local topography, as well as the skill of fishermen in that area. However, even by assuming r was the same in all three areas, he reduced the number of parameters to be estimated from 9 to 7.

Figure 8.16 shows the data and the best fit. Of particular interest is how far the fishery was away from equilibrium during this development phase; the effort in 1986 was generally twice as high as optimal. The futility of trying to use some form of equilibrium assumption on these data should be obvious.

8.5. Summary and Critique

Biomass dynamic models are perhaps the most abused stock-assessment technique in the tool kit of fisheries scientists. A very high proportion of applications that appear in agency reports or published papers use equilibrium-fitting methods, which we know are biased, unreliable, and simply should be thrown out. This problem is exacerbated by handbooks and manuals that continue to teach the equilibrium methods. Our position on this point is unequivocal; anyone dealing with fisheries data now has access to some form of computer that will allow use of the dynamic fitting methods, and they should be used. It appears highly likely that time-series fitting methods should be used instead of the dynamic-regression methods such as Schnute (1977), but there is undoubtedly much more to be learned about the relative merits of these approaches.

Another persistent problem in the use of biomass dynamic models is the widespread belief that they are methods for analyzing catch and effort data. Most applications have used CPUE as an index of abundance; however, it is widely accepted that CPUE may not be proportional to abundance. This poses no major problem for biomass dynamic methods - they work on any measure of abundance, including absolute abundance surveys from fishing gear or hydroacoustics. Biomass dynamics models should be considered the simplest assessment method to consider the net effects of recruitment, growth, and mortality when some measure of abundance is available. Biomass dy-

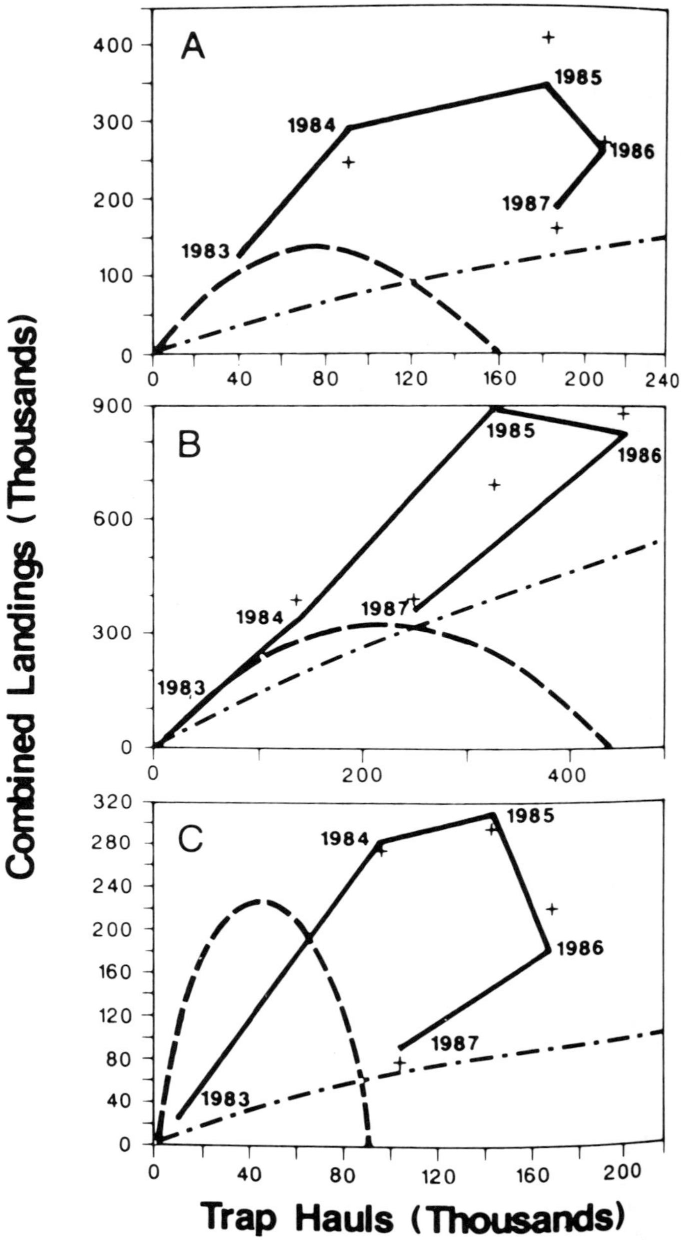

Figure 8.16. Polovina's data and fit for three lobster fisheries. The dashed parabola is the estimated equilibrium curve, the solid line is the observed trajectory. The dot dash line is the predicted catch in 1988. From: Polovina, 1989. Can. J. Fish. Aquat. Sci. 46: Figure 1.

namic models require a measure of abundance and catch data; but they do not need any effort data.

There are two alternative methods for assessing the dynamic response of fish populations to exploitation; biomass dynamic models, and age-structured models. Historically, biomass dynamic models have been the second choice used only when age-structured data were unavailable. The recent work of Ludwig and Walters (1985, 1989) indicate that the choice between the two methods may not be so clear cut and that in many circumstances the biomass dynamic approach may provide a better answer, even when age-structured data are available. It is better to think of the two methods as simply different; if biomass dynamic methods provide a different answer than age-structured methods, then the scientist should try to understand why they are different and analyze the management implications of the different predictions, rather than concentrating on deciding which method is correct.

Bibliographic Notes

The first formal models that can be thought of as biomass dynamic models were proposed by Graham (1935), who used the simple logistic equation that remains the core of the Schaefer model. Schaefer (1954) added the catch relationship and explored various methods of estimating the parameters. Gulland (1961) explored equilibrium assumptions for estimating the parameters and Garrod (1969) considered some variations on the theme.

Pella and Tomlinson (1969) provided the first dynamic time-series fitting method, and Walters and Hilborn (1976) and Schnute (1977) presented linear-regression methods. Quantitative analyses of the performance of these methods were done by Hilborn (1979), Uhler (1979), Roff and Fairbairn (1980), and Mohn (1980).

The most thorough testing involved looking at over 100 different methods of estimating parameters and uncertainties (Punt 1988). The only weakness of Punt's work was that he considered only one particular perturbation history; one that involved a long period of increasing effort and declining CPUE, followed by a reduction in effort and a slight recovery.

9

Delay Difference Models

9.1. Introduction

Until the early 1980s, most fish stock assessment was based on two alternative levels of biological description or realism. At one extreme were very simple biomass dynamic models that attempted to predict changes in total biomass and production in relation to stock biomass, using only a few aggregate parameters for maximum population growth rate and carrying capacity. At the other extreme were dynamic pool models based on population age structure, which attempted to predict changes in numbers and biomass in terms of parameters for natural mortality, growth, and recruitment. The simpler models were attractive in that they could be fit to aggregate data on relative abundance, fishing effort, and catch to give a rough idea of surplus production, even in cases where it had been impractical or uneconomical to gather more detailed biological information. On the other hand, the more realistic age-structure models could take better account of various time delays and changes in population structure associated with growth and recruitment and hence seem likely to make more precise predictions under conditions of rapid population change. Also, the more detailed models allow predictions about more refined management measures than just total quotas or effort limits, such as size limits and restrictions on gear mesh sizes and types. A conceptual bridge between the two levels of description would be useful in several regards. For some stocks, the two levels of modelling gave substantially different predictions about sustainable yields, and it would be useful to know whether this disparity was due to the surplus production model being structurally too simplistic, or to errors in estimating particular parameters of the more detailed model. For simulation and policy-gaming studies, it is often useful to have a relatively simple and computationally efficient model that can be solved or simulated quickly over hundreds or even thousands of case situations in terms of parameter values, policies, and future patterns of random (unpredictable) environmental changes. When mainly aggregate data on relative abundance, catch, and effort are available, it would be comforting and perhaps more accurate to analyze the data with simple models whose parameters make more direct biological sense than the aggregate parameters of surplus production models, especially if some of the model parameters could be directly estimated from bits of auxiliary information on growth, survival, and perhaps recruitment. Such a conceptual bridge was discovered by Deriso (1980) and elaborated considerably by Schnute (1985,1987) and Fournier and Doonan (1987). These authors showed that the basic biomass dynamics of age structured populations can often be predicted directly over time, without keeping track of or simulating the whole population age structure (number of animals in every age class), by using a delay-difference model that predicts this year's biomass directly from the last few years biomass and parameters for survival, growth, and recruitment. The delay-difference equation is a biomass dynamic model, but with biologically meaningful (directly measurable) parameters and accounting for basic time delays due to growth and recruitment. It can be compared directly (mathematically) to simpler biomass dynamic equations, can be fitted directly to simple time-series data when various kinds and qualities of auxiliary information are available, and can be used to build very efficient population simulations without keeping separate track of the numbers and sizes of animals in many age groups. This chapter shows how the basic delay-difference models were derived, how their predictions differ from simple surplus production equations, and how to use them in relatively simple ways for parameter estimation and simulation. More elaborate estimation and simulation schemes can be developed using them, particularly to make use of information on population size distributions (Fournier and Doonan 1987, Schnute 1987); we will not discuss these schemes further, but urge the advanced reader to find out more about them.

9.2. Derivation of Delay-Difference Models

The delay-difference models for population biomass over time are derived by using specific process models for (assumptions about) survival, growth, and recruitment, and then substituting these models into a general equation defining population biomass. Since the substitutions and algebraic manipulations that follow from them do not involve any approximations or further simplifying assumptions beyond those made in the specific process models initially, the final delay-difference equations are exact biomass predictors provided the process models hold exactly in the first place. In other words, the quality of the delay-difference predictions is completely dependent on the process assumptions; no extra hidden assumptions or approximations are added to make the calculations simpler.

Basic assumptions

The delay-difference derivations generally make three key assumptions about growth, survival, and harvesting: (1) growth in mean body weight at age can be described by the linear relationship

$$w_a = \alpha + \rho w_{a-1} \tag{9.2.1}$$

where a is age, and α and ρ are empirical constants (i.e., intercept and slope of a Ford-Walford plot) over the ages $a = k, k + 1, \ldots$ that are fully vulnerable to fishing; (2) all fish aged k and older are equally vulnerable to fishing (i.e., "knife edge" selection at age k); and (3) all fish aged k and older have the same annual natural mortality rate. Assumptions (2) and (3) imply that all animals aged k and older have the same annual total survival rate s_t, where s_t reflects the combined effects of fishing and natural mortality.

It is generally (but not necessarily) assumed that the total survival rate in any year, s_t, can be written as the product of a constant natural survival rate $\mathcal{S}$ (not dependent on time or stock size) and a time-varying survival through harvesting, $1 - h_t$, where h_t is the annual harvest rate:

$$s_t = \mathcal{S}(1 - h_t) \tag{9.2.2}$$

This assumes that harvest takes place in a short season at the beginning (or end) of each year; more realistic assumptions and equations for fisheries that are spread over the year can be found in Schnute (1985).

The annual recruitment rate R_t, as number of fish reaching age k (or more generally, reaching weight w_k), can be any arbitrary time series. For studies of long-term productivity and sustained yield, R_t is generally assumed to be some function of the spawning stock in year $t - k + 1$, where the spawning stock S_{t-k+1} is usually taken to be the total biomass of age k and older animals after harvest in year t. The recruits R_t reaching weight w_k can also be assumed to come from spawning in several previous years (i.e., to depend on several $S_{t-1}, S_{t-2}, \ldots$ spawning stocks), reflecting variation in individual growth rates and/or ages when the fish move into areas where they are vulnerable to fishing.

The assumptions about growth and mortality lead to three model parameters that must be estimated or assumed: the growth parameters α and ρ and the natural survival rate $\mathcal{S}$. The age at recruitment, k, could also be viewed as an unknown in situations where no growth data are available. Additional model parameters come from whatever assumptions are made about the relationship between spawning stock size and recruitment; for example, two additional parameters are added if R_t is predicted from S_{t-k+1} by using the Ricker or Beverton-Holt stock-recruitment equations (see Chapter 7). When recruitment to weight w_k is assumed to involve animals of several calendar ages (produced by several past spawning stocks), then there are additional recruitment parameters specifying the proportions of the total recruits from each spawning year t that reach weight w_k at different calendar ages or lags.

General derivations

Suppose we call B_t the total stock biomass of fully vulnerable fish (age k and older, weight w_k and larger) at the beginning of year t, the total number of fish in this stock N_t, and the numbers by age $N_{a,t}$ for ages $a = k, k + 1, \ldots$. The biomass B_t can be written in terms of the numbers and body weights w_a as

$$B_t = \sum_{a=k}^{\infty} N_{a,t} w_a \tag{9.2.3}$$

This is just the definition of biomass for age k and older animals. B_t can also be written as the biomass of age k recruits, $w_k R_t$, plus the sum of biomass contributions over all ages $k + 1, k + 2, \ldots$:

$$B_t = \left[\sum_{a=k+1}^{\infty} N_{a,t} w_a \right] + w_k R_t \tag{9.2.4}$$

Notice that each of the $N_{a,t}$ in the sum can be written as $N_{a,t} = s_{t-1} N_{a-1,t-1}$ (i.e., as survivors from last year) and each of the w_a can be written as $w_a = \alpha + \rho w_{a-1}$. Substituting these survival and growth equations into the biomass definition gives

$$\begin{aligned} B_t &= \Sigma s_{t-1} N_{a-1,t-1}(\alpha + \rho w_{a-1}) + w_k R_t \\ &= s_{t-1}(\alpha \, \Sigma N_{a-1,t-1} + \rho \Sigma N_{a-1,t-1} w_{a-1}) + w_k R_t \end{aligned} \tag{9.2.5}$$

where the sums are over age $k + 1$ to ∞. Notice above that factoring out the terms that do not depend on age (s_t, α, etc.) results in sums over ages k and older for year $t - 1$, which are the defining sums for total numbers and biomass in year $t - 1$. Hence B_t can be written in terms of these totals as

$$B_t = s_{t-1} \alpha N_{t-1} + s_{t-1} \, \rho B_{t-1} + w_k R_t \tag{9.2.6}$$

This equation, along with the total numbers equation

$$N_t = s_t N_{t-1} + R_t \tag{9.2.7}$$

gets us down from many equations for the entire population age structure, age by age, to just two equations for the total biomass and the total numbers. Obviously this is a computational and conceptual advantage in thinking about the effects of age structure. Notice that the above two equations for B_t and N_t make no assumptions at all about recruitment variation; they apply for any sequence of values of R_t over time. Thus they will work to describe total biomass and numbers even if the population age structure is grossly distorted or erratic over time due to cyclic changes in recruitment, dominance of a few age classes, random changes in recruitment, or various recruitment trends.

The pair of equations for biomass and numbers above can be used to develop quite efficient simulation schemes. But these equations can be combined together to generate a single delay-difference equation for biomass alone, so that population numbers can be ignored entirely. Note that the term αN_{t-1} in the equation for biomass B_t can be written in terms of recruits and survivors from N_{t-2}, as

$$\alpha N_{t-1} = \alpha s_{t-2} N_{t-2} + \alpha R_{t-1} \tag{9.2.8}$$

Then in this equation, the term $\alpha s_{t-2} N_{t-2}$ can be written using the biomass equation for predicting B_{t-1} from B_{t-2} and N_{t-2} as

$$\alpha s_{t-2} N_{t-2} = B_{t-1} - \rho s_{t-2} B_{t-2} - w_k R_{t-1} \tag{9.2.9}$$

Substituting this result into the equation for αN_{t-1}, and that result in turn into the basic equation for B_t, we finally get (after some tedious algebra)

$$\begin{aligned} B_t = {} & s_{t-1} B_{t-1} + \rho s_{t-1} B_{t-1} - \rho s_{t-1} s_{t-2} B_{t-2} \\ & - (w_k - \alpha) s_{t-1} R_{t-1} + w_k R_t \end{aligned} \tag{9.2.10}$$

This is the basic delay-difference model. From the growth equation, notice that α can be written as $\alpha = w_k - \rho w_{k-1}$ where w_{k-1} is the body weight 1 year prior to recruitment. Substituting this expression for α into the delay-difference equation gives the form suggested by Schnute (1985):

$$\begin{aligned} B_t = {} & s_{t-1} B_{t-1} + \rho s_{t-1} B_{t-1} - \rho s_{t-1} s_{t-2} B_{t-2} \\ & - s_{t-1} \rho w_{k-1} R_{t-1} + w_k R_t \end{aligned} \tag{9.2.11}$$

This messy looking equation is not as formidable as it appears. It says in essence that biomass in year t can be predicted from three basic terms involving past biomasses and recruitment: (1) surviving biomass from last year, $s_{t-1} B_{t-1}$, (2) growth of surviving individuals from last year, represented by the complicated term $\rho s_{t-1} B_{t-1} - \rho s_{t-1} s_{t-2} B_{t-2} - s_{t-1} \rho w_{k-1} R_{t-1}$, and (3) biomass of new recruits added to the stock, $w_k R_t$. The growth effect term is complicated because it must keep track of shifts in mean size of animals, and hence growth rate changes, that are associated with recruitment variations.

It is important to recognize at this point that the delay-difference structure provides an accounting framework for the effects of survival and growth on biomass, given any arbitrary sequence of recruitments R_t. The framework is likely to perform very well at this task, considering that growth and natural survival rates tend to be fairly stable over time in fishes. The framework becomes much more difficult to justify in situations where the recruitment rate must also be predicted from past stock sizes and/or other factors such as environmental correlates of juvenile survival. To construct a complete population dynamic or biomass production model, we must make some substitution of a recruitment equation $R_t = f(S_{t-k+1})$ into the delay-difference structure. The recruitment function $f(\ )$ chosen is then likely to dominate the overall model relationship between B_t and past biomasses.

Alternative representations and extensions

So far we have presented two ways of predicting biomass over time for age-structured populations that obey a simple body growth model ($w_a = \alpha + \rho w_{a-1}$) and have age-independent survival rates. First, we can use a pair of equations for the biomass B_t and the total numbers N_t, expressing each in terms of growth and survival parameters and numbers of recruits R_t. Second, B_t can be predicted directly from B_{t-1}, B_{t-2} and past spawning biomass S_{t-k+1}, by using the delay-difference equation structure. Schnute (1987) and Fournier and Doonan (1987) have shown that other useful predictions can be obtained if the delay-difference representation is avoided in favor of examining other relationships between biomass and numbers. An obvious and important relationship between biomass and numbers is that they define the mean body weight of animals in the population, $\bar{w}_t$. This mean is just the ratio of total biomass to numbers making up that biomass

$$\bar{w}_t = \frac{B_t}{N_t} \tag{9.2.12}$$

This is the same thing as saying that $N_t = B_t/\bar{w}_t$, that is, the numbers must be the biomass divided by the average weight of animals making up that biomass. If we substitute this expression for N_{t-1} into the basic biomass-numbers relationship of the previous section, we get

$$B_t = \frac{s_{t-1}\alpha B_{t-1}}{\bar{w}_{t-1}} + \rho s_{t-1} B_{t-1} + w_k R_t \tag{9.2.13}$$

Notice that this substitution eliminates N_t from the biomass equation, without resorting to delay-difference algebra except as it may be needed to predict $\bar{w}_t$ and R_t. It also shows that the biomass dynamics will follow a "simple" relationship

$$B_t = G_{t-1}B_{t-1} + w_kR_t \tag{9.2.14}$$

where G_{t-1} represents a combined growth-survival factor for the effects of B_{t-1} on B_t, and w_kR_t represents biomass of new recruits. The growth-survival factor G_t for each year t is given by

$$G_t = \rho s_t + \frac{s_t\alpha}{\bar{w}_t} \tag{9.2.15}$$

Notice how this growth factor will tend to decrease as the average age and size of fish in the population increases (more older, slower growing fish), and to increase when the average size of fish is smaller (more younger, faster growing fish).

Size distribution moments

Information about the population size distribution, namely the first moment or mean of that distribution, is represented by $\bar{w}_t$. It is also possible to predict higher moments (variance, skewness, kurtosis, etc.) of the size distribution (if some additional assumptions are made) and to predict the impact of changing growth, survival, and recruitment rates on these rather easily measured size statistics. Schnute (1987) assumed that all individuals in the exploitable stock grow deterministically according to Equation 9.2.1, whereas Fournier and Doonan (1987) assumed that the variance of size at age is constant.

Schnute (1987) has shown that in the framework of his model, the propagation of the second and higher moments carry unique information only about the parameter ρ, whereas changes in the first moment $\bar{w}_t$ provide information about changes in survival and recruitment rates. Early equations for estimating total mortality (or survival) rate from mean size of fish in the catch (Beverton and Holt 1956, see also the review of others by Pauly and Morgan 1988) are based on looking at the mean size, while assuming constant recruitment and survival rates. In essence, the equations presented in the previous section provide a generalization to dynamic, changing situations of the early equilibrium survival equations.

For details of the higher moment calculations, see Schnute (1987) and Fournier and Doonan (1987). The calculations involve setting initial conditions for the moments, then predicting changes in them over time by equations analogous to the basic equations above for predicting biomass and numbers changes.

Stock-recruit relations

The delay-difference model derivations shown above demonstrate that the effects of growth and survival on total stock size will likely be relatively simple (i.e., can be described by simple difference equations), even when the population has a complicated age structure. This means that any serious complications or errors in predicting stock size are likely to arise from difficulties in predicting recruitment rates. The assumptions used to predict R_t will dominate the delay-difference model predictions of stock behavior over time and response to harvest rate changes, except for very long-lived fishes whose survival rate s_t is high (say 0.8 or larger) and the recruitment rate R_t is small compared to the total harvestable population N_t.

The most critical modelling decision in developing a delay-difference prediction scheme is whether to treat recruitment rate R_t as depending somehow on the parental stock size S_{t-k+1}. If R_t is treated as independent of stock size, the delay-difference model will be able to capture only growth overfishing effects, along with temporal patterns of change in stock and harvest associated with removal (and rebuilding) of accumulated older fish. For long-lived fish that recruit to the fishing gear at advanced age (large k), it may in fact be correct to treat recruitment as independent of stock size over at least the first k years of predictions, since the recruitments for these years have already been established by historical spawning stocks.

For long-term predictions that require some assessment of the risk of recruitment overfishing, that is, the predictions require the use of some assumption about the effect of low spawning stocks S_{t-k+1} on recruitments R_t, there are two obvious choices to use as initial hypotheses concerning the form of the stock-recruitment relationship. First, one may assume that recruitment is independent of stock size unless stock size is very low; this leads to the Beverton-Holt recruitment equation

$$R_t = \frac{aS_{t-k+1}}{b + S_{t-k+1}} \tag{9.2.16}$$

where the parameter a is the maximum recruitment rate when S_{t-k+1} is large and b is the spawning stock size S_{t-k+1} needed to produce a recruitment of half the maximum (i.e., to produce $R_t = a/2$). Second, one may assume that recruitment will decline at both low and high stock sizes (dome shaped recruitment curve), according to a Ricker form of recruitment equation

$$R_t = S_{t-k+1} \exp(a' - b'S_{t-k+1}) \tag{9.2.17}$$

where the parameter a' measures the maximum productivity per spawner when the spawning stock is low and b' measures density-dependent reductions in productivity as stock size increases. Schnute (1985) and Fournier and Doonan (1987) recommend using the more complex recruitment equation suggested by Deriso (1980), which can assume both Beverton-Holt and Ricker shapes depending on a third parameter besides a and b or a' and b'; in our experience, it will rarely be possible to say anything about that third parameter based on empirical data, except whether the recruitment curve is dome-shaped in the first place. Thus we recommend using one of the two simpler forms.

When growth is variable, so that animals from a given year class may reach the recruitment body weight w_k over a range of calendar ages, say k_1 to k_2, then a simple protocol is to calculate R_t in two stages. First, predict the total recruitment R'_t resulting from each spawning stock S_t. Then, calculate the realized recruitment in year t, R_t, as a sum of proportions of the total recruits R'_t from several spawning years:

$$R_t = \sum_{j=k_1}^{k_2} p_j R'_{t-j} \tag{9.2.18}$$

where p_j is the proportion of the total recruits R'_t that will recruit to size w_k after j years. This formulation shows that the effect of variable growth (or ages at movement into locations where the fish are vulnerable to the fishing gear) is to "smooth out" the recruitment rate R_t into the harvestable population. That is, R_t can be viewed as a running average of the past recruitment rates R'_t, which may be individually quite variable due to environmental factors; running averages of variable quantities are generally less variable (i.e., smoother over time) than the individual quantities.

9.3. Equilibrium and Dynamic Predictions

Delay-difference models as described above can be used either to estimate average equilibrium relationships between harvest rates and stock sizes or to simulate stock changes over time under variable harvest regimes. This section first describes how to calculate equilibrium stock size and harvest as a function of the harvest rate and delay-difference model parameters (i.e., how to calculate the average relationship between surplus production and harvest rate). Then it describes simple simulation techniques for making sure that the equilibrium calculations make sense in more realistic settings where harvest and recruitment rates may vary over time.

Form of the surplus production relationship

It is often useful to predict the equilibrium relationship between stock size or catch and the annual harvest rate h, while pretending that h can be held constant over enough years for an equilibrium to be reached. This relationship usually indicates at least the direction of stock size change to be expected after changes in exploitation levels, and helps identify a target level of exploitation that would produce the highest long-term yields.

In the delay-difference model format, finding the equilibrium stock size

B_e for a given harvest rate h involves substituting B_e for B_t and $h\mathcal{S}$ for s_t everywhere in the delay difference equation, then trying to solve the resulting equation algebraically for B_e. Then the equilibrium catch C_e is predicted simply from B_e and h, as $C_e = hB_e$. The algebraic manipulations to predict B_e are relatively simple but tedious, and the result depends strongly on what stock-recruitment model is assumed. The basic result for the recruitment models discussed above is

$$B_e \begin{cases} = \dfrac{a}{\mathcal{K}} - \dfrac{b}{(1-h)} & \text{Beverton–Holt} \\ = \dfrac{[a' + \log(1-h) - \log\mathcal{K}]}{[b'(1-h)]} & \text{Ricker} \\ = \dfrac{R^*}{\mathcal{K}} & \text{constant at } R^* \end{cases} \tag{9.3.1}$$

where $\mathcal{K}$ is the complicated growth-survival constant:

$$\mathcal{K} = \frac{[1 - (1+\rho)\mathcal{S}(1-h) + \rho\mathcal{S}^2(1-h)^2]}{w_k - \rho w_{k-1}\mathcal{S}(1-h)} \tag{9.3.2}$$

Thus, calculating the equilibrium relationship between stock, harvest, and harvest rate involves three steps for each value of harvest rate h to be considered: (1) calculate the $\mathcal{K}$ value as a function of growth and survival parameters; (2) predict B_e using $\mathcal{K}$ and the recruitment parameters; and (3) predict equilibrium harvest from h and B_e. Incidentally, the equilibrium recruitment rate R_e associated with each h can be obtained by substituting the equilibrium spawning stock prediction, $R_e = (1-h)B_e$, into whatever stock-recruitment equation is used, or more simply $R_e = B_e\mathcal{K}$.

It may be useful to know the equilibrium relationship between mean body size and total survival rate as a means of checking survival rate estimates against body size data or of obtaining direct estimates of survival rate from body size. Note from above that the relationship between the equilibrium biomass B_e and recruitment rate R_e is given by $B_e = R_e\mathcal{K}$, where $\mathcal{K}$ is the complex growth-survival constant. The equilibrium number of animals N_e in the population is defined by $N_e = \mathcal{S}(1-h)N_e + R_e$ (i.e., that value of N such that $\mathcal{S}N + R$ just gives N again the next year), implying that N_e is just $N_e = R_e/[1 - \mathcal{S}(1-h)]$.

The equilibrium body weight, W_e, is the ratio of B_e and N_e, that is

$$\begin{aligned} W_e &= B_e/N_e \\ &= R_e/\mathcal{K}/R_e/[\mathcal{S}(1-h)] \\ &= [1 - \mathcal{S}(1-h)]/\mathcal{K} \end{aligned} \tag{9.3.3}$$

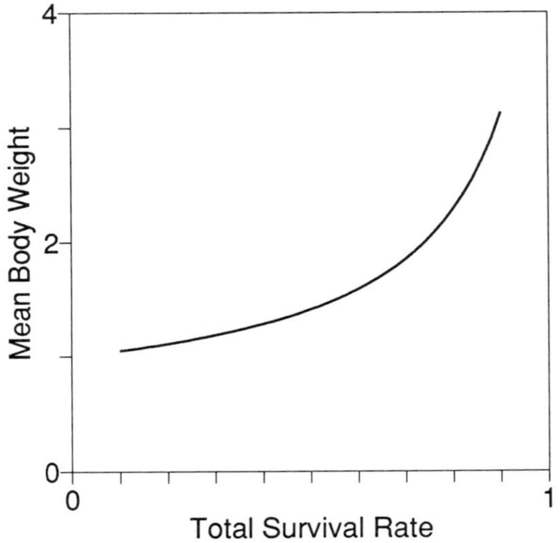

Figure 9.1. Equilibrium relationship between mean body weight and total survival rate for a hypothetical population with growth parameters $w_{k-1} = 0.5$, $w_k = 1$, and $\rho = 0.9$.

If we call s' the total survival rate $\mathcal{S}(1 - h)$, which is the same as $s' = e^{-Z}$ in traditional fisheries notation, the expression for W_e can be written in terms of s' and the growth parameters as

$$W_e = \frac{(1 - s')(w_k - \rho w_{k-1}s')}{[1 - (1 + \rho)s' + \rho s's']} \tag{9.3.4}$$

A simple way to use this equation is to calculate W_e for a range of values of s', then plot the resulting relationship as in Figure 9.1.

Such figures can be read either directly as predictions of the effect of survival rate changes on body weights in the stock, or inversely to estimate survival rate given an observed (and temporally stable) value of the mean weight. There is one key point of caution in using the relationship: it is valid only if the stock is large enough to display a sustainable recruitment rate R_e; stable mean body weight will also occur if the stock is declining exponentially toward extinction with recruitment just proportional to stock size, but the mean weight under this disastrous circumstance is not given by the above equation.

Example relationships between equilibrium stock size and harvest rate are shown in Figure 9.2, for a relatively long-lived ($\mathcal{S} = 0.8$) fish. The biomass equilibrium B_e is partitioned into new recruits R_e and older fish. Note how

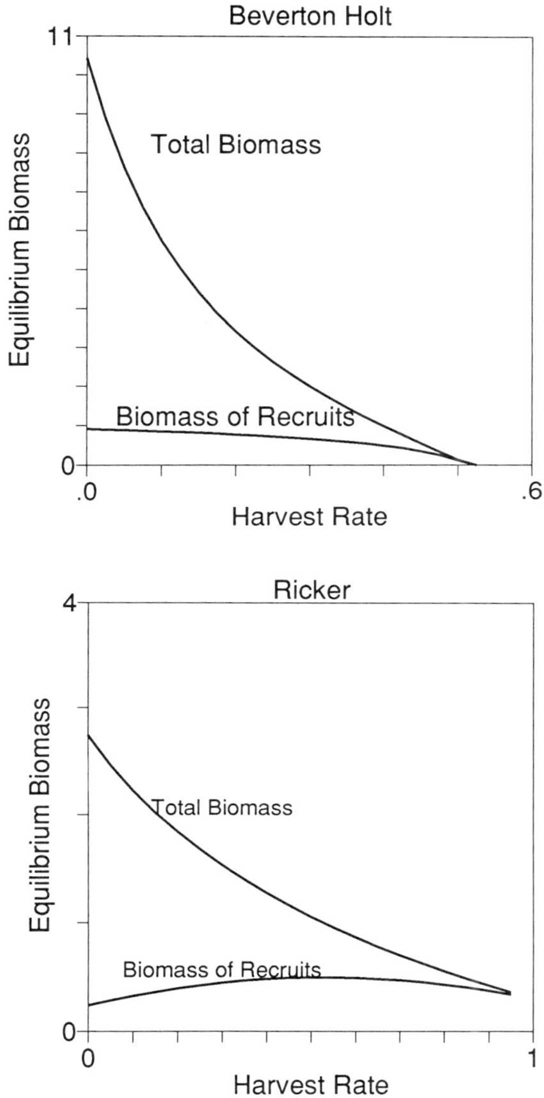

Figure 9.2. Equilibrium stock size as a function of equilibrium harvest rate for a hypothetical fish stock with $\mathcal{S} = 0.8$, $w_{k-1} = 0.5$, $w_k = 1$, and $\rho = 0.9$. For the Beverton-Holt case, $a = 1$, and $b = 0.1$. For the Ricker case, $a' = 2$ and $b' = 1$.

the equilibrium biomass of older fish decreases as the harvest rate is increased (an obvious expectation), whereas the biomass of new recruits looks like a reversed image of the stock-recruitment curve. If a Beverton-Holt recruitment curve is used, the recruitment component of B_e remains nearly constant as h is increased, until a high enough h is reached to cause R_e to fall enough to result in lowered recruitment. If a Ricker recruitment curve is used, increasing h can actually result in increasing recruitment, as R_e is lowered; but at high h, recruitment falls due to recruitment overfishing as in the Beverton-Holt case.

The equilibrium relationships between exploitation rate and stock measures can be converted into relationships between fishing effort and the stock measures, by providing an estimate of the relationship between fishing effort and exploitation rate. Without this additional relationship, the predictions involve at least five parameters (two for growth, one for survival, and at least two for stock recruitment and body size at recruitment). If the effort-exploitation relationship has only one parameter (for example, $h = 1 - e^{-qE}$ where E is effort and q is a catchability parameter), then the overall model will involve at least six parameters. This is in contrast to the simplest two-parameter surplus production model based on the equilibrium relationship between catch and effort for logistic population growth (see Chapter 8).

Transient Behavior in Relation to Time Lags

The delay-difference models imply that harvestable biomass in any year t will depend on biomasses in at least two previous years, $t - 1$ and $t - 2$, and, if recruitment depends on stock size, at least a third year $(t - k + 1)$. The growth-survival dependence on years $t - 1$ and $t - 2$ generates modest time lags (or delayed responses in mean body weight) that can in practice be ignored for short-term forecasting. However, the time lags involved in recruitment responses and recovery from recruitment overfishing can be much more substantial and may make it practically meaningless to provide only equilibrium or very short-term predictions of response to changes in harvest policy.

Consider for example the predicted response to reduced fishing pressure of a stock that recruits to fishing at age 4 and is currently believed to be suffering from recruitment overfishing. If fishing is curtailed immediately, the first increases in stock size due to improved recruitment (by the spawners saved from harvest this year) will occur after 4 years and further effects of these recruits on improving recruitment (through their own spawning contributions) will take an additional 4 years. Thus the overall effect of the immediate fishing reduction may not be seen for 8 years or more.

Simulation procedures

It is simple to simulate transient biomass dynamics and changes in mean body weight for age-structured populations using the pair of equations for B_t and N_t from the previous section. The delay-difference format can be used to predict just biomasses over time, but this is no more efficient than keeping track of both biomass and numbers. To begin the simulation, you must set (1) an initial stock size B_1, (2) total numbers N_1 (or an initial mean body weight $\bar{w}_1$ and calculate $N_1 = B_1/\bar{w}_1$); and (3) the first k years of recruitments $R_1 \ldots R_k$ which were produced by spawning populations prior to time 1. Alternatively, you can set initial spawning stocks for times $-k - 1, \ldots, -1, 0$, and set the initial recruitments from these by using a stock-recruitment relationship. Notice that setting a series of initial recruitments or past spawning stocks is equivalent to partially specifying the initial population age structure for animals aged *1* to k years old. This structure must be retained somehow in order to account for time delays from spawning to recruitment at age k.

Then for each year $t = 1,2, \ldots$ in the simulation, you will need the follow these steps: (1) set an exploitation rate h_t and total survival rate $s_t = \mathcal{S}(1 - h_t)$; (2) calculate the harvest $H_t = h_t B_t$ and spawning stock $S_t = B_t - H_t$; (3) calculate and store the recruitment to occur in year $t + k$, R_{t+k}, from S_t and a stock-recruitment relation; (4) calculate the mean body weight $\bar{w}_t = B_t/N_t$; (5) predict next year's biomass as

$$B_{t+1} = s_t[\alpha N_t + \rho B_t] + w_k R_{t+1} \tag{9.3.5}$$

and (6) predict next year's numbers as

$$N_{t+1} = s_t N_t + R_{t+1}. \tag{9.3.6}$$

The results of steps (5) and (6) give the starting values for the next year's cycle of calculations.

A sample simulation is shown in Figure 9.3, for a case in which an initially stable stock with $s_t = 0.8$ is subject to a sudden increase in exploitation rate to a level (0.3) that causes a reduction in recruitment after a delay of $k = 4$ years. Notice that the mean body weight $\bar{w}_t$ provides a clear signal of the change in exploitation rate, and stabilizes quickly at a new lower level over about the same time scale as it takes for the harvest to stabilize; the delayed decline in recruitment does not cause any marked change in the pattern of decline in $\bar{w}_t$.

9.4. Estimation of Parameters

A number of procedures have been suggested for estimating population parameters by comparing delay-difference model predictions to observed

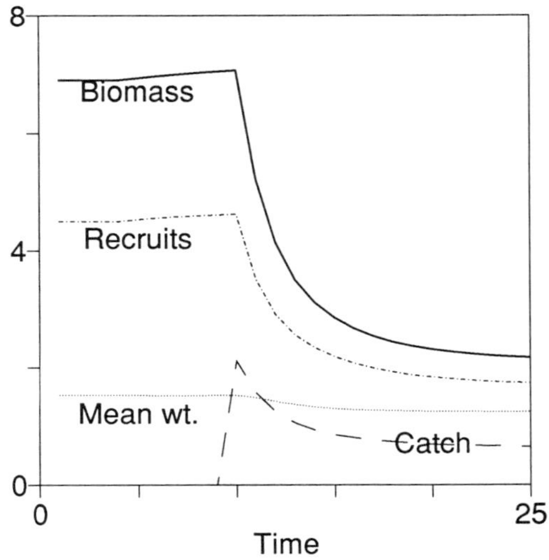

Figure 9.3. Simulation of a sudden increase in exploitation rate, which leads to subsequent decline in recruitment.

patterns of change in relative abundance and average body size (Deriso 1980, Schnute 1985, Walters 1987, Fournier and Doonan 1987, Jacobson et al. 1987). The general concept behind these procedures is to find a set of parameter values (for growth, survival, recruitment, and relative-versus-absolute abundance) that will minimize some criterion for goodness of fit between predicted and observed patterns.

In the simplest procedure (Walters 1987), it is assumed that the only data available are total catches and relative abundances over time, where the relative abundances (such as CPUE's) are assumed to be proportional to stock size (i.e., CPUE $= qB_t$) where the proportionality constant q is unknown. Then a nonlinear estimation algorithm (Chapter 6) is used to seek parameter estimates that give a minimum sum-of-squares deviations between observed and predicted relative abundances. For any combination of parameter values, the model predictions are obtained by simulating biomass over time while subtracting the observed catches for each year, then taking the predicted relative abundance each year to be the catchability or relative abundance scaling parameter q times the simulated biomass for that year.

Experience with fitting only relative abundance time-series data suggests that this procedure will permit estimation of only one or two, or occasionally three, of the six or more delay-difference model parameters. Usually it is necessary to estimate the growth parameters (α,ρ,w_k) from independent age-weight data, and to provide an independent estimate of the natural survival rate $\mathcal{S}$. Then the fitting procedure is used to estimate only stock-recruitment parameters and the catchability parameter. For typical fisheries time series, which do not contain observations of response over a wide range of spawning stock sizes, only one stock-recruitment parameter can usually be estimated. It is then necessary to assume some prior knowledge of either the slope of the recruitment curve for low stock sizes (maximum recruits per spawner at low stock) or of the maximum recruitment rate for the unfished stock.

More elaborate fitting procedures make use of further auxiliary information besides independent growth and survival parameters. For example, Fournier and Doonan (1987) assume a time series of mean body weights and other moments of the population size distribution, and they use these data to improve the estimates of growth parameters, survival rates, and variations in recruitment rate over time. Fournier and Doonan, along with Ludwig and Hilborn (1983), use fishing effort data to predict changes in catches and survival rates from the assumption that $h_t = 1 - e^{-qE_t}$. The delay-difference fitting procedure of Walters (1987) allows the user to enter a variety of auxiliary information, ranging from time series of relative recruitment rates to point estimates of stock size in particular years. The various point estimates and their predicted values are included in a weighted sum-of-squares fitting criterion. By setting high weights on particular auxiliary observations, the user can then force the fitting procedure to choose parameter combinations (or restrict its search to parameter combinations) that are consistent with those observations.

As an example of a fit of the delay-difference model, consider Figure 9.4, the catch and CPUE series for Pacific cod (*Gadus macrocephalus*). The stock exhibits a strong cyclic trend, which the Schaefer model is totally unable to capture. The best Schaefer fit to these data is a flat line. The delay-difference model, with a lag of 3 years to recruitment and a Ricker stock-recruitment curve, can capture the cyclic dynamics quite well. This fit was performed assuming that biomass was CPUE/q for the first 4 years.

Judgement and bias in fitting delay-difference models

The general experience in fitting delay-difference models is that simple time-series data (CPUE, catches, efforts) can usually be explained equally well by (i.e., fit equally well to) a wide variety of parameter combinations. These parameter combinations may imply very different responses to policy changes. For example, it is often not possible to distinguish whether the data came from a large, unproductive stock that has been exploited only lightly, or instead from a small, productive stock that has been heavily exploited. Often such ambiguities remain even when the fitting procedures are

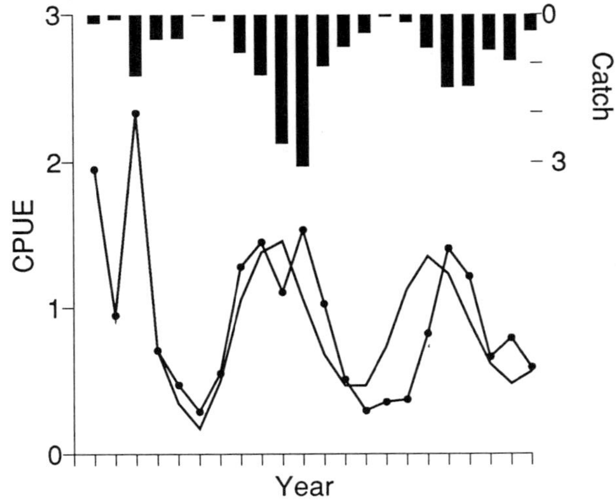

Figure 9.4. Observed catch and CPUE data for Hecate Stait Pacific cod and the best fit series of predicted CPUE.

provided with considerable prior information on growth and survival parameters. Simpler assessment procedures have attempted to side step these ambiguities by estimating only aggregate surplus production parameters while assuming that the stock size has remained near equilibrium over the entire period of historical data.

When one attempts to use delay-difference models to analyze surplus production, the ambiguities must be confronted and dealt with somehow through explicit judgmental decisions about the various parameters that cannot be estimated directly. In one sense, it is healthy to use delay-difference models, because the user is forced to think carefully about various parameters and how to estimate them from the available data. This thinking may lead to substantial improvements in data-gathering programs and to policies that are more cautious or deliberately experimental than would be employed if simpler surplus production model results were trusted for policy formulation. The user is at least forced to drop the pretense that a good statistical fit of his data to any particular assessment model is indicative that he has found the "right" model for policy prediction.

On the other hand, there are two major risks in fitting delay-difference models. First, they contain enough parameters as "fudge factors" to allow the analyst to obtain good looking fits to almost any data set, even when the best fitting parameter estimates are biologically meaningless or misleading. Thus an unscrupulous user, or one committed to some particular hypothesis about the stock, can usually find a good fit to justify almost any prior belief about the productivity and potential of the stock. The search for biological realism in a model structure can easily turn into an exercise in wishful thinking and/or circular reasoning about the available data, to the point that it would have been better to rely on simpler models whose best fitting parameter estimates are at least obvious from the data.

Second, the fitting procedures can give badly biased parameter estimates, even while giving a good visual fit to the data. Ludwig and Walters (1985, 1988) have shown that better estimates of optimum fishing effort can often be obtained by assuming simpler production models that are known to be incorrect, than by assuming the biologically correct delay-difference structure. In such circumstances, the essential problem is that the delay-difference fitting procedures are forced (by the model structure) to separately estimate the effects of growth versus recruitment as components of production; when these effects are not clearly distinguishable in the data (a common circumstance), the fitting procedures will tend to give very poor estimates of stock-recruitment parameters. In particular, they will tend to set the recruitment equation parameters so that the simulated recruitment rate is independent of stock size and hence lead the user to underestimate the risk of recruitment overfishing (overestimate the optimum fishing rate and/or optimum effort level).

Insight about likely estimation biases and sensitivity to auxiliary information about parameters can be obtained by Monte Carlo simulation techniques. However, these techniques will not help to detect biases due to treating key quantities that may be changing over time as though they had been constant parameters. Key parameters that aren't in the delay-difference structure are likely to be (1) the age of entry to the harvestable population k, (2) the catchability or abundance index coefficient q, and (3) parameters of the stock-recruitment relationship. The age at entry k often decreases during fishery development, then may increase later as regulations are introduced to improve the yield per recruit. The catchability q is likely to increase over time and to be inversely related to stock size. Stock-recruitment parameters are likely to change as discussed in Chapter 7. In fitting delay-difference models to simple time-series data, parameter changes can often be mimicked by constant parameters of the wrong form of equation; for example, decreasing k over time during a stock decline may make recruitment rate appear to increase during the decline, and this pattern might be well fitted (and misinterpreted) by assuming a dome-shaped (Ricker) recruitment relationship.

9.5. Summary and Critique

Delay-difference models are extremely useful for understanding age-structure effects and for efficiently simulating temporal changes in exploited stocks.

However, these models must be treated with extreme care when used as devices for extracting information about population dynamics parameters from simple time-series data. They can give better estimates than simpler surplus production models when used wisely and honestly in conjunction with auxiliary information, provided their predictions are not judged solely by goodness of fit to historical data.

Bibliographical Notes

Deriso (1980) developed the delay-difference approach to capturing the dynamics of age structure in a set of simple equations. Deriso's model was generalized and related to a large number of other models by Schnute (1985). Fournier and Doonan (1987) and Schnute (1987) extended the model further, examining the use of length data in conjunction with the index of abundance used in the Deriso (1980) and Schnute (1985) formulation.

10

Virtual Population Analysis

10.1. General Principles

The most common methods currently used in temperate fisheries stock assessments rely on catch-at-age data. While these methods are often referred to collectively as catch-at-age methods, we like to think of them as divisible into two classes of models. The first, commonly called virtual population analysis (VPA) or cohort analysis methods are recursive algorithms that calculate stock size based on catches with no underlying statistical assumptions. These methods are covered in this chapter. The second class of catch-at-age methods rely on formal statistical models and we consider these in the next chapter (Chapter 11) entitled "Statistical Catch-at-Age Methods." Megrey (1989) provides an excellent review of all of these methods.

Virtual population analysis (VPA) calculates past stock abundances based on past catches. Once stock sizes are calculated, fishing size-selectivity as well as changes in vulnerability over time can be determined. The stock size estimates, which include recruitment estimates for each year, can be used for stock and recruitment analysis. VPA, also known as cohort analysis, is one of the most powerful techniques available for the analysis of fisheries data and forms the heart of many current assessment methods where catch-at-age data are available. When catch-at-age data are not available, the related approach known as *stock reduction analysis* can also be very helpful in reconstructing stock histories (Kimura et al. 1984)

Virtual population analysis calculates the number of fish alive in each cohort for each past year. It is also called cohort analysis because each cohort is analyzed separately. VPA relies on a very simple relationship *for each cohort*

$$\begin{pmatrix}\text{Number alive}\\ \text{at beginning}\\ \text{of next year}\end{pmatrix} = \begin{pmatrix}\text{number alive}\\ \text{at beginning}\\ \text{of this year}\end{pmatrix} - \begin{pmatrix}\text{catch}\\ \text{this}\\ \text{year}\end{pmatrix} - \begin{pmatrix}\text{natural}\\ \text{mortality}\\ \text{this year}\end{pmatrix} \quad (10.1.1)$$

Table 10.1. Life history events for British Columbia herring (*Clupea harengus*) and Ontario Lake trout (*Salvelinus namaycush*).

Period	Events
British Columbia herring	
January 1—mid March	natural mortality
mid March—mid April	fishery and spawning
April—December	natural mortality
Ontario Lake trout	
January 1—December 31	natural and fishing mortality

Note that recruitment does not enter into Equation 10.1.1 because we are dealing with a single cohort.

If we knew the initial cohort size, and the natural mortality rate, we could use Equation 10.1.1 to calculate the number alive each year. Unfortunately, we rarely if ever know the initial number alive; this is in fact one of the things we want cohort analysis to tell us.

Do we ever know the number of fish alive at any age for most stocks? Yes! There are few, if any, 100-year-old fish in most exploited stocks. In fact, we are reasonably safe in assuming that there are essentially no fish older than the oldest fish we have ever caught. This will not be true if our gear is size selective and old (and perhaps big) fish can avoid the gear. In general though, especially when stocks are intensively harvested, there will be so few fish older than our oldest catch that we can ignore them.

Equation 10.1.1 can be rewritten as

$$\begin{pmatrix}\text{number alive}\\ \text{at beginning}\\ \text{of this year}\end{pmatrix} = \begin{pmatrix}\text{number alive}\\ \text{at beginning}\\ \text{of next year}\end{pmatrix} + \begin{pmatrix}\text{catch}\\ \text{this}\\ \text{year}\end{pmatrix} + \begin{pmatrix}\text{natural}\\ \text{mortality}\\ \text{this year}\end{pmatrix} \qquad (10.1.2)$$

If we are willing to assume that at some age there are none alive (or that we know the number alive at some terminal age) and that we know the natural mortality rate, we can use Equation 10.1.2 to iteratively calculate the number alive each year, starting from the oldest ages and moving backward to the youngest.

In VPA, we must be very careful to outline the annual time sequence of events in the fish's life. Table 10.1 shows the timing of life history events for two species: herring in British Columbia and lake trout in Ontario. The two events of interest are the periods of fishing mortality and natural mortality. In the British Columbia herring fishery, fishing takes place over a few days which coincide with spawning. Natural mortality takes place year round. In the Ontario lake trout fishery, fishing and natural mortality occur simultaneously all year long. The significant difference between these two life histories is that we can ignore the natural mortality that occurs during the fishery in British Columbia herring, but we must calculate the simultaneous mortalities for lake trout. The calculations for herring are therefore simple but are more complex for lake trout.

10.2. Discrete Fisheries

In any fishery that has a relatively short fishing period, we can assume that the natural mortality during this period is negligible. The life history of British Columbia herring can be considered in two steps: one of natural mortality from the end of the fishery in one year until the beginning of the fishery the next year, and second the period of fishing. Equation 10.1.2 then becomes

$$N_t = N_{t+1} + C_t + D_t \tag{10.2.1}$$

where N_t is the number alive after the fishery last year, C_t is the catch during this year's fishery, D_t is the number dying from natural mortality this year, and N_{t+1} is the number alive after this year's fishery. The number dying is assumed proportional to the number alive at the beginning of the period of natural mortality (N_t) or

$$D_t = N_t(1 - s) \tag{10.2.2}$$

where s is an annual survival rate. We can modify and rearrange Equation 10.2.1 as

$$N_t - D_t = N_{t+1} + C_t \tag{10.2.3}$$

or

$$N_t s = N_{t+1} + C_t \tag{10.2.4}$$

or

$$N_t = \frac{N_{t+1} + C_t}{s} \tag{10.2.5}$$

This gives us a useful equation, because everything on the right-hand side is known, and the left-hand side can thus be calculated.

What are we assuming? Principally that we have some age at which $N_{t-1} = 0$, and that we know the survival rate s. The survival rate need not be the same for all ages, and if we had estimates of age-specific s's we could use them in Equation 10.2.5.

Table 10.2 shows the calculations for the 1970 cohort of the Georgia Strait stock of British Columbia herring. We can theoretically calculate cohort

Table 10.2. Cohort analysis of 1970 brood of Georgia Strait stock of British Columbia herring ($s = 0.6$).

Year	Age	Catch	Cohort size at start of year	Cohort size before fishery	Harvest rate	Instantaneous fishing mortality
1979	9	0	0	0	—	—
1978	8	1050	1750	1050	1.00	0.00
1977	7	2640	7317	4390	0.60	0.92
1976	6	6182	22498	13499	0.46	0.61
1975	5	14145	61071	36643	0.39	0.49
1974	4	8333	115674	69404	0.12	0.13
1973	3	7813	205811	123487	0.06	0.07
1972	2	1286	345162	207097	0.01	0.01

(Data from Stocker et al. (1983))

strength back to time of birth, but since natural mortality during the first few ages is thought to be high and is difficult to estimate, the normal practice is to calculate cohort size back only to the age of first recruitment to the fishing gear. Table 10.2 also shows the cohort size at the beginning of the fishery and the harvest rates and instantaneous fishing mortality rates for each year. It is easily demonstrated that

$$\begin{aligned} N_t' &= N_t s \\ h_t &= C_t/N_t' \\ F_t &= -\ln(1 - h_t) \end{aligned} \tag{10.2.6}$$

where N_t' is the number alive at the beginning of the fishery, h_t is the harvest rate (fraction captured during fishery) and F_t is the instantaneous fishing mortality rate from the equation

$$C_t = N_t'(1 - e^{-F_t}). \tag{10.2.7}$$

If we had reliable measures of fishing effort (which we do not for the British Columbia herring fishery), we could calculate age-specific vulnerability q by the following equation:

$$q_t = \frac{F_t}{E_t} \tag{10.2.8}$$

where E_t is the effective fishing effort. If we perform cohort analysis on all cohorts, we generate Table 10.3. We start with catch at each age and assume a survival rate and that no fish are alive after age 9. We then calculate numbers at each age and fishing mortality by age.

Table 10.3. VPA for Georgia Strait stock of B.C. herring.

Year	Age 2	3	4	5	6	7	8	9
				Catch				
1971	2,607	8,037	6,984	971	644	0	0	0
1972	1,286	14,382	16,930	9,519	1,755	191	13	1
1973	818	7,813	5,665	8,040	2,445	687	89	0
1974	60	856	8,333	6,371	3,254	713	135	0
1975	212	4,302	13,311	14,145	4,866	1,187	286	8
1976	28	324	17,728	20,678	6,182	1,930	302	58
1977	1,479	24,000	21,051	31,577	10,654	2,640	444	96
1978	126	12,319	22,658	17,835	20,572	7,064	1,050	325
1979	89	333	11,681	24,762	5,424	2,169	535	0
			Population at end of last years fishery					
1971	258,835	147,920	59,824	10,818	2,016	36	3	0
1972	345,162	152,694	80,715	28,910	5,520	566	22	2
1973	—	205,811	77,234	31,499	7,827	1,557	148	0
1974	—	—	115,674	40,676	10,859	2,251	247	0
1975	—	—	—	61,071	18,034	3,262	638	13
1976	—	—	—	—	22,498	5,955	770	97
1977	—	—	—	—	—	7,317	1,643	160
1978	—	—	—	—	—	—	1,750	542
1979	—	—	—	—	—	—	—	—
			Population just prior to fishery					
1971	155,301	88,752	35,894	6,491	1,210	22	2	0
1972	207,097	91,616	48,429	17,346	3,312	339	13	1
1973	—	123,487	46,341	18,899	4,696	934	89	0
1974	—	—	69,404	24,405	6,516	1,351	148	0
1975	—	—	—	36,643	10,821	1,957	383	8
1976	—	—	—	—	13,499	3,573	462	58
1977	—	—	—	—	—	4,390	986	96
1978	—	—	—	—	—	—	1,050	325
1979	—	—	—	—	—	—	—	0
				Harvest rate				
1971	0.017	0.091	0.195	0.150	0.000	0.000	0.000	0.000
1972	0.006	0.157	0.350	0.549	0.530	0.563	1.0	0.000
1973	—	0.063	0.122	0.425	0.521	0.735	1.0	0.000
1974	—	—	0.120	0.261	0.499	0.528	0.910	1.0
1975	—	—	—	0.386	0.450	0.607	0.747	1.0
1976	—	—	—	—	0.458	0.540	0.654	1.0
1977	—	—	—	—	—	0.601	0.450	1.0
1978	—	—	—	—	—	—	1.0	1.0
1979	—	—	—	—	—	—	—	.000

10.3. Continuous Fishing and Natural Mortality

When we have continuous fishing and natural mortality, we must use a different set of relationships. The basic assumption is that the survival of the cohort during the year is governed by

$$\frac{dN}{dt} = -N(F + M) \tag{10.3.1}$$

where F and M are the instantaneous fishing mortality and natural mortality rates. Solving for the number surviving the year we find

$$N_{t+1} = N_t e^{-(F+M)} \tag{10.3.2}$$

The total loss during the year is $N_t - N_{t+1}$ so the catch during the year is

$$C_t = \frac{F}{F + M}(N_t - N_{t+1}) \tag{10.3.3}$$

which simply says that the catch is the total loss during the year, times the proportion of the total mortality rate $(F + M)$ that is due to fishing. If we try to solve for F using Equation 10.3.2, we get

$$\begin{aligned} \frac{N_{t+1}}{N_t} &= e^{-(F+M)} \\ F &= -\log\left(\frac{N_{t+1}}{N_t}\right) - M \end{aligned} \tag{10.3.4}$$

We are trying to find an equation analogous to Equation 10.2.5, which expresses N_t in terms of the known quantities N_{t+1}, C_t, and M. If we substitute the equation for F from Equation 10.3.4 into Equation 10.3.3, we only have these terms left.

$$C_t = \frac{-\log(N_{t+1}/N_t) - M}{-\log(N_{t+1}/N_t) - M + M}(N_t - N_{t+1}) \tag{10.3.5}$$

which simplifies to

$$C_t = \left[1 - \frac{M}{\log(N_t) - \log(N_{t+1})}\right](N_t - N_{t+1}) \tag{10.3.6}$$

Let us now stop and ask if Equation 10.3.6 makes any sense. The $N_t - N_{t+1}$ term on the right is the total loss during the year, so everything else on the right-hand side is $F/(F + M)$ or the proportion of total mortality that is due to fishing. If $M = 0$, everything on the right-hand side except the $N_t - N_{t+1}$ term is 1.0, and Equation 10.3.6 then says that catch is equal to total loss, which makes sense when $M = 0$. If M is large, less of the total loss will be due to fishing. As $\log(N_t) - \log(N_{t+1})$ approaches M, all loss is due to natural mortality and $C_t = 0$. Equation 10.3.6 does makes sense, but it does not allow a simple solution for N_t as did Equation 10.2.5. Equation 10.3.6 is transcendental; it has no direct solution. It must be solved by iterative methods. This is where Gulland (1965) left cohort analysis, and it was not used too much because computers were not really available at the time and most people were daunted by having to do iterative solutions on hand-crank calculators.

There are two solutions to the problem. First, modern calculators and microcomputers make iterative solutions very easy. Newton's method will solve Equation 10.3.6 in two or three iterations. Newton's method finds iterative improvements in the value of some variable x which will make $f(x) = 0$. The iterative equation to find x is

$$x_{\text{new}} = x_{\text{old}} - \frac{f(x)}{f'(x)} \tag{10.3.7}$$

where $f'(x)$ is the derivative of $f(x)$ with respect to x. In our case, x is N_t, and $f(x)$ is Equation 10.3.6 transformed so everything is on the right-hand side; $f'(x)$ is therefore:

$$f'(x) = -1 - \frac{(\log(N_t) - \log(N_{t+1}))M - \dfrac{N_t - N_{t+1}}{N_t}M}{(\log(N_t) - \log(N_{t+1}))^2} \tag{10.3.8}$$

The second alternative is to approximate Equation 10.3.6 by a simpler model that does not result in a transcendental equation for N_t. Pope (1972) showed that N_t can be approximated by the following equation:

$$N_t = N_{t+1}\, e^M + C_t\, e^{M/2} \tag{10.3.9}$$

which looks very much like Equation 10.2.5 and is solved the same way. Pope's approximation assumes that all of the catch takes place in an instantaneous fishery, which happens midway during the year. Equation 10.3.9 is derived by using the handy approximation

$$\frac{Z(1 - e^{-F})}{F(1 - e^{-Z})} \approx e^{M/2} \tag{10.3.10}$$

where Z in this case is $F + M$. Table 10.4 shows the relationship between the two sides of the above approximation for Z between 0.1 and 1.7 (as-

Table 10.4. The reliability of Pope's approximation used in cohort analysis.

Z	$[Z(1 - e^{-F})]/[F(1 - e^{-Z})]$	$e^{M/2}$
0.1	1.025	1.025
0.3	1.075	1.078
0.5	1.124	1.133
0.7	1.173	1.191
0.9	1.221	1.252
1.1	1.268	1.317
1.3	1.314	1.384
1.5	1.358	1.455
1.7	1.401	1.530

suming $M = F$). When Z is less than 0.7, the approximation is very good; it is 9% off when $Z = 1.7$. This may not seem like a very large error, considering the types of uncertainty in catch and age data; however, remember that this approximation is used several times in the analysis of a cohort, so the errors accumulate. The effect of this error is to overestimate N_t in Equation 10.3.9. In general, it is probably safe to use the approximation so long as $Z < 1.0$, but, if $Z > 1.0$, you should use the exact solution. However, considering the wide availability of computers today there is really no reason for not solving the equation exactly. On the other hand, few fisheries now have a 12-month fishing season, and you might as well realistically model the length of the fishing season.

Table 10.5 shows solutions for a single cohort of Flack Lake trout using Pope's approximation and the exact method. The natural mortality rate M is assumed to be 0.4 in all cases. Table 10.6 shows the complete VPA for Flack Lake trout, using the exact method. Once we have solved for the cohort sizes, we can also find the fishing mortality rates from Equation 10.3.4. These are also shown in Table 10.6.

Once we have fishing mortality rate, we can solve for the catchability coefficient q. Given that $F = Eq$, we get

$$q = \frac{F}{E} \tag{10.3.11}$$

This gives us a reasonably straight forward solution for q, which was one of the purposes in doing cohort analysis. From this, we can plot age-specific q's as in Figure 10.1.

Terminal F assumptions

A problem with cohort analysis as introduced above is the assumption that there are no animals alive at a certain age, which means we can only analyze

Table 10.5. Comparison of 1965 cohort from Flack Lake using Pope's method and the exact method of VPA.

Age	Catch	Cohort size	
		Exact	Pope
3	13	5335	5536
4	169	3566	3700
5	606	2253	2342
6	644	1023	1074
7	116	184	193
8	22	33	34
9	4	5	5

Flack Lake data from Ontario Ministry of Natural Resources.

cohorts that have passed completely through the fishery; we can say nothing about cohorts still present. This leaves us with the blank triangle in the lower left corner of Table 10.6. Unfortunately, these cohorts are also the ones we are often most interested in because they are still in the fishery. We must find some way to estimate their abundance if we want to use VPA to estimate current stock size for use in developing current fishing plans.

There are two possible solutions to the problem of incomplete cohorts. Probably the best solution is to use the formal statistical methods for catch-at-age analysis that are discussed in the next chapter. A substitute, but one that is currently more often used, is to make some independent estimate of the fishing mortality rate on cohorts currently being fished and use this to estimate the cohort sizes now present. There are two ways to estimate the current cohort sizes: (1) direct estimates based on surveys, mark-recapture, or some other population estimation method; or (2) by assuming a current value for F. This assumption, often called the *terminal F assumption*, relies on the following relationship:

$$N_t = \frac{C_t}{(1 - e^{-Z_t})}\left(\frac{F_t + M}{F_t}\right) \qquad (10.3.12)$$

So how do we get an F_t to substitute into this equation? Again there are two approaches: (1) get a direct estimate of F by some method such as tagging, or (2) estimate F from effort data while assuming that q is known (using $F = Eq$).

An obvious approach is to use the q's for specific ages from complete cohorts to estimate the q's for each age of cohorts still present, and then use Equations 10.3.12 and 10.3.11 to estimate the current numbers at age of the present stock. This approach relies on the q's being stable over time.

Table 10.6. VPA for Flack Lake trout (*Salvelinus namaycush*) using only complete cohorts.

Year	Age 3	4	5	6	7	8	9
				Catch			
1968	13	129	646	954	99	19	4
1969	19	169	416	1,031	243	47	18
1970	40	354	606	479	152	18	7
1971	32	606	1,424	644	157	23	17
1972	0	226	1,178	1,156	116	16	5
1973	2	165	593	982	428	22	11
1974	53	209	560	410	30	0	4
1975	0	105	674	446	16	2	2
1976	46	422	838	726	70	4	4
1977	3	310	1,224	1,068	65	0	0
1978	14	354	1,264	1,172	69	0	6
1979	6	429	1,222	1,067	192	0	0
				Population size			
1968	5,335.8	2,955.8	3,067.7	1,652.6	217.4	55.6	5.0
1969	11,279.6	3,566.1	1,876.6	1,536.0	359.5	67.2	22.1
1970	8,158.0	7,545.5	2,253.3	923.1	229.9	52.6	8.8
1971	3,674.4	5,436.0	4,770.6	1,023.8	240.9	36.1	20.9
1972	3,978.0	2,437.0	3,153.3	2,056.4	184.1	39.4	6.3
1973	5,252.2	2,666.6	1,450.5	1,174.0	470.4	32.9	13.7
1974		3,519.0	1,653.6	500.4	49.8	3.7	5.0
1975	—	—	2,189.3	660.7	26.1	9.9	2.5
1976	—	—	—	927.6	97.1	5.0	5.0
1977	—	—	—	—	69.3	11.2	0.0
1978	—	—	—	—	—	0.0	7.5
1979	—	—	—	—	—	—	—
				Catchability coefficient			
							Average
1968	0.000000	0.000006	0.000030	0.000114	0.000078	0.000053	0.000108
1969	0.000000	0.000006	0.000034	0.000165	0.000167	0.000180	0.000166
1970	0.000001	0.000009	0.000059	0.000144	0.000221	0.000080	0.000156
1971	0.000001	0.000015	0.000046	0.000138	0.000148	0.000142	0.000140
1972	0.000000	0.000013	0.000063	0.000115	0.000141	0.000070	0.000116
1973	0.000000	0.000008	0.000067	0.000279	0.000448	0.000150	0.000324
1974	—	0.000010	0.000066	0.000326	0.000155	0.000000	0.000310
1975	—	—	0.000058	0.000191	0.000158	0.000035	0.000188
1976	—	—	—	0.000195	0.000157	—	0.000192
1977	—	—	—	—	—	0.000000	0.000000
1978	—	—	—	—	—	—	0.000000
1979	—	—	—	—	—	—	0.000000
Average	0.000000	0.000010	0.000051	0.000167	0.000228	0.000110	0.000000
				Instantaneous fishing mortality			
							Average
1968	0.003	0.054	0.292	1.125	0.774	0.524	1.068
1969	0.002	0.059	0.310	1.499	1.522	1.639	1.508
1970	0.006	0.058	0.389	0.943	1.452	0.524	1.022
1971	0.011	0.145	0.442	1.316	1.411	1.351	1.334
1972	0.000	0.119	0.588	1.075	1.320	0.658	1.088
1973	0.000	0.078	0.664	2.761	4.433	1.483	3.205
1974	—	0.075	0.517	2.552	1.216	0.000	2.431
1975	—	—	0.459	1.517	1.760	0.279	1.490
1976	—	—	—	2.195	—	—	2.153
1977	—	—	—	—	—	0.000	.000

Table 10.6. (Continued)

	Age						
Year	3	4	5	6	7	8	9
1978	—	—	—	—	—	—	0.000
1979	—	—	—	—	—	—	0.000
Average	0.004	0.084	0.448	1.545	2.123	0.998	0.000

(Data from Ontario Ministry of Natural Resources). Catchability coefficients and fishing mortality rates are not included for age 9. Entries are left blank if they cannot be calculated or are meaningless.

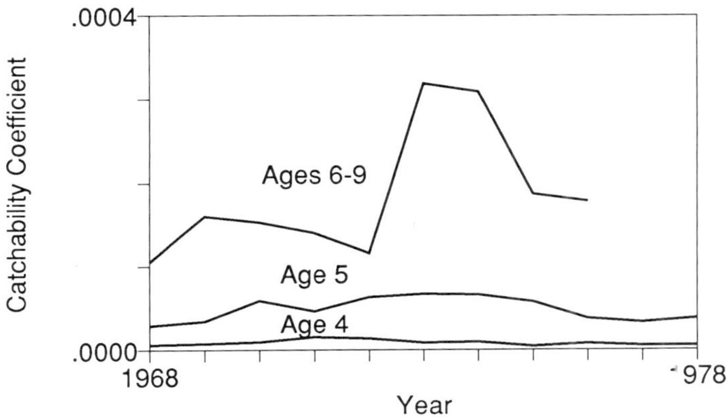

Figure 10.1. Catchability coefficients for Flack Lake trout for several ages in each year.

A variation on the theme, sometimes called *iterated* or *tuned* VPA (Laurec and Shepherd 1983, Pope and Shepherd 1985), is to first use the q's from completed cohorts, then use a new set based in part on the q estimates that then result for the incomplete cohorts.

There are two major schools of thought on how to treat incomplete cohorts, and the relative merits of full statistical catch-at-age methods as opposed to tuned VPA. In general, the North American scientists have favored statistical methods, whereas the Europeans have opted for tuning methods. In 1988, ICES held a workshop in which these methods were compared using *artificial* data from simulation models (ICES 1988). Although the results are not definitive, the general indications were that the tuning methods did quite well, and several of them consistently outperformed many of the more sophisticated statistical catch-at-age methods. There was general agreement among participants in this workshop that more testing was needed, and it may be a long time before we really understand the merits of the different methods, but there does not appear to be any method that is clearly superior.

In Table 10.7, the average fishing mortalities for ages 6+ were calculated and given on the right-hand side of the page. Similarly, the average fishing mortalities and catchabilities were calculated for each age (across all years) in the completed cohorts. If we use the average F's by age in 1979, we can estimate the cohort size of each age in 1979 using Equation 10.3.12. We can also get around the assumption that there are no age 10 individuals alive by using the average F for age 6+ fish as the estimated F for age 9. The results of this for Flack Lake are shown in Table 10.7. Note that the estimated terminal F's are now slightly different, and, as noted above, one might iterate by redoing the analysis with these *revised* terminal F's.

10.4. Pacific Salmon and Other Fishes that Die After Spawning

Pacific salmon have a life history that is different from those discussed earlier, but nevertheless are quite amenable to cohort analysis. Being semelparous, they spawn once and then die. Usually there is some direct estimate of the number of fish that spawn, that is, of the terminal number alive at the spawning age(s). Some species, such as pink salmon (*Oncorhynchus gorbuscha*), all mature at the same age. Other species, particularly chinook (*Oncorhynchus tshawytscha*) and chum (*Oncorhynchus keta*), mature at different ages. Cohort analysis can be used to calculate not only cohort sizes and harvest rates, but also maturation rates. Fishing and spawning take place over an extended period of several months, but we generally assume discrete fishing periods to simplify the calculations, because the annual natural mortality rates are generally low.

The annual life history of chinook salmon can be simplified as shown in Table 10.8. Some stocks of chinook are fished year round, but most are fished during an intense summer fishery of a few months.

The basic cohort analysis equation for chinook is similar to Equation 10.2.5:

$$N_t = \frac{N_{t+1} + C_t + S_t}{s} \tag{10.4.1}$$

where S_t is the number maturing at age t ("spawners") and s is again the annual natural survival rate. Given fishing efforts and the knowledge that $C_t/(N_t s)$ is the proportion of the stock alive at the beginning of the fishery that is caught (the harvest rate), we can calculate the vulnerabilities (q's) as

$$q_t = \frac{-\log\{1 - [C_t/(N_t s)]\}}{E_t} \tag{10.4.2}$$

Table 10.7. VPA for Flack Lake trout using terminal F's from previous analysis.

Year	Age 3	4	5	6	7	8	9	Average
				Numbers				
1968	5,335.2	2,947.9	3,082.8	1,676.2	227.8	62.7	7.0	—
1969	11,291.1	3,565.7	1,871.3	1,546.1	374.7	74.0	26.8	—
1970	8,160.7	7,553.2	2,253.0	919.5	236.2	62.1	13.0	—
1971	3,676.1	5,437.8	4,775.7	1,023.7	238.6	40.0	27.2	—
1972	3,981.7	2,438.1	3,154.5	2,059.8	184.0	38.0	8.8	—
1973	5,339.2	2,669.0	1,451.2	1,174.8	172.6	32.9	12.7	—
1974	6,843.7	3,577.4	1,655.2	500.9	50.2	4.4	5.0	—
1975	9,744.9	4,544.4	2,228.4	661.8	26.4	10.1	3.0	—
1976	9,169.3	6,532.2	2,960.9	953.6	97.8	5.2	5.2	—
1977	9,646.9	6,109.0	4,036.4	1,312.4	84.3	11.6	0.4	—
1978	9,627.9	6,464.1	3,843.4	1,724.9	72.9	6.9	7.8	—
1979	1,823.3	6,442.4	4,045.7	1,565.0	248.2	0.0	0.0	—
				Catchability coefficient				
1968	0.000000	0.000006	0.000029	0.000111	0.000073	0.000046	0.000112	0.000105
1969	0.000000	0.000007	0.000034	0.000163	0.000154	0.000147	0.000165	0.000160
1970	0.000001	0.000009	0.000059	0.000144	0.000209	0.000065	0.000152	0.000153
1971	0.000001	0.000015	0.000046	0.000138	0.000151	0.000117	0.000137	0.000140
1972	0.000000	0.000013	0.000063	0.000115	0.000141	0.000074	0.000117	0.000116
1973	0.000000	0.000008	0.000067	0.000278	0.000431	0.000151	0.000323	0.000319
1974	0.000001	0.000009	0.000066	0.000325	0.000153	0.000000	0.000306	0.000309
1975	0.000000	0.000004	0.000057	0.000191	0.000155	0.000034	0.000189	0.000187
1976	0.000001	0.000007	0.000037	0.000180	0.000154	0.000187	0.000187	0.000178
1977	0.000000	0.000005	0.000032	0.000177	0.000150	0.000000	—	0.000000
1978	0.000000	0.000005	0.000037	0.000115	—	—	0.000157	0.000000
1979	0.000000	0.000007	0.000038	0.000131	0.000179	—	—	0.000000
Average	0.000000	0.000010	0.000051	0.000165	0.000218	0.00096	0.000000	—
				Instantaneous fishing mortality rate				
1968	0.003	0.054	0.290	1.098	0.724	0.450	1.1	1.034
1969	0.002	0.059	0.311	1.479	1.397	1.339	1.5	1.458
1970	0.006	0.058	0.389	0.949	1.376	0.426	1.0	1.005
1971	0.011	0.145	0.441	1.316	1.438	1.117	1.3	1.333
1972	0.000	0.119	0.588	1.072	1.322	0.693	1.1	1.086
1973	0.000	0.078	0.664	2.754	4.266	1.489	3.2	3.154
1974	0.009	0.073	0.517	2.543	1.198	0.000	2.4	2.420
1975	0.000	0.028	0.449	1.512	1.227	0.271	1.5	1.483
1976	0.006	0.081	0.414	2.026	1.731	2.100	2.1	1.998
1977	0.000	0.063	0.450	2.490	2.100	0.000	—	0.000
1978	0.002	0.069	0.498	1.539	—	—	2.1	0.000
1979	0.004	0.084	0.448	0.550	2.120	—	—	0.000
Average	0.004	0.084	0.446	1.521	2.032	0.865	0.0	—

although we often use an approximation for q_t derived as follows:

$$C_t = N_t s E_t q$$

$$q_t = \frac{C_t}{N_t s E_t} \tag{10.4.3}$$

Table 10.8. Annual life history sequence for chinook salmon (*Oncorhynchus tschwytscha*).

January—April	natural mortality
March—August	fishery mortality
September—October	maturation
October—December	natural mortality

The q's derived from Equations 10.4.2 and 10.4.3 will be nearly identical as long as the natural mortality rate is low ($s > 0.3$).

Note that we use $N_t s$ in both equations because only s of the cohort that started the year will be around at the start of the fishery. Similarly, to calculate a maturation rate g_t, we must take the numbers that mature and divide that by the numbers alive at the beginning of the period of maturation:

$$g_t = \frac{S_t}{N_t s - C_t} \tag{10.4.4}$$

An interesting type of cohort to examine in Pacific salmon is one formed of hatchery fish released with coded-wire-tags implanted in their noses. The catch and escapement are sampled and estimates produced of how many of these fish were caught in different fisheries or escaped. Typical data for such a cohort are shown in Table 10.9 as are the results of cohort analysis. Cohort analysis of coded-wire-tagged chinook salmon has become an integral part of the policy evaluation and formulation process by the Pacific Salmon Commission for the United States and Canada and is regularly performed for dozens of stocks (Starr et al. 1986, also see Kope 1987).

10.5. Forward VPA

The secret to cohort analysis as we have presented it so far is in doing the calculations (reconstructing stock sizes) backward in time from some known endpoint. A little algebra will show that this can be switched around to go forward instead. Suppose we use the basic equation

$$N_t = \frac{N_{t+1} + C_t}{s_t} \tag{10.5.1}$$

where s_t is the natural survival rate from t to $t + 1$. We can write this repetitively as

$$N_t = \frac{N_{t+1} + C_t}{s_t} \tag{10.5.2}$$

Table 10.9. Cohort analysis of a representative tag group of chinook salmon.

Age	Catch	Maturing	Cohort Size	s	Maturation Rate	Harvest Rate
5	12	5	18.9	0.9	1.00	0.71
4	22	50	88.6	0.8	0.73	0.82
3	1047	50	2979.4	0.7	0.09	0.74
2	1096	0	6792.4	0.6	0.00	0.27
1	0	0	13584.8	0.5	0.00	0.00

Data from Paul Starr, Canadian Department of Fisheries and Oceans.

$$N_{t+1} = \frac{N_{t+2} + C_{t+1}}{s_{t+1}}$$

We can substitute the equation for N_{t+1} into Equation 10.5.1 to yield

$$N_t = \frac{N_{t+2} + C_{t+1}}{s_t s_{t+1}} + \frac{C_t}{s_t} \tag{10.5.3}$$

or expanded to many years

$$N_t = \frac{C_t}{s_t} + \frac{C_{t+1}}{s_t s_{t+1}} + \frac{C_{t+2}}{s_t s_{t+1}\, s_{t+2}} + \cdots + \frac{C_{t+n}}{s_t s_{t+1} \cdots s_{t+n}} \tag{10.5.4}$$

Our cohort size at the beginning time t is all future catches inflated to correct for natural mortality. If we assume constant survival (as is usual in VPA), we obtain the even simpler expression

$$N_t = \sum_{i=t}^{i=t+n} \frac{C_i}{s^{i-t+1}}. \tag{10.5.5}$$

Equations (10.5.4) and (10.5.5) should frighten you a little: they say that the VPA estimate N_t is just a weighted sum of catches from the cohort, but with each catch weighted by the inverse of a product of numbers less than 1.0 (the survival rates); these weights go up very rapidly for older ages if you assume survival rates much less than 1.0 (for example, if you use s = 0.5, you get $1/s = 2$, $1/s^2 = 4$, $1/s^3 = 8$, etc.).

10.6. Summary and Critique

Virtual population analysis has been very widely used, but not so widely criticized. It is arguably the least criticized technique used in fisheries data analysis. We believe the main reason for this is that there are very few instances where an independent estimate of stock size or recruitment is available to evaluate estimates obtained by VPA. In many cases, VPA is taken to provide the ultimate authority on cohort sizes. With nothing to compare it to, it is difficult to find fault with the technique.

The major exception to this has been evaluation of VPA's made using terminal F assumptions, done by using later data from VPA's conducted after the cohorts had completely passed through the fishery. In these cases, discussed below, we often find that the VPA estimates were very bad.

Another reason VPA appears to be so reliable is that it is really very simple, relying only on the following assumptions:

1. There are no fish alive at some age
2. The natural mortality rate is known
3. There is no net immigration or emigration

By fisheries standards, these are rather modest assumptions. Their implications and difficulties are discussed below.

Terminal F assumptions

The place that VPA has most often been found to be wrong is when catchability has increased while the stock was declining. In these cases, the assumption that the terminal F has not been changing in the most recent few years leads to a systematic overestimation of the stock size. In such cases, catches tend to remain high as the stock declines, leading one to believe that the stock is not declining as fast as it really is. Sinclair et al. (1985) documented this problem for several herring stocks in the northwest Atlantic. Figure 10.2 shows the estimated biomass of herring for VPA's performed from 1978 to 1982. If we accept the 1982 results as correct, we see that the 1978–1980 estimates were all seriously overestimating the stock size.

This has been commonly found in declining stocks; it is an almost universal feature of clupeoid fisheries that catchabilities increase rapidly when the stock is declining. There is no simple solution. To estimate the current cohorts present in the fishery, some terminal F assumption must be made. However, one should now be well aware that for many types of fish, catchabilities will change, and a terminal F assumption that reflects anticipated changes in catchabilities should be used.

Using the wrong *M*

The use of an incorrect mortality rate will have a major impact on the reconstructed stock sizes in VPA. If the M used is too large, the estimated cohort sizes will be larger than they should be, and, if the estimated M is smaller than the true one, the estimated cohorts will be too small. Since one is often looking for trends, rather than absolute numbers, in such analyses,

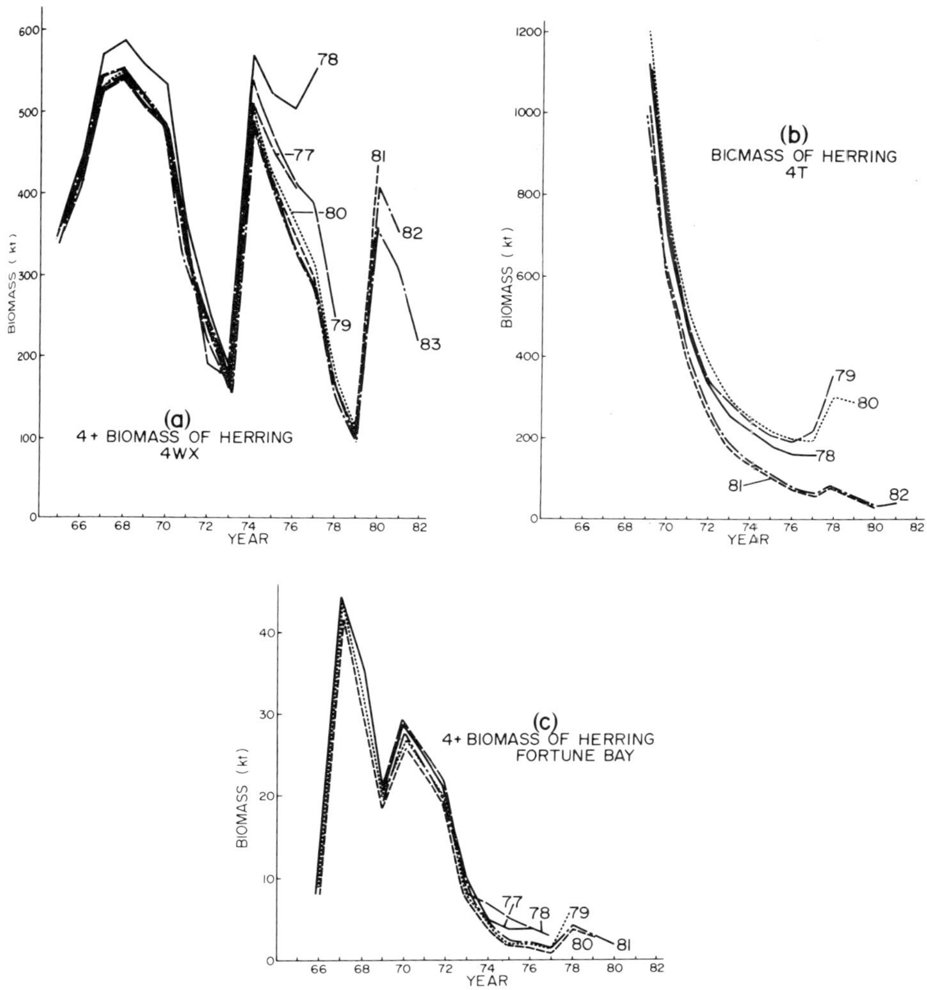

Figure 10.2. Estimated biomass of herring from VPA performed in different years for three herring stocks. From Sinclair et al., (1985). Can. J. Fish. Aquat. Sci. 42: Figure 2.

these biases may not be serious (but see Bradford and Peterman 1989, and LaPointe, et al. 1989).

Constant M

If the true M is constant, then the size of the bias caused by an incorrect estimate of M will depend primarily on the fishing mortality in relation to

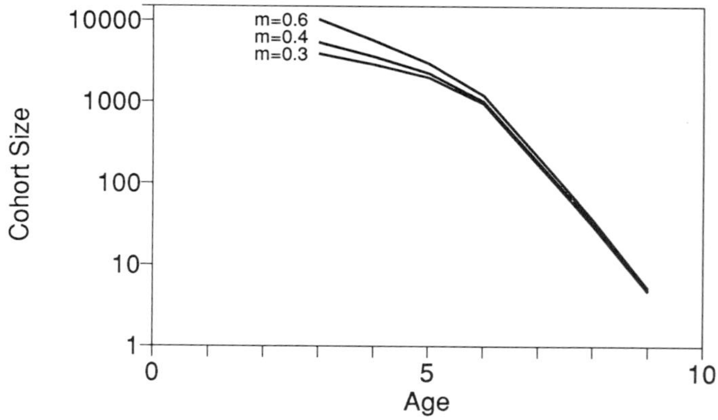

Figure 10.3. The effect of assumed natural mortality on cohort size for the 1965 cohort of Flack Lake trout.

the natural mortality. If fishing mortality rate is high relative to natural mortality, an incorrect M will have almost no impact on the estimates, whereas, if F is small relative to M, the impact will be much larger. Figure 10.3 shows estimated cohort sizes for different M's for the Flack Lake trout data presented earlier.

Changing M

An additional assumption is that M is constant. It is generally hard enough to get one estimate of M; the idea of trying to measure changes in M is quite frightening. Unfortunately, there are a number of reasons to believe that M might change over time, among them changes in food availability, predators, or other environmental factors. Major efforts are being made to include these effects in VPA, mainly by including food chain effects in what is called *multispecies* VPA (discussed under multispecies approaches). The data requirements are so great that such approaches are likely only to be used in a few of the most intensively studied fisheries.

Aging errors

The foundation of VPA is a catch-at-age time series. If there are errors in the determination of age, this will introduce systematic biases in the results of VPA. As a simple example assume that there is an 80% chance of determining the age of a fish correctly, a 10% chance of misclassifying it as 1 year older than actual and a 10% chance of misclassifying it as 1 year younger than actual. For many types of aging these would be considered very good classifications.

If all cohorts are roughly equal, the consequences of such misclassification are not at all serious, but, if some cohorts are much stronger than others, the weak cohorts will be systematically overestimated. Consider for instance a cohort that is one tenth the size of the cohort 1 year older. If there are 1,000 individuals at age 4 in the older cohort there will be 100 individuals at age 3 in the younger one (assuming $M = 0$ for simplicity). One hundred of the age 4 fish will be misclassified as age 3, whereas 10 of the age 3 will be misclassified as age 4. This will cause us to overestimate by a factor of two the number of age 3 fish and, in general, the strength of the weak cohort.

This overestimation would be even worse if a strong cohort followed a weak cohort, and there was a large M. The age classes 1 year behind the weak cohort would be much larger because their cohort was both bigger, and had seen 1 year less mortality. Any misclassification would have major impacts.

Aging errors will tend to mask recruitment variability. The bias introduced by such aging errors will have the most significant impact on stock-recruitment analysis. If there is a good relationship between stock and recruitment, aging errors will tend to mask the stock-recruitment relationship. Small spawning stock sizes, which in fact produced small recruitments, will tend to appear to have produced larger recruitments than they really did. This will make the stock appear to be more productive at small stock sizes than it really is and thus overestimate the optimal harvest rate and underestimate the optimal stock size.

Unit stock assumption

Virtual Population Analysis also assumes that there is no net immigration or emigration from the stock; the only two possible sources of loss are from natural or fishing mortality. Any immigration into the stock will cause the number of fish alive to be increased, which when worked backward into the VPA equations will cause the cohort sizes to be overestimated. This will be particularly bad if the immigration takes place at older ages, where each individual that entered the stock and was then caught will be magnified a number of times as the VPA equations are solved backward. Again, this type of error can cause problems in stock and recruitment analysis because small cohorts will be magnified proportionally more by any amount of immigration.

Emigration is potentially less serious because it should be related to population size and thus may be considered a form of natural mortality. If emigration is perfectly proportional to density, it will act exactly like natural mortality, but, if it occurs in a highly random way, it may act more like white noise.

Discarding, unreported catch, and fishing-induced mortality are all potential problems with VPA. Each of these can be thought of as underreporting of catch. Many fisheries agencies now attempt to estimate the extent of these problems and correct the VPA accordingly.

Bibliographic Notes

Megrey (1989) has reviewed the history of age-based methods, as well as the methods themselves and includes an extensive discussion of the evolution of the terms VPA and cohort analysis. Any readers seriously interested in the history of these terms should read his paper. The idea of summing up catches and inflating by natural mortality has been in use at least since Ricker (1948). Ricker (1948) used Baranov's catch equation in backward sequential computation for halibut. Jones (1961) explored the convergence and divergence properties of such estimators depending upon what was used as the known stock size. Murphy (1965) developed a computer program for doing these calculations. Gulland (1965) attached the label virtual population analysis to this technique and was credited by Pope (1972) as the popularizer of VPA. Pope (1972) suggested the approximation that simplifies calculations, as was discussed earlier. Pope also suggested the use of the term *cohort analysis* as a distinct term for his method, but this had been previously used and thus VPA and cohort analysis are used interchangeably.

Methods for fine tuning VPAs are discussed in Laurec and Shepherd (1983) and Pope and Shepherd (1985). Rivard (1989) provides an overview of the effects of different types of errors on VPA.

11
Statistical Catch-at-Age Methods

11.1. Introduction

In the last chapter, we saw how historical stock sizes and fishing mortality rates could be estimated from catch-at-age data and an assumed natural mortality rate using VPA. There are two major problems with VPA: it can give unreliable estimates for cohorts that have not completely disappeared from the fishery, and it requires an assumption about the natural mortality rate.

The first problem is more important; fisheries managers generally want to know the relative strength of cohorts still being fished, and although VPA will provide such estimates, they are obtained by various forms of ad hoc manipulation, generally called tuning. The statistical catch-at-age methods described in this chapter provide more formal methods for estimating the current abundance of cohorts still being fished. Statistical catch-at-age methods should, in principle, be better than tuned VPA, but few comparisons have been made. The ICES workshop and its followup mentioned in the previous chapter are the most rigorous comparisons performed to date.

Statistical catch-at-age methods also provide a method for estimating the natural mortality rate provided the data contain clearly contrasting levels of fishing effort and total mortality rate. However, our experience has been that the estimates of natural mortality rate are highly unreliable, even when there have been major variations in fishing mortality.

Statistical catch-at-age analysis very elegantly synthesizes many pieces of fisheries theory that have been well known for some time, in particular, catch curve analysis and VPA.

11.2. Catch Curves

It has been noted for a long time, going back at least to 1908 (Edsen 1908, Heinke 1913, Wallace 1915, Baranov 1918), that there are usually more smaller (and therefore younger) fish in commercial catches than there are large or old fish. Assuming fish can be aged, if we plot the number of

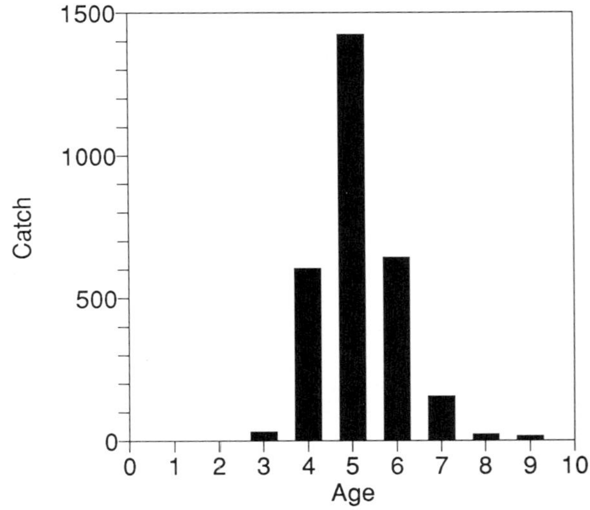

Figure 11.1. Age distribution for 1971 catch of Flack Lake trout (*Salvelinus namaycush*) in Ontario. Data from Ontario Ministry of Natural Resources.

fish caught at each age in a given year, we often get the pattern shown in Figure 11.1. Few fish of the youngest ages are caught, the catches peak for a certain age, and then the catch of older fish drops.

Two processes are at work. First, the younger fish are less vulnerable to the fishing gear, but as they age they become more and more vulnerable, until they are said to be *fully recruited* at a certain age. Lake trout, for example, are thought to be fully recruited at age 5 or 6. Second, catches of older (fully recruited) ages decline simply because there are fewer of them due to fishing and natural mortality; also, older fish may become less vulnerable to the fishing gear for various reasons. For fish older than the age of full recruitment, it is often assumed that catch is proportional to abundance of the age class, that is, that they do not become less vulnerable with age. The higher the natural and fishing mortality, the steeper the decline will be in catch of older aged fish, as seen in Figure 11.1.

We can quantify this observation by calculating the survival as

$$\text{survival from age } a \text{ to } a + 1 = \frac{\text{catch-at-age } a + 1}{\text{catch-at-age } a}$$

If these catches are taken in the same year, that is, they represent cohorts born a and $a + 1$ years ago, then we are making the implicit assumption

Table 11.1. Catch-at-age for lake trout in Flack Lake, Ontario, 1971.

Age	Catch	Survival	Instantaneous Mortality Rate
3	32	—	—
4	606	—	—
5	1424	45%	0.79
6	644	24%	1.41
7	157	14%	1.92
8	23	74%	0.30
9	17	—	—

that the two cohorts were of the same initial size (constant recruitment over time) and had similar mortality histories. If we are looking at catches of the same cohort in 2 successive years (year a and $a + 1$), then we are implicitly assuming that fishing effort and catchability were the same in the 2 years. Table 11.1 shows the 1971 catch-at-age and calculated survivals of lake trout in Flack Lake, Ontario. If we dismiss age 9 because of small sample size, the estimates of survival decrease from 45% to 24% to 14%. We might take the average and use 28% as an average estimate of total survival.

For lake trout, which are fished year round in Flack Lake, it is best to assume that there is continuous natural and fishing mortality, so that

$$N_{a+1} = N_a e^{-Z} \tag{11.2.1}$$

which means that the total instantaneous mortality rate is

$$Z = -[\log(N_{a+1}) - \log(N_a)] \tag{11.2.2}$$

The calculations for Z are also shown in Table 11.1. When dealing with instantaneous rates, a constant mortality rate will appear as the slope of a straight line when the catches or numbers are plotted on a logarithmic scale. Figure 11.2 shows the catch plotted for each age on a logarithmic scale. We are now able to do a regression using ages 5 through 8 to obtain an estimate of Z, which will be the slope parameter of the regression.

This may be more clear if we write the regression as

$$\log(C_a) = b - Za \tag{11.2.3}$$

where C_a is the catch-at-age a, b is the intercept, and Z is the instantaneous mortality rate. The intercept b says what the catch of age 0 fish would have been if they were as vulnerable as fully recruited fish. Exactly where did Equation 11.2.3 come from? Although it may be intuitively obvious to you, we admit to jumping a few steps. Let us go back to explore some simple relationships.

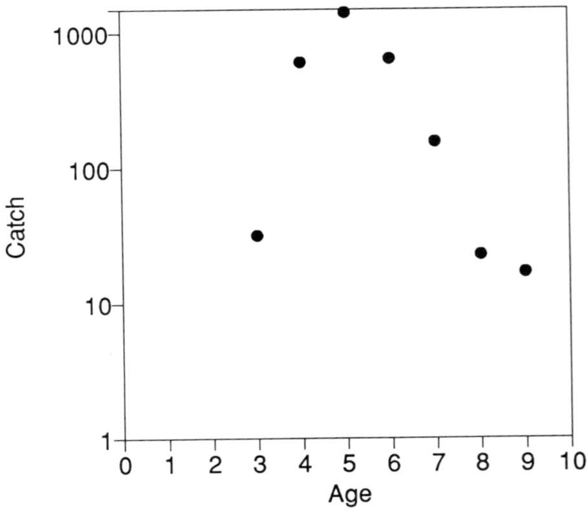

Figure 11.2. The catch plotted against age for Flack Lake trout caught in 1971. Note that the Y axis is scaled logarithmically.

$$N_1 = Re^{-Z}$$

$$N_2 = N_1e^{-Z} = Re^{-Z}e^{-Z}$$

$$N_3 = N_2e^{-Z} = Re^{-Z}e^{-Z}e^{-Z}$$

$$N_a = R(e^{-Z})^a = Re^{-Za} \tag{11.2.4}$$

where R is the recruitment to the cohort, and N_1, N_2, N_3, and N_a are the number alive at 1, 2, 3, and a years of age. If we take logs of the last Equation in 11.2.4, we get

$$\log(N_a) = \log(R) + (-Z)a \tag{11.2.5}$$

If we assume that catch is proportional to abundance by the relationship

$$C = Nv \tag{11.2.6}$$

where v is the vulnerability, then Equation 11.2.5 becomes

$$\log\left(\frac{C_a}{v}\right) = \log(R) + (-Z)a$$

$$\log(C_a) - \log(v) = \log(R) + (-Z)a$$

$$\log(C_a) = \log(Rv) + (-Z)a \tag{11.2.7}$$

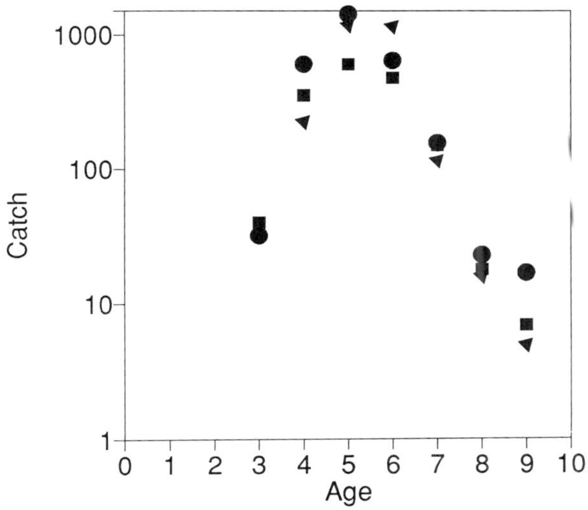

Figure 11.3. The catch (log scale) plotted against age from 1970 to 1972 for Flack Lake trout.

This is just like Equation 11.2.3, except we now know the intercept b of equation 11.2.3 is the log of the recruitment times the scaling factor v.

Equation 11.2.7 can be rewritten as

$$\log\begin{pmatrix}\text{catch}\\ \text{at}\\ \text{age } a\end{pmatrix} = \log\begin{pmatrix}\text{recruitment}\\ \times\\ \text{vulnerability}\end{pmatrix} - \begin{pmatrix}\text{instantaneous}\\ \text{mortality rate}\\ \times\text{ years of mortality}\end{pmatrix} \quad (11.2.8)$$

This equation simply says that as fish get older, more of them will have died. It should be clear that equations 11.2.3, 11.2.7, and 11.2.8 all say the same thing and describe how the catch at a given age will depend upon the age. So long as R and Z are constant, we can deal with ages within a single year or cohorts between years. Thus far, we have only dealt with data from a single year, but we can do the same analysis on data from several years. Figure 11.3 shows several years of data for Flack Lake trout.

This does not look nearly as nice as Figure 11.1. There are several reasons for this messiness. First, for Figure 11.1 , we chose 1971 as a clean example. Second, several assumptions would have to be met for the data to look nice. These assumptions are: (1) recruitment is constant from year to year (or lognormally distributed about an average recruitment), (2) fishing and natural mortality are constant, and (3) vulnerability to fishing gear is constant above a given age.

When using catch curves, few people would accept assumption 1, whereas assumptions 2 and 3 are more often believable. Many fish stocks show trends in recruitment over time, thus assumption 1 will not hold. A low point on a graph such as Figure 11.3 would most likely be ascribed to a weak year class. It could, however, be due to lower vulnerability in that year or to lower fishing effort. Also note that there is apparently little mortality between ages 5 and 6. This is undoubtedly due to the fact that age 5 fish are not fully recruited, a phenomenon we found when we used VPA on the same data.

11.3. Catch Curves on Cohorts

So far, we have been plotting catch-at-age for individual years. We can eliminate assumption 1 above (constant recruitment) by plotting the catch-at-age for cohorts. Equation 11.2.7 becomes

$$\log(C_{aj}) = \log(R_j v) + (-Z)a \tag{11.3.1}$$

where C_{aj} is the catch of cohort j at age a and R_j is the initial recruitment (at age 0) of cohort j. This not only lets us eliminate assumption 1 (constant recruitment), but it also lets us estimate relative recruitments for each cohort, as we promised at the beginning of this chapter. Figure 11.4 shows the catch for three cohorts of Lake Erie perch (*Perca flavescens*), where year class strength is much more variable than for Flack Lake trout. The 1962 cohort was very large, the 1966 cohort was smaller, and the 1971 cohort was the smallest of all. For each of these cohorts we can estimate both the initial relative cohort size and the total mortality rate. Eyeball fits of a straight line for each cohort show that the slope is relatively consistent for the three cohorts shown; we have a pretty good idea of the total mortality rate Z.

We are unable to separate natural mortality from fishing mortality; the cohorts of Lake Erie perch could be disappearing quickly because they are fished very hard or because they have a high natural mortality rate. We also do not know the actual stock sizes, because we do not know the scaling factor v. If we assume a natural mortality rate M, then we can calculate actual stock sizes and vulnerabilities using VPA. Unfortunately, the catch curve analysis provides only a maximum value of the natural mortality rate and this maximum is the total mortality rate Z. Thus we have little useful information from cohort-based catch curves. For a more powerful technique, we must move on to full catch-at-age analysis.

In summary, cohort-based catch curves assume that

1. Fishing and natural mortality are constant
2. Vulnerability to fishing gear is constant above a given age.

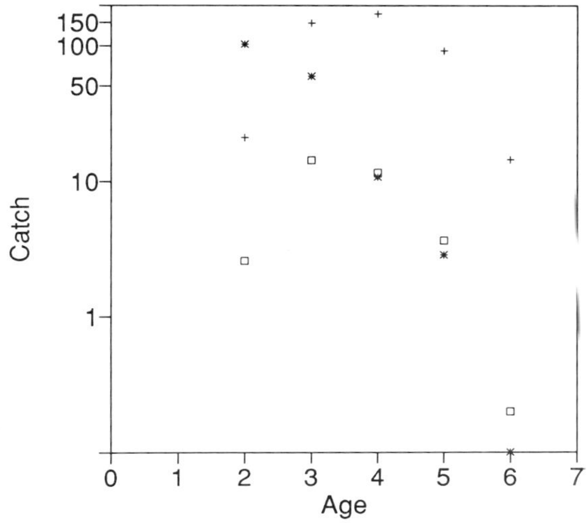

Figure 11.4. Catch-at-age data for three cohorts of Lake Erie perch (*Perca flavescens*). + = 1962 cohort, * = 1966 cohort, □ = 1971 cohort. Data from Ontario Ministry of Natural Resources.

11.4. Paloheimo's Catch-at-Age Analysis for Multiple Cohorts

Cohort-based catch curves do not allow us to estimate natural mortality rate or vulnerability to fishing gear. Because we have assumed that both are constant, we have been prevented from distinguishing between natural and fishing mortality. Imagine closing Flack Lake to fishing for a number of years and plotting the catch for a new cohort. This will not work because without fishing there is no catch. However, since we are already imagining (a popular past-time for fisheries modellers), assume that we could indeed plot the log of numbers versus age for two cohorts, one which had been fished and one which had not. Figure 11.5 shows such an imaginary plot. Both cohorts started out at about the same size, but the cohort that was fished declined faster. We could draw a regression through each cohort's data and estimate the total mortality Z for each. Since total mortality is equal to fishing mortality (F) plus natural mortality (M), and we have assumed no fishing mortality for one of our cohorts, the estimate of Z we derive from this cohort is our estimate of the natural mortality rate. In Figure 11.5, the slope for the cohort which was not fished is -1 so $Z = M = 1$. For the cohort which was fished, the slope is -2, so $Z = M + F = 2$ and therefore $F = 1$.

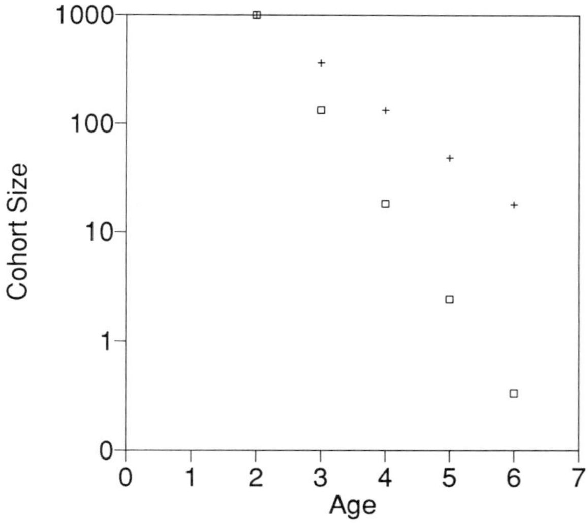

Figure 11.5. Hypothetical numbers at age for a cohort that was fished for several years, and another that was not fished.

We can see that in principle we could separate fishing from natural mortality if we had one cohort that was fished and one that was not. We can split fishing from natural mortality and rewrite Equation 11.3.1 for numbers instead of catch as follows:

$$\log(N_{aj}) = \log(R_j) - Ma - Fa \tag{11.4.1}$$

This can be rewritten as

$$\log\begin{pmatrix}\text{number}\\\text{alive}\end{pmatrix} = \log\begin{pmatrix}\text{initial}\\\text{cohort}\\\text{size}\end{pmatrix} - \begin{pmatrix}\text{cumulative}\\\text{natural}\\\text{mortality}\end{pmatrix} - \begin{pmatrix}\text{cumulative}\\\text{fishing}\\\text{mortality}\end{pmatrix} \tag{11.4.2}$$

In practice, it is most unlikely that there will be one cohort that is constantly fished hard and another cohort that is not fished at all. More likely there might be 1 or 2 years of unusually low (or high) fishing rates. Figure 11.6 shows another hypothetical cohort which normally was fished hard, but was not fished at all for 1 year (age 4). Again, cohort size is plotted against age. The total mortality rate for all years is 2 except for age 3, in which $Z = 1$. We could thus calculate M as 1.

This method is great in principle, but it does require accurate estimates of abundance for each age and a period of no fishing. If we really had a year of no fishing, all cohorts in the fishery would have 1 year of different

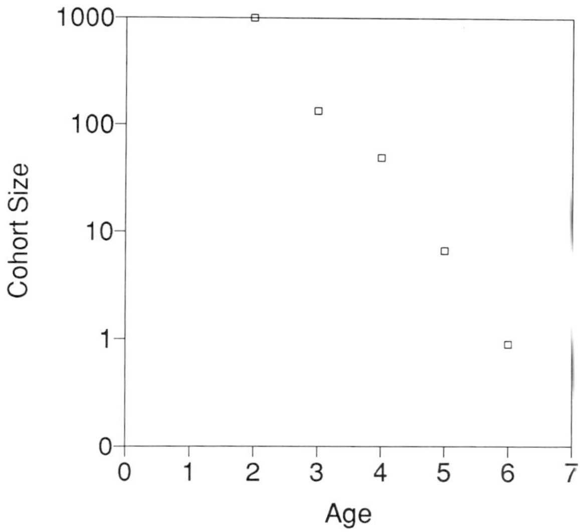

Figure 11.6. Hypothetical numbers at age for a cohort that was unfished for 1 year (age 4).

slopes (occurring at different ages of course), and we would actually see a number of separate estimates of the fishing mortality rate. However, it is most unlikely that there will be a period of no fishing or that accurate estimates of actual abundance will be available. Furthermore, we have been plotting cohort size because catch curves assume constant exploitation rates. In short, all the preceding approaches are included to prepare you for some painful algebra, which tries to deal with more realistic situations.

The following section is not necessarily designed to cure you of any desire to do catch-at-age analysis. We know that it is not as bad as it looks, proof being that it has been understood by the occasional graduate student. Let us go back to first assumptions and rederive Equation 11.4.1 in a little more explicit fashion, and in a way that lets us look at catches instead of just numbers. The following catch-at-age model was first derived by Doubleday (1976), but it was Paloheimo (1980) who inserted a devious trick that makes the problem much simpler to solve and widely accessible to anyone with a large linear regression program. Doubleday's method was actually quite general, and we present here a special case of it.

We start with three basic relationships. The first relates catch to current abundance and fishing and natural mortalities

$$\text{catch} = \text{numbers} \times \text{fishing mortality} \qquad (11.4.3)$$

and we shall assume that fishing mortality acts independently of natural mortality and takes a fraction $F/(F+M)$ of the total mortality so

$$C = N \frac{F}{F + M} [1 - e^{-(F+M)}] \qquad (11.4.4)$$

The second relates the number at any age a to the initial cohort strength (R) and the cumulative fishing and natural mortalities:

$$N_a = Re^{-\Sigma F - \Sigma M} \qquad (11.4.5)$$

If Equation 11.4.5 is confusing, it may help you to think of it as

$$N_a = Re^{-(F_1+M_1)}e^{-(F_2+M_2)} \ldots e^{-(F_{a-1}+M_{a-1})} \qquad (11.4.6)$$

where F_a and M_a are the fishing and natural mortality rates at age a. Each $e^{-(F+M)}$ in this equation takes the number alive at the start of 1 year and produces survivors at the start of the next year.

The last basic equation we need is the relationship between fishing effort (E) and fishing mortality

$$F = Eq \qquad (11.4.7)$$

You should remember this from VPA and biomass dynamics models, where q was called the catchability coefficient. If we now substitute the relationship for abundance N_a in Equation 11.4.5 into catch Equation 11.4.4, we get

$$C_a = Re^{-\Sigma F - \Sigma M} \frac{F}{F + M} [1 - e^{-(F+M)}] \qquad (11.4.8)$$

(Here we are omitting some subscript notation for simplicity.) Rewriting Equation 11.4.8 in words we get

$$\begin{pmatrix} \text{catch} \\ \text{at} \\ \text{age } a \end{pmatrix} = \begin{pmatrix} \text{initial} \\ \text{cohort} \\ \text{size} \end{pmatrix} \times \begin{pmatrix} \text{survival} \\ \text{to} \\ \text{age } a \end{pmatrix} \times \begin{pmatrix} \text{fishing} \\ \text{mortality} \\ \text{at age } a \end{pmatrix} \qquad (11.4.9)$$

with the two key relationships being survival to age $a = e^{-\Sigma F - \Sigma M}$ and

$$\text{fishing mortality at age } a = \frac{F}{F + M} (1 - e^{-(F+M)}). \qquad (11.4.10)$$

Now, if we substitute our relationship for F from Equation 11.4.7, into equation 11.4.8 we get

$$C_a = Re^{(-q\Sigma E - \Sigma M)} \frac{Eq}{F + M} [1 - e^{-(F+M)}]. \tag{11.4.11}$$

The only mildly tricky part of this was to write ΣqE as $q\Sigma E$. We are assuming that q is constant and independent of age. If we divide both sides by E and replace $F + M$ by Z, we get

$$\frac{C_a}{E} = Re^{(-q\Sigma E - \Sigma M)} q\left(\frac{1 - e^{-Z}}{Z}\right) \tag{11.4.12}$$

If we then take logs of both sides, we get

$$\log\left(\frac{C_a}{E}\right) = \log(Rq) - q\Sigma E - \Sigma M + \log\left(\frac{1 - e^{-Z}}{Z}\right) \tag{11.4.13}$$

Should you be wondering how we got Rq in one log, rewrite $\log(R) + \log(q) = \log(Rq)$. Equation 11.4.13 may look like a lot of algebraic gibberish, but translated back into English, we find

$$\log\begin{pmatrix}\text{catch}\\ \text{per}\\ \text{effort}\end{pmatrix} = \log\begin{pmatrix}\text{initial cohort}\\ \text{size} \times\\ \text{catchability}\end{pmatrix} - \begin{pmatrix}\text{cumulative}\\ \text{fishing}\\ \text{mortality}\end{pmatrix} - \begin{pmatrix}\text{cumulative}\\ \text{natural}\\ \text{mortality}\end{pmatrix} + \begin{pmatrix}\text{something}\end{pmatrix} \tag{11.4.14}$$

Ignoring the "something" on the right, we see that the catch per effort should be higher if initial cohort size is large or catchability is high, and lower the more fishing or natural mortality the cohort has suffered prior to age a.

Equation 11.4.13 also looks pretty much like a linear regression. Ignoring the "something" again, and assuming M is constant over a years, we can rewrite Equation 11.4.13 as

$$\log\left(\frac{C_a}{E}\right) = \log(Rq) - q\,\Sigma\,E - M(a - 1) \tag{11.4.15}$$

which is a multiple regression of the form

$$Y = b_0 + b_1X_1 + b_2X_2 \tag{11.4.16}$$

where b_0, b_1, and b_2 are parameters, Y is a dependent variable, and X_1 and X_2 are independent variables. X_1 is ΣE, the total fishing effort the cohort has endured up to age a; and X_2 is $a - 1$, the number of years of natural mortality on the cohort after the age of recruitment. Notice the very similar structures of Equations 11.4.15 and 11.4.1; we have simply rewritten things in such a way that we will arrive at a linear estimation method. Notice also that the regression intercept parameter b_0 is a measure of the initial recruitment of the cohort a years ago, but scaled by the catchability q (compare Equation 11.4.15 to Equation 11.3.1).

To get rid of the ugly "something" on the right hand side of Equation 11.4.13 and turn it into a nice multiple regression, Paloheimo (1980) used a rather sneaky trick. It was the same slight of hand that Pope used for VPA, namely

$$\log\left(\frac{1 - e^{-Z}}{Z}\right) \approx -\frac{Z}{2} \tag{11.4.17}$$

As you remember, this is a good approximation when Z is not too big.

Since $Z = M + F$, we can rewrite equation 11.4.13 as

$$\log \frac{C_a}{E} = \log(Rq) - q\Sigma E - M(a - 1) - \frac{M}{2} - \frac{F}{2} \tag{11.4.18}$$

A somewhat intuitive assessment of the $-M/2$ and $-F/2$ terms is that the C_a/E not only depends on the starting cohort size at the beginning of the year, but also on how quickly the age class is depleted during the year. Noting that $F/2 = qE/2$, Equation 11.4.18 can be rewritten as

$$\log\left(\frac{C_a}{E}\right) = \log(Rq) - q\left(\Sigma E + \frac{E}{2}\right) - M\left(a - \frac{1}{2}\right) \tag{11.4.19}$$

Since we have been a bit untidy with subscripts, we should note that $\Sigma E + E/2$ is the sum of all efforts this cohort has seen up to this year plus half the effort it will face this year (at age a).

Equation 11.4.19 with all the subscripts looks like

$$\log\left(\frac{C_{aj}}{E_j}\right) = \log\,(R_{j-a}q) - q\left(\sum_{k=j-a}^{j-1} E_k + \frac{E_j}{2}\right) - M\left(a - \frac{1}{2}\right) \tag{11.4.20}$$

where C_{aj} is the catch of age a fish in the jth year, E_j is the fishing effort in the jth year, R_{j-a} is the number of fish recruited (age 0) in year $j - a$ (the recruits of the cohort caught in C_{aj}), and q and M are the catchability and natural mortality rate.

Equation 11.4.19 (or Equation 11.4.20) is a multiple regression of the form described in Equations 11.4.15–11.4.16; it can be used with standard multiple-regression packages to produce estimates of Rq, q, and M from a set of catches-at-age and efforts. Table 11.2 shows how data for a single

Table 11.2. Catch and effort data for a single cohort (1971) of Lake Erie perch.

Age	Catch	Effort	C/E	Y $\log(C/E)$	X_1 $\Sigma E + \frac{E}{2}$	X_2 $\text{age} - \frac{1}{2}$
2	103	15.9	6.5	1.9	7.9	0.5
3	59	15.4	3.8	1.3	23.5	1.5
4	11	13.5	0.82	−0.2	37.9	2.5
5	3	12.6	0.24	−1.4	50.9	3.5

$$Y = 2.37 + 0.22X_1 - 4.34X_2$$

$$\log(\hat{R}q) = 2.37$$

$$\hat{q} = -0.22$$

$$\hat{M} = 4.34$$

Parameter Correlations

$$\begin{array}{c} \\ Rq \\ q \\ M \end{array} \begin{pmatrix} Rq & q & M \\ 1 & & \\ -0.71 & 1 & \\ -0.69 & -0.999 & 1 \end{pmatrix}$$

cohort of Lake Erie perch can be used to attempt estimation of the parameters Rq, q and M.

The data assembly shown in Table 11.2 is very straightforward except for a little sneakiness regarding X_2, the cumulative years of natural mortality. From Equation 11.4.20, we should have put 1.5 instead of 0 5 in our X_2 column for 2 year olds, because if $a = 2$, then $a - 0.5$ is 1.5. We have subtracted 3 from all ages, which has the pleasant consequence that recruits (Rq) estimates age 2 recruits instead of age 0 recruits. This is the normal procedure in catch-at-age analysis (just as it is in VPA) because, even though we must assume a constant natural mortality rate, we are most comfortable assuming it constant only after the fish have been recruited to the fishery.

The parameter estimates emerging from the regression are not as satisfying as we might wish. The estimated instantaneous mortality rate is 4.3 (roughly 99% annual mortality), and, even worse, the q is negative, which says that more fishing effort increases the population size! This does not make any biological sense. The reason for this unhappy result is our old bugbear, *lack of contrast*. Look at the X_1 and X_2 columns; as X_1 goes up, so does X_2. This is not too surprising because as fish get older, there will be more cumulative natural and fishing mortality, but, because effort is about the same each year, the cumulative fishing effort goes up linearly with age, just as years of natural mortality do.

To separate natural mortality from fishing mortality, we must have had variation in fishing effort (look back at Figure 11.6), but we do not have this contrast in the single Lake Erie perch cohort. This is also shown by the parameter correlations in Table 11.2. Note that the estimates of q and M have a correlation of -0.999, which means that their effects are totally confounded and only the combination $Z = qE + M$ (total mortality) associated with the average effort E can be estimated.

We do know that $Z = M + qE$, so using an E of 14 as an average, we get $Z = 1.26$. At least this makes more sense than $M = 4.34$. From this 1971 cohort, we have a good estimate of Z, but we cannot separate Z and F. Not wanting to discourage you too much, we must still point out that this is a very common result. In our experience, most catch-at-age analyses are plagued by terrible correlations between M and q due to a lack of systematic contrasts in effort levels over time.

11.5. The Experimental Design Table

The above example for a single cohort is relatively easy to understand, since there are only three parameters to estimate. But what happens when there are several cohorts or a whole table of catch-at-age data? We are assuming that M and q apply for all ages and years (an assumption that will be relaxed later), but surely we do not want to assume that recruitment is the same for all cohorts. We want to estimate M and q and separate R's for each cohort. As usual, there is a trick, and it is known as using dummy variables, or an experimental design matrix. If we were trying to use data from three cohorts, our model could be written as

$$Y = b_1X_1 + b_2X_2 + b_3X_3 + b_4X_4 + b_5X_5 \tag{11.5.1}$$

where we now define b_4, b_5, X_4 and X_5 as we did b_2 and b_3 before:

$$\begin{aligned} b_4 &= -q \\ b_5 &= -M \\ X_4 &= \Sigma E + E/2 \\ X_5 &= a - 1/2 \end{aligned} \tag{11.5.2}$$

But, we will now define $b_1, \ldots, b_3$ as recruitment parameters;

$$\begin{aligned} b_1 &= \log (R_1q) \\ b_2 &= \log (R_2q) \\ b_3 &= \log (R_3q) \end{aligned} \tag{11.5.3}$$

Table 11.3. Paloheimo catch-at-age analysis for three cohorts of Lake Erie perch.

Cohort	Year	Age	Effort	Catch	Y	X_1	X_2	X_3	X_4	X_5
1962	1965	3	12.0	148.2	2.51	1	0	0	6.0	0.5
1962	1966	4	17.1	173.2	2.32	1	0	0	20.6	1.5
1962	1967	5	15.9	91.6	1.75	1	0	0	37.1	2.5
1962	1968	6	12.8	14.5	0.124	1	0	0	51.4	3.5
1966	1969	3	15.8	14.3	−0.10	0	1	0	7.9	0.5
1966	1970	4	15.6	11.6	−0.30	0	1	0	23.6	1.5
1966	1971	5	19.4	3.7	−1.66	0	1	0	41.1	2.5
1966	1972	6	16.7	0.2	−4.22	0	1	0	59.2	3.5
1971	1974	3	15.4	59.3	1.34	0	0	1	7.7	0.5
1971	1975	4	13.4	10.8	−0.22	0	0	1	22.1	1.5
1971	1976	5	12.6	2.9	−1.47	0	0	1	35.1	2.5
1971	1977	6	14.8	0.1	−5.0	0	0	1	48.8	3.5

$$Y = 4.7X_1 + 0.95X_2 + 1.73X_3 + 0.1X_4 + -3.08X_5$$

$$\log(\hat{R}_1 q) = 4.7$$
$$\log(\hat{R}_2 q) = 0.95$$
$$\log(\hat{R}_3 q) = 1.73$$
$$\hat{q} = -0.1$$
$$\hat{M} = 3.08$$

Parameter correlations

	$\log(\widehat{R_1 q})$	$\log(\widehat{R_2 q})$	$\log(\widehat{R_3 q})$	$\hat{q}$	$\hat{M}$
$\log(\widehat{R_1 q})$	1.0				
$\log(\widehat{R_2 q})$	0.21	1.0			
$\log(\widehat{R_3 q})$	0.61	0.17	1.0		
$\hat{q}$	0.42	−0.48	0.48	1.0	
$\hat{M}$	−0.48	0.41	−0.54	−0.999	1.0

where R_1, R_2, and R_3 are the initial cohort sizes for cohorts 1, 2, and 3. Using these definitions for $b_1, \ldots, b_3$ the dummy variable trick is simply to have $X_1, \ldots, X_3$ in equation 11.5.1 take on the values 0 or 1, depending on which is present (which recruitment rate applies) for each observation. If we are looking at a Y from cohort 1, we make $X_1 = 1$ and $X_2 = X_3 = 0$; a Y from cohort 2 has $X_1 = 0$, $X_2 = 1$, and $X_3 = 0$. Thus $X_1, \ldots, X_3$ are codes for the presence or absence of the qualitative treatment cohort recruitment $\times$ q. Table 11.3 shows the data input for the multiple regression in Equation 11.5.1 and the output for the 3 cohorts of Lake Erie perch shown previously in Figure 11.4. We have used the catch data from age 3 to age 6.

Examination of Table 11.3 shows that there is obviously going to be good contrast between X_1, X_2, and X_3, which means we will have good estimates

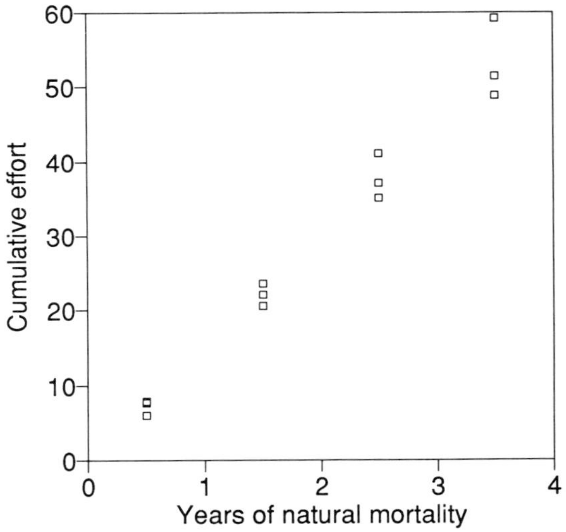

Figure 11.7. Cumulative effort (X_4) plotted against cumulative years of natural mortality (X_5) for three cohorts of Lake Erie perch.

of the relative cohort strengths. Unfortunately, X_4 and X_5 are highly confounded. The extent of this confounding is shown in Figure 11.7, which shows X_4 plotted against X_5 for the twelve observations of Table 11.3. It is little wonder that the correlation between M and q is again very high (-0.999).

We see that even having data from three cohorts is not enough to allow separation of qE and M effects. The estimates of q are still negative, and M is still too high. About all you can do in such cases is to obtain an independent estimate of M, then do the regression without it (move terms multiplying M to the left-hand side of the regression equation to get Ys corrected for natural mortality) or else use VPA.

11.6. Doubleday's Method

Paloheimo's method is actually a special linear case of a more general catch-at-age analysis method developed by Doubleday (1976). Doubleday's method goes back to Equation 11.4.8, which was

$$C_a = Re^{-\Sigma F - \Sigma M} \frac{F}{F + M} [1 - e^{-(F+M)}] \tag{11.6.1}$$

This can be transformed, including all subscripts, to the messy form

$$\log (C_{aj}) = \log (R_{j-a}) - \sum_{k=1}^{a-1} F_{a-k,j-k} - \sum_{k=1}^{a-1} M_{a-k,j-k} + \log\left\{\frac{-F_{aj}}{F_{aj} + M_{aj}} [1 - e^{-(F_{aj}+M_{aj})}]\right\} \quad (11.6.2)$$

where C_{aj} is the catch of age a fish in year j, R_{j-a} is the recruitment in the $j - a$th year and M_{aj} is the instantaneous natural mortality rate of age a fish in the jth year.

If we have n ages and m years of data, then we have $n \times m$ observations of catches at age. Ideally, we would like to estimate the annual recruitments ($m + n - 1$ of them), the fishing mortality rates ($n \times m$ of them) and the natural mortality rates ($n \times m$ of them). We need to estimate a recruitment for the first age for each year of data (m) and a starting cohort size for each age in the first year of data ($n - 1$). This would mean estimating $2(n \times m) + (m + n - 1)$ parameters from only $n \times m$ data points, which is obviously impossible. Also, Equation 11.6.2 is not linear, so a nonlinear estimation method would have to be used. However, some simplifying assumptions can be made to reduce the number of parameters to be estimated.

The first assumption usually made is that all M_{aj} are the same, that is, natural mortality is independent of age or year. Although we have been using this assumption from the beginning, it still does not reduce the problem enough because we have $n \times m$ F's and m R's. The next commonly used assumption is that F is the same for all ages but varies from year to year; then there are only m F's. We now have $m + n - 1$ R's and one natural mortality rate, or $2m + n$ parameters and $m \times n$ data points.

Providing that $n > 2$, we have enough data to try to do the estimation. Doubleday's method has the same general problem as Paloheimo's; if there is no contrast in F's then it will be impossible to discriminate between F and natural mortality. Doubleday's method has the advantage that it does not assume that F is proportional to effort, so it can be used when effort is not available or unreliable. Note that if we assume $F = Eq$, Doubleday's method reduces to Paloheimo's method.

Doubleday's method has the nice feature that you can look at changing vulnerability with age by assuming that $F_{aj} = v_a F_j$ (this is called the "separable model" for age versus time effects on fishing rate). In other words, we have different fishing mortalities for each year and different vulnerabilities for each age. This will add n more parameters to estimate. Paloheimo also presented a nonlinear estimation procedure for getting age-specific vulnerabilities by expanding his basic formulation. We can also look for time-specific changes in mortality rate by breaking the data into blocks of time and estimating different M's for each block.

Table 11.4. Relative strengths and weaknesses of the Doubleday and the Paloheimo methods.

Feature	Paloheimo	Doubleday
Computation	Linear, generally easily solved	Nonlinear, requires lots of computation, potential problems with false minima
Recruitment estimation	Good, generally reliable for relative recruitment	Good, generally reliable for relative recruitment
Assumption about fishing mortality	Assumes $F = Eq$, requires effort data	Requires no assumption, can calculate age-specific vulnerabilities
Assumption about natural mortality	Assumes M constant over all years and ages	Can in principle estimate different mortalities for all ages or time
Ability to discriminate fishing mortality from natural mortality	Requires major contrast in effort, generally poor	Requires contrast in F, generally poor

In principle, we can estimate lots of parameters. In practice, we can usually find the relative R's and age-specific vulnerabilities, but it is usually difficult, if not impossible, to get good estimates of F's and M's. Even if M is fixed, there still is a correlation between the R's and the F's. Table 11.4 outlines the relative strengths and weaknesses of the Doubleday and Paloheimo methods.

11.7. Explicit Errors and Auxiliary Information

Fournier and Archibald (1982) extended the application of catch-at-age analysis by explicitly considering how errors enter into the model. The error structures of both the Paloheimo and Doubleday approaches were totally arbitrary; the models were derived as if there was an underlying deterministic process and the estimation equations were developed on that basis. Fournier and Archibald pointed out that there are errors in both the assessment of ages of fish and the measurement of catch. In addition, there is natural variation in recruitment and mortality.

The paper by Fournier and Archibald is quite rich, and we present here a major taste, but probably not enough of it to satisfy serious readers. Much of the material from Fournier and Archibald is summarized in Deriso et al. (1985), who have also prepared a public-domain package for catch-at-age analysis. Anyone seriously interested in doing catch-at-age analysis would be well advised to obtain the package from Deriso.

Fournier and Archibald's approach uses three components: catch-at-age observations, fishing effort observations, and an assumed form for the stock-recruitment relationship. Each of these components leads to a collection of prediction errors (deviations from catches at age, deviation of fishing rates from qE predictions, and deviations of estimated recruitments from stock-recruit curve). They formulate a likelihood for each of these collections of deviations, and then allow the analyst to provide relative weightings that reflect his beliefs concerning how large the relative deviations of each type should be.

Catch relationship

The first step in their approach is the equation for predicted catch of a certain year class in a specified year (Equation 11.6.2). They define the sum of squares associated with catch (SSQ_{catch}) as:

$$\text{SSQ}_{\text{catch}} = \sum_{a}\sum_{j}(\log(C'_{aj}) - \log(C_{aj}))^2 \tag{11.7.1}$$

where a is the age, j is the year, C' is the predicted catch from Equation 11.6.2 and C is the observed catch. This by itself is exactly the same as Doubleday's method, and leaves unstated the definition of the mortality rates and how they are aggregated by year or age.

Fishing mortality

The second step in Fournier and Archibald's approach is to define a function between age and vulnerability in any year. The normal approach of Paloheimo's method is to assume that $F = qE_t$ for fully recruited ages. Fournier and Archibald suggest using a two- or three-parameter function relating fishing effort to mortality by age. Instead of allowing for a different vulnerability parameter for each age, they compress the relationship into two or three parameters. They used one version that also included time-dependent changes in the vulnerability/age relationship.

From the vulnerability relationship one obtains a predicted fishing mortality F' on fully recruited ages (qE_t is normally used), and the observed fishing mortality for any specific set of estimated R's and other parameters. In an ideal situation we could find some value of the parameters that make the F' exactly equal the observed F. However, the deviation between computed fishing mortality and that expected based on effort is

$$\epsilon_{1t} = \log[F(t)] - \log[F'(t)] \tag{11.7.2}$$

and we define another sum of squares to minimize as

$$SSQ_{\text{effort}} = \lambda_1 \Sigma (\epsilon_{1t})^2 \tag{11.7.3}$$

where λ_1 is the weight assigned to the SSQ_{effort} and is related to how variable we believe the effort-fishing mortality relationship should be. If λ_1 is small then the effort data has little effect, but if λ_1 is large then the nonlinear estimation will try very hard to find parameters that make the apparent fishing mortality closely related to effort.

Stock and recruitment

In the Paloheimo and Doubleday methods, we allowed the recruitment for each cohort to be a totally free parameter to be estimated. We may wish to force the estimated recruitments toward fitting some stock-recruitment curve. In the extreme case, we might actually have estimates of recruitment in each year.

The assumption used by Fournier and Archibald was that the recruitment relationship should approximate a Ricker stock-recruitment curve, or that the predicted recruitment for year t should be

$$\log (R_t') = \log (S_t) + a - bS_t \tag{11.7.4}$$

where a and b are the parameters of the Ricker stock-recruitment relationship and S_t is the spawning stock size in year t. The deviation between the predicted recruitment and the estimated recruitment is therefore

$$\epsilon_{2t} = \log (R_t') - \log (R_t) \tag{11.7.5}$$

and the sum of squares due to the recruitment deviations is

$$SSQ_{\text{spawn}} = \lambda_2 \Sigma(\epsilon_{2t})^2 \tag{11.7.6}$$

Putting catch, fishing mortality and recruitment together

The final step is to use nonlinear parameter estimation to find parameters that minimize

$$SSQ_{\text{catch}} + SSQ_{\text{effort}} + SSQ_{\text{spawn}} \tag{11.7.7}$$

which, of course, depends upon the data and the values assigned to λ_1 and λ_2. This brief overview does not really do justice to the new ideas in Fournier and Archibald's paper. In particular, we have hardly mentioned their analysis of errors in aging. However, we hope it has given enough of a taste to interest readers in further exploration of their work.

Auxiliary information

In addition to the basic idea of examining how errors enter into the catch-at-age analysis process, the major contribution of Fournier and Archibald was to introduce in a more formal way the use of data other than catch in catch-at-age analysis. Of course, effort data were an integral part of Paloheimos's method. Deriso et al. (1985) provide an excellent review of the use of auxiliary information. Besides effort data, and the stock-recruitment relationship mentioned above, auxiliary information might include estimates of recruitment or fishing mortality from totally different studies.

Deriso et al. explored the importance of auxiliary information using data for the Pacific halibut (*Hippoglossus stenolepis*). They found it was almost essential to include effort data and very helpful to include a spawner-recruitment curve. Methot (1989) has developed a method he calls *stock synthesis*, that is probably the most sophisticated method for dealing with catch-at-age and other data in a single computational framework. Methot allows biomass survey, CPUE, catch-at-age, and other forms of data to be used simultaneously to estimate stock abundance, fishing mortality, and gear/selectivity curves.

11.8. Summary and Critique

Catch-at-age analysis, particularly with auxiliary information, is an extremely powerful and elegant synthesis of VPA, catch curves, selection curves, and stock and recruitment. When proper consideration is given to the errors introduced in data collection, and to natural variability in the recruitment and mortality processes, catch-at-age analysis is the *state of the art* in analysis of fisheries data.

Although it is difficult if not impossible to obtain catch-at-age data in many fisheries, there are a surprising number where it is possible, and the more widespread availability of computer software to perform catch-at-age analysis will undoubtedly lead to its wider use.

The main risk in using catch-at-age analysis is the same as in using VPA to estimate the abundance of incomplete cohorts, namely that the catchability coefficient and/or fishing rate may be increasing rapidly in the most recent years if a stock decline is underway. For the most recent cohorts entering the data, the regression methods will be unable to tell whether the cohorts are small and are being fished hard or are large and are being subjected to lower fishing rates.

Bibliographic Notes

The use of catch curves can be traced back to Edser (1908) and Heincke (1913) and has seen broad application in fisheries for over 50 years. The widespread acceptance of VPA stimulated more intense use of catch-at-age data, but the real major breakthrough was by Doubleday (1976). Paloheimo (1980) provided a computational shortcut which is both historically and mathematically analogous to Pope's (1972) shortcut for VPA. Fournier and Archibald (1982) provided the next major step, with explicit consideration of errors and use of auxiliary information. Megrey (1989) has reviewed all catch-at-age methods.

12

Depletion Estimates of Population Size and Turnover

12.1. Introduction

An important role of stock assessment is to provide estimates of the total harvestable stock size, especially early in fishery development when it is important to have some idea of the potential for development. There are four general approaches to estimating total stock: (1) direct counts and surveys; (2) reconstruction from historical catch data (VPA, catch-at-age analysis); (3) mark-recapture and change in ratio methods (Peterson, Schnabel estimates, etc.); and (4) depletion estimators. All of these topics are discussed briefly in Chapter 5 and there are thorough external references on direct counts and mark recapture statistics. We devote two chapters to VPA and catch-at-age analysis. This chapter discusses the fourth of these approaches, depletion estimates, which under some circumstances can be substantially cheaper and more effective than the others.

The general concept behind depletion estimators is to examine how measured removals of fish (by regular or experimental fishing) influence the relative abundance (CPUE, other abundance index) of fish remaining in the total stock or in a designated depletion study area. In the simplest case, for a closed stock with no recruitment or natural mortality, the concept is essentially to predict how large the total (cumulative) removal would have to be in order to drive the relative abundance to zero; this predicted total removal is then an estimate of the initial stock size before removal began. In more complex situations where there is recruitment (and/or immigration) and mortality (and/or emigration) from the study area during the experiment, the concept is extended to try to estimate how much recruitment and/or mortality must have occurred in order to have produced the observed pattern of change in relative abundance.

It is important not to confuse depletion estimation with deliberate overfishing, and to reject the approach out of hand as being too risky from a conservation standpoint. Often, depletion estimation models can be applied to historical data, especially from early in fishery development, to provide estimates of the virgin biomass before (often uncontrolled) development began; knowing virgin biomass is often useful in determining whether overfishing has occurred during the development. In planning for data gathering during early fishery development, the possibility of using depletion estimators after a few years is strong motivation for investment in survey fishing and other field programs that will give a time series of relative abundances that are not so messy to interpret as CPUE data (which reflect all sorts of changes in technology and fishing skill, as well as fish abundance changes during early development). In other cases, deliberate depletion experiments can be designed to safely measure abundance in restricted local areas (bays, reefs, particular substocks), without jeopardizing the overall stock; these local totals can then be used to "calibrate" relative abundance indices that are applied to other areas.

12.2. Derivation of Depletion Estimators

Depletion estimators are derived by combining two basic "submodels" or collections of assumptions. First, there is a population dynamics submodel that predicts population size N_t at any time t given an initial population N_1, catches C_1, C_2, . . ., C_{t-1} taken prior to time t, and (optionally) parameters for recruitment, survival, and growth. This model may also use fishing effort data E_1 E_2, . . ., E_{t-1}. The key requirement is that the predicted values N_t be functions of the unknown parameter N_1, which can hopefully be estimated.

Second, there is an observation submodel that defines an assumed relationship between the population size N_t and an abundance index y_t. Usually it is assumed that the index is measured in such a way that it will be proportional to N_t. Under this assumption, the observation submodel is just $y_t = qN_t$. The parameter q (and any other parameters of more complex observation models) is one of the unknowns to be estimated (along with N_1); assuming some prior knowledge of q would be equivalent to saying that total population size was already known (i.e., $N_t = y_t/q$).

Depletion estimation procedures are then derived by substituting the population submodel equation(s) into the observation submodel equation, so that the observed index values y_t are predicted as a direct function of the unknown parameters q, N_1, and so on. Given this direct prediction model, statistical estimation procedures are used to find parameter values that give a best fit of predicted y_t's to the observed sequence y_t, in a least-squares or maximum-likelihood sense.

12.3. Leslie and DeLury Estimates for Closed Populations

Derivation

Suppose you have a population that is completely closed, so that it gains no new recruits or immigrants and loses no animals to natural mortality or emigration. The behavior of this population under harvesting can be described by the very simple model

$$N_t = N_1 - K_{t-1} \tag{12.3.1}$$

where K_{t-1} is the cumulative catch taken prior to time t, that is, $K_{t-1} = C_1 + C_2 + \ldots + C_{t-1}$ for $t > 1$, and $K_0 = 0$. If it can be assumed that the catch rate is directly proportional to an effort measure f_t, that is, $dK_t/dt = qf_t$, and the cumulative effort (integral of f_t) up to time t is E_t, then an alternative to Equation (12.3.1) is the continuous or exponential fishing model

$$N_t = N_1 e^{-qE_t} \tag{12.3.2}$$

where $E_1 = 0$. Notice that the population submodel (12.3.1) makes no assumption whatever about how the cumulative catch K_t arises or varies over time, while the alternative submodel (Equation 12.3.2) makes a very strong assumption about the distribution of fishing effort (all fish equally vulnerable, random search) and your ability to measure it.

The *Leslie estimate* (Leslie and Davis, 1939) of N_1 is obtained by assuming Equation 12.3.1, whereas the *DeLury estimate* (DeLury, 1947) is obtained from Equation 12.3.2. In both cases, the simple observation model $y_t = qN_t$ is assumed. For the Leslie case, the abundance y_t may be measured in any way and may be completely independent of the fishing process that generates K_t; for example, y_t might be an acoustic index of school sizes and densities, whereas K_t is taken from purse seine fishing. For the DeLury case, y_t must be CPUE, measured over short enough periods t so that C_t/f_t is approximately equal to qN_t.

For the Leslie method, substitution of Equation 12.3.1 into the observation model $y = qN$ gives

$$\begin{aligned} y_t &= qN_t \\ &= q[N_1 - K_{t-1}] \\ &= qN_1 - qK_{t-1}. \end{aligned} \tag{12.3.3}$$

Notice that the last expression is in the form of a linear regression $y = a + bX$, where the intercept parameter a is qN_1, the slope parameter b is q, and the regression X is $-K_{t-1}$. Thus if a and b are estimated by regression of y_t on $-K_{t-1}$ as in Figure 12.1, the slope should estimate q and the ratio a/b (intercept of regression line on the X axis) should estimate N_1.

In simpler terms, you can think of the Leslie method as drawing a line through the plot of relative abundance versus cumulative catch, then looking to see at what cumulative catch the relative abundance would be predicted to drop to zero.

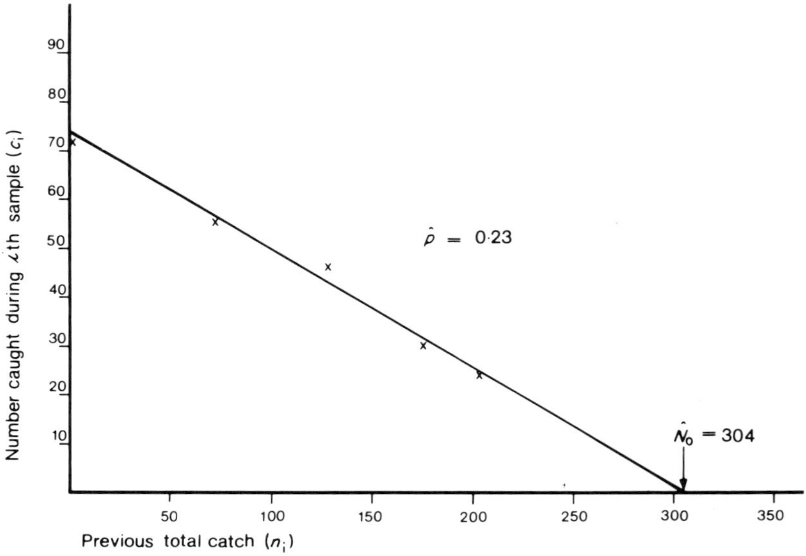

Figure 12.1. Closed population estimate of trout from Afon Dulas at Pentre, June 19, 1979. From Cowx (1983). Reprinted by permission of Blackwell Scientific Publications Limited.

For the DeLury method, substitution of Equation 12.3.2 into the observation model $y = qN$ gives

$$\begin{aligned} y_t &= qN_t \\ &= qN_1 e^{-qE_t} \end{aligned} \quad (12.3.4)$$

Taking logarithms, we then get

$$\log[y_t] = \log[qN_1] - qE_t \quad (12.3.5)$$

This direct prediction model is again in the form of a linear regression, but with the regression dependent variable being $\log[y_t]$ and the independent variable being the cumulative effort E_t prior to time t. The regression slope is the same as for the Leslie method, whereas the regression intercept is $\log[qN_1]$. An estimate of N_1 is obtained from the regression intercept a and slope b as $N_1 = e^a/b$.

Figure 12.1 shows plots of CPUE (y_t) versus cumulative catch for a fishing experiment, performed on a trout population that was sequentially electrofished five times on June 21, 1979 (Cowx 1983). The catches in each electrofishing episode were 72, 56, 46, 30, and 24. The total catch was, therefore 228. Cowx fitted a line by eye and obtained an X intercept of 304, which is the estimated initial population size.

Sources of error and bias in closed population depletion experiments

Whereas both the Leslie and DeLury regression estimates of q and N_1 are approximately unbiased provided all fish are equally vulnerable to fishing and K_t or E_t is measured exactly (Seber 1982, Ricker 1975), there is much practical and Monte Carlo simulation experience to indicate that the estimates must be treated with considerable care (see e.g. Bohlin and Sundstrom 1977, Mahon 1980, Cowx 1983, Peterson and Cederholm 1984).

Warning: Be sure to estimate the reliability of depletion estimators by Monte Carlo methods for your specific problem.

Especially for the DeLury estimator, errors in measurement of the regression-independent variable (K_t or E_t) cause the estimate of q to be biased downward and N_1 to be overestimated; this problem is likely to be less serious for the Leslie estimator, which can also be used under a greater variety of circumstances in the first place, because the relative abundance index can be measured independently of the process that removes fish to give K_t. Thus the Leslie estimator is generally preferable.

Ricker (1975) points out that "Inconstant catchability is perhaps the greatest potential source of error in applying methods of estimation based on secular change in catch per unit of effort." It is often found that the first few units of removal effort cause a rapid depletion of more vulnerable (stupider, slower, more accessible) fish, with accompanying rapid change in catch per effort or other abundance index. After this initial removal, the remaining fish have effectively lower q values, so that q declines progressively as the depletion proceeds. There may even be a large pool of animals with $q = 0$ for some reason, and this pool will not be sampled at all by the depletion process. Thus the general effect of varying catchability among individuals is to bias the estimate of q upward (only animals with higher q are seen, at least in short experiments) and to bias the estimate of N_1 downward. This effect is likely to be much larger than the upward bias due to statistical errors mentioned in the previous paragraph, so the depletion estimate of N_1 is likely to be too low; underestimates in the order of 30% to 50% are not at all unlikely. The presence of large numbers of fish with low catchability may be indicated by curvature (flattening) of the depletion regression, but only if the experiment is carried out for many (at least 10) time periods.

In chapter 5, we discussed the relationship between CPUE and abundance; all of these problems will occur in depletion methods that attempt to use CPUE as an index of abundance. We suggest readers using depletion methods should carefully examine Section 5.4.

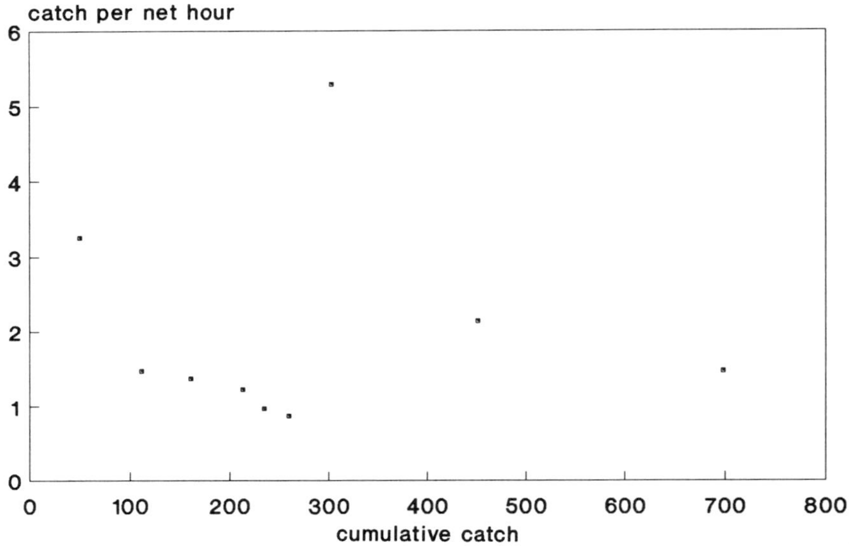

Figure 12.2. CPUE versus cumulative catch for Wonder Lake. Data from Hall (1991).

Figure 12.2 shows the CPUE versus cumulative catch data for brook trout in Wonder Lake, a small alpine lake in California (Hall 1991). The first six data points were obtained during a 4-day period, and the next three points were obtained 8 weeks later. Wonder Lake is a closed system, with no possibility of immigration. Hall concludes that there was an enormous change in catchability between the end of the first fishing period and the second episode 8 weeks later. He suggests two possible mechanisms, (1) either territoriality so that the territorial individuals were removed in the first fishing period or (2) major behavior changes in feeding mode. For the purposes of depletion methods, this experiment serves as a strong warning: any abundance estimate based on the first fishing episode would have drastically underestimated the population size.

If the same fishing process is used to both deplete the stock and to provide an abundance index for each depletion period, then random variations in the fishing catchability coefficient will lead to nonindependence in the y_t values and to correlations between y_t and K_t (or E_t) that are not modelled in the basic linear regression equations. Such correlations are partly responsible for overestimation of q and N_1 when all fish have equal catchability and will lead to higher statistical confidence in the estimates (narrower confidence limits, etc.) than is warranted by the data. These problems can be dealt with by using maximum-likelihood estimation (Schnute 1983, Zippin 1956), but are likely to be small in comparison to biases caused by unequal catchability.

12.4. Depletion Estimates for Open Populations

It is often impractical to design depletion experiments over time and space scales such that the experimental population remains closed to migration, recruitment, and mortality as required for the simple Leslie and DeLury methods. Imagine for example trying to estimate the "stock" of tuna aggregated near a fish attracting device (FAD) somewhere in the South Pacific; obviously this stock will be subject to recruitment (immigration) and loss (emigration) even on time scales of a few hours or days. Fortunately, it is possible to develop depletion estimators, for a variety of time scales, that can provide estimates of both initial population size and combined rates of immigration-recruitment and survival-emigration.

Derivation

The development of open population depletion estimators proceeds with the same general structure as for closed populations, namely a population submodel and an observation submodel. However, we now want to have the population submodel account for gains and losses besides the catches C_t. To make the derivation as general as possible, let us assume that the stock size N_t may be measured in biomass units and that body growth may occur between depletion periods; let us assume, as in the delay-difference models (Chapter 9), that body size of fish that have been in the stock for i periods (e.g. ages if the periods are 1 year each) can be predicted from the body size of fish one period younger ($i - 1$) from the Ford-Walford model $w_i = \alpha + \rho w_{i-1}$. Then, as shown in Chapter 9, we can write a general dynamic model for the stock size as

$$N_t = G_{t-1}[N_{t-1} - C_{t-1}] + R_t \qquad (12.4.1)$$

This model just says that N_t is given by a factor G_{t-1} times the number of survivors $N_{t-1} - C_{t-1}$ from the last depletion period, plus a number of new recruits R_t. If N_t is measured in biomass units, then G_t represents a combined growth-survival factor of the form $G_t = s[\rho + \alpha/\bar{w}_t]$ and R_t measures recruitment of new biomass (numbers times weight w_k at the age of recruitment). When N_t is measured in numbers units, G_t represents only the natural "survival" rate s (proportion of animals that neither die or emigrate from the stock $N_t - C_t$ remaining after harvest in period t), and R_t represents the net number of new recruits added to the population (through reproduction and/or immigration) between time $t - 1$ and t. In fact, G_t also contains (represents) any component of recruitment that is proportional to $N_{t-1} - C_{t-1}$; if we wrote a model $N_t = bN_{t-1} + sN_{t-1}$, where b is birth rate and s is survival rate, this model would be indistinguishable in its predictions from the simpler form $N_t = gN_{t-1}$ where g is the combined parameter $b + s$.

Various special cases of Equation 12.4.1 have been used as the basis of depletion estimators (Allen 1966, Chapman 1974, Sainsbury 1984). For a review of these cases and discussion of various estimation procedures, see Seber (1982, pp. 541–545). The population submodel (Equation 12.4.1) can obviously represent a wide variety of circumstances in terms of recruitment, survival, migration, and growth, depending on how the stock size is defined and on the duration of each depletion period t. However, to make further use of it for depletion estimation, we will need to make further assumptions about the G_t and R_t factors. In particular, let us assume that: (1) there is an independent estimate of each G_t, i.e. of the survival rate s and (if appropriate) the growth parameters ρ and α and mean body sizes $\bar{w}_t$ over time; and (2) recruitment rate R_t is constant over time at a value (or randomly varying around a value) R that is independent of stock size. Later we will see that, under some circumstances, it is possible to relax the first of these assumptions somewhat and to estimate at least the survival rate component s of G_t. Fixing survival rate s at some independent estimate is essentially the same idea as is used in obtaining virtual population estimates (cohort analysis).

By successively substituting Equation 12.4.1 into itself to solve for N_2, N_3, and so on in terms of N_1 and the parameters, it is easily shown that Equation 12.4.1 can be written in a more convenient form when recruitment is constant ($R_t = R$):

$$N_t = K'_{t-1}N_1 - K''_{t-1} + K'''_t R \tag{12.4.2}$$

where the growth-survival-catch factors K', K'', and K''' are defined recursively as follows:

$$\begin{aligned} K'_1 &= G_1; & K'_t &= K'_{t-1}G_t & &\text{for } t > 1 \\ K''_1 &= G_1C_1; & K''_t &= [K''_{t-1} + C_t]G_t & &\text{for } t > 1 \\ K'''_1 &= 0; & K'''_t &= K'''_{t-1}G_{t-1} + 1 & &\text{for } t > 1 \end{aligned} \tag{12.4.3}$$

These factors depend only on the growth-survival factors G_t, and they measure the cumulative effects of past G values on the population present at time t.

Notice how Equation 12.4.2 splits up the prediction of N_t into three terms. The first, $K'_{t-1}N_1$, measures essentially the contribution to N_t of the initial population N_1; if no growth effects are included in the model, K'_t is the probability s^t of surviving t periods for a fish that has not been caught. The second term K'_{t-1} is analogous to the total catch K_t of closed populations, except that it is a weighted sum of past catches; catches further in the past are given less weight since the growth-survival factors G_t will generally be less than 1.0. The third term, K'''_t, measures the cumulative effect of recruitment on N_t, weighting each past recruitment contribution R by the odds of surviving to period t.

Let us note one further generalization before proceeding. Suppose that the recruitment rates R_t are unknown, but that a time series of relative recruitment rates $r_t = R_t/R$ is known, where R is an unknown base or mean recruitment rate. Then if we could estimate R, the recruitments R_t would be given by $R_t = r_tR$. In some cases we might know for example that recruitment due to migration is zero for some time periods due to periodicity in movement behavior (e.g. for some hours of the day during a diurnal experiment). In other cases, there might be a sampling program that has provided a relative recruitment index over time. A time series of known r_t values can easily be incorporated into Equation 12.4.2, just by calculating the factor K_t''' using the r_t values rather than 1.0 at each recursive step: $K_t''' = K_{t-1}'''G_{t-1} + r_t$.

Suppose now that we substitute equation 12.4.2 into the observation model $y_t = qN_t$, to obtain a model that predicts y_t only from the unknown parameters q, N_1, and R. The result is

$$y_t = [qN_1]K_{t-1}' - [q]K_{t-1}'' + [qR]K_t''' \tag{12.4.4}$$

where the bracketed terms collect together the parameters that multiply each K term. Notice that we now have a model of the form $y = b_1X_1 + b_2X_2 + b_3X_3$, that is, a multiple linear regression, with the "independent variables" $X_1 - X_3$ being the three K factors for each y and the regression coefficients being $b_1 = qN_1$, $b_2 = -q$,and $b_3 = qR$. Thus if the growth-survival factors G_t are assumed known in advance, so that the K factors can be calculated as above, we can do a simple multiple linear regression, comparable to the closed population depletion estimation procedures, in order to estimate q, N_1 (as $-b_1/b_2$), and R (as $-b_3/b_2$). You must be a bit careful about doing this regression with package linear regression programs, since many of these programs assume an intercept parameter b_0; you must either use a program that allows you to constrain the intercept to zero or else write your own program to solve the linear model estimation equations $\hat{b} = (X'X)^{-1}X'Y$.

Alternative estimation procedures

So far, we have derived a simple multiple-regression procedure that will provide estimates of q, N_1, and R for open populations, when an independent sequence of growth/survival rates G_t and perhaps relative recruitment rates r_t is provided. The basic procedure can be thought of as a variation on the theme of virtual population analysis. Given an independent estimate of the survival rate s and the historical catches C_t, what values of q, N_1, and R best describe (fit) the historical relative abundance sequence y_t? The best fitting parameter estimates may or may not be unique; especially for short time series, there may be many combinations of q, N_1, and R estimates that fit equally well; only in longer time series with strong depletion will the effects of R be readily discernable.

Suppose now that there are no credible independent estimates of the survival or turnover parameter s that is needed to calculate the G_t values. It is possible to do a sequence of linear regressions, each giving estimates of q, N_1, and R, for a range of different estimates of the survival rate s (and hence different G_t estimates). In principle, this sequence may allow you to find a best fitting (minimum-regression-error sum of squares) estimate for all four parameters q, N_1, R, s. However, in practice, it is unusual for more than two of the parameters to be well-defined (data equally well explained by many combinations of the four parameters).

Thus it is usually necessary to use at least some independent auxiliary information or assumptions about the parameters, beyond what can be obtained from the multiple regressions. One possibility is to assume that the stock was at equilibrium before the depletion started (Chapman, 1974); in this case the recruitment rate must have just balanced mortality, that is, $R = [1 - G_1]N_1$. Using this assumption as a constraint among the parameters allows you to eliminate one of the parameters from the estimation. For example, you can substitute $[1 - G_1]N_1$ for R in Equation 12.4.3 to obtain a reduced regression equation containing only N_1, q, and the survival/growth parameters.

Examples of open population depletion estimates

Figure 12.3 shows fits of model Equation 12.4.4 to three sample data sets. These data sets illustrate how the model may be applied to data from very different time scales. The first data set represents catch and catch per set for two days of beach seining for small bait fishes in a small bay of Nuku Hiva (data provided by J. Ianelli, personal communication). Four seine sets were made one morning (about an hour apart), then four more were made two mornings later. A clear depletion pattern was seen in each visit, and the stock substantially recovered (presumably due to immigration of new schools into the bay) over the 2-day rest period. In this case, it was reasonable to assume that the stock was initially at equilibrium (movement into the bay balancing movement out), with no growth, so $G_t = s$ and $R = [1 - s]N_1$. The basic time step for t was taken to be 1 hour. When the regression fit with these conditions was performed for a range of different s values, a clear best fit was found at $s = 0.95$, which implies a turnover rate of around 5% per hour for the stock in the bay.

The second data set represents catch and catch-per- purse seine set for seiners working on seven successive days around fish attracting devices (FADs) in the Indian Ocean (Hallier 1985) catching skipjack (*Katsuwonus pelamis*)

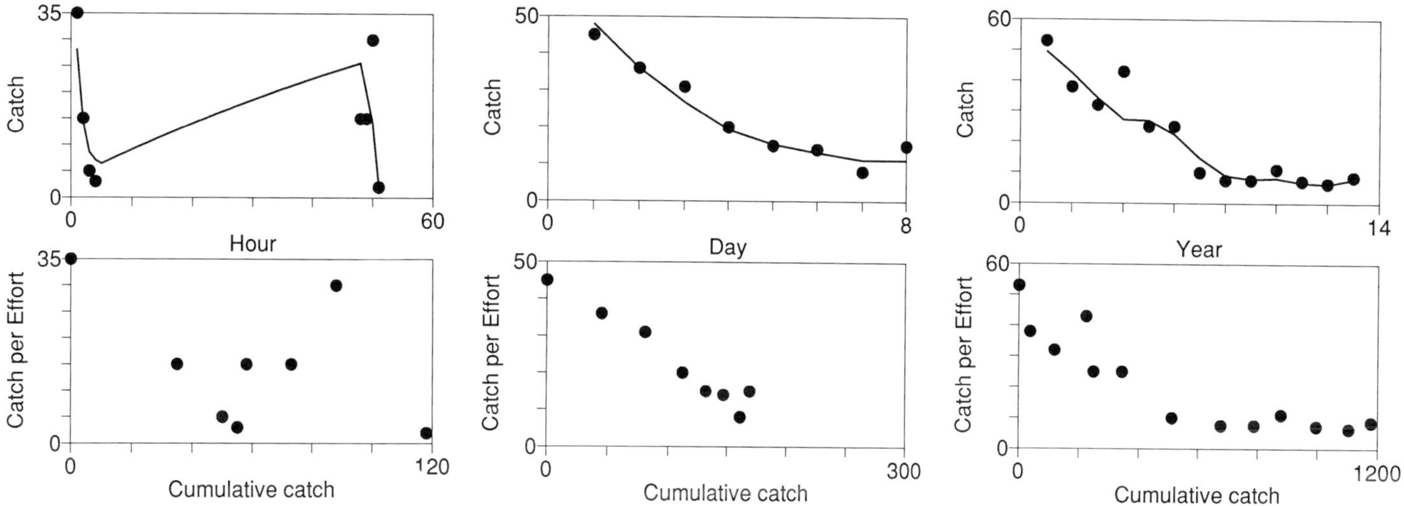

Figure 12.3. Depletion experiments from baitfish in a bay on Nuka Hiva, Fiji (a), Purse seiners catching tuna at a FAD (b) and for Barramundi (*Lates calcarifer*) (c). For sources of data see text.

and yellowfin tuna (*Thunnus albacares*). The data for several FADs were averaged to obtain the smooth depletion pattern shown in Figure 12.3.b. Again it was assumed that body growth was negligible over the depletion period and that immigration initially balanced emigration so $R = [1 - s]N_1$. The basic time step in this example is 1 day. Again a clear best fit was found when linear regressions were done for a range of s values, again at $s = 0.95$. In this case, $s = 0.95$ implies a turnover rate of around 5% per day, implying a recovery time of approximately 3 weeks after a FAD is hit hard by seiners. It would be interesting to determine by tagging experiments whether the average residence time of tuna around a FAD is in the order of 20 days as predicted from the best-fitting parameter estimates.

The third data set represents 14 years of commercial gill net catch and CPUE from a barramundi (*Lates calcarifer*) stock in the Mary River, Northern Territories, Australia. This fascinating fish, with its complex life cycle (catadromous, hermaphroditic) and high value (large body size, good sport and table fish), was rapidly depleted in abundance by fisheries in several Northern Territories rivers. The basic depletion time step in the data is 1 year. Here we were willing to assume that the stock was initially at equilibrium, but not that growth is negligible. We had no time series of mean body weights for the stock, so we used a simulated time series obtained from the delay-difference equations of Chapter 9 to calculate G_t values. By iterating back and forth between the depletion estimation and mean weight simulations (using the parameter estimates obtained from the estimation to get new mean weight simulation parameters, then putting the simulated mean weight results back into the depletion estimation), we obtained a very good fit to the historical CPUE data at an estimated survival rate $s = 0.5 - 0.6$; independent age composition studies later gave survival rate estimates of that magnitude as well (R. K. Griffin, Northern Territories Department of Fisheries and Development, personal communication). Note that the good fit was obtained while assuming recruitment rate independent of stock size over the entire stock decline; had there been recruitment overfishing during the decline, we should have obtained a poor fit to the later years in the series since poor recruitments in those years should have led to continued decline (rather than the observed pattern of low but stable CPUE in later years). The Mary River apparently contains a distinctive enough stock to have few new recruits as immigrants from other rivers (Davis 1987), so constant recruitment in this case likely represents improved reproductive success by the declining spawning stock rather than immigration of new recruits from other river systems.

Bias in open stock depletion estimates

With more parameters to estimate from time series that will generally be relatively short, the open stock depletion estimators are more prone to both random estimation error and to bias than are the closed population procedures. As for the simpler estimators, Monte Carlo studies indicate that errors in y_t and C_t measurement will generally lead to underestimates of q and overestimates of N_1. R may also be overestimated, though one tendency we have noted in some data sets and simulations is for the regression to fit best with low q and high N_1, but a low or negative estimate of R. Systematic biases will of course also be introduced by incorrect assumptions about s (and ρ and α in cases where growth is considered) in the calculation of the G_t series.

More complex and interesting effects can be expected in the presence of varying catchability among individuals, so that the more vulnerable ones are removed early in the depletion sequence. In this case, q will be overestimated and N_1 will be underestimated as in the closed population case. However, R may be overestimated. This is because fish of low catchability will appear to recruit to the stock later in the depletion sequence, and to hold the y_t values higher than would be predicted from early in the sequence. For short time series, there is in principle no way to distinguish from the y_t and C_t data alone whether the depletion is proceeding more slowly due to removal of fish from a low catchability reservoir, or removal of new recruits from immigration and growth into fishable sizes. Eventually the effects of recruitment must become apparent, as any low catchability pool is slowly depleted, but this may well mean an impractically long experiment in terms of repeated fishing and/or maintaining a standardized abundance indexing program.

The parameter estimates must be interpreted carefully when recruitment rate may be correlated with stock size. For instance, we obtained a very good fit of Equation 12.4.4 to catch-effort data on Leiognathids from the Gulf of Thailand (Pauly 1986), as shown in Figure 12.4. Assuming constant annual recruitment, the good fit was obtained with G_t held constant at $G = 0.8$. But Leiognathids are small, very short lived fish, and it is doubtful that G_t is greater than 0.3. What likely happened in this case is that recruitment rate declined during the fishery development, so that R_t could be approximated over the development period by the linear relation $R_t = r_1 + r_2(N_{t-1} - C_{t-1})$. Substituting this relation for R_t into Equation 12.4.1 and rearranging, we get

$$N_t = [G_{t-1} + r_2](N_{t-1} - C_{t-1}) + r_1 \tag{12.4.5}$$

Notice in this equation how the growth factor $G_{t-1} + r_2$ contains the effect of past stock size on recruitment; if G is treated only as a survival/growth factor, its best estimate will be too high by a factor r_2, whereas the recruitment parameter R will contain only the recruitment rate intercept term r_1.

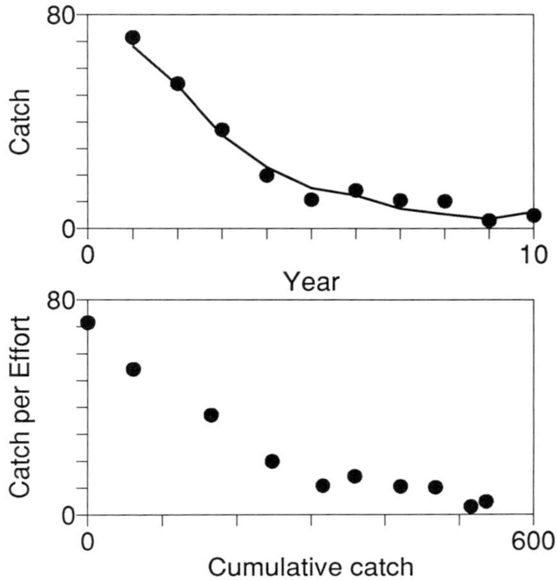

Figure 12.4. Depletion estimates for Gulf of Thailand data on Leiognathids. Data from Pauly (1986).

12.5. Age-Structured Depletion Estimates

So far, we have assumed that only aggregate catch and relative abundance data are available for the depletion estimation. Suppose now that the basic depletion time step is 1 year and that several years of data on age composition of the numerical catch and of a numerical relative abundance index are available. Now we have tables of information $C_{a,t}$ on catch, and $y_{a,t}$ on relative abundance by age class. Let us assume further that (1) the relative abundance data have been collected by a survey sampling method that is not age-selective, so that $y_{a,t} = qN_{a,t}$ with the same index catchability q for all ages a; and (2) the annual survival rate s is known and is the same for all ages of fish. Under these assumptions, we can construct depletion estimators for the initial cohort sizes (recruitments) for each cohort present in the age composition tables (Walters and Collie 1988, Johnson and Walters 1991, Collie and Sissenwine 1983).

Under the above assumptions, each cohort in the age-time data table will decline according to Equation 12.4.4 with recruitment set at zero,

$$y_{a,t} = [qN_{1,c}]K'_{a,c} - [q]K''_{a,c} \tag{12.5.1}$$

Here, the index c represents a cohort index number for the animals aged a at year t. $N_{1,c}$ is the initial number of animals present in cohort c in the first year when it was sampled, K' represents cumulative survival s (number of years -1 in the table for cohort c), and $K''_{a,c}$ represents cumulative catches weighted by survival for the years and ages that cohort c appears in the table prior to age a and year t. Here the initial numbers $N_{1,c}$ represent numbers at the youngest age for all cohorts that first appear on or after the first sample year and numbers at later ages for those cohorts that are already present in the population at the first sample year.

As for the depletion estimators for aggregated populations, Equation 12.5.1 is in the form of a multiple regression. Each relative abundance $y_{a,t}$ is predicted to result from two factors: (1) a recruitment parameter $qN_{1,c}$ for the initial cohort size leading to age a animals in year t, multiplying an independent variable $K'_{a,c}$; and (2) a catchability parameter q, multiplying an independent variable $K''_{a,c}$ representing cumulative catch from the cohort. Note that the multiple regression has more than two parameters; although the q is shared, there is a different $[qN_{1,c}]$ recruitment parameter for each cohort present in the data. If there are n years of data for m ages, then there will be $n + m - 1$ recruitment parameters, which along with q makes a total of $n + m$ parameters to be estimated.

The multiple regression equations can be solved efficiently using a recursive procedure (Johnson and Walters 1991) that adds each new year's survey and catch data as these data become available. Note that the regression procedure makes no assumptions whatever about the pattern of change in the recruitment rates $N_{1,c}$ over time. Random changes, trends, cycles, or whatever pattern may occur in the recruitment rates, without affecting the accuracy of the estimation procedure. That is, the accuracy of the procedure does not depend in any way on the absolute or comparative values taken by the $N_{1,c}$ parameters. Thus age structured depletion methods are potentially a very valuable tool for measuring recruitment changes and patterns.

Sources of error and bias

Since age-structured data involve essentially a whole series of depletion experiments, with each cohort providing some information about q (and cross-comparisons of apparent q values among cohorts present at the same time), the q parameter is likely to be much more accurately estimated than in aggregate depletion experiments. Improved accuracy in q leads to more accurate and less biased estimates of the $N_{1,c}$ recruitment parameters as well, at least for cohorts that are well represented (over several ages) in the data table. Monte Carlo simulation tests (Johnson and Walters 1991) indicate that random variation in q from year to year (due to changes in the effectiveness of the survey gear) will also not bias the estimates and will lead to serious inaccuracy only for cohorts that are represented in the data table by only a few age-year catches (e.g., youngest ages in most recent years, oldest ages in earliest years).

There are two serious sources of bias to be concerned about with the age-structured estimates. Both of these concern the index catchability coefficient q. First, it may be impossible to standardize the survey or commercial CPUE index so that q is constant over time. In particular, increases in q over time as a stock is depleted will make the $y_{a,t}$ remain high, and hence will cause overestimates of the most recent recruitment rates $N_{1,c}$.

Second, age selectivity in the survey gear can cause severe biases in the estimates of q and $N_{1,c}$. In particular, incomplete recruitment of younger ages to the survey gear can cause gross underestimates of q, and corresponding overestimates of $N_{1,c}$. For example, if the first and second ages have survey catchabilities 80% and 90% as high as older ages, then for a 10-age, 10-year data base with 1,000 fish aged each year the estimate of q will be 25% to 40% too low, and the recruitment rates (and total stock size) will be overestimated by 20% to 40%. If the youngest age has a catchability less than 50% as high as for older ages, it is not unlikely that the estimate of q will be negative. In such cases, it is necessary to either analyze fewer (older) ages or to use a much more complex catch-at-age analysis procedure that allows estimation of age-specific catchabilities.

Temporal and age-related changes in q can sometimes be detected by examining the age-time table of residuals from the regression predictions of $y_{a,t}$. In particular, individual years of high or low catchability should be clearly visible, although longer term trends will be absorbed (as bias) in the estimates of recruitment rates. Age-related catchability changes will usually be apparent through residuals being all of one sign over time, particularly for younger ages.

12.6. Planning For the Use of Depletion Experiments

The most critical requirement for depletion estimators is to have a relative abundance index y_t or $y_{a,t}$ that is proportional to the stock size present at any time during the depletion. Misleading changes in the index parameter q are likely if the index is constructed only from commercial or sport catch and effort data. It is especially likely that the stock size will be underestimated if the same gear is used to both deplete the stock and measure relative abundance; in this case, y_t will drop more rapidly than it should as more vulnerable fish are removed from the population. Another serious concern in longer term data sets or experiments is change in q due to technology, fisherman learning, and shifts in the distribution of both fish and fishing effort.

The only way to be sure of avoiding various bias problems with the abundance index data is to design survey procedures that measure relative abundance independently of the catch removal process, and are carried out in a consistent pattern over time. This same admonition applies to estimation procedures based on time-series data in general, for example, in the analysis of surplus production patterns using simple or delay-difference models for biomass dynamics. It is a major challenge for the fishery scientist to design an indexing procedure such that $y_t = qN_t$ over a wide range of N_t. The procedure will likely have to deal with substantial changes in the distribution and behavior of the fish as N_t changes. For example, density per survey area may stay constant as N_t declines, while the total area occupied by the stock contracts. Fish may learn to avoid not only commercial, but also survey, gears. Depth distributions and diurnal behavior patterns may change as the fish seek to avoid fishing activity or as fish with particular preferences and patterns are depleted more rapidly. It may be prohibitively expensive to conduct a survey that is sure to detect and account for all of these changes, so that a number of judgements will have to be made about how to limit the survey to a manageable size and perhaps how to extend it through selective use of commercial catch and effort statistics.

Depletion estimation procedures can of course be used just for modest space-time experiments on local research areas, where various economical procedures can be used for abundance indexing. However, it is wasteful during the early development of a fishery to restrict the indexing only to local experimental situations, without taking advantage of the depletion process caused by the overall fishery. Large-scale overall abundance indexing during fishery development provides a unique opportunity to collect information about the impact of fishing, and no amount of intensive and expensive sampling later in the development can replace this information.

The deployment of a consistent, survey-based procedure to provide a relative abundance index over time should be a top priority during the early development of any fishery that appears likely to be of significant economic importance. Within a few years (4–5) after the survey is initiated, and provided the fishery develops rapidly, the index data (along with total catches) will permit relatively accurate assessments of the initial and current stock sizes. These stock size estimates will then be of substantial value in estimating the potential for further development and in avoiding overcapitalization.

Where it would be uneconomical or impractical to set up an indexing procedure that is completely independent of the fishery, it may still be possible to obtain some systematic survey information from voluntary or required survey work by the fishermen (Walters and Collie 1988). This essentially means convincing (or forcing) the fishermen to devote some percentage of their fishing activity to sampling fish densities in a systematic or representative pattern over space and time. Their basic reward for this activity early in fishery development would be an insurance program against grossly overestimating stock size and thereby permitting too many competitors to enter the industry (overcapitalization). Other incentives for fishermen to provide survey information, such as direct payment or payment in increased catches, should also be considered. For a further discussion of such cooperative management activities, see Chapter 18.

12.7. Summary and Critique

When it is possible to fish a population intensively over a short period of time, depletion methods may provide a powerful technique for assessing initial abundance. When the fishing down extends over a longer period, the assumption that the population is closed may need to be relaxed and depletion methods that incorporate recruitment may be used. As the assumptions are relaxed, the depletion methods start to look more like biomass dynamic models than depletion models. If one were to extend the depletion models so that recruitment were represented as a function of stock size, you would actually have a biomass dynamic model. If the perturbation is informative, the *open-population* depletion models may prove very powerful.

The major impediment to depletion models is the need for an index of abundance that is truly proportional to stock size; this is the same problem that plagues most fisheries analysis. A second problem can be spatial structure; if the fishery extends spatially as the depletion continues, the index of abundance may not drop as quickly as the population. It may be that there is a sequence of populations that are being sequentially depleted. Bockstoce and Botkin (1983) present data from the western arctic bowhead whale fishery and analyze the data using a depletion estimator. However, they graphically show that the fishery began south of the Bering Strait and gradually extended north of the Strait, apparently sequentially depleting a series of areas. This could seriously bias an estimator based on the assumption of a homogeneous population.

Depletion estimators are similar to many fisheries statistical methods; although theoretically elegant, the assumptions are rarely perfectly met. The ability to estimate parameters from these methods depends primarily on how informative the data series has been, and, if the data series contain strong enough contrasts in abundance, then the methods may be very powerful and robust. However, one should use depletion estimators only after Monte Carlo analysis of the robustness to violation of assumptions for the particular data set under analysis.

Bibliographic Notes

According to a review by Ricker (1975), the first depletion methods were used by Helland (1913–1914) on harvest statistics on bears, and Hjort et al. (1933) on whale stocks. Leslie and Davis (1939) used depletion methods on rat populations, and Smith (1940) examined a starfish removal experiment. DeLury (1947, 1951) presented his methods, which became widely known in the fisheries community.

Allen (1966), Chapman (1974), and Sainsbury (1984) all extended the closed population model in various ways to allow for open populations. Cowx (1983) provides a recent review of depletion methods.

13

Analysis of Body Size and Growth Data

13.1. Introduction

The analysis of body size and growth information has been a major preoccupation of fisheries scientists throughout most of this century. Growth in body size is obviously a major component of biological production, and the problem of "growth overfishing" (taking too many fish when they are too small) cannot be ignored in many fisheries. Beyond the valid concern with growth as a component of production, large volumes of size information are often gathered simply because they are cheap and easy to get, and might contain useful information about longevity, mortality, and variation in recruitment rates (e.g., presence of strong modes in the length frequency distribution).

In this chapter, we review a few of the key models and methods used in analysis of body-size information for growth and mortality rate estimation. We begin the discussion with some strong words of warning about size frequency sampling, since some of the methods (especially for mortality estimation) require a standard of representative sampling that cannot be achieved with many stocks except at great expense. Then we look at growth models, which are a central part of population models for stocks where growth is a significant component of production. We briefly review methods for extracting growth information from length frequency samples, for stocks where tagging studies and direct methods of age determination are not available. Finally we review methods for reconstructing population age structure from size structure, and for estimation of mortality rates directly from size information.

13.2. Difficulties in Gathering Representative Samples

Often the stock assessment specialists are presented with size information gathered from fishery or survey sampling, and to make use of it they must pretend that the information is somehow representative of the size distribution of the stock. Usually the size information will show various modes that might be indicative of individual year classes or pulses of recruitment. Usually there are fewer large fish than small ones, which might be indicative of mortality processes. Usually there is at least some indication of decreasing growth rate with increasing body size (large mode in the size distribution where several ages are apparently clustered, decreases in length increments between successive modes of size distribution, etc.).

It is irresistible to use these patterns to try to quantify key dynamic parameters. However, before embarking on such an exercise, it is important to look very carefully at exactly how the data were gathered. Most size information, especially from commercial catch samples and demersal fish species, will be grossly misleading if treated as representative of the population size/growth structure. The gathering of representative information on population size structure is in fact extremely difficult and costly for most fish stocks, because of three key problems: gear selectivity, size-related changes in fish distribution, and difficulties in getting representative subsamples from catches. It may sound cheap and easy to gather size information, but do not be misled! There are some really nasty sampling difficulties hiding behind those simple shipboard and port sampling procedures. If these difficulties were honestly admitted and confronted more often, we would likely see less emphasis by fisheries agencies on size sampling, and more on the development of reliable age-determination procedures.

Gear selectivity

For mobile organisms, there is no such thing as a sampling gear that is completely nonselective with respect to body size. Nets are obviously selective in terms of mesh size and nonretention of smaller fish. Gillnets are obviously selective as well with regard to fish whose girth is far larger than the mesh size. Hook-and-line gear is selective with respect to mouth size relative to hook size.

But the most difficult components of gear selectivity to estimate are those related to fish movement and gear avoidance. By virtue of higher swimming speeds, larger fish are more likely to escape from mobile gears (trawls, purse seines), and on the other hand are more likely to encounter fixed gears such as longlines and gill nets. Bigger fish generally have higher cruising speeds when searching for food (which would make their vulnerability to hook and line or gillnet gear larger), but on the other hand are often more wary of attractants such as baits. Behavioral interactions between smaller and larger fish within localized aggregations or schools can also lead to differences in vulnerability; for example larger individuals may better defend territories or preferred feeding stations, making them more or less vulnerable to particular gears depending on how the gears are sited or move relative to the preferred locations. Smaller fish are often pushed into more marginal or dangerous microhabitats (e.g., off the bottom or away from protective cover), making them more vulnerable to open-water fishing gears.

Gear selectivity is generally evaluated by comparing size distributions of catches from different gears fished in the same locations. Then some guesses are made about which gears are most likely to have given representative information about particular sizes of fish, and the relative fishing powers of the other gears are adjusted accordingly. Better information about selectivity can be obtained by doing depletion (removal) experiments, while recording size information so as to estimate differential depletion rates and separate population estimates for various size categories. However, short of draining the pond, there is no generally reliable way to correct for size selectivity.

Size-related changes in distribution

As fish grow, they face a changing suite of opportunities for feeding, threats from predators, and requirements for mating and successful production of offspring. It is not surprising, therefore, that almost all fish show some pattern of ontogenetic migration, at least among habitat types. Even slow-moving species such as abalone show such changes: smaller abalone adopt cryptic behaviors (hide under rocks) presumably to avoid predators, whereas larger individuals emerge to forage for better algal sources on exposed rock surfaces. Many marine fishes have adopted reproductive tactics (spawning locations, buoyancy patterns of eggs and larvae, etc.) that will result in their juveniles being carried into shallower, more protected coastal waters, and these juveniles gradually move offshore into deeper waters (often over several years) to feed and mature. Anadromous and catadromous migrations to exploit spawning opportunities are common and have led to much concern about the protection of rivers and estuaries as fish nursery areas.

Figure 13.1 shows length distributions of two species of tropical snappers at a series of depths. In both cases, the fish tend to be smaller in shallow water and larger in deeper water. It is absurd to expect that any commercial fishery would provide a representative sample of these size distributions.

What these tidbits of fisheries biology mean in terms of size frequency sampling is that every dip of the net or set of the line will take a highly selective sample of fish. It is difficult enough to design a sampling program that will take fish from a representative set of dips, even from a fishery or survey program that thoroughly covers the full ontogenetic distribution of the stock. But most fisheries are not even representative of that distribution in the first place, and their selectivity is likely to change from year to year as the fishery develops. Some fisheries develop by first taking the most accessible (often smaller or inshore) components of the size distribution,

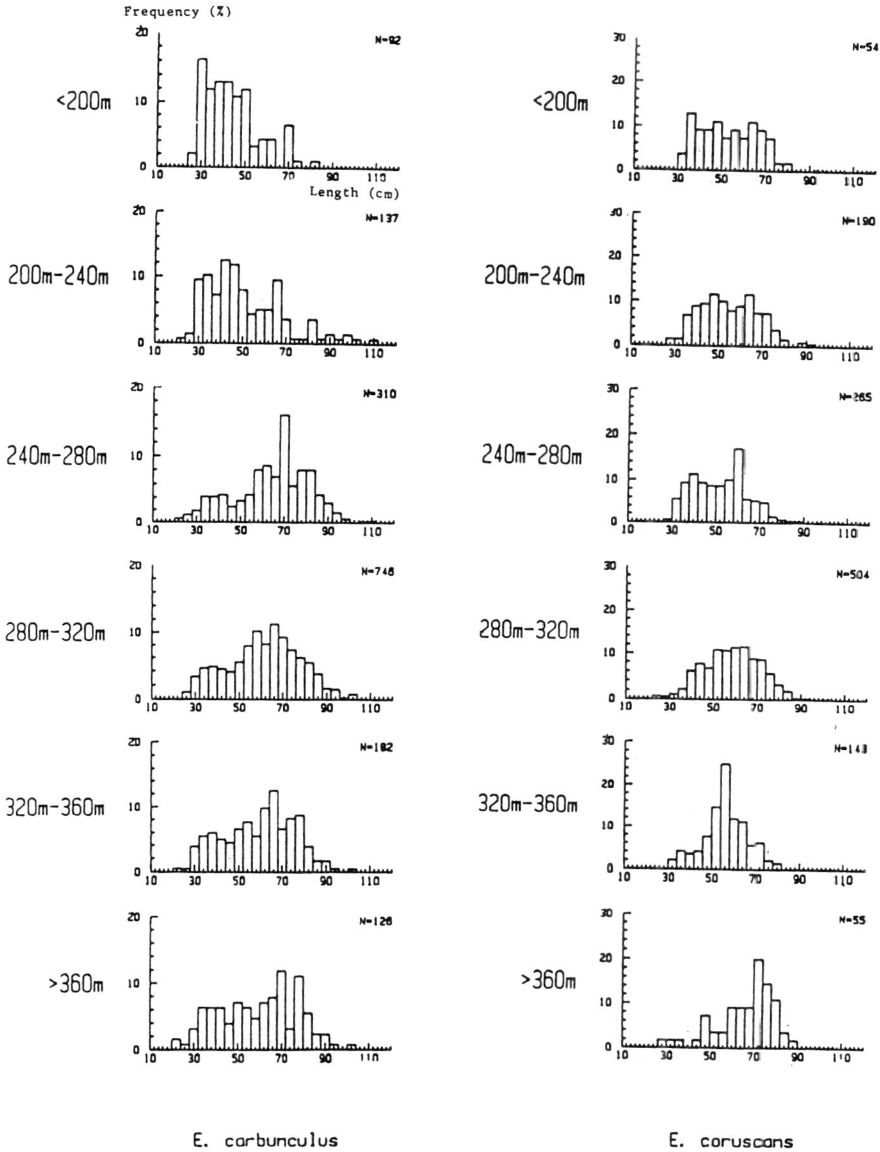

Figure 13.1. Length distributions of *Etilis carbunculus* and *Etilis coruscans* at different depths. From Brouard and Grandperrin (1985).

then shift to other components as the accessible ones are depleted or protected by management measures. This has occurred for example in some artisanal fisheries for small pelagic bait fishes, where juveniles are concentrated in shallows and bays and are easily accessed by artisanal gears. Then later the larger, offshore gears (purse seines) have pursued schools of mature individuals for reduction processing.

Other fisheries focus first on areas where the largest, most valued individuals are concentrated (often on spawning aggregations), and later move (often inshore) to take smaller individuals. This occurred for example with the Pacific salmon fisheries, where river nets and traps for maturing migrant fish have given way (in part) to offshore troll, seine net, and recreational fishing for immature fish. The fishery for southern bluefin tuna (*Thunnus maccoyii*) developed first through Japanese fishing of spawning aggregations (over 8 year-old fish) off Australia, then spread to the Southern Ocean and eventually to the coastal waters of Australia, where Australian fishermen now take substantial numbers of juveniles 1–4 year old (Murphy 1977, Hampton and Majkowski, 1986).

To find out about size selectivity associated with changes in distribution with fish age and maturation, there is no substitute for (1) survey and tagging programs that thoroughly cover the stock distribution, and (2) detailed logbook programs that permit precise spatial location of catches for catch sampling design and interpretation. When ontogenetic movements involve changes in basic habitat type, so that different sampling gears must be used in each type (e.g., from rocky to smooth bottoms), then there may be no effective way of obtaining a broadly representative size sample except by estimating absolute abundances separately in the various habitat types (e.g., by depletion experiments).

Subsampling catches

Most fish are slippery and slimy, and, even when they are not, there is a nasty tendency for the smaller individuals to be shuffled toward the bottoms of holds, tanks, and sample buckets. It is no mean feat to define a field sampling regime, and to train field workers properly, to obtain representative samples of the size distributions of catches.

Particularly for schooling fishes, it is not unusual to see holds filled with fish of relatively uniform size that were taken in one or a few net sets. This means that the catch sampling program has to be careful not only with how the catches from each vessel are sampled, but also with the selection of vessels from which to take samples. The fact that this care is not always taken is evidenced in some fishery data sets by the sudden appearance and disappearance of size frequency modes, rather than the movement of modes through the size distribution over time.

In seasonal fisheries, especially for spawning fish (e.g., salmon and herring), there can be substantial day-to-day changes in the size and age composition of catches. These changes are associated with size-related differences in spawning times and/or migration speeds and, in some cases, with seasonal changes in fishing tactics (mesh sizes, gear set locations, bait types, etc.). These comments lead to an admonition for the stock assessment specialist: before you use size information for assessment, try to spend some time in the field and have a close look at how the sampling program is conducted. Your advice on how to improve that program may be far more valuable than the bad estimates you might otherwise provide of growth and mortality parameters, based on hopelessly biased historical samples.

13.3. Growth Models

Prediction of the growth component of fish production generally involves specifying some relationship between body weight and age, that is, a growth model. The simplest and often safest such model is just a table of average body weights over time (by years, or by shorter intervals for seasonal fisheries where seasonal growth rate changes may be important), obtained from analysis of size and age data with correction if possible for size selectivity in the sampling gear(s) used to obtain the data.

Analysis of length data should always begin with an examination of the length versus age relationship as shown in Figure 13.2. Shown here are the age as estimated by counts of daily otolith rings and fish length. The first panel shows growth data for millet-seed butterflyfish, a small, rapidly growing short-lived fish. These data show a reasonably consistent pattern of length versus age, but where there are numerous samples of fish of similar age, as between roughly 125 and 175 days, note that the length may range between roughly 35 and 55 mm. This is better illustrated by the second panel, for the tropical snapper (*Etilis carbunculus*), where individuals of roughly 2,000 days of age are found between 30 and 60 cm.

There is a common tendency to plot only mean length at age and to not plot the actual data. Indeed, neither of the two most common fisheries reference books, Ricker (1975) and Gulland (1983) show any data for individual fish length versus age: they only show average length versus age. If more people would actually plot length (or weight) versus age, there would be a lot fewer people willing to use length as a surrogate for age.

For some assessment procedures it may be necessary to fit the tabled weight data to some particular growth model equation. Much analysis of length data falls into the same trap we discussed regarding stock and recruitment; people become obsessed with the length-age curve rather than the length-age relationship. The models discussed below are simply convenient curves that attempt to represent the average length at age, and you simply must remem-

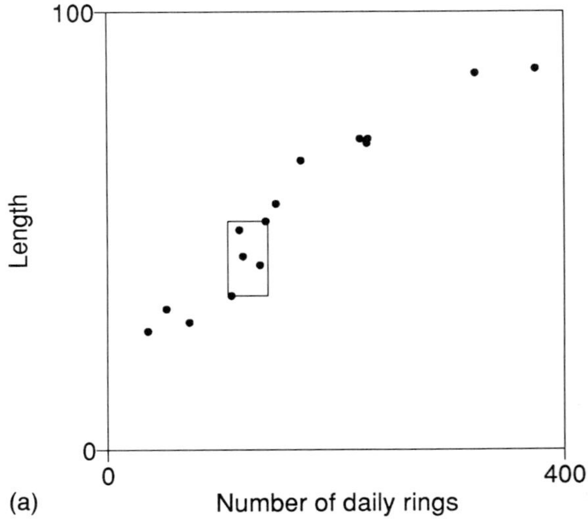

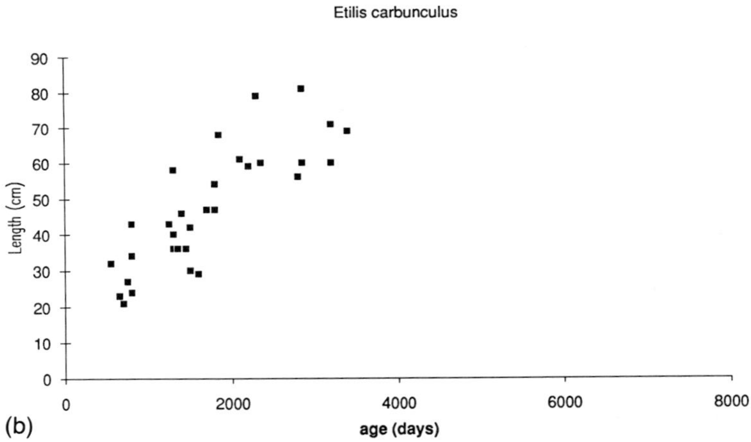

Figure 13.2. Individual fish length versus estimated age from counts of daily rings in otoliths. (a) Data from Table 1 of Ralston (1976) for millet-seed butterflyfish (*Chaetodon miliaris*). (b) From Brouard and Grandperrin (1985) for a snapper (*Etilis carbunculus*).

ber that every fish of the same age is not the same size. Where weight sampling is difficult or impractical, a common procedure is to estimate relationships between length versus age and weight versus length, then combine these relationships to predict weight versus age.

For length-based virtual population analysis (see Section 13.8), it is necessary to provide an estimate of the average time spent in each sampled length interval. To obtain this time, we essentially invert the growth curve so as to predict the change in age from the change in length. But an important point is that in doing this inverse calculation, the appropriate growth curve is not the curve defining mean length as a function of age, but rather the curve defining mean age as a function of length (Jones 1984). The curve of mean age versus length is also a growth curve, but one that does not necessarily reach a clear asymptote when the curve is plotted with age on the x axis, because those fish remaining alive to be sampled at large sizes will not in fact have correspondingly high ages. Instead they will be the fish that have grown more rapidly than predicted by the mean length versus age curve.

Growth in length and length-weight relationships

The most common model used to describe growth in fish length was developed by von Bertalanffy (1938):

$$l_t = l_\infty[1 - e^{-K(t-t_0)}] \tag{13.3.1}$$

In this three parameter model, l_∞ defines an asymptotic or maximum body size, K is called the *Brody growth coefficient* and defines growth rate toward the maximum, and t_0 shifts the growth curve along the age axis to allow for apparent nonzero body length at age zero. Combining this equation with a power model for the length-weight relationship, $w_t = a(l_t)^b$, and defining the asymptotic maximum weight to be w_∞, we get the von Bertalanffy growth curve for body weight

$$w_t = w_\infty[1 - e^{-K(t-t_0)}]^b \tag{13.3.2}$$

which has the same parameters as for length growth, plus the allometric growth parameter b.

We do not have much to say about length-weight relationships and the allometric growth parameter b. Such relationships are plotted in a multitude of fisheries papers, we think largely because they are one of the few fisheries relationships that look nice statistically. Length-weight analysis is a good thing to have your teenage children do as a way of learning about ideas of correlation and regression, and you might find the results mildly useful in estimating average weight of fish caught from samples of lengths of fish caught. If your teenager is having trouble understanding how to estimate b for you, it may be of some comfort to know that you will not likely go far wrong by just assuming $b = 3$.

There are a variety of methods for estimating the von Bertalanffy parameters l_∞, K, and t_0, although often either l_∞ or t_0 is poorly determined due to

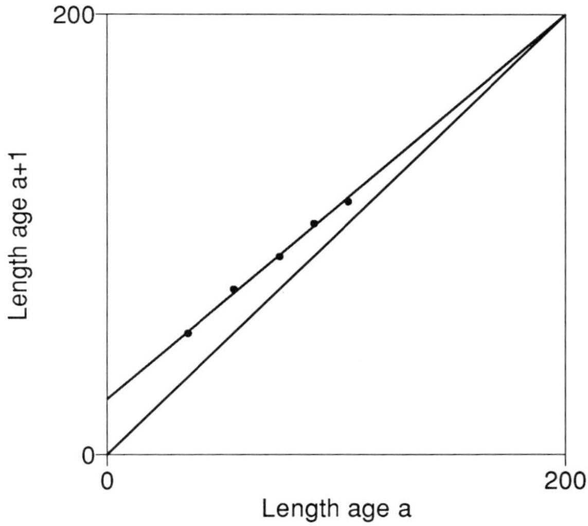

Figure 13.3. A Walford plot for Atlantic yellowfin tuna data (*Thunnus albacares*) from Postel (1955), cited in Pauly (1984). The lower line is $y = x$.

lack of data for very old or very young fish. When t_0 is not needed, the simplest method is the Walford plot (Ricker 1975) of length at age against length at one age younger (Figure 13.3). This plot should be linear if the model is correct, have slope equal to e^{-K}, and have intercept with the $l_{t+1} = l_t$ line equal to l_∞. When t_0 is needed, it can be obtained by an iterative procedure (Ricker 1975, p. 227) using the Walford plot estimates as a starting point.

Nonlinear estimation procedures (Schnute 1981a, b, Ratkowsky 1983) can also be used to estimate the length growth parameters. If you bother to use such procedures, you should take care to define precisely what you want to assume about variability in individual growth rates around the average curve (Sainsbury 1980, Francis 1988). Generally the best assumption will be that individual variation leads to a similar coefficient of variation for all ages (i.e., standard deviation of length at age equal to the coefficient times mean length, with the same coefficient for all ages).

Length growth parameters can be estimated from tagging data by using nonlinear estimation procedures with the transformation (Fabens 1965)

$$C_i = (l_\infty - l_i)(1 - e^{-Kt_i}) + e_i \qquad (13.3.3)$$

where C_i is the change in weight from tagging to recovery of the ith fish recovered, l_i is its length at tagging, t_i is its time between tagging and recovery, and e_i is its deviation from the mean growth curve. For alternative assumptions about the distribution of e_i, see Sainsbury (1980) and Francis (1988), and, for an example of fitting under the assumption that l_∞ varies among individuals, see Kirkwood and Somers (1984).

Models for growth in weight

A variety of equations have been used to describe weight growth patterns in fish. The most common are the von Bertalanffy described above, the Richards equation, the Gompertz equation, and the logistic equation (Ricker 1979). However, Schnute (1981a) has shown that all these models are special cases of a single general model, where the growth rate dw/dt is the product of w times a specific growth rate (rate per weight) z, where z varies over time according to the differential equation

$$\frac{dz}{dt} = -z(a + bz) \tag{13.3.4}$$

Specific models are obtained by making further assumptions about a and b. For example, $a > 0$, $b > 0$ results in the von Bertalanffy model, whereas $a > 0$, $b = 0$ results in the Gompertz model.

Schnute (1981a) not only provides a way of comparing alternative growth equations (by fitting his general model with various restrictions), but also shows that his parameterization avoids various nonlinear estimation difficulties (convergence, wandering of parameter values to wild extremes, etc.) that are regularly encountered with growth curve fitting. He notes that the general model involves estimating four parameters: a, b, and mean body weights at any two arbitrarily chosen ages (for example, age zero and age 1). We recommend using his nonlinear estimation package (Schnute 1981b, Mittertreiner and Schnute 1985) for doing the fitting and comparing the various special case models.

13.4. Extracting Growth Information from Length Frequencies

For almost a century, fisheries scientists have been trying to use length frequency information to study growth, beginning perhaps with the work of Petersen (1892). The basic idea is that length frequencies generally have modes, at least at smaller sizes, that presumably represent single cohorts or recruitment events. Once a set of modes is found, the lengths at these modes are taken to be the mean lengths for the associated cohorts, and these mean lengths can then be used as a tabled growth model or for fitting growth curves. Two approaches can be used to analyze for location of the modes, depending on whether single or multiple length frequency samples are available.

Before proceeding with this section and the one that follows it, the reader should look again at the section above on difficulties with size frequency sampling. In particular, you should be aware of how difficult it is to obtain representative size frequency samples over any range of fish ages and of how localized sampling problems can create apparent size modes or mask any real ones that may be present. In short, be very wary of the methods that we review in the remainder of this chapter.

Analysis of modes in length frequencies

When only a single sample is available, the basic idea is to use various statistical techniques to detect modal positions and to correct for overlaps between the frequency distributions of fish around the modes. The techniques are of two types: (1) simple graphic analyses (Bhattacharya 1967, Cassie 1954), and (2) fitting procedures based on assuming that the size distribution consists of a collection of overlapping frequency curves with unknown mean and variance parameters to be estimated (Macdonald and Pitcher 1979, Fournier and Breen 1983). The fitting procedures work best when an underlying growth curve (mean length versus age) can be assumed *a priori,* so that only the parameters of this curve (rather than every mean length at age) need be estimated from the data.

A key problem with all the single sample methods is that generally only the first few fish ages will produce clear size modes, while the older, slower growing fish will be "piled up" into a single broad mode. This is illustrated in Figures 13.4 and 13.5, which show size distributions of coral trout (*Plectropomus leopardus*) from Queensland and abalone (*Haliotis rubra*) from Tasmania. Some information about growth and mortality can be extracted from the broad mode for older ages if it can be safely assumed that (1) all ages and sizes of fish were equally vulnerable to the size sampling gear and (2) all older fish had the same total mortality rate (see Fournier and Breen 1983, for further details.)

Analysis of repeated length frequency samples

Better information about growth patterns can often be obtained when a set of length frequency samples over time is available. In this case, samples can be cross compared to correct for spurious modes created by sampling problems, and strong modes can be followed through time to provide direct reconstructions of historical growth patterns.

The most widely used methodology for analysis of growth patterns from length frequency samples is the ELEFAN program developed by Daniel Pauly.

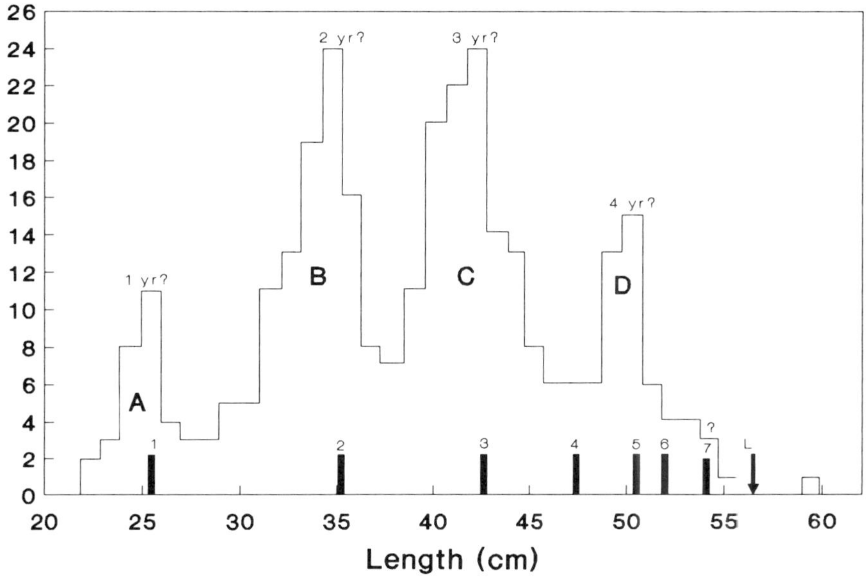

Figure 13.4. Length frequency of coral trout (*Plectropomus leopardus*). Redrawn from Goeden (1978).

For reviews of this methodology and its applications, see Pauly and Morgan (1987). ELEFAN provides objective methods for detecting size frequency modes and also allows the user to set modal sizes based on subjective analysis of the data. Connections between modes over repeated samples can be used to provide very useful reconstructions of growth histories for different cohorts.

One of the best examples of using multiple size frequency samples is the analysis of Peruvian anchoveta length frequencies by Pauly et al. (1987) (Figure 13.6). Monthly samples of length frequency from purse seine catches were collected. These monthly samples show very clear cohorts, and Pauly et al. estimated growth rates by following the changing distributions over time. They were also able to estimate recruitment by year from the data. The anchoveta data are a particularly good example because they illustrate both the strength and weakness of length frequency data. The facts that recruitment occurs over a short period, and that growth is rapid assure that the annual cohorts are easily separated; compare this to the coral trout graph shown earlier. In the anchoveta case, one hardly needs a sophisticated computer algorithm to tell you how old an individual fish is; it is obvious by looking at the graph. Second, the fitted growth curve shown in this figure essentially follows the modes, and allows for seasonal changes in growth rate, which are common to many if not most fish. However, as we continue to caution, the observed size distribution is a product of gear selectivity and

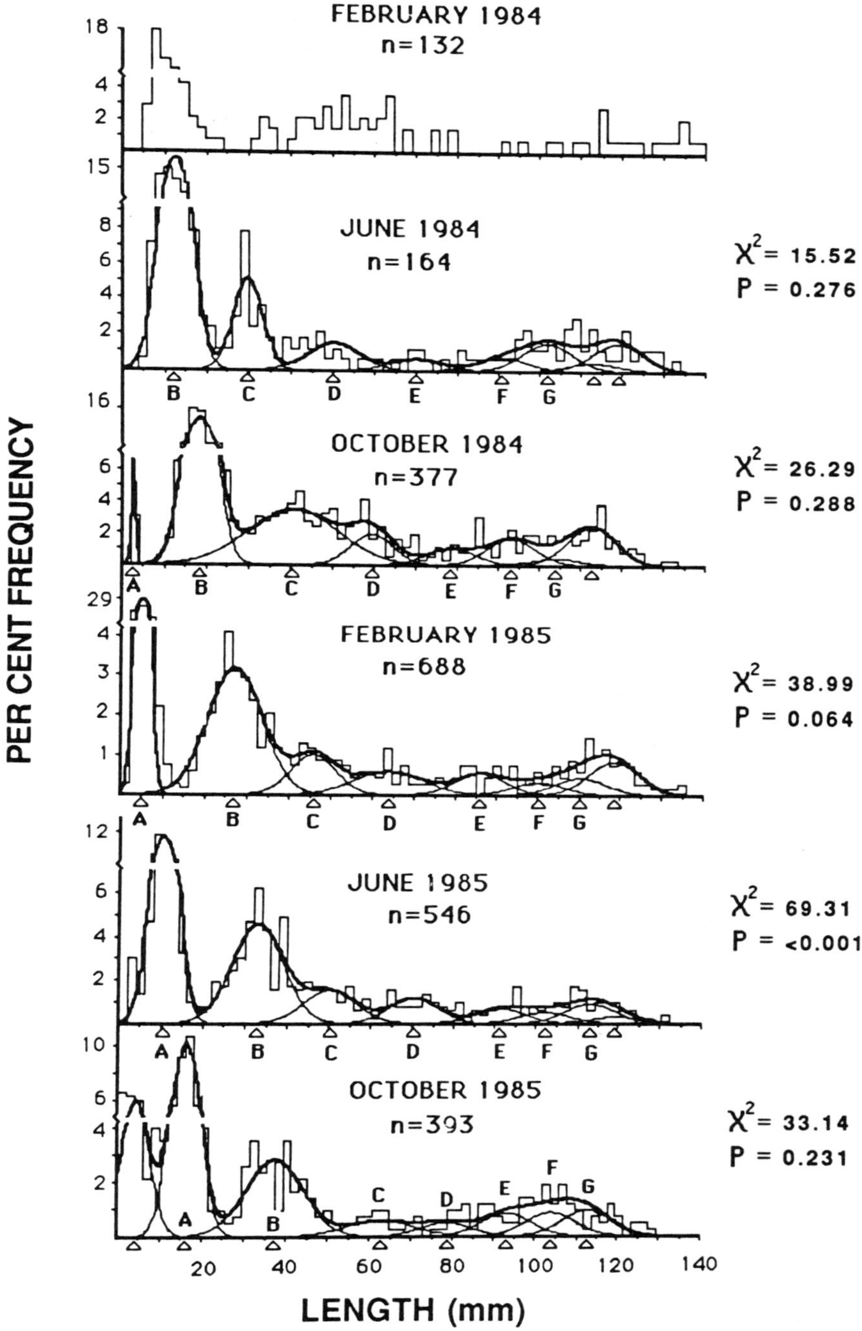

Figure 13.5. Size distribution of abalone in Tasmania. Reproduced with permission from Prince (1989).

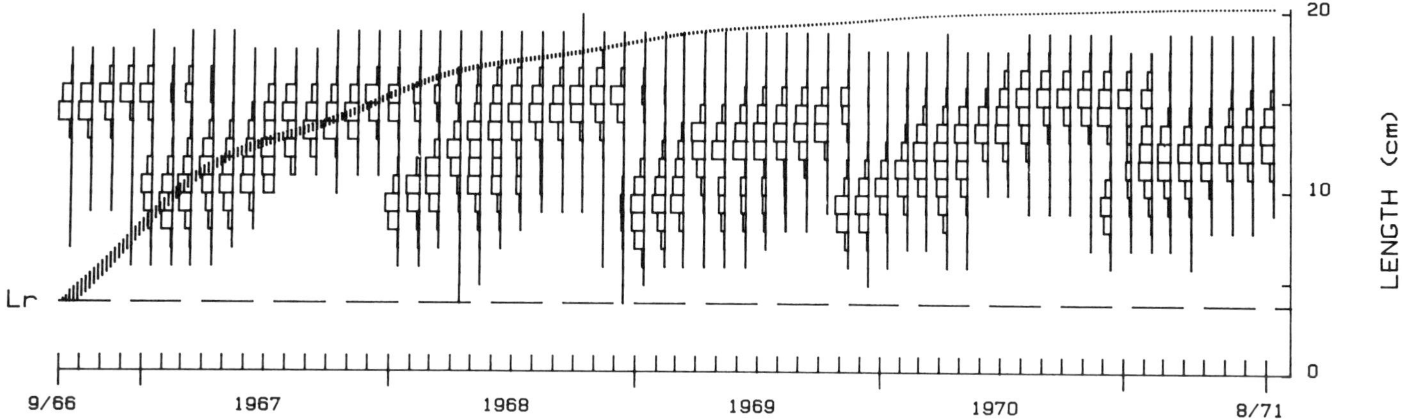

Figure 13.6. Size distribution of catch of the Peruvian anchoveta and fit growth curve. Reproduced with permission from Pauly et al. (1987, p. 146).

growth, and the best "fit" growth curve to size frequency data will only be the actual growth curve if the gear is totally unselective.

Besides ELEFAN, two major microcomputer programs have recently become available for extraction of growth and abundance information from repeated length frequency samples. These are the MULTIFAN system of Fournier et al. (1990) and the CASA program of Sullivan et al. (1990). Both of these programs work by assuming a general dynamic model for how the length distributions are generated as a function of recruitment, survival, and growth parameters (including variability in growth rates among individual fish), then using nonlinear estimation methods to vary the model parameters so as to fit the observed distributions. This is an ambitious undertaking in both computation and statistical terms, and it is not clear as yet just how good the programs will be — particularly at recovering information on abundance and recruitment variation.

13.5. Mortality Information from Length Samples

Direct estimates of mortality from mean length

Provided recruitment rate is essentially independent of stock size, we expect the average length of fish in a population to decrease as the total mortality rate (fishing plus natural) increases. A number of models for the equilibrium relationship between mean length and mortality rate have been proposed, based on different assumptions about growth variability.

The simplest model, and one that is good at recovering mortality rates from simulated data even when its assumptions are rather badly violated, was proposed by Beverton and Holt (1956). They showed that when (1) the stock is at equilibrium (constant recruitment, mortality), (2) all fish above size l_{crit} are equally vulnerable to fishing, (3) growth follows the von Bertalanffy model, and (4) all fish have the same total instantaneous mortality rate Z, then Z is related to mean length $\bar{l}$ and to the von Bertalanffy parameters by

$$Z = \frac{K[l_\infty - \bar{l}]}{[\bar{l} - l_{\text{crit}}]} \qquad (13.5.1)$$

Several elaborations on this equation have been developed; for reviews see Hoenig et al. (1983) and Pauly and Morgan (1987).

The Beverton-Holt equation above and its relatives should be used only as a last resort, when just a single size frequency sample is available or when a rough check is needed for other procedures that use multiple samples. The equation depends critically on the stock being at equilibrium, on representative sampling of all fish sizes above l_{crit}, and on having a good independent estimate of l_∞. Note that the total mortality rate estimate is directly proportional to l_∞ above. We hope we have convinced readers that few populations will be at equilibrium, and that representative length frequency sampling is difficult — in other words only the most naive biologist will use this method with any confidence.

13.6. Predicting Dynamics of Mean Length with Delay-Difference Models

The machinery of delay-difference models (Chapter 9) can be used to gain some understanding of how violations of the equilibrium assumption will affect estimates of total mortality rate based on average body size (length or weight). Recall from Chapter 9 that when (1) all fish aged k or older have the same total survival rate s, and (2) body size w_a at age a (w_a may be either weight or length) follows the Brody equation $w_a = \alpha + \rho w_{a-1}$ where α and ρ are the intercept and slope of the Walford plot of w_{a+1} versus w_a, then the dynamics of population biomass (or total length) B_t and numbers N_t are given by the two equations:

$$B_t = s\alpha N_{t-1} + s\rho B_{t-1} + R_t w_k \qquad (13.6.1)$$
$$N_t = sN_{t-1} + R_t$$

The average body weight (or length) is given from these equations by $\bar{w}_t = B_t/N_t$.

At equilibrium, R_t will be constant at $(1 - s)N^*$, where N^* is the equilibrium number of animals in the population. The biomass (or total length) at equilibrium, B^*, will then be defined by

$$B^* = s\alpha N^* + s\rho B^* + (1 - s)N^* w_k \qquad (13.6.2)$$

and the average body weight $\bar{w}$ will be given by $\bar{w} = B^*/N^*$. Dividing the equation for B^* by N^*, and substituting $\bar{w}$ for the B^*/N^* ratio, we get an equilibrium relationship between survival s and $\bar{w}$:

$$\bar{w} = s\alpha + s\rho\bar{w} + (1 - s)w_k \qquad (13.6.3)$$

Solving this equation for s while noting that $\alpha = (1 - \rho)w_\infty$, where w_∞ is the asymptotic maximum body weight (or length), we get an equation similar to the Beverton-Holt equation:

$$s = \frac{\bar{w} - w_k}{\bar{w} - w_k + (1 - \rho)(w_\infty - \bar{w})} \qquad (13.6.4)$$

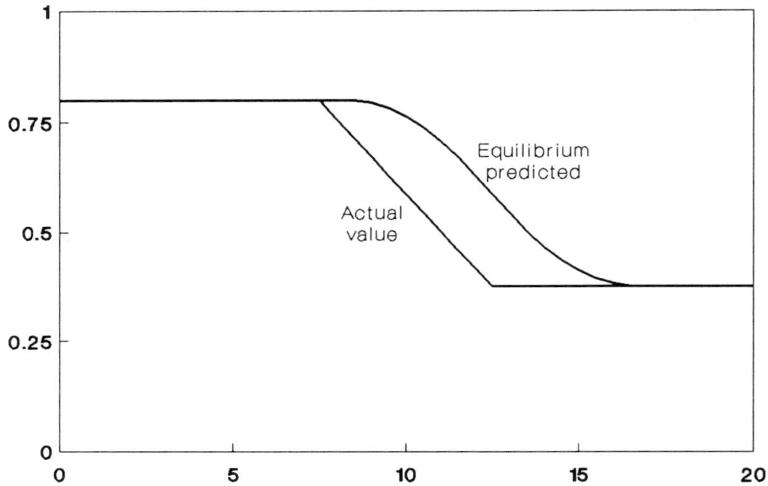

Figure 13.7. Actual and estimated survival rates from a simulation.

This equation makes intuitive sense: s can be large only if $\bar{w}$ is substantially larger than w_k, the weight at recruitment; if $\bar{w}$ is observed to be near w_k, the survival rate must be very low.

Figure 13.7 compares this equilibrium prediction of s to the actual value of s over time for a simulated population where s decreases over 5 years due to fishery development, then stabilizes at a lower value. For the simulation, we took $R_t = 1.0$ for all t (constant recruitment), $\rho = 0.8$, $w_\infty = 1.0$, and $w_k = 0.3$. Initially, s was set to 0.8, then decreased over the fishing development to $s = 0.4$.

Note in Figure 13.7 that the equilibrium estimate of s does track the dynamic change in s during the development period, but with a substantial lag (roughly 5 years); thus using the equilibrium calculation to analyze data from a developing fishery would give a misleading picture of how much the survival rate has decreased as of any point in the development (i.e., effect of fishing on survival would be underestimated until the decline was almost over).

13.7. Conversion of Length to Age Frequencies

Rather than using simple equations relating mean length to total mortality rate, the more common attack is to analyze the length frequency information in two steps: (1) use an age-length key to estimate age frequencies from the length frequencies, then (2) analyze the resulting age structure estimates using the variety of techniques available for age-structure data (see Chapter 11). The key difficulty is the first of these steps, where you attempt to reconstruct what the age structure must have been from a length-frequency sample that will generally involve overlapping length distributions for at least the older ages, that is, each age will be smeared over several length categories and each length category will contain fish of several ages.

Deterministic age-length keys

The simplest approach to age-length keys is to ignore growth rate variation and overlapping length distributions, and instead to associate each length category with a relative biological age computed from a model for the mean length at age. For example, the von Bertalanffy growth model can be solved for age t given length l, as

$$t = t_0 - (1/K)\log[1 - l/l_\infty] \qquad (13.7.1)$$

The term $log[1 - l/l_\infty]$ is sometimes referred to as the *relative age*; note that the relative age is not defined for fish larger than l_∞. The full computation of age t from t_0, K, and relative age is referred to as the age-length key.

The simplest way to use relative age is in the analysis of a single length frequency sample. If you plot log (number at length) against the age index $\log[1 - l/l_\infty]$ for a representative sample of lengths from a population at equilibrium, the result will be a "catch curve" with slope $1 - Z/K$ (Pauly 1984). In Monte Carlo tests where simulated data are generated with growth rate variation of 10% to 20% around the von Bertalanffy curve, we have found that this method actually works quite well provided that the equilibrium and nonselective sampling assumptions are not violated, and l_∞ is known (only the data for lengths less than l_∞ are used in the regression).

More elaborate age-length keys can be constructed by solving the above equation for ages t at the lower and upper bounds of each length-frequency interval, and using the results to allocate all fish from the length sample to integer age categories. Fish from length intervals whose corresponding age limits span two integer age categories are assigned to the integer ages in proportion to how near the integer age is to one or the other computed age limit. Also, it is possible to correct at least partly for stochastic variation in growth rates. Additional properties and elaborations of age-length keys are discussed in Bartoo and Parker (1983) and Hoenig and Heisey (1987).

An interesting approach to cohort analysis for a single length frequency sample can be developed by assuming that relative ages computed from a mean growth curve are a good indication of how long it takes for fish to grow through each length interval in the length distribution (Jones 1984).

Suppose the length data are summarized as numbers of fish falling in length intervals of width Δ_l. If l is the length at the start of one of these intervals, and $l_{\text{end}} = l + \Delta_l$ is length at the end of the interval, then it takes the average fish some time t_l to grow through the interval. This average time is predicted by the von Bertalanffy growth curve to be

$$t_l = \frac{1}{K}[\log(1 - l/l_\infty) - \log(1 - l_{\text{end}}/l_\infty)] \qquad (13.7.2)$$

t_l is the difference in average age between fish of length l versus fish of length l_{end}; presumably t_l increases with increasing l, as the growth curve decelerates. Now, if the instantaneous natural mortality rate M per time is constant, then the expected natural survival rate of fish that pass through the length interval l to l_{end} is just $S_l = e^{-Mt_l}$; note that S_l will decrease for increasing l, since t_l will increase. Provided t_l is reasonably short (Δ_l is small compared to the annual growth increment), then the number of fish $N_{l\text{end}}$ from a typical cohort that survive to length l_{end} can be predicted from the number N_l that reach length l, the catch of fish C_l of lengths l to l_{end}, and S_l:

$$N_{l_{\text{end}}} = [N_l - C_l]S_l \qquad (13.7.3)$$

Solving this equation for N_l leads to the backward recursion

$$N_l = \frac{N_{l_{\text{end}}}}{S_l} + C_l \qquad (13.7.4)$$

which can be used to estimate numbers alive at the start of each length interval given a terminal number N_L reaching some maximum size L. The simplest procedure would be to take $N_L = 0$. In this procedure, it is not clear how to deal with fish having lengths l greater than the l_∞ from the growth curve; one option is to use an artificially high l_∞ (e.g., slightly larger than the largest fish in the sample), which is conservative in that it will result in an overestimate of growth rate and hence an overestimate of the survival rates S_l, leading to an underestimate in N_l.

Use of a size distribution matrix

Instead of converting from length to age using a growth equation to provide the age-length key, a more complicated procedure is to use growth information to estimate the size distribution matrix $\mathbf{P}$, where P_{ij} is the proportion of fish of age j that are expected to be in length interval i. Each column of $\mathbf{P}$ must sum to 1.0, that is, all the fish of age j must fall in some size interval i. All ages with identical columns (identical size distributions) should be treated as a single aggregate age class, since the length data will contain no information upon which to disaggregate them. Differences among columns represent growth, but these differences need not follow any particular growth equation.

Using the size distribution matrix **P**, the total number of fish n_i that are in length interval i is given by

$$n_i = \sum_j N_j P_{ij} \tag{13.7.5}$$

where N_j is the (unknown) number of fish of age j in the sample. In matrix form, the above equation can be written as

$$\mathbf{n} = \mathbf{PN} \tag{13.7.6}$$

An estimate of the age structure vector **N** is then provided by $\mathbf{N} = \mathbf{P^*n}$ where $\mathbf{P^*}$ is a generalized inverse of **P** (generally, **P** will not be square, and, if square, will often not have an inverse due to higher ages having nearly the same P_{ij} values). One might view the matrix $\mathbf{P^*}$ as a key for going from length to age, but note that it will generally assign fish from a range of size categories n_i to each age category N_j.

The difficulty in using this approach is to estimate the **P** matrix, by using tagging data and/or information about the distribution of lengths around an average growth curve. Each column of the **P** matrix must be constructed to define the proportions of fish of the column age that are expected to be in the length intervals represented by the rows of the matrix. If the fish of age j are assumed to be normally distributed in length around some mean for the age, then each P_{ij} will represent the area under the normal density curve between the two limits of length that are used to define the ith size frequency category. However, you need not assume normality in constructing the **P** matrix; for example, the elements (proportions) in any column may be chosen to reflect presumed cumulative effects of size-selective mortality on one part or another of the size distribution.

When dealing with fish that have highly variable growth rates, so that each length frequency interval is likely to contain fish of a substantial range of ages, it is safer to estimate numbers at age from even a crude **P** matrix than to assume a deterministic relationship (key) between biological age and length. Otherwise, you take the risk of trying to extract too much information from the size distribution, essentially by confusing variability in length at age with declining numbers (mortality) associated with increasing age alone.

Solution of the matrix equation $\mathbf{n} = \mathbf{PN}$ can be accomplished efficiently using the singular value decomposition (SVD) of **P** (Walters 1986). Numerical recipes for SVD are generally available and result in expressing **P** as **P**

= **USV′**, where **U**, **S**, and **V** are three matrices. **P*** is given from these matrices by **P*** = **VS*U′**. Here S^*_{ij} is a diagonal matrix whose ith diagonal element is equal to $1/S_{ii}$ if the singular value S_{ii} is nonzero, or 0.0 if S_{ii} = 0. Using this generalized inverse **P***, fish will be distributed evenly across all ages N_j for which the columns of **P** are nearly identical (i.e., all ages with the same size distribution).

13.8. Length-Based VPA

Jones (1984) introduced a quite powerful method for estimating recruitment and abundance-at-length by using *length-based cohort analysis*. This method has been further evaluated by Lai and Gallucci (1988). The method is critically dependent on having a length frequency distribution that is either from a population at equilibrium, or represents an average distribution over some period of time for which recruitment and exploitation rates have been stable on average (no significant trends in either).

The method is very simple. Suppose C_l is the annual number of animals caught in length interval l, where the interval consists of lengths l_1 to l_2. The average time required for animals to pass through the interval is given by

$$\Delta t = a_{l_2} - a_{l_1} \tag{13.8.1}$$

where a_l is the mean age for animals at length l (from the inverse of the growth curve obtained by plotting length of animals versus mean age), and assuming that over this time they will be exposed to a known (or assumed) natural mortality rate M. Then the average number of animals entering length class $l + 1$ (N_{l+1}) can be predicted from the number entering length class $l(N_l)$, the catches by length interval (C_l), the natural mortality rate (M), and the time spent in the interval by using the standard exponential equations for continuous fishing and natural mortality. The cohort analysis then proceeds recursively by calculating N_l from N_{l+1}, using Pope's (1972) approximation. Jones (1984) shows that this recursion can be written conveniently as

$$N_l = N_{l+1}(A_l^{M/K}) + C_l[A_l^{M/(2K)}] \tag{13.8.2}$$

where K is the von Bertalanffy growth coefficient and A_l is calculated from the asymptotic length l_∞ and the width of the length classes $\Delta = l_2 - l_1$ as

$$A_l = \frac{l_\infty - l}{l_\infty - (l + \Delta)} \tag{13.8.3}$$

The average time spent in the interval can be calculated from A_l as Δ =

$\log(A_l)/K$. The fishing mortality rate for animals in the interval F_l, is given by

$$F_l = \frac{\log(N_l/N_{l+1})}{\Delta} - M \qquad (13.8.4)$$

Note that N_l as defined above is the average number entering length class l during a typical year, rather than the number present in the class at the start of the year. Jones (1984) presents a clever derivation to show that the average number present "at sea" in the class during the year, $\bar{N}_l$ is given by

$$N_l = \frac{N_l - N_{l+1}}{Z_l} \qquad (13.8.5)$$

where Z_l is the estimated total mortality rate $F_l + M$.

While length-based cohort analysis is an appealing method for trying to estimate abundance and size selectivity of gear from simple size-frequency data, it is important to note that the method depends critically on the assumption of equilibrium. We recently had a nasty experience with this assumption while looking at length-frequency data for the West Australian pilchard (*Sardinops neopilchardus*). In this case, a fishery developed rapidly over the period 1982–1989. Using length-based cohort analysis for data from 1989, we estimated the stock size to be around 100,000 M tons. This was good news because the catch was only just approaching 10,000 M tons. However, the fishermen reported declining sea bird populations, increased search times to find pilchard schools, and seasonal disappearance of the fish — all danger signs of overfishing. It might well be that the stock is in fact 40,000 tons or even less, and that the apparent higher abundance is due to the length-frequency samples still containing high proportions of large fish that would be eliminated if the fishery had time to reach equilibrium (and if recruitment did not collapse).

13.9. Summary and Critique

Length data are probably the most frequently collected information in fisheries simply because they are so easily gathered. As we said earlier, it is irresistible to try to do something with them. Our advice is generally to resist this temptation. We have seen very few examples where length-based analysis leads to useful management advice. Perhaps the most common use is to provide an estimate of growth rate and natural mortality rate, which can be used for a yield-per-recruit analysis. In some circumstances, such an analysis will demonstrate that the fishing mortality rate is excessive and may provide the only technical justification available for reducing fishing pressure.

At its best, length-based analysis can be used for cohort analysis, to reconstruct recruitments and stock sizes and provide all of the retrospective analytic power available from VPA, stock-recruitment analysis, and so on. The analysis by Pauly et al. (1987) of the Peruvian anchoveta is the only convincing example we have seen of such an application. It requires very clear cohorts; they must in fact be so clear that no formal models are needed, so that you can simply look at an individual and tell what age it is. This circumstance will not work for anything that does not have very seasonal reproduction and a very short life span.

We believe that attempts to use length-based analysis to formulate management advice for species that do not exhibit unambiguous modes is misguided and fundamentally hopeless. Attempts to use length-based analysis for yield-per-recruit depend upon the nonselectivity of the fishing gear and the ability to representatively sample the catch. It is a certainty that the yield-per-recruit analysis will be biased because of these problems. Attempts to use lengths to do VPA are even more dangerous. Length-based VPA is a poor imitation of age-based VPA with large and perhaps enormous aging errors.

One should not confuse length-based analysis with the use of length-age keys. Length-age keys can be constructed by aging fish every year to provide a table giving the probability that an individual of length i will be of age j. Normally far more individuals are measured for length than for age, and the length-age keys are essentially a sampling shortcut. The keys must be constructed every year, because the distribution of the elements of the table depends upon the year class strengths as well as the growth relationship.

Analysis of length data is currently a hot topic in fisheries, particularly for tropical stocks. A recent FAO handbook on stock assessment (Sparre 1985) devoted most of its pages to analysis of length data. Much of our strong cautions on the use of these methods stems from their current widespread popularity. In this chapter, we have provided only the briefest overview of the estimation of parameters from length data. A full length book devoted solely to this topic would be required to treat the subject properly, and both Sparre (1985) and Pauly (1984) cover many of the topics in detail.

Bibliographic Notes

The early history of the use of length data and estimation of growth rates are reviewed in Maier (1906) and Ricker (1975). The earliest work (pre-1950s) was primarily associated with relating length or weight to age. The next step was the analysis of modes to estimate growth rates with Cassie (1954) providing a key technique. At roughly the same time, Beverton and Holt (1956) examined the relationship between mean size and exploitation rate. Schnute (1981a, b) provided a more rigorous look at the curve fitting aspects of length-age data. Direct use of lengths for estimation of other parameters such as cohort size, mortality rate, and so on, was stimulated by Jones (1981, 1984) in two manuals.

Major impetus to the intensive use of length frequency data has come with the widespread availability of microcomputers and the development of software for such computers, particularly the ELEFAN package distributed and taught by ICLARM. A similar growth in analysis of such data has occurred in more rigorous statistical analysis of length distributions in the series of papers by Fournier and Doonan (1987), Fournier et al. (1990), Schnute (1987), and Schnute et al. (1989).

The conference proceeding edited by Pauly and Morgan (1987) illustrates many of the approaches being taken in using length-frequency data and the critiques leveled at such analysis.

14 Multispecies Analysis

14.1. Introduction

The tradition in most fisheries science has been to treat each species and stock as a separate entity to be analyzed and managed. We all know that species in ecosystems interact with one another through processes like predation and competition, yet we normally ignore these interactions.

There are several reasons for ignoring species interactions. Most importantly the connections may not matter very much. If we can reliably predict the dynamics of a stock by examining only the stock size and structure, harvests, and perhaps the physical environment, then why bother about other species? Secondly, even if we believe that stock interactions may be important, there may be nothing we can do about it, that is, we may not be able to estimate the parameters of the stock interaction. Finally, even if we can estimate the parameters of the interaction, if we cannot predict the future condition of the other species nor control them, then it may be best to just treat the species interactions as unexplained noise in the system and rely on controlling the species of interest.

These arguments for ignoring species interaction have serious weaknesses. First, we often do know something about how the interaction works, and this knowledge can be used to improve predictions and/or help to design policies to control (usually via harvesting) the other species so as to provide a more productive community structure. Second, even if there is no biological interaction between species, there is often an interaction caused by fishing gear; several species are often caught in the same type of gear and we simply cannot target effort cleanly on a species-by-species basis. Interaction caused by nonselective fishing is called *technological interaction* and may easily dominate the biological interaction between species. In this chapter, we explore the extent to which biological and technological interactions are significant and we can estimate the parameters of the interaction. We begin by examining how the single-species methods discussed in the preceding chapters can be extended to include biological and technological interactions. Then we step back and take a broader view of the problems and survey the relative merits of different approaches.

14.2. Multispecies Stock and Recruitment

Effects of predation and competition on a stock are likely to be most important for its juveniles. In single-species analysis, the net effects of juvenile production and survival are usually represented in terms of the stock-recruitment relationship. Thus a key starting point for multispecies analysis should be to examine whether the stock-recruitment relationship shows evidence of variation due to changes in the abundance of predator or competitor species.

Ricker model

The standard Ricker stock-recruitment model can be extended to include effects of other species by adding additional terms to the exponent of the equation; this exponent represents variation in juvenile survival rate. For a single hypothesized competitor or predator, the extended Ricker model would be written as

$$R = Se^{a-bS-cX} \tag{14.2.1}$$

where R is the recruitment, S is the spawning stock, X is the stock of the second species, and a, b, and c are parameters. The parameters a and b have the same meaning as in the normal Ricker model and c represents the effect of X on R. If $c = b$, then one unit of X has the same effect on survival to recruitment as one unit of S. If $c > b$ then the second species has a stronger effect, and, if $c < b$, then the second species has a weaker effect. If we have regular measurements of R, S, and X over time, we can estimate the parameters a, b, and c using the regression

$$\log\left(\frac{R}{S}\right) = a - bS - cX + w \tag{14.2.2}$$

where w is the residual. Reliable estimates of a, b, and c would require good contrast in S and X, that is periods of high S and low X, periods of high X and low S, and periods of both species being at high and low abundance. Walters et al. (1986) examined the relationship between stock and recruitment in Pacific herring (*Clupea harengus*), as affected by the abundance of Pacific cod (*Gadus macrocephalus*). Figure 14.1 shows their estimates of recruits per spawner (from VPA) for herring plotted against their estimates of cod biomass. Herring survival to recruitment appears to be negatively related to cod abundance. Walters et al. estimated that the instan-

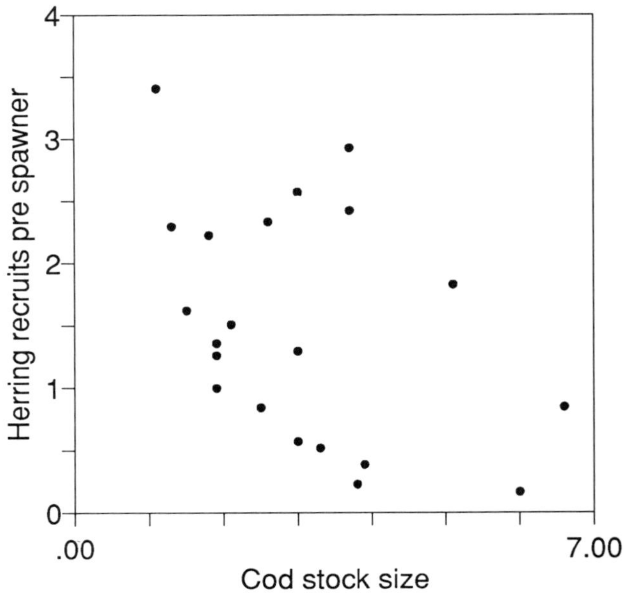

Figure 14.1. Herring recruits per spawner plotted against cod stock size from Walters, Stocker, Tyler and Westrheim (1986). Can. J. Fish. Aquat. Sci. 43: Figure 2.

taneous mortality rate of herring juveniles caused by cod predation was 0.75/year (average value of cX). To produce this rate, each cod would have to consume several hundred juvenile herring per year. Analysis of cod stomach contents is consistent with this level of herring consumption.It is important to have an independent check (such as stomach contents and feeding rate data for predators) on the interaction parameter estimates. Spuriously high estimates can easily be obtained if there is some other factor, such as oceanographic variation, that is causing correlated change in S and X. Also, the mere existence of predation (as evidenced by stomach contents data) does not imply that the predator has an additive effect on juvenile mortality; instead the predators might select juveniles with a higher probability of natural mortality in the first place.

Equation 14.2.1 can be extended to any number of species by the following equation:

$$R = S\, exp\left[a - bS - \sum_i c_i X_i\right] \qquad (14.2.3)$$

where c_i is an interaction coefficient for the ith species and X is the abundance of the ith species. The estimation of all c_i's is theoretically possible but would require contrast between all of the X's and S, which is most unlikely considering the number of observations generally available in fisheries time series.

Other models

It is algebraically quite simple to construct multispecies analogs of Beverton-Holt or other recruitment models, which would simply involve adding additional terms for other species. However, the general lack of data makes this a somewhat uninteresting exercise.

14.3. Multispecies Biomass Dynamics Models

Biomass dynamics models can be similarly modified to account for competitive and predation interaction between species. We will develop the algebra for the discrete case, but as usual the continuous case is similar. For a two-species case, the biomass B at time $t + 1$ is

$$B_{t+1} = B_t + rB_t\left(1 - \frac{B_t}{k} - cX_t\right) - C_t \qquad (14.3.1)$$

where B_t is the biomass of the stock at time t, X_t is the biomass of the competing or predator species, c is a parameter that relates how much a unit of biomass of X_t reduces the relative growth of B, C_t is the catch of the stock at time t, and r and k have the usual Schaefer model meanings. Given the usual catch equation

$$C_t = B_t E_t q \qquad (14.3.2)$$

where E_t is the fishing effort at time t and q is the catchability, the parameters r, k, q, and c can be estimated by

$$\left(\frac{U_{t+1}}{U_t}\right) - 1 = r - \frac{r}{kq}U_t - rcX_t - qE_t \qquad (14.3.3)$$

The problems in this estimation are even worse than for a single-stock biomass dynamic model because in addition to the need for contrast between U and E, we need contrast between X and U and X and E. Furthermore, X would need to be observed directly, not by an index. Needless to say, such an approach is basically hopeless as a realistic management tool.

The generalized extension for many species is

$$B_{i,t+1} = B_{i,t} + r_i B_i\left(1 - \sum_{j=1}^{j=n} b_{ij} B_{j,t}\right) - C_{i,t} \qquad (14.3.4)$$

This is conceptually quite simple and elegant, and a whole generation of ecological modellers has explored the continuous version of this model. The matrix of b_{ij}*'s* is known as the *community matrix*. Because of parameter estimation difficulties, the structure is not going to be very useful for real fisheries work.

Essentially, it is hopeless to attempt to get the b_{ij}*'s* from real fisheries data if you can only observe an index of abundance such as catch-per-effort. The only cases where it has been done successfully have been in the laboratory where each species can be paired with each other species, and the r's and b's estimated separately in experimental trials. Even in the two-species case, the experimental control required between fishing mortality and abundance of the two stocks is so stringent that it is very difficult to ever get data with the needed contrast. Some of this difficulty is due to the fact that we are trying to understand q as well as r, k, and the interaction term (b's). If we knew abundances, then the problem reverts to one of stock and recruitment (couched in biomass dynamic terms), and this certainly is possible, at least in the two-species cases. We then only need contrast between the abundances of the species, and do not need contrast in fishing mortality. The application of this type of theory to fisheries is reviewed in Sainsbury (1989).

Aggregated production models

An alternative to multispecies production models in which each species has an interaction with each other species is to aggregate species and treat the aggregates as stocks to be analyzed via production models. Ralston and Polovina (1982) explored various methods of aggregation for a multispecies tropical handline fishery in Hawaii, and found that on a species-by-species basis the results from production model analysis were highly erratic. However, when the data were aggregated into major species groups, consistent trends in catch rate and yield became apparent. This has often been found in other mixed-species tropical fisheries (Pauly 1979, Pope 1979, cited in Ralston and Polovina 1982). Figure 14.2 shows catch rate and total catch versus total effort for the entire aggregate of Hawaiian species studied by Ralston and Polovina.

14.4. Multispecies VPA

One of the most interesting approaches to multispecies analysis is a variation on VPA in which the natural mortality rate for each species and each

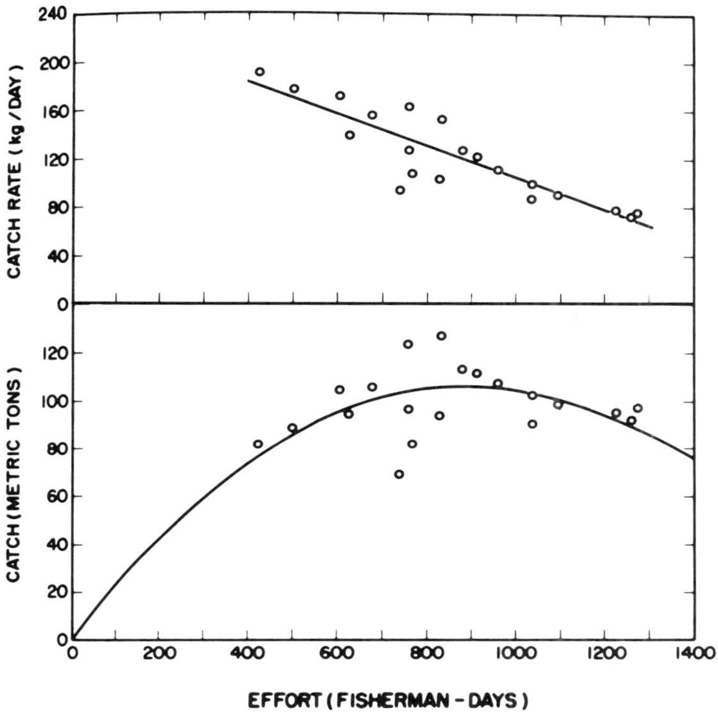

Figure 14.2. Catch and catch rate versus effort for a multispecies tropical fishery from Ralston and Polovina. (1982). Fisheries Bulletin 80.

age depends upon the abundance of the species which prey upon it (Pope and Knight 1982, Daan 1987). The total amount consumed by a cohort (j) of all other cohorts is calculated as

$$U_j = N_j R_j P_j \tag{14.4.1}$$

where N_j is the number of individuals in the jth cohort U_j is the total amount consumed by all individuals of the jth cohort, R_j is the annual ration (in kilograms) of an individual of the jth cohort, and P_j is the proportion of the total ration obtained from species and ages included in the cohort analysis. If $P_j = 0.8$, then 20% of all the food consumed by the jth cohort is assumed to come from species not included in the analysis.

The number of individuals of the ith cohort consumed by the jth cohort is assumed proportional to the prey abundance and a preference table for the ith prey by the jth predator (A_{ij})

$$E_{ij} = \frac{U_j A_{ij}}{W_i} \tag{14.4.2}$$

where E_{ij} is the number of the *i*th cohort eaten by the *j*th cohort, U_j is the total consumption of the *j*th cohort calculated above (Equation 14.4.1), A_{ij} is the proportion of the diet of the *j*th cohort (of U_j) made up of the *i*th cohort, and W_i is the weight of an individual of the *i*th cohort. Note that cohorts *i* and *j* may belong to the same species, and in this case the calculations represent cannibalism.

Once the E_{ij}'s are determined, the numerical cohort analysis equations used are

$$N_{i,t} = N_{i+1,t+1}\, e^{M} + (C_{i,t} + E'_{i,t})\, e^{M/2} \tag{14.4.3}$$

where $N_{i,t}$ is the number of individuals in the *i*th cohort at time *t*, $N_{i+1,t+1}$ is the number 1 year older 1 year later, *M* is the instantaneous natural mortality rate that is natural mortality from causes other than predation by species included within the cohort analysis, $C_{i,t}$ is the catch of the *i*th cohort and $E'_{i,t}$ is the total number of $N_{i,t}$ individuals eaten by all other cohorts ($E'_{i,t} = \Sigma_j E_{ij}$). If we believe that all mortality is due to predation by cohorts within the analysis or to fishing (within the ages we are considering), then $M = 0$ and Equation 14.4.3 becomes

$$N_{i,t} = N_{i+1,t+1} + C_{i,t} + E'_{i,t} \tag{14.4.4}$$

Pope used an iterative method to solve this equation over all cohorts; iteration is required because the *E*'s depend on other *N*s. Pope suggested using a starting estimate of N as

$$N_{i,t} = (N_{i+1,t+1} + 0.5C_{i,t})\, e^{M/2} \tag{14.4.5}$$

and then iterating through the analysis until convergence is reached.

The conceptual simplicity of this approach is its major attraction. It suffers from the assumption that the prey consumption proportion A_{ij} must either be independent of the abundance of the prey or must be measured each year. Considering the difficulty of obtaining even single-time estimates of the *A* matrix (presumably by stomach contents sampling), the technique is obviously very data intensive.

Table 14.1 shows the natural mortality rates estimated by Pope for a multispecies complex in the North Sea. A nonpredator mortality component (*M* of Equation 14.4.3) of 0.1 was included for each species. Daan (1987) reviewed the North Sea multispecies VPA (MSVPA) project and concluded that, whereas the methods are applicable to many fisheries in the world, the re-

Table 14.1. Estimated natural mortality rates for six species from multispecies cohort analysis.

	Ages							
Species	0	1	2	3	4	5	6	7
Cod	0.975	0.206	0.104	0.1	0.1	0.1	0.1	0.1
Haddock	0.847	0.403	0.242	0.115	0.108	0.104	0.102	0.1
Whiting	1.114	0.405	0.149	0.131	0.131	0.131	0.118	0.109
Saithe	0.1	0.1	0.1	0.1	0.1	0.1	0.1	0.1
Herring	0.407	0.271	0.260	0.220	0.218	0.219	0.193	0.143
Plaice	0.1	0.117	0.109	0.103	0.1	0.1	0.1	0.1

From Pope and Knight 1982.

quirements for reliable catch and stomach contents data are so stringent that "it may not be possible to apply MSVPA methods elsewhere." Pope (1990) also reviewed the North Sea MSVPA project and said that the MSVPA had caused a complete rethink of the understanding of the North Sea but that MSVPA was probably not the appropriate tool for further investigation. Both Daan and Pope suggest that single-species management is inadequate, but we do not really yet have the theory, tools, and resources to do multispecies management.

14.5. Multispecies Yield-Per-Recruit Analysis

Murawski (1984) developed a logical extension of Beverton and Holt's yield-per-recruit analysis, to account for many species fished by several gears. The discrete time equation for yield per recruit is

$$Y_i = \sum_{t=r_i}^{t=m_i} D_{it} \left(\frac{F_i p_{it}}{M_{it} + F_i p_{it}} \right) W_{it} \tag{14.5.1}$$

where r_i is the age at recruitment for the ith species, m_i is the maximum age for species i, D_{it} is the total number of fish dying at age t, $p_{i,t}$ is the fraction recruited at age t, $M_{i,t}$ is the natural mortality rate, and $W_{i,t}$ is the mean weight of fish at age i in year t. The relative recruitment proportion p_{it} can be adjusted by gear selectivity, whereas F can be adjusted by effort changes. We would normally assume that fishing mortality (F) is proportional to fishing effort by some catchability:

$$F_i = Eq_i \tag{14.5.2}$$

We can then substitute this relationship into Equation 14.5.1 to obtain

$$Y_i = \sum_{t=r_i}^{t=m_i} D_{it} \left(\frac{Eq_i p_{it}}{M_{it} + Eq_i p_{it}} \right) W_{it} \tag{14.5.3}$$

We need two more equations, one to calculate the initial cohort size, and one for the number dying. Numbers over time are given by

$$N_{i,t+1} = N_{i,t}\, exp[-(Eq_i p_{it} + M_{it})] \tag{14.5.4}$$

where $N_{i,t}$ is the number alive in the ith cohort at time t. The deaths are given by

$$D_{i,t} = N_{i,t+1} - N_{i,t} \tag{14.5.5}$$

or

$$D_{i,t} = N_{i,t}(1 - exp[-(Eq_i p_{it} + M_{it})]) \tag{14.5.6}$$

The initial size of the cohorts is needed to determine the total yield. Unlike single-species yield per recruit, where the optimal policy is independent of average recruitment, in multispecies yield per recruit, we need the relative average recruitments of the different species. Murawski suggested that we assume that the average initial cohort size for each species is $N_{i,1}$ is obtained from other data such as surveys.

Murawski then extended the analysis to n fishing gears using the relationship:

$$F_i = \sum_{j=1}^{n} q_{i,j} E_j \tag{14.5.7}$$

where F_i is the fishing mortality rate on the ith species, q_{ij} is the vulnerability of the ith species to the jth fishery, and E_j is the effort in the jth fishery. If we let subscript i denote age as well as species, then we can include size-specific vulnerability to gear. Figure 14.3 shows the total yield per recruit and the species compositions for a mixed stock fishery on cod (*Gadus morhua*), haddock (*Melanogrammus aeglefinus*), yellowtail flounder (*Limanda ferruginea*) and winter flounder (*Pseudopleuronectes americanus*) from Murawski's paper.

Pikitch (1987) used Murawski's formulation, and combined it with economic data on the value of different species, to analyze different fishing policies shown in Figure 14.4 for a three-species trawl fishery off the Oregon coast. The policy shown at the far left, called MSYS, is the theoretical maximum obtainable if each species could be managed individually.

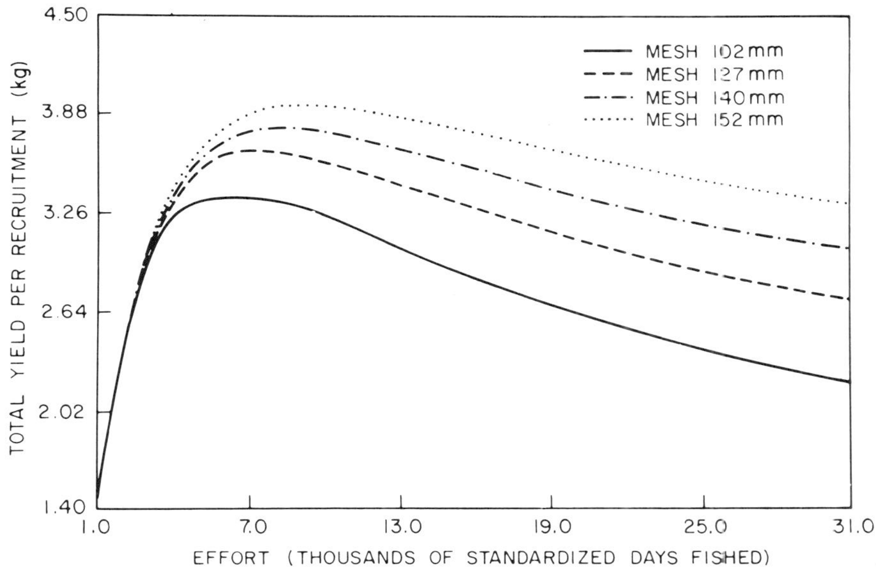

Figure 14.3. Total yield of four demersal species (Atlantic cod, haddock, yellowtail flounder, and winter flounder) at various standardized fishing effort levels. From Murawski. (1984). Can. J. Fish. Aquat. Sci. 41: Figure 4.

14.6. Equilibrium Yield From Several Stocks Harvested Jointly

It is computationally straightforward to calculate equilibrium yields for any number of stocks, so long as we can describe the dynamics of each stock by a simulation model. We simply simulate for some specific fishing mortality rate until the stocks come to equilibrium. This approach has been used since computers were first available to fisheries scientists. Figure 14.5 shows such a calculation for three stocks of Pacific salmon from Paulik et al. (1967).

The authors point out that the yield from a mixed-stock fishery will always be less than the yield from each stock harvested separately. How much yield is lost will depend upon the differences in productivity between the stocks. In Figure 14.5, the optimum harvest rate for the total stock is approximately 80%, with stock 3 dominating the yield. The small unproductive stock 1 is driven to extinction at such high harvest rates, but the increase in yield from stocks 2 and 3 more than makes up for the loss of stock 1. It is a common

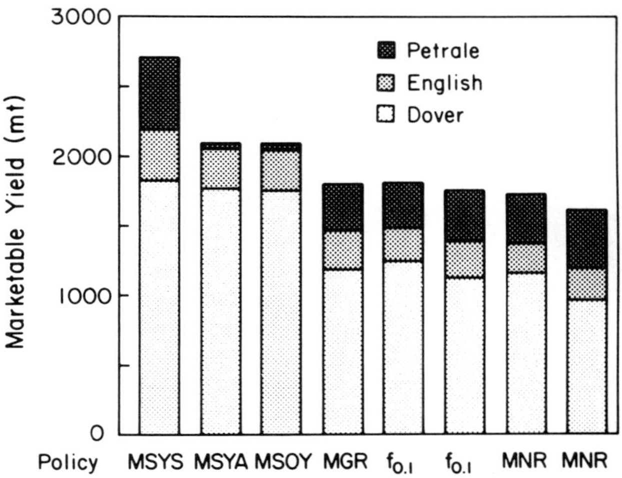

Figure 14.4. Species composition of the marketable yield that would result from the implementation of various management policies in a three-species trawl fishery off the Oregon coast. MSYS= sum of the single-species maximum sustained yields (MSYS); MSYA = MSY for the three-species aggregate; NSOY = mesh size only policy, maximum yield; MGR = maximum gross revenues policy; $F_{0.1}$ = $F_{0.1}$ policy; MNR = maximum net revenues policy. From Pikitch. (1987). Can. J. Fish. Aquat. Sci. 44 (Suppl 2), Figure 2.

result of mixed-stock fisheries that unproductive stocks will be lost when the harvest rates are chosen to maximize total yield.

Figure 14.6 shows a plot of combined yield from two chum salmon stocks (*Oncorhynchus keta*), a natural stock, and one from a hatchery. Note that the less productive natural stock is larger, so that the optimum harvest rate for the mixed stock fishery is close to the optimum harvest rate for the less productive stock. In this case, we should forgo yield from the productive enhanced stock to maintain the large potential yield from the less productive stock.

Even when all the stocks or species in a mixed fishery have the same productivity (same optimum harvest rate), the yield from a mixed-stock fishery will be less than the potential yield from separate fisheries because of environmental variation. To maximize biological yield for a single stock, the harvest rate should be adjusted according to the stock size. This cannot be done for each stock in a mixed stock fishery. The more uncorrelated and

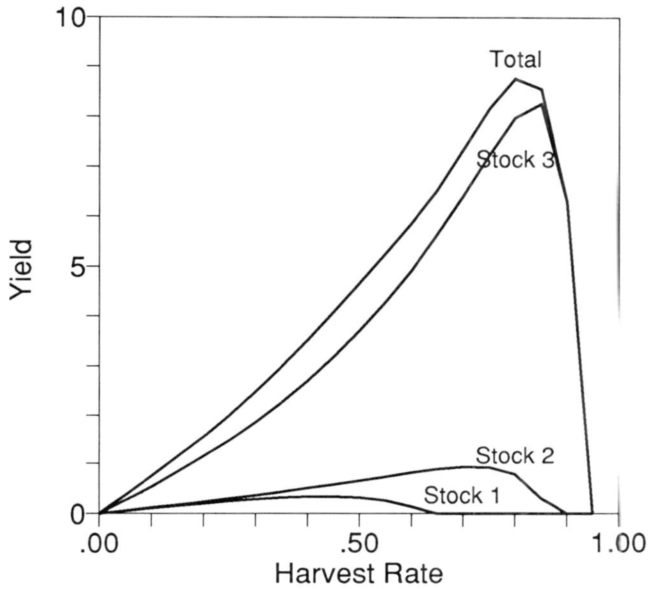

Figure 14.5. Total equilibrium catch at various rates of exploitation in a hypothetical fishery on three mixed stocks. Redrawn from Paulik et al. 1967.

larger the random variability of the stocks, the more yield is foregone by a mixed-stock fishery.

14.7. Dynamic Harvesting in Mixed Fisheries

The analyses of mixed-stock fisheries reviewed above looked only at equilibrium harvest policies. When we have only two or three stocks or species, we can calculate an optimal feedback control policy for all stocks using stochastic dynamic programming (see Chapter 16). Figure 14.7 shows contours of optimal harvest rate for a two-stock mixed fishery, from Hilborn (1976).

In the example, the two stocks are governed by a Ricker stock-recruitment curve; stock 1 has an *a* value of 1.0, stock 2 has an *a* value of 2.2. When either stock is virtually extinct, then that stock is largely ignored and the harvest rate will be determined solely by the remaining stock. When both stocks are reasonably abundant, the rules are a bit more difficult to describe.

Three simple alternative policies could be used to manage a mixed stock fishery. We look at these alternatives and see how close they come to the computed optimal policy shown in Figure 14.7.

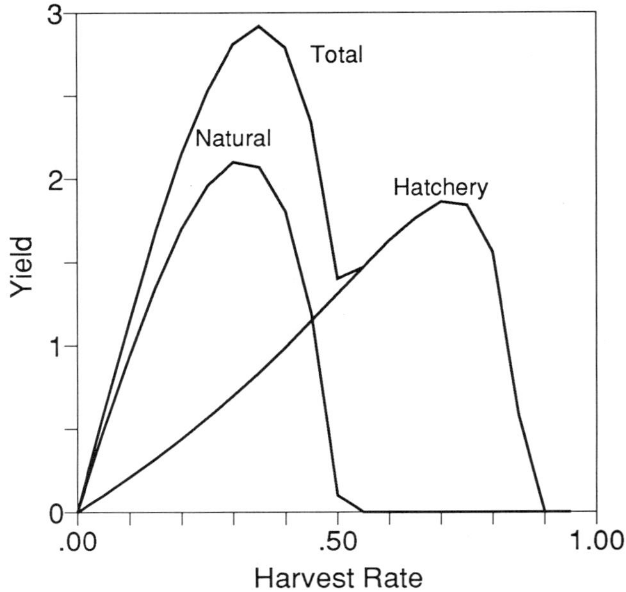

Figure 14.6. Total yield from a mixed fishery on a hatchery and natural stock of Pacific salmon.

Manage to the weakest stock This rule says to use the harvest rate that permits the weakest stock to meet its optimum escapement. Implicit in this argument is that underescapement is worse than overescapement. When one pushes this policy to the extreme case with dozens of stocks, some of which are very unproductive, it is clear that total yield would be very low. Figure 14.5 shows such a case; if we managed to the weakest stock in that example, the large potential yields from stocks 2 and 3 would be foregone.

Maintain constant escapement for the total mixed stock. An alternative to trying to meet the optimum escapement for all stocks is to try to maintain a constant escapement for all stocks combined. That is, if the escapement goal for all stocks summed to 1 million, we would attempt to allow a constant escapement of 1 million fish, regardless of the stock composition. The dashed lines in Figure 14.7 shows how the harvest rate contours would look for such a policy. Note that the optimal policy looks very much like these contours except when the relative strengths of the two stocks are very different.

Use a constant harvest rate. The third option is to use a constant harvest rate. It is difficult to see how this would perform from Figure 14.7, but Hilborn (1985d) explored constant harvest rate policies for mixed-stock fisheries and found rather good performance. In most multispecies fisheries,

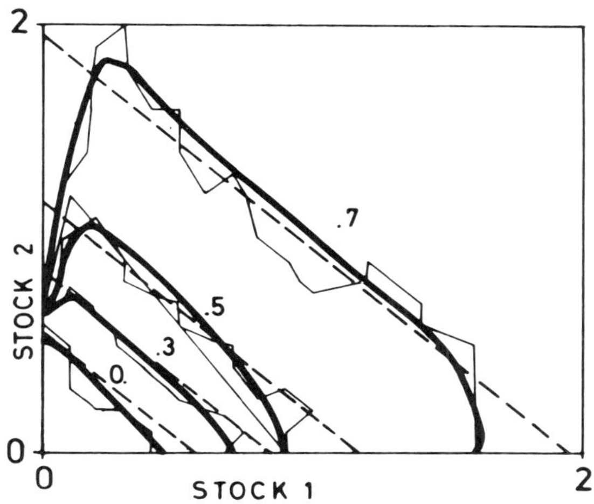

Figure 14.7. Optimal harvest rate isoclines for a mixed fishery on two salmon stocks. Both stocks have the same Ricker *b* values (1.0); for stock 1, $a = 1.0$, and, for stock 2, $a = 2.2$. Note that except when the stocks are greatly different in abundance, the optimum policy is a constant total escapement. From Hilborn. (1976). J. Fish. Res. Board Can. 33: Figure 1.

constant harvest rate would really mean constant effort, since the harvest rates would be different for each species. Chapters 15 and 16 consider methods for evaluation of such policies.

14.8. Summary and Critique

Previous sections were straightforward extensions of single-species approaches. At this point, we should stand back a bit and look at the alternative ways of thinking about, modelling, and managing multispecies fisheries. Research on biological and technological interactions are almost totally different; as Pikitch (1988) has pointed out, biological interaction is difficult and expensive to study and takes many years, whereas technological interaction is much simpler and cheaper to study, and useful results can be obtained rapidly. Nevertheless, the biological orientation of most fisheries scientists means that much more effort has been expended in the analysis of biological interaction. In this section we review the strengths and weaknesses of different approaches to analysis of biological and technological interaction.

Biological interaction

There are six basic approaches to the analysis of biological interaction:

Multispecies biomass dynamics models. These approaches were discussed in Section 14.3, and although they may prove useful in some rare two-species interactions, in general, the data requirements for such types of analysis are so demanding that we see little if any practical utility of such approaches in formulating fisheries management advice.

Multispecies age-structured models. These include multispecies VPA and multispecies stock-recruitment models. Although far from proven, the work on the North Sea stocks, as illustrated in Pope and Knight (1982), shows that the methods hold considerable promise, with a very expensive price tag.

Food web and trophic level models. This approach can consist of very complex multispecies food web models, such as Laevastu et al. (1982), or can involve derivation of simple and general empirical relationships between community structure and potential biological yield (Ryder et al. 1974, Marten and Polovina 1982). We believe the food web modelling approach is hopeless as an aid to formulating management advice; the number of parameters and assumptions required are enormous. Such large model-building exercises are perhaps of some utility in designing research programs, but will not help manage fisheries. The use of trophic relationships to estimate general yield potentials also seems unlikely to produce anything that fishery managers can use in the near future, although such relationships may become clearer in the long term as data are accumulated from a wide variety of fisheries.

Direct empiricism. Sainsbury (1989) has proposed that managers set up experimental fishing regimes that are tried and evaluated in the multispecies context. Sainsbury's proposal is an extension of the adaptive management concept (Walters and Hilborn 1976, Chapter 17), and we not surprisingly find it very attractive. The essence of Sainsbury's arguments is that the only way to find out how to manage multispecies systems is to try different regulations, hopefully replicated and controlled in different spatial sites, and then evaluate them and see what works. There are many practical impediments to the success of direct empiricism (discussed in Chapter 17), but the potential is also very attractive.

Ignore biological interaction. This is a very viable alternative to spending a lot of time and effort trying to understand biological interaction, and represents a sort of default assumption. Even though it may represent a "stick your head in the sand" approach, it may often prove to be the best thing to do.

Lump the species together. This is the approach adopted by Ralston and Polovina (1982) and one that may prove the best general strategy. Whereas the previous method just went ahead with individual single-species management measures, in this approach we treat the entire mix of species as a single stock to be managed (presumably for economic yield not biomass yield) and hope that species interaction effects are captured in the overall empirical relationship between yield and effort. This is likely to be the only alternative to the direct empirical approach for tropical fisheries with dozens of species.

Technological interaction

Technological interaction involves understanding (1) gear effectiveness on each species, (2) fishermen's targeting on species, and (3) discarding. Long ignored, this field has recently received some much deserved attention (IFREMER 1987, Pikitch 1987, 1988). The multispecies yield-per-recruit method of Murawski (1984) is also an analysis of technological interaction.

The beauty of technological interaction is that we know it is important, and unlike biological interaction it can be rather easily studied. We anticipate considerable progress in this field over the next few years.

Bibliographic Notes

Multispecies analysis is quite a young field and we have presented the potential techniques rather uncritically. The symposium edited by Mercer (1982) provides the single largest series of papers on the question of multispecies analysis, and the rather unrelated nature of most of the papers in that symposium reflects the undeveloped nature of the field. In October 1989, ICES held a workshop on multispecies models, which contains many papers reviewing different multispecies approaches. At the time of this writing, the proceedings from this conference are not yet available, but by the time this book is published, the ICES proceedings should provide a valuable reference document. Kerr and Ryder (1989) also review multispecies approaches, and categorize approaches into four categories (1) descriptive multivariate, (2) dynamic multivariate, (3) multivariate system, and (4) integral systems. Shepherd (1988) and Saila and Erzini (1987) are recent papers suggesting new approaches to the analysis of multispecies data.

Walters et al. (1985) developed the question of multispecies stock and recruitment by combining both abundance and food consumption estimates to present a consistent story.

Multispecies production models are almost totally unused, except when species are aggregated. Pope and Knight (1982) in the Mercer symposium, and a number of later papers e.g. Daan (1987), have developed multispecies VPA quite extensively, but the potential for widespread use will be very limited because of the intense data requirements. Multispecies yield per recruit and dynamic harvesting has been almost totally untouched outside of the papers by Murawski, Paulik et al., and Hilborn discussed in the text.

Part IV

Managing Fisheries

15

Harvest Strategies and Tactics

15.1. Introduction

In a well-run fishery, all of the key players (fishermen, biologists, and managers) should be able to state in unambiguous terms what harvest strategy is used for the fishery. A harvest strategy is a plan stating how the catch taken from the stock will be adjusted from year-to-year depending upon the size of the stock, the economic or social conditions of the fishery, conditions of other stocks, and perhaps the state of uncertainty regarding biological knowledge of the stock. Many stocks in the Canadian Atlantic fishery are exploited with a constant harvest rate. Such constant-exploitation-rate strategies are becoming quite popular and have been adopted in the eastern Pacific for Pacific halibut and for herring in British Columbia. Many Pacific salmon stocks are managed on the basis of an escapement goal, by which managers attempt to adjust the catch every year so that the same number of fish reach the spawning grounds. This type of harvest strategy is also known as a constant-stock-size strategy. Another type of harvest strategy involves taking a constant catch, or fixed quota, from the stock every year. In the Tasmanian abalone fishery, for many years every licensed fisherman was permitted to harvest 28 tons of abalone each year.

A harvest strategy is not a set of annual regulations; rather it is a plan that should be robust to the unpredictable and/or uncontrolled biological fluctuations that are expected from the stock. Thus, any well-designed harvest strategy should not need modification due to an unusually poor or unusually good year class, nor should it need modification due to normal levels of price change. The tactics needed to implement the strategy each year will usually need to be adjusted annually. If the strategy is a constant exploitation rate implemented by doing an annual stock size assessment and setting an annual *Total Allowable Catch* (TAC) to be a constant fraction of this assessment, then obviously the TAC will be adjusted from year-to-year and the tactics of monitoring and enforcement may vary each year depending on how long it is expected that the fishermen will take to achieve the TAC.

The more stable and predictable a fishery, the longer should be the life of a harvest strategy. It may be quite difficult to formulate a strategy for a newly developing fishery, and the strategy may need updating every year or two, whereas in a long established fishery, the strategy may need updating only every 5–10 years.

Harvest strategies are also not a vaguely worded set of fishery objectives stating, for instance, that "the purpose of the fishery is to maximize benefits to the fishermen subject to biological constraints" or some other mumbo-jumbo that provides no explicit guidance to formulation of annual regulations. Harvest strategies must be explicit and quantitative, saying how much catch should be attempted to be harvested under what circumstances.

A harvest strategy involves biological, economic, social, and political decisions, and should be formulated at the highest political levels with the active participation of fishermen and processors. Normally, harvest strategies implicitly make major decisions about tradeoffs between average yield and year-to-year variability and major decisions about how to deal with biological uncertainty in stock assessments. In our view, these are exactly the types of questions that fishermen and processors are best equipped to discuss, and they should be involved in the choice of a harvest strategy. Ideally, biologists and economists should perform a technical role in determining the likely outcomes of alternative harvest strategies, but the actual choice of a harvest strategy should be made by politicians, fishermen, and processors.

The purpose of this chapter is to review the types of harvest strategies that are currently used in many fisheries and to consider tradeoffs between alternative strategies. More importantly, because the tradeoffs are almost always quite specific to individual fisheries, we introduce some rather simple techniques for comparing the benefits of alternative harvest strategies for specific fisheries. These tools can then be used to look at a wide variety of cases. Rather than try to make generalizations, we wish to provide the readers with the ability to analyze their own situations.

Once a harvest strategy has been adopted, a detailed set of annual regulations must be put in place to try to implement the strategy. These annual regulations are called harvesting *tactics,* and include measures such as annual catch quotas, season limits, and gear restrictions. In a later section of this chapter, we consider types of tactics, and the interrelationship between specific strategies and specific tactics.

15.2. Stock-size-dependent strategies

The most common and simplest harvest strategy is to set the target catch for each year to be a linear function of population size, as shown in Figure 15.1. Within this class of strategies, constant-stock-size strategies are a straight line with a slope of 1 and intercept equal to minus the desired escapement.

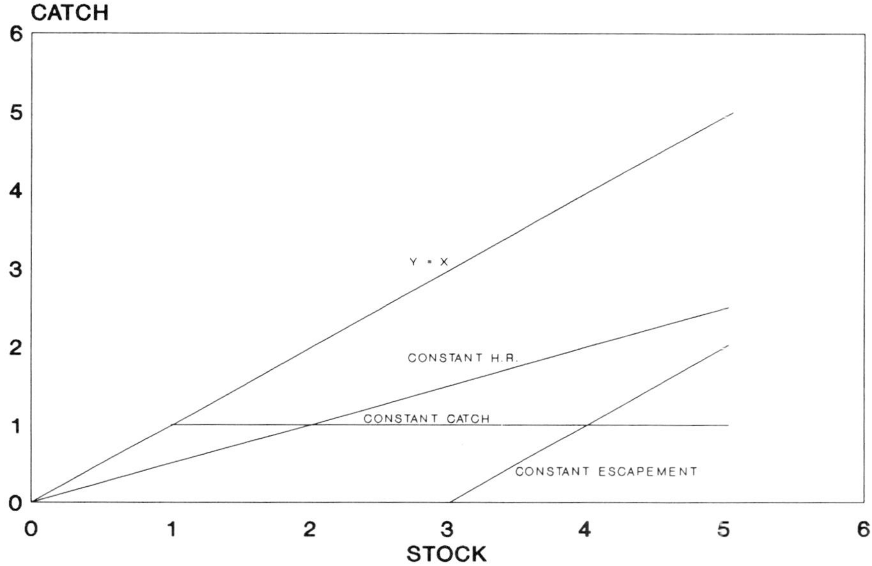

Figure 15.1. Simple stock-size-dependent strategies, with desired catch as a linear function of stock size.

Constant-exploitation-rate strategies have an intercept of zero, and a slope equal to the exploitation rate. Constant-catch strategies have a slope of zero and an intercept equal to the catch.

The simplest way to compare harvest strategies is by a combination of simulation and graphic presentation of the simulation results. We will rely on simulation of alternative strategies to see what happens when we apply the strategy to a specific model of population dynamics. For instance, we can take a simple Schaefer production model ($r = 0.4$, $k = 1{,}000$) and try alternative exploitation rates and constant-catch policies, simulating each for 100 years. Figure 15.2 shows the results of such simulations. We know from the analytic results of Chapter 8 that the optimum escapement will be $k/2$ or 500, and the optimum exploitation rate will be $r/2$ or 0.2. The deterministic sustainable yield will be 100. We can see from Figure 15.2 that the best yield is slightly over 100 and is obtained with a constant escapement of 400. Our graphic optimization technique relies on simulating over 100 years, starting from unfished conditions. This means that the average yield is a little higher than 100, because of the initial unexploited biomass. This also explains why very high harvest rate policies still have some average catch. If we wish to consider only long-term yields, we can first simulate for 100 years to let the stock come to equilibrium, and then simulate another 100 to calculate average yield.

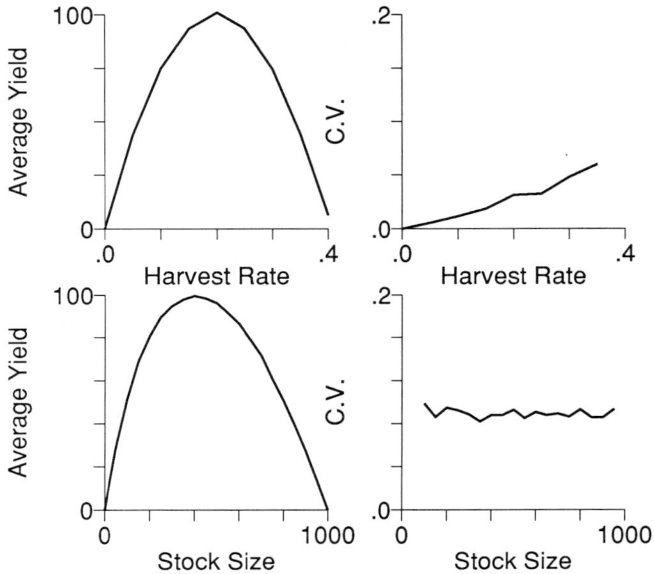

Figure 15.2. Average yield and variability of yield for constant-stock-size and constant exploitation-rate harvest strategies.

The most important results of this simple graphic example are that constant stock sizes between 400 and 600 produce roughly the same average catch, and constant harvest rates between 0.15 and 0.25 also produce nearly the same average catches. Note, however, that the coefficient of variation of the catch is much lower for the constant-harvest-rate policies, that is, there is much less interannual variation in catches.

We can adopt a more general graphic view for the comparison of alternative harvest strategies if we recognize that any harvest strategy that involves catch as a linear function of stock size can be written as

$$\text{catch} = \text{intercept} + \text{slope} \times \text{stock size} \tag{15.2.1}$$

We can then plot the average yield for different intercepts and slopes as shown in Figure 15.3. Note that the results for constant-stock-size policies are along the right-hand edge of the plot, constant-harvest-rate results are along a line where the y intercept is zero, and constant-catch policy results are along the upper left-hand edge. The main benefit of this type of plot is that it enables us to see that many alternative policies will produce quite similar average yields and forces us to decide what other factors, such as variability of yield, will be used to determine which harvest strategy is best for a particular situation.

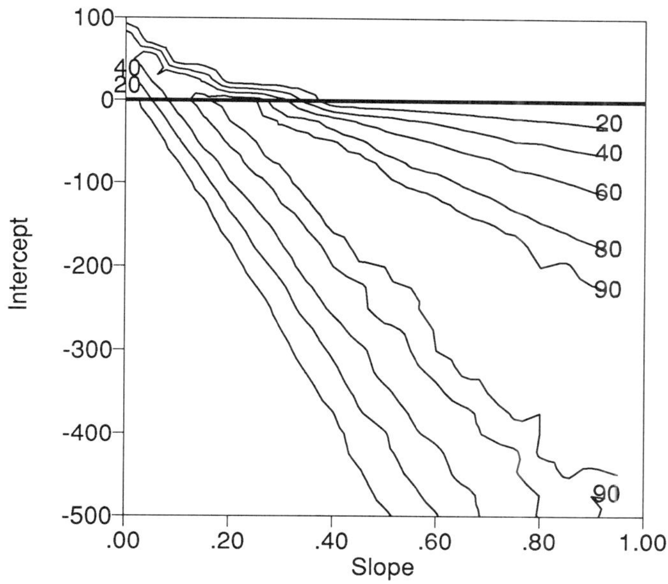

Figure 15.3. Average yield plotted as a function of the slope and intercept of the linear harvest strategy.

A second benefit of the graphic method is that it permits any model to be used. Figures 15.2 and 15.3 use a simple production model, but a full age-structured model can be easily used, as can a complex multistock model, or for that matter any computational procedure that produces an average (or total) outcome when supplied with a specific harvest strategy rule.

Stock-dependent harvest strategies need not be linear functions of the stock size. Curves, lines with breakpoints, and so on, are all perfectly legitimate harvest strategies. However, for practical purposes, it is not often that greatly improved yields are obtained by going to more complex strategies. Nevertheless, we encourage you to try your own strategies and see how well you can do. In one policy analysis exercise with salmon fishermen who pursue sockeye salmon in Rivers Inlet, British Columbia, we found that the fishermen preferred to specify a smooth, sigmoid-shaped curve relating target harvest rate to stock size. Their reasoning was not that the curve would produce the best harvest pattern in a statistical sense, but rather that the smooth shape would help to avoid arguments about on which side of break points or thresholds the stock size actually is each year, in other words, it would minimize quibbling about assessment procedures.

There does exist a rather substantial literature on optimal exploitation of fish populations that is relevant to our current discussion of harvest strategies. However, with few exceptions this literature has been devoted to looking at models with three very restrictive assumptions: (1) that the objective is maximization of average discounted catch, (2) that the stock is a single homogeneous unit, and (3) that the model parameters are perfectly known. In general, formal analysis, beginning with Ricker's simulations (1958b) and continuing on to some quite sophisticated mathematical analyses (Clark 1976, 1985, Mangel 1985, Reed 1978, 1980), has shown that under the restrictions mentioned above the optimal harvest strategy is usually the constant-stock-size strategy.

Unfortunately, this result has proved more of a hinderance than a help to the formulation of good harvest strategies, although, in truth, most management agencies have ignored it. In fact, there are few fisheries where any of the three assumptions are met. Deriso (1985) showed that when the objective is to maximize the logarithm of the catches (a risk-averse utility function as discussed in Chapter 2), the optimal policy can be shown to be fixed exploitation rate for some specific models. Similarly, Hilborn (1985d) showed that the optimal harvest strategy is constant exploitation rate for certain mixed stock conditions. The key point is not what harvest strategy is generally best, but rather that it is a simple matter to *compare* strategies yourself, rather than rely on results someone else has obtained.

There are, however, some general attributes that normally emerge from such comparisons. First and foremost, there is almost always a tradeoff between average yield and variability of yield. For the simple models, the constant-stock-size policy that maximizes average yield also maximizes the variance of yield. The policy choice is therefore often a matter of choosing how much average yield to give up in order to obtain less variability over time. Figure 15.4 shows a graph of average catch and coefficient of variation CV for all policies examined in Figure 15.3. For the specific model and alternative harvest strategies examined, a wide range of alternative yields are possible. Of primary interest are policies that lie on the boundary of the feasible and infeasible regions. If average yield and CV are the only two indicators of interest (ignoring management cost, etc.), then there is no reason to choose a policy that is not on this border. Any policy on the interior of the feasible region is dominated by a policy on the border that has a higher average yield and a lower CV. The choice of strategies than becomes making a tradeoff between CV and average yield.

Constant-stock-size policies are particularly uninformative. Stock assessment requires understanding the relationship between stock size and subsequent recruitment, and a harvest strategy that holds the stock size constant will provide little information about stock and recruitment. Constant exploitation rate strategies maximize the time-series bias in the analysis of stock-recruitment data and minimize contrast in fishing mortality rates. Holding exploitation rate constant is quite inconsistent with trying to determine the effects of fishing by many analytic techniques such as catch-at-age analysis.

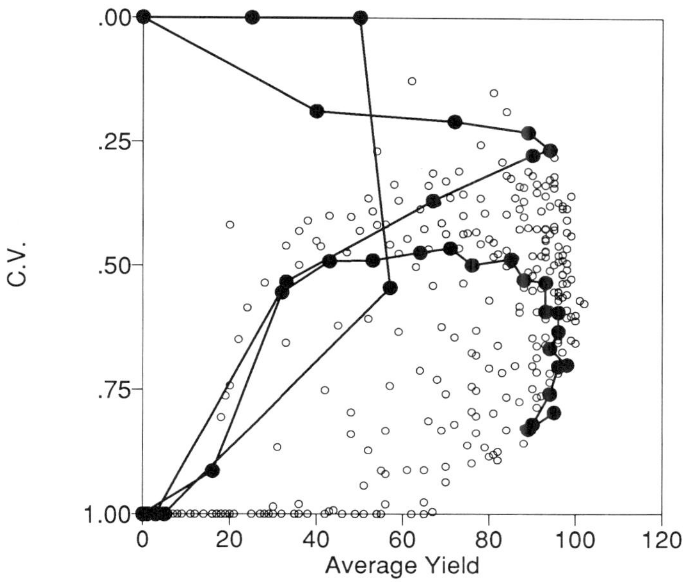

Figure 15.4. Average catch and coefficients of variation for the policies of Figure 15.3. The three connected sets of solid dots represent fixed-quota, fixed-escapement and fixed-harvest-rate strategies.

$F_{0.1}$ strategies

In the 1980s, a number of major fisheries have been managed by a type of strategy called $F_{0.1}$. The $F_{0.1}$ strategy is a constant exploitation rate strategy with the fishing mortality rate (denoted $F_{0.1}$) set equal to the value of F where the slope of the yield per recruit function is 0.1 times the initial slope. This is illustrated in Figure 15.5. The fishing mortality rate that maximizes yield per recruit is often denoted F_{max}, and occurs where the slope of the yield per recruit function is 0.0 times the initial slope; thus F_{max} strategies could also be denoted as $F_{0.0}$ strategies. $F_{0.1}$ will always be less than F_{max}, and thus in general a little more economically efficient.

The calculation of $F_{0.1}$ policies is done numerically using the methods shown in chapter 3. The age-specific weights and vulnerability to fishing gear, as well as a natural mortality rate are required; then solution for yield as a function of F is quite simple, and the slopes can be easily calculated.

$F_{0.1}$ policies have become very popular in eastern Canadian fisheries (Doubleday et al. 1984) and are also seeing consideration and use elsewhere (for example, Andrew and Butterworth 1987). Probably the most important aspect of $F_{0.1}$ strategies is that they are totally ad hoc. That is, there is no

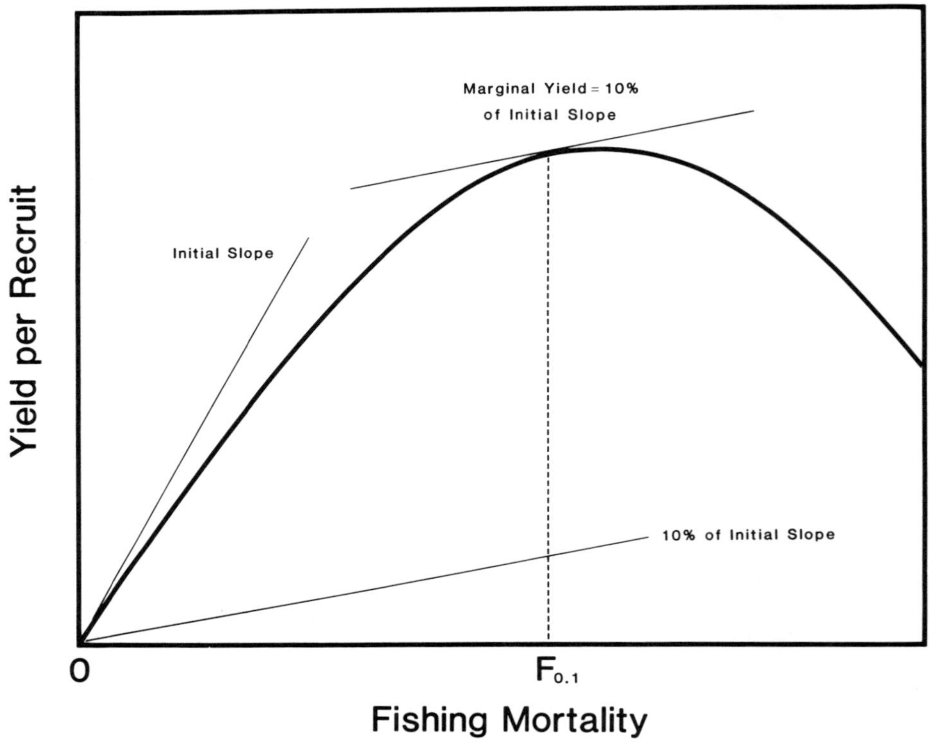

Figure 15.5. Location of the $F_{0.1}$ level of fishing mortality in yield-per-recruit analysis. From Deriso. (1987). Can. J. Fish. Aquat. Sci. 44:(suppl. 2) Figure 1.

theoretical reason that yield from a fishery will be optimized or maximized by applying a fishing mortality rate that is $F_{0.1}$. $F_{0.1}$ is an essentially arbitrary choice of fishing mortality rate, which often appears to be in the right ball park.

Deriso (1987) has shown that for a broad range of models of stock dynamics the $F_{0.1}$ policy does not unduly reduce the spawning abundance, and this suggests that $F_{0.1}$ policies may be rather robust to alternative stock-recruitment relationships. This remains to be examined for a wide variety of actual stocks.

The acceptance of $F_{0.1}$ policies despite their ad hoc status stems from a recognition that for most fish stocks, some type of constant exploitation rate strategy is useful—it provides for a stability that is widely deemed important. Having gained acceptance in some fisheries, $F_{0.1}$ policies are now "legitimate"; one can justify their use because they are used elsewhere.

$F_{0.1}$ policies may be one of the most significant changes in fisheries harvesting practice since the earlier widespread acceptance of MSY. They are significant not because of any theoretical breakthrough, or any intrinsic elegance, but simply because they provide a replacement for F_{max} and MSY and appear to often be robust.

15.3. Periodic Harvest Strategies

An alternative fisheries harvesting strategy is "pulse" or periodic harvesting in which the stock is not fished every year, but only periodically. Such harvesting strategies were first considered for wildlife by Walters and Bandy (1972) and for fisheries by Walters (1969) and Clark et al. (1973). Hannesson (1975) used the general results of Clark et al. and applied them to a specific cod fishery, while De Klerk and Gatto (1981) and Botsford (1981) considered extensions and generalizations of the earlier work.

The essential result of the above work is that periodic harvesting will be optimal when the economics of yield (rather than biomass) are considered and one of the following holds true.

1. There are large economies of scale, so it is economically more efficient to take a large catch every few years than a smaller catch each year.
2. The economic value of older individuals is much higher than the economic value of younger individuals.

Another condition that can lead to periodic harvest giving the highest average yield even in purely biological terms, is for there to be no way to avoid taking fish that are far smaller than yield-per-recruit analysis would suggest is the best minimum size (Walters 1969). For example, in some clam fisheries there is no way to selectively dig only for larger clams (small clams make holes that look the same to diggers), and digging may kill the smaller ones or expose them to predation; in this case it may be best to divide a large clam beach into sections, and fish each section only once every 2 years or more. That way the yield from every section is maximized without introducing economically damaging variation from year-to-year in the total yield.

In practice, periodic harvest strategies are seldom adopted by management agencies, but may be much more commonly employed in cases where some form of individual choice and control has evolved. For instance, in commercial fisheries that involve diving, such as for abalone and some clams, divers normally rotate their effort, periodically visiting different beds. Similarly, most forms of aquaculture involve periodic harvesting. Many subsistence and artisanal fishermen rotate their effort, visiting different beds, or fishing grounds, periodically and letting the stock rebuild in between.

These examples mostly reflect economies of scale. Subsistence fishermen travelling by foot or small boat have considerable "set up costs" involved in getting to a fishing ground. They will prefer to rotate their visits rather than visit each site for a short time. This is also true of abalone divers, who must travel to a location and set up their gear in anticipation of a dive.

We would expect pulse fishing to evolve naturally in situations where there are few participants, and sole ownership fisheries (such as a mussel or salmon farm) provide the most obvious example. Periodic harvesting certainly increases the year-to-year variance in catch (unless the fishing area can be subdivided like the clam beach mentioned above) and would therefore seem an unlikely strategy for areas with dependent users, unless the spatial scale of the beds is considerably smaller than the spatial scale of the dependent users.

Although there have been no general prescriptions for how to calculate optimal pulse fishing policies for many fisheries models, it is rather easily done using either graphic or fixed-form optimization, as discussed in Chapter 16. Figure 16.4 shown later shows an example, where the two axes are the stock size at which a harvest takes place and the exploitation rate used when harvesting occurred. An alternative approach, which results in a non-feedback policy, would be to graphically search over possible rotation periods and F values. We suspect that yields would normally be better by using a fixed stock "trigger," although this would obviously be more expensive to implement for a manager.

15.4. Strategies that Include Uncertainty

Simple stock-size-dependent strategies assume that the state of the fishery can be described by a single number: the stock size. Obviously, there are many things that should in principle affect the harvest decision made each year, and these might include the level of confidence we have in our understanding of the stock dynamics. Walters and Hilborn (176) first considered harvesting strategies that explicitly consider uncertainty, and Chapter 17 on *Designing adaptive management policies* discusses this subject further.

15.5. Sex-Specific Harvesting

Sexes are frequently harvested at different rates. This may be due to sex-specific growth and size-specific vulnerability, or more commonly, sex-specific retention by the fishing gear. Many crustacean fisheries involve release of all, or at least berried females. Recreational fisheries often involve release of gravid females. Many if not most big game harvests are male only, fur seal harvests are primarily males not involved in the harem breeding; whaling regulations often banned taking of females with calves. In short, sex-specific harvesting is rather widespread. This practice has the theoretical foundation that females make a disproportionate contribution to the successful reproduction of the stock. Although appealing from a superficial biological perspective, there is actually little theory or data to justify this practice.

Such a theory would require an explicit model of the relationship between number of female and male spawners and subsequent recruitment success. There are some theoretical models that could be applied (Das Gupta 1972, Caswell and Weeks 1986), but these have not been used in practical policy design. There is reason to be suspicious of any argument that proposes "more females, more recruits" uncritically. For example, many of the world's major crab fisheries are selectively targeted on larger males, but male-male and male-female behavioral interactions are thought to be complex; there is little understanding of how distortion in sex ratios may affect cannibalism, mating patterns, or molting and growth of females.

15.6. Size-Limit Strategies

Another type of strategy is to set the size limit greater than the size at which females first begin reproduction, thus assuring the presence of an "invulnerable" group of females. This is widely used in invertebrate fisheries, and the remarkable persistence of some of these fisheries despite enormous fishing efforts argues that this can be a successful strategy. This strategy requires, of course, some form of very size selective fishing gear and is normally used in trap fisheries where undersized individuals can be returned to the water with a minimum of incidental mortality.

In several Australian rock lobster fisheries, for instance, the legal size limit is such that 2, or perhaps 3 years of reproduction occur before females reach legal size. Yield per recruit analysis indicates that the yield would be improved by a smaller size, but managers have preferred to be cautious, and to maintain spawning stock by keeping the larger size limit.

In many recreational fisheries, size limits have become a de facto harvest strategy, particularly where the primary source of recruitment to the stock is hatchery plantings. The fisheries are often regulated by a size limit (in conjunction with seasons and gear restrictions) simply because it is tradition: no one has ever done any formal analysis to compare alternative size limits, but the system works and managers prefer not to tamper with it.

However, the lack of formal analysis of size limits can lead to some peculiar regulatory structures. A minimum size limit is a common prescription to maximize yield per recruit and is probably the most common form of size limit. However, in many fisheries both minimum and maximum size limits have been imposed, to protect large mature individuals as well as young rapidly growing ones. Such regulations have sometimes been justified by using quantitative models (Argue et al. 1983, for instance). Other fisheries have exactly the opposite policy, where all the fish in a middle range of sizes must be released! It is possible that given the objectives, biology and socioeconomics each of these two types of policy might be best in different situations. It is equally possible that they are arbitrary policies that simply seemed like a good idea to managers without any quantitative analysis.

15.7. Economics and Harvest Strategies

Our discussion of harvest strategies thus far has considered only biological yield as the output of the fishery. Fisheries consist of fish and fishermen, and evaluation of alternative harvest strategies should normally include explicit consideration of the impacts of the strategy on fishermen—indeed this is one of the reasons we believe that fishermen should be key actors in selecting a harvest strategy.

In Chapter 2, we considered a number of indicators of performance for a fishery, including economic and social indicators. These indicators cannot be forgotten in comparing harvest strategies. It is particularly important to consider the costs associated with harvest strategies, including the following:

Fixed harvesting costs. These will generally be proportional to fleet size and include depreciation, insurance, and interest payments.

Variable harvesting costs. These include fuel, crew expenses, food, and some maintenance. Variable costs will normally be proportional to fishing effort, or to catch if the crew works on a venture-share basis.

Management costs. The costs to the management agency of running the fishery, including fixed costs of the staff, and variable costs, such as vessel charters and overtime.

We should also compare revenue, which in its simplest form is just price times the tons landed. However, in many fisheries, the price depends greatly on the average weight of size of fish, or on the gear used. A good example occurs in the yellowfin tuna fisheries of the world, where a large yellowfin caught by longline gear and sold on the sashimi market may be worth 5 or 10 times as much as the same fish caught by a purse seiner and sold for canning. Indeed, price differentials by gear, size, and so on, are the rule rather than the exception in the world's fisheries and should be incorporated into comparison of harvest strategies.

Investment as a second variable

Stock-dependent strategies consider harvest in relation to one variable, size of the stock. Implicit is the assumption that the fishery is capable of taking any catch. In many fisheries, particularly developing ones, the fleet

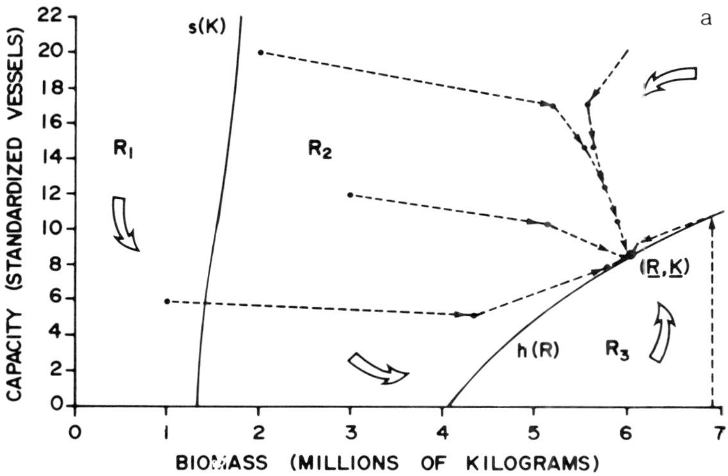

Figure 15.6. Optimal management policies where both catch and number of vessels built are control variables. From Charles. (1983). Can. J. Fish. Aquat. Sci. 40: Figure 1.

may not be capable of large catches, and harvest decisions should be based not only on the size of the stock, but also on the size of the fleet. Charles (1983) examined such situations in a general sense, looking at both harvest and investment decisions.

Figure 15.6 is a typical result from Charles' analysis for a prawn fishery. The size of the stock is on the x axis and the size of the fleet is on the y axis. The large arrows indicate the general direction the fishery should follow. The solid line indicates the equilibrium where fleet and stock are approximately balanced. When the stock is large, and the fleet is also large (upper right), no investment should take place, and the fleet will decline while the stock is also fished down. When the fleet is very small and the stock is large, investment should take place while the stock is maintained near its optimum level. At small stock sizes, the fleet should be reduced by depreciation and lack of investment while the stock is allowed to rebuild.

15.8. Harvesting Tactics

A harvest strategy is not of much value unless a mechanism or set of tactics can be devised to implement it in the field. It is easy to decide we want to keep fishing mortality constant at some specific level, but it is often very hard to determine what fishing mortality actually is and to regulate it. It might be nice in a theoretical sense to maintain a fixed spawning stock, but the short-term (in-season, annual) monitoring and management for achieving the target each year might be prohibitively expensive. In this section, we consider the merits of some tactics, and consider the relationship between the tactical tools to be used and the strategy we are attempting to implement.

Alternative tactics

The following is a list of tactical tools that are commonly used in various fisheries.

Gear restrictions	mesh size, net length, trap size, vessel size, etc.
Season lengths	
Gear Limitations	number of traps, number of vessels
Effort limits	control on number of trips
Annual catch quotas	for entire fleet or individual vessels
Size limits	

Most fisheries employ more than one of the above tactics; we can think of very few fisheries where there is absolutely no form of gear restriction. Season lengths were historically the most common type of regulation, but it often found that shorter and shorter seasons did not really reduce the catch. Recently, managers have been turning more and more to gear limitation and annual catch quotas (often both) to reduce fishing mortality on heavily exploited stocks. There is a general preference for annual catch quotas, because they put a ceiling (in principle at least) on total catch, whereas gear limitation has often been found to be ineffective at reducing harvesting pressure.

Fishermen's behavior is an integral part of the effectiveness of a tactic. For instance, many managers have naively assumed that reducing the fishing season by 30% will reduce the fishing mortality rate by 30%. Generally, reducing the fishing season has little if any impact on fishing mortality rate; the fishermen simply fish more during the open season than they did before. The extreme example of this is the Pacific halibut fishery, which now exerts as much fishing pressure in 2 days of fishing (the entire season), as it used to exert in 6 months. Similar experiences can be cited for attempts to reduce fishing mortality (or prevent its growth) by gear restriction.

Given the range of tactical tools, how do we best implement a certain harvest strategy? The choice of tools for implementation is very dependent upon the nature (physical and temporal structure) of the fishery, and requires specific local knowledge. As an example, conside a fishery in which an $F_{0.1}$ strategy is used; the managers seek a way to achieve the target exploitation rate. A common approach is to use catch-at-age analysis or surveys to estimate the current stock size, and then multiply the stock size times the desired exploitation rate to determine a total allowable catch for the year. If we estimate that we have 700,000 tons of fish, and we want a 12% exploitation rate, then we set the TAC at 84,000 tons. This is quite simple and understandable, but the biomass of 700,000 tons is just an estimate, and in most fisheries is subject to considerable error. Similarly, while the quota may be set at 84,000 tons, the actual catch (including fishing-induced mortality and discarding) may be quite different.

Considering errors in estimation of the current biomass, if we underestimate the biomass we will set the TAC too high, and this may have a prolonged effect on future stock sizes and catches. If biomass is quite difficult to estimate, so that potentially damaging overharvests can be expected to occur with some regularity, we might be better off with a simpler policy that involves fixed season length or fixed fishing effort; these alternative tactics will at least tend to keep the harvest rate stable, unless catchability increases sharply with decreased stock size.

Consider, for instance, how fishermen are likely to behave under a fixed season length, if there are alternative fisheries or employment opportunities. Ideally, a fixed season length would produce a constant fishing effort and a constant fishing mortality rate. However, this would rarely be the case. If stock abundance were low, price were poor, or vulnerability were low, then fishermen would be drawn to other fishing and nonfishing activities, and the realized fishing mortality rate would be lower than normal. Similarly, if stock abundance was unusually high, or price was good, then fishermen would tend to fish more, and the exploitation rate would be higher than "expected." In this situation, there would be a desirable relationship between realized exploitation rate and the stock size; if the stock size was high it would be harvested harder. This is just the opposite of what would be obtained by using TACs.

Table 15.1 summarizes the general relationship between strategy and tactics when there is only uncertainty in stock size. Unfortunately, we are often uncertain about vulnerability and expected effort as well as stock size. If stock size were known perfectly, then a TAC would be a better tactic than either effort or season length limits, no matter what strategy we wished to follow.

Table 15.2 summarizes the effects of uncertainty in stock size, vulnerability, and expected effort on various harvest tactics. When stock size is well known, fishing to a TAC is an effective tactic for any harvest strategy. When vulnerability is perfectly known, then effort limits are the most effective. When effort is perfectly known, it is more difficult to prescribe the optimum tactic because it still depends upon the uncertainty in stock size and vulnerability.

In real fisheries, stock size, vulnerability, and expected effort are all unknown. The optimum tactic will depend upon the relative uncertainties in

Table 15.1. Relative merits of tactics and strategies when only stock size is unknown

Strategy	Tactic: TAC	Tactic: Time Limits	Tactic: Effort Limits
Fixed Escapement	worst	best	medium
Fixed Harvest rate	worst	medium	best
Quota	best	worst	medium

these quantities, and some simulation analysis should be performed to see how they compare.

Implementation

Different harvest tactics have different implementation requirements and costs, as well as differences in their intrinsic reliability. Annual catch quotas or TACs require the most information and are the most expensive to implement. Normally, the total catch to date is estimated by sampling the landings for average catch and multiplying times the total number of landings. An estimate is also made of the number of vessels currently at sea and the amount they have on board. Once the current landings plus the estimated amount on board are equal to the TAC, the fishery is closed. This entire estimation procedure requires a rather elaborate port and vessel sampling program, which can be very expensive.

Errors in the reliability of TACs can arise from each element involved. Average catch may be misreported or landings not randomly sampled; estimates of the total number of landings may be in error and the average catch of vessels currently at sea may be higher or lower than the average reported for vessels already landed. When many vessels are involved, or many ports are receiving catch, the sources of error and costs of implementation can be enormous.

Effort limitation is easier to implement. Even though the same monitoring of fishing effort is required (boats at sea, boats already landed, etc.), effort limitation does not require sampling of the catch per landing. Therefore it is intrinsically less expensive and subject to fewer sources of error than TACs.

Season lengths are even more simply implemented since they require only enforcement of the fishing season. This probably accounts for their widespread use in many of the worlds fisheries. In some cases, they are combined with crude and relatively inexpensive monitoring of fishing effort to guard against unexpected changes in timing and extent of fishing pressure during each season.

In some cases, it is possible to achieve relatively precise exploitation-rate

Table 15.2. Effects of uncertainty in stock size, vulnerability, and effort on different tactics

Effect of Uncertainty in	Effect of Uncertainty on		
	TAC	Time Limits	Effort Limits
Stock size	lower stock higher harvest	lower stock lower harvest	constant harvest rate
Vulnerability	none	higher vulnerability higher harvest rate	higher vulnerability higher harvest rate
Effort	none	higher effort higher harvest rate	none

targets by making use of biological characteristics of migration patterns in conjunction with time/area fishing seasons. For example, guantlet fisheries for migrating Pacific salmon have long used the approach of closing the fishery for several days each week, and restricting the size of fishing areas, to insure that a fraction of the migrating fish escape to spawn. For fish and invertebrates with seasonal or developmental migrations, fishing areas and times can often be located to protect smaller juveniles (nursery areas) and/or maturing adults. However, tactics that rely on space-time regulations related to animal movements will generally produce variable results due to environmental factors that cause changes in migration/dispersal patterns (for example, stalling in salmon migrations due to low freshwater flows at river mouths), so these tactics often need to be supplemented with monitoring and emergency regulation protocols.

15.9. Summary and Critique

The most important aspect of harvest strategies and tactics is to explicitly think about them! All too often there is no formal comparison of different strategies, nor is there any consideration of the relationship between implementation and the performance expected from the harvest strategy. A strategy that will work well only if stock size can be perfectly measured and catch perfectly regulated is simply dangerous. Search for strategies that are robust not only to biological uncertainty, but also to the practical difficulties of implementation.

We offer no prescriptions about what strategies are best, since the objective and tactical opportunities for regulation of fisheries are so variable that it is simply impossible to make any firm generalizations. However it does seem clear that constant stock-size strategies will rarely be appropriate, despite the body of theory showing they are "optimal." Variability is often as important an indicator as average catch, and the assumptions that MSY is the objective and stock dynamics are perfectly known are rarely if ever met. No simple replacement for constant stock size has emerged, but $F_{0.1}$ does seem to be emerging as a default option. Do not accept $F_{0.1}$ blindly; compare alternative harvest strategies for different biological assumptions.

Pulse harvesting, sex-specific harvesting, and size-specific harvest strategies are all probably underutilized. These strategies offer the possibility of being very robust, and, when the biology and economics are appropriate, these may be excellent strategies.

Bibliographic Notes

The literature on harvest strategies is rather spotty: the standard fisheries reference books do not explicitly deal with comparison of alternative harvest strategies. Ricker (1958b) simulated different stock dependent strategies and examined the tradeoffs between average yield and variability of yield. This work was extended by Larkin and Ricker (1964), Tautz et al. (1969), Mendelssohn (1980) and Sissenwine et al. (1988). Doubleday (1976) compared yield and variability of yield for constant stock size, constant quota, and constant effort regulation. Beddington and May (1977) showed that variation in yield increased as yield approached MSY. Sissenwine (1977), Shepherd and Horwood (1977), May et al. (1978), Ludwig (1981), and Murawski and Idoine (1989) all examined how variability in stock production affected the potential production from the stock. Beddington and Taylor (1973) examined age specific harvest strategies. Beddington (1978), Beddington and Grenfell (1979), and Garrod and Horwood (1979) considered the risks associated with alternative harvest strategies.

Clark et al. (1973), Clark (1976), and Reed (1978, 1979, 1980) explored several biological models and showed that average yield is maximized by constant escapement strategies. Walters (1975) showed how stochastic dynamic programming (Chapter 16) could be used to solve very general problems for optimal harvest strategies. Walters (1975), Mendelssohn (1982), and Deriso (1985) all examined the importance of objectives and discounting in the nature of the optimal harvesting strategy.

Most of the above work was theoretical, and did not really have a very large impact on actual fisheries management — the theoretical work generally indicated that fixed escapement policies were "optimal" and few fisheries managers wanted to try to manage their fisheries on this basis because of the high year-to-year variability in catch associated with such policies.

The relationship between harvest tactics and strategies is discussed in Sissenwine and Kirkley (1982), and Hilborn (1986).

The concept of $F_{0.1}$ was first proposed by John Gulland and has now become very widely accepted. $F_{0.1}$ represents, in our view, a rather realistic compromise between variability of yield and average yield. $F_{0.1}$ is discussed along with other "biological reference points" in Sissenwine and Shepherd (1987), Andrew and Butterworth (1987), and Deriso (1987).

Hilborn (1979) used simulation models to test combinations of estimation systems and harvest strategies, a methodology that is currently intensively used by the International Whaling Commission (see Punt and Butterworth 1989 for a recent example).

Hilborn (1985) developed the graphical method of examining all linear stock-size-dependent strategies, and this has been subsequently used by Hightower and Lenarz (1989).

16 Optimization

16.1. Requirements

Optimization methods are used to find the best fishing plan given (1) a specified quantitative objective, (2) a model of the stock dynamics, and (3) specific management alternatives. A typical fisheries optimization problem would be to calculate optimum mesh size and fishing effort, so as to maximize the biological or economic yield from a stock. This chapter discusses a number of specific techniques for calculating optimum fishing policies given a specified set of objectives. One will rarely take the results of a specific optimization and simply call them management recommendations; more practically one uses the results of formal optimization to help guide the development of management policies. When used most powerfully, optimization methods can be used to generate new management alternatives that had not been previously considered. The use of optimization results is discussed further in Chapter 17. Here we concentrate on optimization methods.

16.2. Equilibrium or Feedback

Optimization produces a set of rules about what actions to apply. These rules may be in the form of an equilibrium policy where the rules do not depend on the status of the stock, or they may be feedback rules where the actions do depend on the stock.

An equilibrium solution to the mesh-size/fishing mortality problem would be one mesh size and one fishing mortality rate, to be applied for all possible stock sizes. Yield-per-recruit analysis traditionally provides this type of result.

A feedback policy for this same problem would specify the mesh size and fishing mortality rate as a function of the stock size. It is likely that an optimum policy to maximize average catch would specify little or no fishing when the stock is small. The best known feedback policy resulting from optimization calculations in fisheries is the constant stock size or "fixed escapement" policy, which says not to harvest when the stock is below an optimum stock size and to harvest all fish surplus to this optimum stock size when the stock is greater than it.

Equilibrium policies are normally much easier to compute than feedback policies because they do not attempt to account for possible reactions to changes in the stock size over time. Using methods reviewed in this chapter, we can compute the optimum equilibrium policy for almost any possible objective, model, or alternative. We can usually compute the optimum feedback policies only for very simple models.

Because of the highly dynamic nature of the biological, economic, and social elements of fisheries, equilibrium policies have many shortcomings. Here we will place most of our emphasis on methods for finding optimum feedback policies; we feel that it is better to deal in a relatively crude way with the right problems (dynamics and change) than to deal in detail with the wrong problems (equilibrium).

16.3. Analytic Methods

The most elegant method for calculating an optimal fishing policy is by mathematical analysis. This was essentially the only method available until the emergence of modern computers. When a problem is tractable to mathematical analysis, the results are often quite general and the implications are easily understood. For equilibrium policies, analytic methods calculate which combination of actions produces the best value of the objective by looking at the derivatives of the yield with respect to the possible actions. The optimum action set is one where these derivatives are all zero. If the objective is simple and the model straightforward, algebraic solutions for the optimum are often possible. For more complex models or objectives, the numerical methods of Sections 16.5 and 16.6 are more often required. A frequent result of mathematical analysis is an equation that then requires some numerical computation.

Analytic solutions for feedback policies are difficult. Clark (1976, 1985) and Mangel (1985) provide the best summary of the available solution methods, which are based on the theory of optimal control and dynamic programming. However, be warned: the methods are mathematically hard to understand, and they will usually provide solutions only for very simple problems. Analytic methods can generally only solve problems where the status of the stock can be described by a single number and where the stock dynamics are deterministic. Problems with explicit age structure, more than one stock, or economic capital as well as fish stock size do not often have analytic solutions. As a practical guide, we think it is highly unlikely that analytic techniques will prove a useful tool for most people working in fisheries stock assessment and management, except to provide general guidance about the structure (functional form, shape) of optimal policies. Most of the problems that are simple enough to solve analytically and realistic enough to be meaningful to managers have been solved. The literature over the last 5 years shows very few optimization results being obtained by analytic methods. Most optimization problems in fisheries are now tackled with graphic and numerical methods.

16.4. Graphical Methods

When the number of possible actions is one or two, we can plot the value of the objective against the action, given any computational procedure for predicting what the value of the objective will be given any action. Figure 16.1 shows a yield-per-recruit diagram in which the average yield-per-recruit is plotted against fishing mortality and mesh size. Such graphic presentations have several advantages.

1. They can be computed for very complex models
2. The results are easily understood
3. The results for alternative objectives can be easily compared; several indicators of performance can be presented and weighted by the viewer.

With a modern digital computer we can easily build models of very complex fish and fishery dynamics. For each control variable we can simulate the outcome dozens of times to compute average yields for each possible action. We can add any biological or economic elements we wish. As long as we have the computer time we will get an answer. For instance, to generate a yield isopleth diagram as shown in Figure 16.1, we might try 10 levels of fishing mortality and 10 mesh sizes. We might calculate the average yield over a 50-year time horizon and repeat the calculations 10 times for each set of actions to allow for different random sequences of recruitment. We would therefore need to run our model for a total of $10 \times 10 \times 50 \times 10 = 50{,}000$ time steps or years. Even with a complex age-structured model, this will take only a few seconds on a large computer and a few hours on the older microcomputers. Such computational resources are available to almost every fisheries biologist.

Most people understand graphs. If care is taken in presenting the axes, the results can easily be explained. We would suggest, for instance, that, in presenting a yield isopleth diagram to decision makers or fishermen, the x axis should be labelled *fishing effort,* or *annual harvest rate,* rather than *instantaneous fishing mortality* or simply F. The mathematical tradition of obscure graph labels, particularly Greek symbols, is a major obstacle to communicating optimization results.

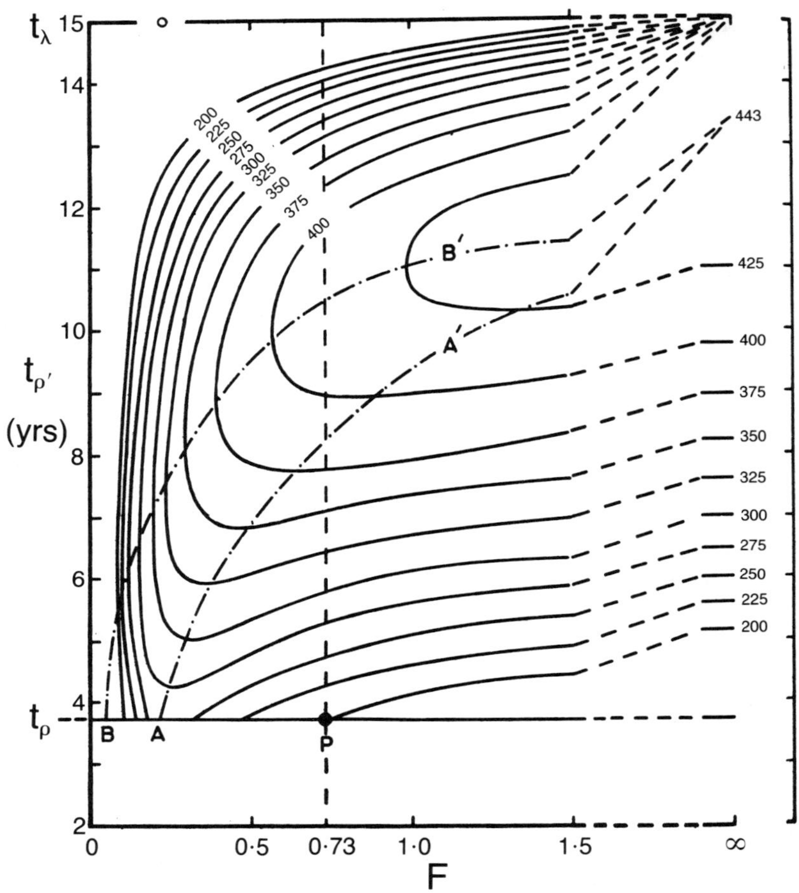

Figure 16.1. Yield-per-recruit diagram showing average yield per recruit for different fishing mortalities and age at entry. Redrawn from Beverton and Holt (1957).

With graphic methods, the objective does not need to be specified prior to analysis. These graphs are not truly optimizations, because they do not determine the maximum; they simply present a view of the results, and the viewer can choose the best point. When more than one indicator of fishery performance is important, graphic methods let you present each indicator on a separate graph.

Figure 16.2 shows 20 indicators of performance for a sport and commercial fishery in British Columbia. Sport and commercial size limits are compared to see how different indicators respond. J. Gross (unpublished.) and Peterman (1975, 1977) developed this graphic technique further by mak-

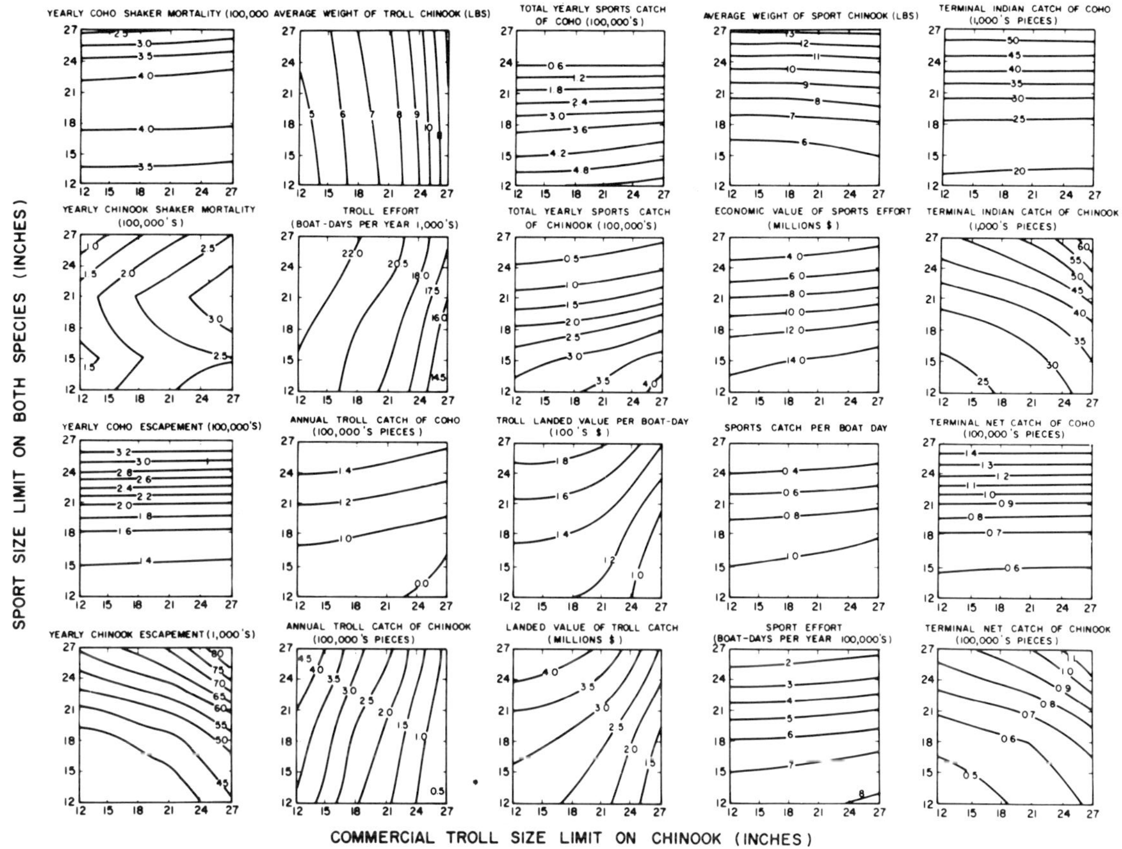

Figure 16.2. Twenty indicators of performance in the British Columbia chinook and coho fishery for different sport and commercial size limits. Results generated by a large simulation model described in Argue, et al. (1983). Can. Bull. Fish. Aquat. Sci. 211: 50, Figure 10–3.

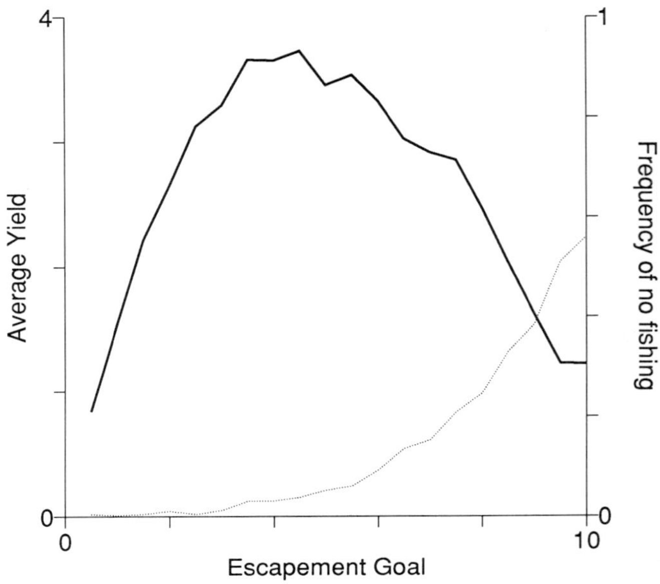

Figure 16.3. Average yield and number of years with no fishing plotted against escapement goal.

ing plastic transparent overlays with an *x* at the origin for each graph. One can then move the overlay (representing movement among policy choices) and view the consequences on each indicator.

Multiple indicator graphs allow you to add constraints by shading unacceptable regions (combinations of actions) on the indicator graphs. By overlaying all the unacceptable regions we can find acceptable ones.

Graphic methods generally do not produce feedback policies, unless the graph axes are parameters of some equation used to represent a feedback relationship. The examples in Figures 16.1 and 16.2 are equilibrium policies. The mesh size and fishing mortality in Figure 16.1 do not depend upon stock strength; the same fishing and mesh size are used irrespective of stock condition.

We can use graphic methods to evaluate alternative feedback policies, if we are clever in defining the axes. Instead of using fishing mortality as an axis, we could use the *target stock size*. This would let us evaluate alternative *fixed escapement* strategies on the *x* axis. Figure 16.3 shows average yield and number of years with no fishing plotted for a single stock using *target stock size* as the axis. This is a graphic feedback optimization. Each point on the average yield curve was generated by doing many runs of a Monte Carlo simulation over time, while taking the harvest at each time to be the excess of stock size over the target.

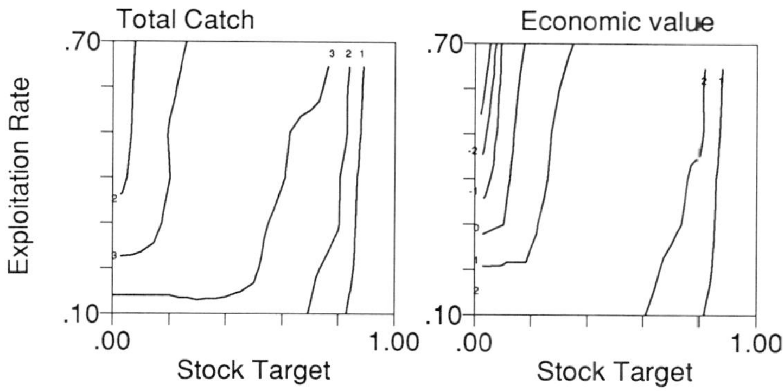

Figure 16.4. Average catch plotted for periodic harvest policies defined as a minimum stock size for exploitation, and the exploitation rate used when harvest takes place.

Graphic representation of alternative feedback policies usually requires more than one graph axis. For example, Figure 16.4 shows contours of total catch and net value of catch (catch minus relative cost of fishing effort) from 50-year simulations of an age-structured model, where the fish are assumed to become vulnerable to fishing over ages 1–3, but have accelerating body growth up to age 4. In this case, the best harvest strategy may be periodic, where there is no fishing for several years in a row followed by 1 year of high harvest rate (see Chapter 15 on harvest strategies). One feedback policy to represent this possibility would be the simple decision rule: "for any year, harvest with exploitation rate X if the stock size that year exceeds Y, and take no harvest if the stock is less than Y." Here we have two policy variables, X and Y, and the rule is a feedback policy because the stock size in any year is used in deciding whether to harvest.

The major constraint on graphic methods is the number of axes. We can really only show two control variables at a time. Thus any policy can only be graphically analyzed on a "two-at-a-time" basis. If we had three control variables, perhaps sport and commercial size limits as well as a season length, we could produce a number of graphs. One approach would be to produce a different graph for each season length with sport size limit on one axis and commercial size limit on the other. Such complex graphic presentations lose much of the intended simplicity and, therefore, benefits. For more complex problems, the numerical fixed-form optimization becomes appropriate.

16.5. Fixed-Form Optimization

When the number of control variables is greater than two, graphic optimization methods become more difficult; we must resort to multiple dimensional displays and multiple panels. An alternative is fixed-form optimization. In fixed-form optimization, you specify a functional form for effects of the control variables, and then use non-linear search methods to find the values of the control variables that maximize the objective measured from multiple simulation runs of the model.

Another way to think about this approach is to think of policies as rules for prescribing specific actions to take under different circumstances. That is, prescribed action is a function of circumstances. When you define such a function, you must define its form (equation, shape) as well as a set of quantitative parameter values for how the function is to be scaled in terms of the variables that are used to measure changing circumstances. These parameter values become the control variables for fixed-form optimization.

For instance, for simple stock recruitment or biomass dynamic models the optimal catch is often a continuous function of the stock size. For fixed escapement and fixed exploitation rates, it is even a linear function of stock size. If we specify that

$$catch = a + b \times stock + c \times (stock)^2 \tag{16.5.1}$$

and specify our objective (long term average catch for instance), we can then search for the values of a, b, and c that maximize the objective using standard nonlinear optimization techniques. This is totally analogous to parameter estimation, except that the objective is not to maximize a goodness of fit, but rather to maximize management objectives.

There is no true guarantee that such fixed-form optimization will give the best possible result, unless you already know the basic form of the optimal control law. However, when dealing with simple biological models of fishery dynamics (ignoring uncertainty), we know that the optimal policy is usually a reasonably simple function of biomass. The advantages of fixed-form optimization are that it is computationally straightforward and allows for any complexity in the model and form of decision rules. Simulation time is the only limiting factor.

The main disadvantage is that you have to have some idea of what functional form to use. Walters and Hilborn (1978) used stochastic dynamic programming on a very simple model to find the functional form and then used fixed-form optimization on a more complex and more realistic model.

16.6. Nonlinear Search Procedures

There are a variety of very general procedures for trying to find the maximum value of an objective Y that depends on a series of control variables $x_1, \ldots, x_n$, where there are simply too many x's to permit a simple graphic or tabular search for the optimum. Here the x's might be parameters of a fixed-form decision rule as discussed above; they may be annual harvest rates to be applied to a stock for the next few years in an open loop (no feedback) management policy; or they might be the minimum size limit and fishing effort.

The procedures fall into three general categories: (1) procedures that use derivatives of Y with respect to the x's; (2) derivative-free, but deterministic, procedures that search the function in some systematic way for an optimum; and (3) random-search procedures that test different x combinations until an optimum is found. The basic requirement is only that you have a computational algorithm (equation, simulation, etc.) that associates a particular value of Y with every possible value of the xs.

The derivative-dependent methods are mainly variations on what is called Newton's method, and they differ from one another mainly in how they constrain or modify the optimization search so as to avoid making jumps in the x values that are too large from one search step to the next. We recommend starting first with Newton's method, then going to more elaborate procedures, such as the Marquardt algorithm, only if necessary. Numerical recipes for these procedures are widely available in books on numerical analysis (Press et al. 1986).

The derivative-free methods are mainly variations on what is called the *simplex method*. Schnute (1982) has developed microcomputer software for applying this method to a variety of fisheries problems.

If you really want to be sure of finding a global optimum for a function that may have several local maxima, about the only sure way is the random search procedure known as the *Matyas algorithm*. Here you simply start from some initial set of x values, and compute Y at several dozen randomly chosen new x points, with the x's perturbed by random normal deviates of reasonably high variance. Whichever of these points is best becomes a new initial point around which another set of random points is chosen. After a few moves of initial point, you begin to reduce the variances of the normal deviates, to home in on the maximum. This procedure generally takes approximately 300–500 function evaluations (calculations of Y) to get reasonably close to the maximum. But on the other hand, it is very simple to program and is sure to eventually find the right answer.

An approach that blends these procedures is known as *stochastic approximation* (Robbins and Monro, 1951). Here the idea is again to choose an x starting point and some other x points near this and to get a stochastic estimate of Y at each of these points (e.g., by several Monte Carlo trials at each of the new points). Then you use the results to find an approximate derivative g_x of Y with respect to the xs, and change to a new starting point according to the rule $x_{\text{new}} = x_{\text{old}} + a_k g_x$. The a_k is a sequence of progressively smaller multipliers that you apply from iteration to iteration. For a fisheries example, see Ruppert et al. (1983).

16.7. Stochastic Dynamic Programming

A very powerful numerical optimization tool now widely used in fisheries is stochastic dynamic programming. Given a specified objective, dynamic model, and list of possible actions, stochastic dynamic programming enables us to calculate the best action to take for all possible states of the system. That is, it calculates the optimal feedback policy. Its major limitation is that the state description and possible actions must not be too complex. Generally the state of the system should be describable by one or two variables, and the possible actions should be similarly limited.

Thus, stochastic dynamic programming is widely used for determining optimal harvest policies for single stocks described by a single measure, such as biomass, but not for determining the optimal feedback policy for full age-structured models. It was first used in fisheries by Walters (1975) for a single stock, and Hilborn (1976) applied it to mixed stock problems. It is currently the most commonly used optimization technique for fisheries harvesting problems.

General overview

Let us pose a general optimization problem, where the fish stock size S may take on any of n possible discrete values $S_1 \ldots S_n$, the harvest in any year H is specified in catch and may be any of m discrete values $H_1 \ldots H_m$ subject to the constraint that catch is less than or equal to stock, and there is some objective ΣY_t (the catch summed over all years) to be maximized that depends upon the catch in each year, $Y_t = f(H_t)$. Initially, let us also assume that we simply want to maximize the sum of Y_t's over time. Further, let us assume that we have a Markov transition matrix that tells us the probability of going from any state after harvest $S' = S - H$, to any new stock size as of the next year. We will call this Markov transition matrix **P**. Notice that **P** here provides a very flexible and general model for the dynamics of transition (survival, growth, reproduction) from the spawning time one year (or generation) to the stock available for harvest the next. No equations are involved, but rather just a general tallying of the frequency with which we expect different stock sizes S to occur following any starting spawning stock S'. We might of course generate **P** from equations, such as a stock-recruitment curve relating S to S', but the optimization procedures turn out not to depend in any way on how we get the **P** matrix.

The above are the conditions required to use stochastic dynamic programming to solve a harvesting optimization problem. Let us now specify these general conditions with some real numbers. Table 16.1 gives the Markov transition matrix for Skeena River sockeye salmon, with spawning stock and recruitment broken into discrete intervals of 150,000 fish. This table is de-

Table 16.1. Markov transition table for Skeena River sockeye salmon stock and recruitment.

Resulting Stock Size	Spawning Stock Size 0–150	150–300	300–450	450–600	600–750	750–900	900–1,050	1,050–1,200	1,200–1,350	1,350–1,500
3,600–3,750	0.0	0.0	0.0	0.0	0.0	0.25	0.20	0.0	0.0	0.0
3,450–3,600	0.0	0.0	0.0	0.0	0.0	0.0	0.0	0.0	0.0	0.0
3,300–3,450	0.0	0.0	0.0	0.0	0.0	0.0	0.0	0.0	0.0	0.0
3,150–3,300	0.0	0.0	0.0	0.0	0.0	0.25	0.0	0.0	0.0	0.0
3,000–3,150	0.0	0.0	0.0	0.0	0.0	0.0	0.0	0.0	0.0	0.0
2,850–3,000	0.0	0.0	0.0	0.10	0.0	0.0	0.0	0.0	0.0	0.0
2,700–2,850	0.0	0.0	0.0	0.0	0.0	0.0	0.0	0.0	0.0	0.0
2,550–2,700	0.0	0.0	0.0	0.0	0.0	0.0	0.0	0.0	0.0	0.0
2,400–2,550	0.0	0.0	0.0	0.0	0.0	0.0	0.20	0.25	0.0	0.0
2,250–2,400	0.0	0.0	0.0	0.0	0.40	0.0	0.20	0.0	0.0	0.0
2,100–2,250	0.0	0.0	0.0	0.0	0.0	0.25	0.0	0.0	0.0	0.0
1,950–2,100	0.0	0.0	0.0	0.0	0.0	0.0	0.0	0.0	0.0	0.0
1,800–1,950	0.0	0.0	0.0	0.0	0.0	0.0	0.0	0.0	1.0	1.0
1,650–1,800	0.0	0.0	0.11	0.0	0.0	0.0	0.0	0.25	0.0	0.0
1,500–1,650	0.0	0.0	0.11	0.20	0.0	0.0	0.0	0.0	0.0	0.0
1,350–1,500	0.0	0.0	0.0	0.10	0.20	0.0	0.20	0.0	0.0	0.0
1,200–1,350	0.0	0.14	0.22	0.20	0.0	0.0	0.0	0.25	0.0	0.0
1,050–1,200	0.0	0.0	0.11	0.20	0.0	0.0	0.0	0.0	0.0	0.0
900–1,050	0.0	0.0	0.22	0.0	0.20	0.0	0.0	0.0	0.0	0.0
750–900	0.0	0.0	0.0	0.10	0.20	0.25	0.20	0.25	0.0	0.0
600–750	0.0	0.43	0.11	0.10	0.0	0.0	0.0	0.0	0.0	0.0
450–600	1.0	0.28	0.11	0.0	0.0	0.0	0.0	0.0	0.0	0.0
300–450	0.0	0.0	0.0	0.0	0.0	0.0	0.0	0.0	0.0	0.0
150–300	0.0	0.14	0.0	0.0	0.0	0.0	0.0	0.0	0.0	0.0
0–150	0.0	0.0	0.0	0.0	0.0	0.0	0.0	0.0	0.0	0.0

scribed in more detail in Chapter 7. The midpoints of the intervals are therefore our S's with $S_1 = 75{,}000$ $S_2 = 225{,}000$, and so on. We will consider possible harvests H's to take the same discrete values. If we assume that the objective is to maximize the expected average catch, then $Y = H$.

Numerical steps

The secret to solving stochastic dynamic programming problems is to work backward. Rather than go into any sophisticated theory, we will look at the numerical steps involved, which nicely reveal the intrinsic beauty and elegance of this form of optimization. For another presentation similar to the one below, along with some discussion about the more general theory, see Mangel and Clark (1988).

Let us assume that we know the future long-term value of being at any possible stock size (S_i) at some time t, and let us call this V_{it}. In principle, the value should be the net present value of the stock size i in year t, and if the resource was private and could be sold, the market value might be a good assessment of V_{it}. Naturally we would expect that V_{it} would be larger for larger values of the stock size i. In a limiting case, where we intended to harvest all the fish at time t, and ignoring any harvesting costs, $V_{it} = S_{it}$. For most fish species, we would generally expect the value of the spawning stock to be greater than its value as catch since the fish left to spawn will produce more fish in the future. As a starting estimate, let us assume that each fish left in the spawning stock is worth only the value of a fish in the catch, or $V_{it} = S_{it}$, for some distant future time t. Note that we are considering value in units of fish, and to convert to dollars all V's would need to be multiplied times the average price of a fish.

Having specified the value of the stock at time t, we then want to work backward in time, and ask what will be the best harvest action at time $t - 1$. That is, what action should we take at time $t - 1$ to maximize the value of the catch at $t - 1$ plus the value of the stock at time t? We can call this total value $V_{i(t-1)}$. Note that it depends on the stock size i at time $t - 1$, and that it can be written as

$$V_{i(t-1)}|H_k = H_k + V_{jt} \tag{16.7.1}$$

where H_k is a harvest of level k, and the subscript j indicates the stock level j that we would end up with if we took harvest level k from stock size i at time $t - 1$.

We can now do a local optimization to find the maximum $V_{i(t-1)}$ that can be obtained for each stock size level i that might arise by time $t - 1$. We simply try each harvest level k, to find the one that produces the highest value of $V_{i(t-1)}$, where for each possible choice k this value is computed using the above equation to combine H_k and V_{jt}. For illustration, we will look in detail at one possible state. For the possibility of having 825 (thousand) fish at time $t - 1$ ($i = 6$), and using our discrete catch options H_k's, then we can take 0, 150, 300, 450, 600, or 750 fish in the catch. If we take a catch of zero, $V_{6(t-1)}$ is simply V_{jt}. Since the stock-recruitment process is stochastic, we do not know for sure what j value we will have at time t. Here is where the population model, in the form of the state transition matrix **P**, enters the analysis. If we have 825 at time $t - 1$, and do not harvest, then, from Table 16.1, we have a 0.25 probability of having 825 fish at time t, and a 0.25 probability of having 2,175 or 3,225 or 3,675 fish. Let us assume that the best way to combine these possible outcomes is to take a weighted average of the values associated with them, with the weights being the probabilities of occurrence from the **P** matrix. The total value of taking no harvest at time $t - 1$ if we have 825 fish at time $t - 1$ is then the sum of probabilities of ending up with the different stock sizes at time t times the values of having those stock sizes. This is summarized below.

Resulting Stock Size	Probability	$V_{St} \times \Pr(S_t)$
825	0.25	206
2175	0.25	544
3225	0.25	806
3675	0.25	919

Thus we see that the expected value of taking no harvest at $t - 1$ with a stock of 825 is $206 + 544 + 806 + 919 = 2475 = V_{6,t-1}|(H = 0)$.

The computational procedure of stochastic dynamic programming is basically just a matter of pure brute force: for every possible state i at time $t - 1$, try (compute expected value for) each possible action and find which one has the highest expected value. Below we show the calculations for a harvest of 600 from the stock of 825, which produces a spawning stock at time $t - 1$ of 225.

Resulting Stock Size	Probability	$V_{St} \times \Pr(S_t)$
225	0.143	32
525	0.286	150
675	0.429	290
1275	0.143	182

The value of taking a harvest of 600 at time $t - 1$ is the catch of 600 plus the expected value of the remaining stock which is $32 + 150 + 290 + 182 = 654$. $654 + 600 = 1254$ is considerably less than the value of taking no harvest at time $t - 1$, which was 2,475. In fact, when we compare all possible catches for a stock of 825, we find that a harvest of zero provides the highest expected value.

We have now calculated the best action and the best expected value at time t for a stock of 825 at time $t - 1$. We then repeat this calculation procedure for every other possible stock size at $t - 1$. The resulting optimum catch and expected values are given in Table 16.2. Note that the optimal action is to not fish at all unless the stock is at least 975, and then to increase the catch to keep the stock size after harvest at 825. Also notice what has happened to the expected values at t. Although we assigned V_{it} to be simply the stock size at time t, at time $t - 1$ the expected value rises very sharply for the first 825, and then increases only at a rate of 1:1. This is obviously because at stock sizes less than 825 the fish left in the spawning stock are very valuable. In the next generation, they will have produced two to four times as many individuals, but any fish over 825 only contributes by being available for catch.

Table 16.2. Optimum catch and expected value at time $t - 1$.

Stock Size (S_i)	Optimum Catch	Expected value $V_{i,t-1}$
225	0	654
375	0	1124
525	0	1380
675	0	1575
825	0	2475
975	75	2550
1125	225	2700
1275	375	2850
1425	525	3000
	⋮	
3675	2775	5250

Solving for $t - 2$ and earlier

So far, we have not solved for the optimal feedback policy in any general sense. We have only determined the best thing to do at time $t - 1$ given an arbitrary value of the stock at time t. The next step is to move back to time $t - 2$ and calculate the optimal action at that time, given the expected value at time $t - 1$ for being in each possible state i that might be reached at $t - 1$ starting from different states at $t - 2$. The computational procedures are exactly the same as for finding the best choices at $t - 1$ while recognizing the values of resulting states at t.

Once we have solved for $t - 2$, we move back to $t - 3$, then $t - 4$, and so on. Again, each of these solutions is not necessarily the optimal policy for an infinite time horizon but, in fact, the optimal policy choice for each state usually stops changing after a few backward time steps and we have converged to the true optimal feedback policy. Technically, we say that the feedback policy becomes stationary (independent of t), and this almost always happens quickly in harvesting problems. In the example, the convergence takes place in one time step, although this is not generally true.

It may be useful for you to think about similarities between simulation and dynamic programming and also how these tools differ in very fundamental ways. In doing a simulation, you apply rules to a single state at time t in order to generate a new state at time $t + 1$. In dynamic programming, you apply rules to a function of the state at time t (the value function V), in order to generate the function at time $t - 1$. In dynamic programming, when you admit stochastic variation in the rules for building the function (i.e., in the local optimizations that give points along the function), you are not even thinking about any particular sequence of states over time as you would if you were doing a simulation. In doing a simulation, you have to hypothesize a particular rule for relating action to state; in dynamic programming, this rule emerges from the computations as a consequence of the local optimizations.

Adding discounting

We often want to consider discounting of future catches. This is easily done by modifying Equation 16.7.1 as follows:

$$V_{i(t-1)} = Y_{t-1} + V_{jt}\left(\frac{1}{1 + d}\right) \qquad (16.6.2)$$

where d is the discount rate. This has the effect of reducing the future value in relation to the current yield.

When discounting is employed, the vector V will eventually converge, that is, it will stop changing over time because the amount being lost by the discounting term to the future value balances the value being added from this year's catch. At this point, one can be assured that the optimal actions will also have converged.

Continuous problems

Stochastic dynamic programming can also be used for problems where the state dynamics are defined by continuous models instead of Markov transition patterns. However, the computational procedure then involves discretizing the states to approximate the value function, and we have generally found it is easier to convert a continuous model (such as the Schaefer production model) into a Markov transition matrix for computation right from the start.

There are a few problems where you cannot safely convert to a Markov representation right at the start. These are mainly problems in adaptive control, where the state of the system includes stock size, as well as quantitative measures of uncertainty about the correct model (probabilities placed on alternative models, variances of parameter estimates, etc.). You still work with discrete levels of these measures, but you have to use rather elaborate interpolation procedures to account for state transitions that end up between the discrete levels. For further details, see Walters (1986).

Larger problems

Mixed stock problems, or ones where stock size and fleet size are involved, are conceptually the same as the simple problem described above, but the state description and list of possible actions becomes more complex. If we have two stocks, each broken into 20 discrete steps, then our state description would have 400 possibilities, and the Markov transition matrix would be 400 × 400, which, while large, is still quite solvable even on a microcomputer. Such a complex state description is often unnecessary since unless there were biological interaction among the stocks, the problem could actually be solved using two 20 × 20 Markov matrices.

Another example where at least two state variables are involved is in harvest regulation by annual quota, if the management agency is constrained in how large a quota change it can make from year to year. In this case, the stock size is one state variable, but last year's quota becomes a second state variable. The decision variable is not the quota, but rather the increment (up, none, down) to make in the quota for this year. The optimum policy then specifies the optimum increment to make for every possible combination of stock size and current (last year's) quota.

16.7. Summary and Critique

We have discussed a number of approaches to optimization, a "bag of tricks" that can be used in various ways for different problems. Quite often, for instance, one might do some mathematical analysis to understand the basic elements of a problem and then resort to stochastic dynamic programming to find the solution for the full problem. Generally, mathematical analysis is preferred for its simplicity but will not solve most problems that are now considered interesting, particularly those involving more than one state variable. Graphic methods are strongest for their communicative power but frequently provide little guarantee of providing a true optimum solution. Stochastic dynamic programming is probably the most powerful technique but is limited in the complexity of problems it can handle.

Bibliographic Notes

Clark (1976) provides the best summary of analytic methods and results for fisheries analysis. Good introductions to numerical procedures are in Mangel and Clark (1988). Beverton and Holt (1957) popularized the use of graphic methods, particularly for yield isopleth diagrams. Peterman (1977) extended this approach to multiple indicators and introduced the use of constraint regions. Walters and Hilborn (1978) first described fixed-form optimization for resource management problems.

Stochastic dynamic programming was invented by Richard Bellman (1961). K.E.F. Watt (1963) showed how it could be used in natural resource management problems, and Walters (1975) first applied it to fisheries.

17

Designing Adaptive Management Policies

17.1. Introduction

It does not take much practical experience with stock-assessment methods and models to realize that the available data usually cannot provide unique and reliable estimates of the best management policy. Assessment calculations based on historical data often reveal gross uncertainties about sustainable yields, optimum effort levels, effects of various stock enhancement measures, and so forth. These uncertainties are not just a matter of annoying imprecision in estimates of a few parameters such as natural mortality rates; usually they reflect a fundamental lack of experience about how the stock behaves under alternative policy options. In other words, the prediction of the best policy choice is usually either a gross extrapolation about how the stock will respond to conditions that have never occurred in the available historical data, or is only one of a collection of alternative predictions that each appear equally credible based on the data.

The uncertainties do not automatically disappear over time as more experience is accumulated and more detailed biological information is gathered about a stock. Consider the eastern Atlantic tropical tuna fishery that has existed for two decades or so and has been sampled primarily for catch and effort statistics, and compare this fishery to the Pacific halibut fishery that has been studied in considerable detail for almost 60 years. In the tuna case, surplus production analyses (plot of catch versus effort, fit of dynamic surplus production model, etc.) reveals that the best fishing effort is apparently somewhat higher than any effort level seen so far (see Figure 17.1 later); here it is obvious how any policy prediction is a gross extrapolation from the limited historical experience available. But are we fundamentally any better off in the halibut case? Deriso et al. (1986) point out that the mass of historical data on halibut are consistent with two very different hypotheses about the relationship between stock size and recruitment and about the impacts of higher exploitation rates than have historically been permitted. Recruitment fluctuations for the halibut may be explained equally well by a model based on environmental fluctuations (in which case, there is a severe risk of recruitment overfishing), or by a dome shaped (Ricker) stock-recruitment relationship (in which case it would be best to fish harder). What is happening with the halibut, and many other fish stocks, is that we are learning a great deal about some aspects of the biology (growth, mortality, migration), but are left with gross gaps in understanding of other processes that are difficult to measure and study, particularly the stock-recruitment relationship. The gaps are not just due to measurement difficulties; management practices often prevent us from even having a chance at measurement by preventing informative stock sizes and harvest rates from occurring in the first place.

In some fisheries, very conservative management policies (very restrictive harvest regulations) have been employed as a hedge against uncertainty. The usual justification for these policies is to avoid taking risks with the stocks until better scientific data can be gathered about their potential productivity. This is a most curious argument; how can scientific data about the potential of a stock be gathered when the stock is prevented from empirically exhibiting that potential? The best that can be expected from data gathering during a period of conservative management is refined estimates of growth and mortality parameters and some reasonable alternative hypotheses (extrapolations) about how recruitment might behave under less conservative harvesting. In other words, more data gathering can lead to a more detailed collection of alternative models for stock response, but cannot in principle define which of these models is correct until some hard empirical experience is obtained.

Major uncertainties are of two types: (1) those that the manager should not or need not learn about to manage a stock well, and (2) those that define untested opportunities to perhaps improve yields and economic performance. An obvious example of an uncertainty of the first type is how recruitment will behave at very low stock sizes for a stock that has already suffered some degree of recruitment overfishing; it might be scientifically interesting to see what would happen if the stock were depressed still further, but the manager should avoid this circumstance if he knows that recruitment overfishing has already occurred (he should be moving the stock in the opposite direction). Our concern in this chapter will be with uncertainties of the second type; how should management respond when the optimum policy is uncertain and the optimum can only be found by testing alternative possibilities (opportunities) through management experience?

In the face of large uncertainties, fisheries management is necessarily an adaptive process where the decisions and policies developed over time may have a profound influence on how rapidly the uncertainties are resolved. The fact that management is an adaptive process raises two key technical questions for practitioners of stock assessment: (1) how can we predict (model) the impact of different decision choices on uncertainty, recognizing that decisions made today will establish a "legacy of uncertainty" for future stock assessment and decision making, and (2) how should management policies be adjusted to take account of the "value of information" associated with decision choices that will help to resolve uncertainties? Obviously we cannot answer the second of these questions unless we can answer the first; unless we can predict how alternative policy choices will affect uncertainty, we cannot judge whether it is worth considering choices that risk or sacrifice short-term fishery performance in favor of getting better information and experience for longer term management.

In this chapter we will first look in general at whether it is worth trying to predict uncertainty over time and to incorporate such predictions into assessments of the relative merits of different policy options. Then we will look at basic steps in the design of "adaptive policies" that do take deliberate account of future uncertainty, and at the key technical problem of how to model uncertainty over time. Finally, we will examine some theoretical and practical impediments to the implementation of adaptive policies.

17.2. Passive Versus Active Adaptive Policies

There are three basic strategies for dealing with uncertainty in the management of dynamic systems over time. These strategies differ in how models based on historical data are used to guide policy choices. First, one can use the available data at each point in time to construct a single "best guess" or "best possible model" based on the data, and then act as though this model were true (or hedge against uncertainty by being more conservative than this model predicts) while counting on any weaknesses or errors in it to reveal themselves in future assessments. This is called a *passive adaptive* strategy, and, in adopting it, one ignores the impact that behaving according to any one model may have on learning about whether that model is correct. Passive adaptive policies can in fact be optimum when uncertainties are small and/or when the passive decision choice is as informative as any other choice would be. However, passive policies may cause the system to be locked in to a narrow range of behavior (e.g., stock size and harvest held near one presumed optimum) without any data ever being gathered to help decide whether the optimum is in fact within this range.

Second, one can simply try a variety of alternative policies, more or less at random, in hopes of accumulating experience about which one is best. This is called an *evolutionary adaptive* or *trial-and-error adaptive* policy. It has the advantage of not requiring the decision maker to make any judgement about which model best fits the data available to him at any point in time. If the manager has considerable flexibility to try a wide range of choices, an evolutionary adaptive approach may lead him to eventually stumble upon some very good choice that would never have been identified or noticed through rigorous but narrow analyses of historical data. However, like natural evolution, a trial-and-error strategy is likely to be very wasteful.

Third, one may deliberately try to construct a range of alternative models that are consistent with historical experience and use these to identify a policy that offers some balance between probing for information (directed experimentation) versus caution about losses in short-term yield and long-term overfishing. This is known as an *actively adaptive* strategy. It involves a great deal more effort in stock assessment and modelling than is required for passive policy design. It may involve testing a much narrower range of best bet policy choices than would be tried in a trial-and-error strategy, but likely a wider range than would be tried in a passive strategy. Thus it is in a sense a compromise between those extremes.

Most fisheries stock assessments are aimed at passive policy design, and there is rarely any attempt to systematically identify alternative models as would be required for the design of actively adaptive policies. However, it is unusual for management agencies to adopt a passive policy directly on the recommendation of assessment experts, without modification. Trial-and-error policy choices often arise in management settings where the results of scientific assessments are discussed and criticized. Often the final decision choice is essentially a trial-and-error one, made by management staff without formal review or reevaluation by assessment staff. In other words, some form of actively adaptive or trial-and-error policy design often creeps into the management process whether or not formal assessment procedures are used; this is sufficient grounds in itself for assessment experts to broaden their perceived mandate to at least look at simple active adaptive policies.

Our emphasis in the remainder of this chapter will be on the design of active adaptive policies. However, this does not mean that the others should be rejected out of hand. The outcome of active adaptive design is sometimes a prescription to just act as though a single best-fitting model were correct (i.e., manage passively), either because the uncertainties are not large enough to justify a riskier policy or because the passive choice is as informative as any other would be. Also, there are times when a pure trial-and-error strategy is worth considering, just to break away from existing management dogmas and habits; in fact, the more adamant the biologists are that they know what the best policy is in some situation, the more important it is to consider some imaginative, trial-and-error choices that are far from their beliefs (witness the success of Pacific salmon stocking into the Laurentian Great Lakes, which was bitterly opposed or scoffed at by many respected fishery biologists in the early 1950s).

17.3. Steps in Adaptive Policy Design

The development and evaluation of actively adaptive policies involves six basic steps:

1. Identification of alternative stock response hypotheses
2. Assessment of whether further steps are necessary by estimating the expected value of perfect information
3. Development of models for future learning about hypotheses
4. Identification of adaptive policy options
5. Development of performance criteria for comparing options
6. Formal comparison of options using tools of statistical decision analysis.

These steps are discussed in detail in Walters (1986); here we provide only an overview of the key ideas and data requirements.

Identification of alternative hypotheses

The first and most critical step in adaptive policy design is to provide a clear and simple description of major uncertainties, in the form of a set of distinct alternative hypotheses or models about how the stock will respond to various policy options. Although this set of alternatives may be defined through various complicated statistical procedures, it is important that it be presented and described initially in very simple terms, that is, as a few extreme alternative models, whose different policy implications are easily understood. It is also important that this set not include any alternatives that differ in terms of biological realism or mathematical formulation, but are essentially the same in their policy predictions. (For example, if a delay-difference equation and a simpler surplus production equation both fit the data well and predict the same optimum harvest policy, then only one of these should be included in the set.)

When only simple catch and effort data are available, the alternatives can usually be described most simply as a set of alternative curves extrapolating beyond the range of available catch versus effort data (Figure 17.1), or as a set of alternative predictions of how catch per effort will change over time if the fishing effort is increased (or decreased). Obviously, one might draw infinitely many curves in such situations. But the key point at this stage in policy analysis is not to identify every possibility, but instead to define a likely range of possibilities and a few intermediate possibilities within this range.

When more complex models and data are used (e.g., delay-difference or age-structured models with mortality, growth, and recruitment parameters), it is usually possible to define a simple range of distinct predictions by the following three-step procedure. First, fix any parameters that are reasonably well known at a set of nominal values (usually possible with natural mortality and growth parameters). Second, pick a leading variable (or two if necessary) from the remaining set of uncertain parameters. Often a good

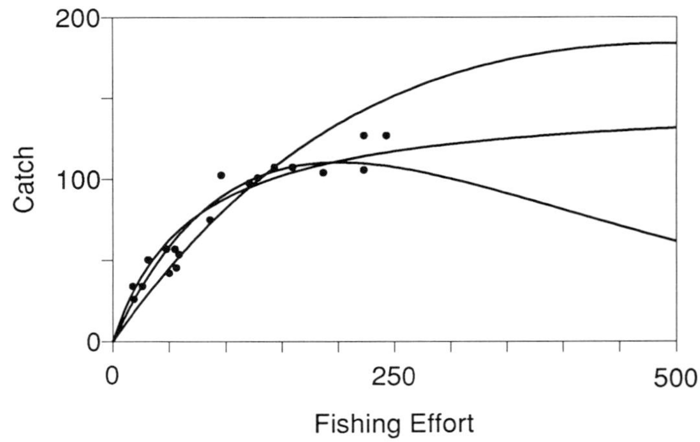

Figure 17.1. Atlantic yellowfin tuna catch effort data with possible extrapolations.

choice here is the initial (unfished) stock size, and/or a stock-recruitment parameter that measures the slope of the stock-recruit relation at low stock sizes. Third, for a reasonable range of values of the leading variable(s), estimate what values of the remaining parameters will give a best fit to the historical data. Each parameter combination of fixed parameter values, leading parameter value(s), and estimated parameter values resulting from these three steps is then taken as a candidate for the final hypothesis set. The final set is obtained by omitting any parameter combinations that are "redundant" in terms of making essentially the same policy predictions.

If all of the uncertainty can be compressed into (represented by variation of) a single parameter, then it is simple to quantify a set of alternative hypotheses. A good example of this is Ludwig's simple one-parameter stock recruitment model (discussed in Chapter 7).

A more elaborate way to define a set of alternatives is to fit some general model to the data (e.g., a delay-difference model), and estimate a joint confidence region for the parameters of this model. The alternatives are then taken to be a set of parameter combinations that systematically span this confidence region. For further details on how to extract parameter sets from multidimensional confidence regions, see the discussion on singular value decomposition in Walters (1986, Chapter 6, pp 177–185). Alternatively, one can bootstrap a large number of alternative parameter estimates (perhaps 20–100), and treat each of these sets of parameter estimates as an alternative hypothesis.

Often the key alternative hypotheses can be defined from the simple example of Table 17.1. This table defines four alternatives:

Table 17.1. Four possible states of the stock.

Recruitment Curve Slope	Initial Stock Size	
	Low	High
Flat	A	B
Steep	C	D

A. The initial stock size is small, and the recruitment curve has a low slope so that recruitment overfishing will occur if the stock size is depressed much.

B. The initial stock size is large, but the stock is sensitive to recruitment overfishing.

C. The stock size is small, but the recruitment curve is steep, so that recruitment overfishing is unlikely.

D. The stock is large and is not prone to recruitment overfishing.

For any particular case, these alternatives must of course be associated with particular parameter values for whatever stock model(s) can be used with the available historical data.

Often the best way to define a good set of alternative hypotheses is to commission or conduct several independent assessments of the available data, then have the authors of these alternative assessments meet to discuss and debate their results in a workshop setting. Such debates have a way of sorting out trivial alternatives ("but your model gives the same answers as mine; it just looks different mathematically") and focussing discussion on basic differences in policy predictions. Note that this is the opposite of how many scientific committees and commission advisory groups try to operate: the emphasis is usually on trying to find concensus and agreement.

Estimating the expected value of perfect information

Once a clear set of alternative hypotheses or stock response models is available, it is worth doing a simple calculation of the expected value of perfect information (EVPI) in order to determine whether further adaptive policy analysis is worthwhile. The essential idea behind this calculation is to find the policy option that would be best if there is no future learning about which hypothesis is correct, then to see how much improvement could be obtained from that nonadaptive baseline if it were known for certain which model is correct, that is, if perfect information were suddenly available. Any adaptive policy is expected to provide a level of performance that is inter-

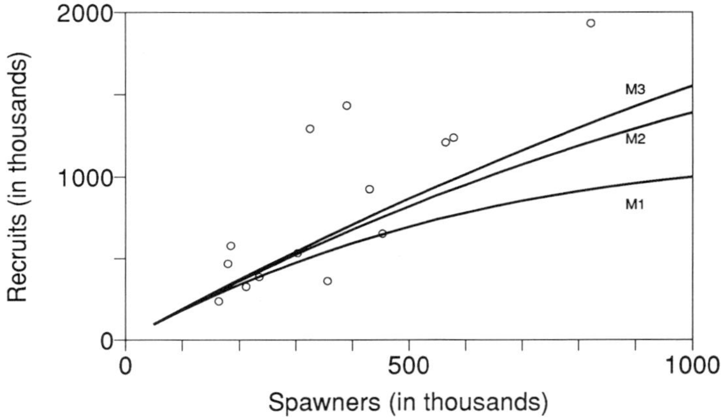

Figure 17.2. Alternative Ricker stock recruitment curves for Fraser river chum salmon data.

mediate between the nonadaptive and perfect information extremes. Often this step in the analysis reveals that there is a "robust" policy that should do very well, no matter which model is correct, so that only minor gains would be expected from having better information. In other cases, the various models imply such similar optimum policies that little improvement can be obtained beyond choosing any one of them as the nonadaptive baseline (that choice is robust in a trivial sense).

The simplest way to estimate EVPI is to construct a decision table that predicts the expected performance for each hypothesis, over a range of policy choices. Then for each model the best choice is read off the table as that giving the highest performance for that model. The best nonadaptive choice is that which gives the best average performance across the models. This average performance across models can be calculated either as an unweighted mean or, more elaborately, as a weighted average where each model is weighted by some prior probability or judgement about how likely it is to be correct based on available historical information. EVPI is then estimated as the difference between the average performance across models, where the value for each model is taken to be the best that could be obtained if the model were known to be correct, minus the expected performance (averaged across models) for the best nonadaptive baseline policy.

Figure 17.2 shows stock-recruitment data for the Fraser river chum salmon that provide no useful information about the potential size of the stock; over the range of available data the recruitment appears to be proportional to the spawning stock size. Any number of stock-recruitment curves could be fit to these data, but for purpose of illustration let us consider three alternatives: each of them involves a different assumption about the parameters of the

Table 17.2. The yield expected from different escapement levels given different biological hypotheses for Fraser River sockeye. The * indicate the optimum escapement level for each hypothesis.

Escapement Level	Hypothesis M_1 $B = 1000$	M_2 $B = 2000$	M_3 $B = 3000$	$\bar{M}$
200	146	171	180	166
400	196*	281	312	263
600	180	350	414	315
800	109	381*	489	327*
1000	0	380	538	306
1200	0	348	563	304
1400	0	292	567*	286
1600	0	212	550	254

Ricker model of the form $R = Se^{a(1-S/B)}$. M_1 says that the Ricker B value is 1000, M_2 says the Ricker B is 2000, and M_3 says it is 3,000. For all hypotheses we assume that the Ricker a is 0.66.

Then consider eight possible strategies for spawning stock management, holding the spawning stock size at levels of 200, 400, . . . 1,600. (Note all units are in thousands of fish.)

Given the alternative models and alternative management strategies, we can solve by simulation for the expected value of following each management strategy if each model is true. This requires programming the model assumptions as true and then solving for many Monte Carlo runs of some specified duration to find the catch that is returned, on average, by using a particular management action. Table 17.2 shows the results of such simulations for the Fraser river chum. Each column of the table represents one hypothesis about the stock recruitment relationship. The last column is the average yield (averaged across the three models, with each model taken to be equally likely) for the escapement level. The table has been constructed so that model M_1 is the most conservative (i.e., the low escapement level of 400 is best if M_1 is correct), M_2 predicts that an escapement of 800 is best, and M_3 predicts that 1,400 is best. The best nonadaptive policy is an escapement of 800, giving an average performance of 327 across all models. From this table we estimate EVPI $= (196 + 381 + 567)/3 - 327 = 55$.

Another way to think of EVPI is how much a sole owner would be willing to pay for the knowledge of the true stock-recruitment relationship. When considering the actual numbers in the table, remember these are in units of average catch per year; to express EVPI in dollar terms we would need to multiply them times the price of a fish, and then find the net present value of a time stream of 55 (thousand) fish per year. As a rough dollar guide, these fish are worth about $15 each, which makes the average annual value of EVPI approximately $825,000. At annual discount rate of 10%, the net present value of this annual average works out to about $8,000,000. This provides a rough guide to what knowledge of the true underlying stock recruitment relationship might be worth.

Another way to calculate EVPI is to take the mean of the differences between the best that could be obtained if each model were known to be true, minus what will be obtained for each if the best nonadaptive policy is used. In the above example, this method gives EVPI = [(196 − 109) + (381 − 381) + (567 − 489)]/3 = 55. Note that in this way of doing the calculation, each difference represents the loss associated with using the best nonadaptive policy (escapement of 800); since 800 was best for hypothesis M_2 in the first place, that loss term (381 − 381) is zero.

The construction of a decision table to assess EVPI is useful beyond providing a test of whether further adaptive policy calculations are worthwhile. It forces an evaluation of the best policy implied by each alternative hypothesis and helps define a range of uncertainty about the best policy. It forces a preliminary assessment of what performance measures should be used to compare policy alternatives. With luck, it helps uncover a robust policy that will do well no matter what model is correct.

Models for future learning

If the EVPI calculations indicate that there would be considerable improvement from knowing which hypothesis is correct, the next step in adaptive policy analysis is to construct models for predicting how long it will take to tell which (if any) of the alternative models is correct, under different policy options (and possibly, alternative monitoring programs). We discuss this important step, which is usually not recognized as a key part of stock assessment, in more detail in Section 17.4.

The basic result of this step should be a computer simulation or algorithm that will predict the probabilities that a manager would place on each alternative hypothesis at any time in the future, given any of a broad range of policy choices and given a future data set (simulated data generated from any one of the alternatives). Alternatively, the algorithm may predict how parameter estimates and measures of uncertainty on them (parameter variances and covariances) would behave over time given any sequence of policy choices. Essentially, the idea is to predict how different policy choices will affect future learning, so that the "value of information" associated with each choice can be assessed.

One simple approach to this step is to first identify what assessment procedures will most likely be used by future managers to analyze their data, then set up a Monte Carlo simulation program that can generate artificial data sets for these assessment procedures given any choice of true model (hypothesis) and sequence of policy choices. Such a program can be used to obtain at least a general feeling for how long it will take to discriminate among alternative models (or obtain reasonably precise parameter estimates) under different policy choices. In particular, it can be used to pinpoint particularly uninformative policy choices; for example, for the Fraser chum stock-recruitment example given above, such an analysis would quickly reveal that low escapement levels would provide very little discrimination between alternative hypotheses, whereas high escapement levels would provide good discrimination.

The analysis of learning patterns can be a frustrating and complicated business, particularly when it is not clear how future data will be analyzed or when attempts to test simulated data with various estimation procedures reveal that those procedures are inadequate (e.g., badly biased) in the first place. It is very easy at this step to become sidetracked into messy and time-consuming evaluations of assessment procedures, when in fact all that is really needed is a general feeling for how different policy choices in the short term will affect the ability of managers to discriminate among alternative hypotheses over the longer term. This basic need must be kept in mind at all times.

Identification of policy options

At this point, the adaptive analysis should have proceeded far enough to provide some baseline policy choices that would be optimal under the different hypotheses and if there is no learning (EVPI calculations) and to have provided a general assessment of how policy choices affect the length of time needed to discriminate among the alternatives. The next step is to do a bit of imaginative thinking about whether there are any deliberately experimental or probing policy choices worth considering, which might provide a reasonable balance or compromise between managing safely versus perturbing the system strongly so as to learn more quickly. For example, in a catch-effort situation, it might be worth considering a very high effort level for a short time, for which the alternative hypotheses make very different catch predictions. In a case such as Figure 17.2, where there is a suspicion that recruitment overfishing may have occurred (some hypotheses predict better performance at higher spawning stocks), it may be worth giving up catches entirely for a few years to allow very high spawning stocks (and hence provide a quick and strong test of where the spawner-recruit curve bends over).

We know of no automatic or mechanical procedures for identifying good experimental policy options. Decision tables and simulations of estimation performance under extreme policy choices can help to identify policies for which the alternative models make particularly distinctive predictions (or for which estimation procedures most quickly converge to correct parameter estimates). Also we know in general that an experimental policy has two basic dimensions: severity and duration; less severe (disruptive, extreme) policies will take longer to provide useful discrimination, and the general tradeoff between severity and required duration can be established by Monte Carlo simulation. However, in the end, the analyst must use considerable judgement and imagination to sort out a reasonably small set of alternatives that appear practical and worth comparing to nonadaptive policy choices.

Development of performance criteria

The next step is to define a precise measure of performance for evaluating the impact of any policy choice, nonadaptive or experimental, when applied given any of the alternative hypotheses. This performance measure must account for both the short-term and the long-term impact of applying the policy, because it will be used to compare policies that may sacrifice short-term performance in favor of improved longer term performance due to better information.

The simplest performance measure that may be credible in management decision making settings is the total catch over some fixed planning horizon, for example over 40 years. A better simple alternative may be the discounted sum of catches over a much longer (or infinite) time horizon, which at least accounts for how a given catch taken in the near future is economically more valuable than the same catch deferred to a later time. Generally, higher discount rates tend to favor nonadaptive policies; experimental or probing policies are favored only when the time horizon is long enough (or the discount rate low enough) for the effects of learning on performance to be fully apparent.

The performance measure may also account for expected product prices and costs of fishing, for monitoring costs, and for hardship costs associated with loss of employment and relocation of fishermen under policies that result in sudden declines in catch at some future time(s). However, care should be taken in constructing such economic cost-benefit accounting, since it must be kept clear who will reap the benefits (e.g., fishermen or the public) and who will bear the costs; simple sole owner accounting for all benefits and costs (as though all would accrue to one owner of the resource) is seldom very useful in policy debates among government and industry actors.

The performance measure may also account for risk aversion to very low catches, by assigning a low utility to very low catches and a diminishingly increasing utility to higher ones. One simple way to do this is to take the logarithms of catches as the utility measure (for an example of this approach, see Walters and Ludwig 1987). Again, care must be taken about whose utility function or risk aversion best represents the management agency's (or the public's) interest in the resource; most biologists and fishery managers are personally very risk averse, but it is not at all clear that they serve the best public interest by assuming that their own personal views represent that interest. Usually, risk aversion in the performance measure will make experimental or probing adaptive policies look less worthwhile.

In general, there will be substantial uncertainty about the best performance measure to use. It is important for the analyst to anticipate this uncertainty by providing a set of alternative performance measures and assessments based on them. It is also important to plan for an iterative process of policy formulation, in which the analyst presents results based on one set of measures, then reacts to criticisms of these results by developing other measures and testing performance under them.

Simple learning: the certainty equivalent policy

The simplest model of learning involves passively using whatever management option is best for whatever model is most likely. At each time period, you estimate the relative likelihoods of different models from the data that are then available, and then assume that the most likely model is true. For your true model, find which management action has the highest expected value. This best action is called certainty equivalent (CE) because you are pretending that you are certain about the model. Certainty equivalent policies may end up acting like actively adaptive policies, because they may inadvertently introduce informative variation, but they do not involve explicit consideration of learning over time.

Assume, for instance, that although the true model parameter for the Fraser chum is $B = 3{,}000$, the data that have been collected so far indicate $B = 1{,}000$. This is presumably due to the luck of the draw in random deviations from the long-term average recruitment curve. If we begin with the escapement goal that is optimal for $B = 1{,}000$, we will, however slowly, eventually learn that the real B value is 3,000. As the data points accumulate, the estimated B will slowly rise, and our escapement goal will similarly rise.

Table 17.2 showed only nonadaptive policies; however, we can also calculate the value of a CE policy under each of the possible models. This calculation involves averaging the results of a number of Monte Carlo simulations, where the simulations include annual updating of parameter estimates (i.e., redoing the estimation regression each year) and corresponding annual revisions of the CE policy.

Doing these simulations results in values of 179 given $B = 1{,}000$, 361 given $B = 2{,}000$, and 532 given $B = 3{,}000$. Averaging these values assuming equal prior odds on the correct B value gives an expected value of the CE policy of 357. This is considerably higher than for any of the non-

Table 17.3. The basic format for evaluation of policy options.

Policy	Hypothesis M_1	M_2	. . .	M_n	$\bar{M}$
Baseline					
AP_1					
AP_2			V_{ij}		
⋮					
AP_m					

adaptive policies and reduces the EVPI from 55 to 24. Another way to look at passive learning and CE policies is that the value of a CE policy is 30 above the best nonadaptive policy.

Formal comparison of options

The last major step in adaptive policy analysis is to formally compare a collection of nonadaptive and adaptive policy options. This involves using the results of the previous steps to provide an extended decision table that shows how different policies are expected to perform under different alternative hypotheses, while accounting for future learning. The basic format of the decision table should be as shown in Table 17.3.

The entries V_{ij} in this table should be the expected total performance measure if policy *i* is used and hypothesis *j* turns out to be correct. The "Baseline" policy should be the best nonadaptive policy found in the EVPI assessment. The first few policies AP_1, AP_2, and so on, should be passively adaptive (or certainty equivalent) policies where management is assumed to initially act as though one or other of the hypotheses is correct and is assumed to change policies over time as (or if) information is gathered about which hypothesis is correct. The last few policies AP_{m-2}, . . . AP_m should represent more extreme experimental regimes, where severe disturbances are introduced in hopes of learning more rapidly.

Note that the simulations from which the V_{ij} are estimated must include learning effects and response to learning, that is, the simulations must simulate future data gathering as well as future reactions by managers to the data. Each policy tested AP_i must thus precisely define (1) a set of initial actions over the first few simulated years, and (2) a feedback rule for future reaction to the results of the initial actions. Policies may differ from one another, both in terms of the severity of the initial actions and in the assumed feedback rules for reaction. For example, policy AP_2 may specify a low initial effort level and that effort be kept low unless (or until) the odds placed on a more optimistic hypothesis become very high; in contrast policy AP_3

Table 17.4. Expected yields from non-adaptive CE policy and three different levels of escapement for 6 years followed by CE policies.

Policy	M_1 $B = 1000$	M_2 $B = 2000$	M_3 $B = 3000$	$\bar{M}$
Baseline non-adaptive	109	381	489	327
CE policy	179	361	532	357
6 yrs 400 escapement	179	350	502	343
6 yrs 800 escapement	182	367	527	358
6 yrs 1400 escapement	175	363	537	358

may also start with low effort, but then specify that effort be increased when there is any statistical indication that a more optimistic hypothesis is correct.

The initial decision table may have many rows; it may compare a large number of policies that differ in various details about how the simulated data are analyzed and reacted to as well as about the initial action regime. However, it is generally possible to "prune" most of the rows (eliminate most of the options) from this table as being *dominated policies*. A dominated policy is one for which there is at least one other policy that will perform better under all alternative hypotheses (i.e., has larger V_{ij} entries for all columns j).

We can construct such a table (Table 17.4) for the Fraser chum example by considering as a baseline policy the constant escapement of 800, which was the best nonadaptive policy. We can add as a passive policy the simple CE policy described in the previous section. Finally we can consider three "experimental probing" policies that consist of 6 year experiments of 400, 800, and 1,400 escapement levels followed by the use of a CE policy after the 6-year experimental period. The three escapement levels were chosen because they are the optimal escapements for the three alternative models, and the 6 year experimental period was chosen more or less arbitrarily. In an exhaustive analysis, different experimental periods could be evaluated. The purpose of this type of probing experiment is to provide *informative* data points during the experimental period that will improve the performance of the CE policy once the experimental period is completed.

We see that, whereas all of the adaptive policies provide a significant improvement over the nonadaptive baseline policy, the actively adaptive probing experimental policies do not improve upon the passive CE policy.

The final decision table should represent a relatively compact and understandable summary of the effects of different basic policy choices, under a range of alternative hypotheses about stock response. It can be used to identify a best policy that maximizes the expected (average) performance over the set of alternative hypotheses. However, the calculations often result in

Table 17.5. A simplified representation of the basic biological uncertainties and possible management actions.

Policy	Hypothesis: Pessimistic	Hypothesis: Optimistic	Average
Baseline	1.0	1.0	1.0
Mild Exp.	1.1	1.3	1.2
Harsh Exp.	0.6	2.0	1.3

V_{ij} values such that the best policy (in an expected value sense) may not be easy to justify, as illustrated in Table 17.5.

In this case, the nonadaptive baseline provides a low return no matter which hypothesis is correct (e.g., it may be a very conservative policy), whereas the policy labelled "mild experiment" is expected to do better than (i.e., dominates) the baseline due to eventual learning about which hypothesis is correct. The decision difficulty is with the policy labelled "harsh experiment." Note that this policy has a higher average value (1.3 versus 1.2 or 1.0) than the others (due to detecting sooner whether the optimistic hypothesis is correct), but gives a bad result (0.6) if the pessimistic hypothesis is correct. This bad result would be viewed by many fisheries managers as a risk (or potential embarrassment) that is not worth taking, and it would be difficult to convince them to implement the harsh experiment. Such conflicts between risk and expected performance are characteristic of adaptive decision problems; all the decision analysis and decision table can do is to expose the options and consequences in an objective manner for political debate and decision.

17.4. Modelling Uncertainty Over Time

Assessment of expected performance for adaptive decision choices requires some prediction of how uncertainty will change over time, and how the pattern of change will be influenced by alternative policy choices. In essence, it is necessary to develop a submodel for uncertainty that can be simulated over time in parallel to submodels for population behavior and for management response to population data and uncertainty. As for modelling any dynamic behavior, the development of a submodel for uncertainty involves two key steps: (1) identification of a set of quantitative state variables that measure the condition of the subsystem at any point in time (uncertainty in this case), and (2) formulation of a set of rules for change for the state variables. Rules for change in measures of uncertainty are essentially statistical formulae. The state variables used to measure uncertainty are often called the information state. For a review of general control theory literature on this subject as related to renewable resource problems, see Walters (1985).

Two different measures are commonly used to represent uncertainty in dynamic systems: (1) Bayesian probabilities placed on alternative hypotheses or (2) means and covariances of parameter estimates for a single model structure (called the *wide sense* information state). For most fisheries applications, Bayesian probability measures are simpler to use and allow greater flexibility in representing alternative hypotheses that are not easily expressed as alternative parameter values of a single model.

Bayesian probabilities on alternative hypotheses

The Bayesian probability for hypothesis or model i in a set of $i = 1, \ldots, n$ alternatives, given t years of data, is defined by the product:

$$P_{i,t} = \frac{P_{i,0} L_{i,t}}{K_t}. \tag{17.4.1}$$

The elements of this product are defined as follows. $P_{i,0}$ is a prior probability placed on hypothesis i before any data are gathered on the system or independently of such data; $P_{i,0}$ represents a judgement about how credible model i is based on experience in other systems and can be forced not to have any effect by setting $P_{i,0} = 1/n$ for all hypotheses. $L_{i,t}$ is the likelihood or probability of all the data available on the system at time t, given that (conditional on) model i is true. K_t is the total probability of the data; it is calculated as the sum over all models i of the products $P_{i,0}L_{i,t}$, and can be thought of simply as a normalizing constant that makes the $P_{i,t}$ sum to 1.0. Equation 17.4.1 is known as "Bayes theorem" (see Chapter 6) and a simple way to understand it is to write it as

$$P_{i,t} = \frac{C_i}{\sum_j C_j} \tag{17.4.2}$$

where C_i is the credibility index $P_{i,0}L_{i,t}$ for model i, based on prior judgement and on how likely the data are given model i were true.

Simulation of the $P_{i,t}$ over time is done by updating the likelihoods $L_{i,t}$ as each new year of simulated data is generated, then recalculating the credibility indices C_i and dividing each by their sum. The only difficulty may be in calculating the likelihoods. If all observations over time are treated as independent of one another (the usual assumption), the likelihood $L_{i,t}$ can be written as the product

$$L_{i,t} = l_{y1} l_{y2} l_{y3} \ldots l_{yt} \tag{17.4.3}$$

where l_{y_k} is the probability of obtaining the observed value(s) y_k in time step k if hypothesis i is true. Suppose for example that it is assumed that future observations y_k will be distributed normally around the mean prediction $\bar{y}_{i,k}$ of model i for year k. Then the likelihoods l_{y_k} would be calculated as

$$l_{y_k} = \frac{1}{\sqrt{2\pi\sigma_{i,t}^2}} exp\left[\frac{-(y_k - \bar{y}_{i,k})^2}{(2\sigma_{i,t}^2)}\right] \qquad (17.4.4)$$

which is the normal probability density with mean $\bar{y}_{i,k}$ and variance $\sigma_{i,t}^2$. This probability density should be used as a default when no better information is available about the future distribution of measurement and/or biological process errors. Conservative predictions of learning rates are obtained by setting the predicted variances $\sigma_{i,t}^2$ to larger values than would be predicted from lack of fit of the hypotheses to historical data. Conversely, improved monitoring programs can be simulated by using smaller values of $\sigma_{i,t}^2$ than would be estimated from historical data.

The effect of policy choice on learning is represented in the Bayesian approach through the effects of policy on model predictions. Recall that the likelihood $L_{i,t}$ is defined as the probability of the data, given hypothesis i. Any hypotheses that predict the same data (i.e., assign the same probability to the data) will end up with the same Bayes' probabilities $P_{i,t}$ except for effects of the prior probabilities $P_{i,0}$. Different policy choices will impact the $P_{i,t}$ differently only in so far as the models predict different data patterns (CPUE over time, age compositions, etc.). For example, all the hypotheses might predict high and stable relative abundance (CPUE) under low fishing rates, so low fishing rates would result in similar $L_{i,t}$ and hence no learning (change in the $P_{i,t}$). One (pessimistic) hypothesis might predict rapid decline in abundance under high fishing rates, and this hypothesis would be assigned high $L_{i,t}$ compared to the others only when high fishing rates did indeed result in consistent decline over time in relative abundance.

The main risk in using a Bayesian approach in conjunction with discrete alternative hypotheses is that the assessment will be too optimistic about learning (correct model assigned high $P_{i,t}$ too soon) if too small and distinctive a set of hypotheses is considered. If the analysis only admits a few very extreme predictions and assumes that future outcomes will be measured very precisely, then any simulation of learning will quickly see which extreme has been used to generate the data. This risk can be avoided by admitting more hypotheses to the range considered, and/or by increasing the assumed data variances $\sigma_{i,t}^2$ to account for a range of predictions by each hypothesis. To find a reasonable and realistic representation, it is wise to play with alternative model sets and likelihood functions (few versus many hypotheses, small versus large data variances) so as to find the smallest hypothesis set that will give realistic learning times relative to a many-hypothesis baseline; fortunately, the Bayesian calculations are simple enough to permit this kind of play on even small microcomputers.

Wide sense information state

The wide sense information state at any time t is simply the best fitting parameter estimates, along with their variances and covariances. A simple way to think about this representation is as the confidence limits or joint confidence region for the parameters, since these limits are defined by the parameter estimates and their covariances. One point is worth noting: when two or more parameters are uncertain, it is important to look at more than the individual confidence limits for each parameter. The covariances among parameters define whether uncertainty about them is restricted to a narrow ellipse or region of parameter combinations; when two parameters have high covariance (or are highly correlated) then in essence only one of them is uncertain, as the other is well specified by the data if the one is fixed at some independently determined value. A common example of this situation is in fitting stock-recruitment curves where data are available from only a narrow range of spawning stock sizes; the data may define very well a set of combinations (or confidence region of combinations) that specify recruitment curves that pass through the mean of the data, but it is impossible to tell which of these curves (which particular parameter combination) is correct. This is the case in the Fraser chum data used earlier.

Prediction of the wide sense information state can be done with any statistical algorithm or procedure that might be used to estimate parameters and their covariances from actual data. One special case or estimation algorithm is worth mentioning. Estimation is often done by linear regression, so the observations y_k at time k are predicted as

$$y_k = \sum_i b_i X_{i,k} + e_k \tag{17.4.5}$$

Here the b_i are unknown parameters, $X_{i,k}$ are functions of time alone, and the errors e_k are assumed to be independent. In this case, the regression parameter estimates and their covariances can be found by recursive formulae, which are easy to use in learning simulations (new parameter estimates at each time are simple functions of estimates from last time, new data, and $X_{i,k}$). For reviews of recursive linear regression formulae and recursive approximations to nonlinear regression estimators, see Gelb (1974), Meinhold and Singpurwalla (1983), Young (1974), and Walters (1986, Chapter 7).

Effects of policy change on learning performance are simulated in wide sense information representations through the effects of these changes on the independent variable values of regression procedures. For example, suppose it is assumed that Schnute's (1977) regression procedure (Chapter 8) will be used to estimate surplus production parameters (r, K, and q) from future data on catch and CPUE. Schnute's procedure is in the form of a linear regression, and the regression estimates can be obtained recursively over time (see Walters 1986, pp. 286–291). The regression-independent variables $X_{i,k}$ are the fishing effort over year k and the mean CPUE over that year. These independent variables must assume contrasting values over time in order for all the surplus production parameters to be well estimated. But early in fishery development, as effort increases, CPUE is likely to decline so that there is initially a lack of proper contrast (one cannot tell whether the dependent variable is changing due to increase in effort, or decrease in CPUE). Only when effort is highly variable, or deliberately manipulated so as to give both high and low values when CPUE is both high and low, will all the parameters be well determined. During periods of fishery stability (similar efforts and CPUEs over time), each new observation adds little or no new information about the parameters.

There are two main drawbacks with using changes in wide sense information state as a model for learning. First, as noted earlier it is often difficult or impossible to capture key uncertainties by representing them as differences in parameter values for a single model. Second, most estimation procedures for model parameters involve some type of nonlinear regression procedure; in these cases, the wide sense information state does not fully capture all the information available from the data about parameter values, and it is necessary to either do very cumbersome simulations of how the nonlinear estimation behaves over time as new data are added or else to use approximate filtering formulae. The approximations may badly overestimate or underestimate learning performance, particularly under extreme policy choices, and must be tested extensively through Monte Carlo simulations.

17.5. Opportunities for Experimental Design: Replication and Control

Adaptive management of a single stock over time requires strong assumptions about stationarity of production relationships (Chapter 7) and about the independence over time of statistical deviations from model predictions. When the production relationships may be changing slowly over time, or when deviations from mean relationships may show strong temporal patterns (cycles, trends, etc.) of many years duration, it becomes difficult or impossible to untangle these temporal effects from the effects of experimental policy changes. In scientific terms, it becomes impossible or unacceptably risky to treat historical data as a temporal control against which to measure effects of policy change. Scientists try to guard against misinterpretation of the causes of temporal change by using planned experiments that provide:

(1) replication to determine which changes are predictably related to actions (rather than possibly being due to chance alone) and (2) controls, which are experimental units that it is hoped are subject to all the factors that cause temporal change (and patterning of change) except for the policy change. For example, by fishing heavily on one stock in a region while fishing lightly on another, it might be hoped that changes in the lightly fished stock will reflect the effects of large-scale environmental factors that are also influencing the heavily fished stock (and could be misinterpreted as effects of fishing if that stock were studied as an isolated unit).

Advantages of planned management experiments

Many managed fisheries involve at least some spatial structuring of the stock into substocks that can be treated as experimental units for testing adaptive management policies. The imposition of an experimental design over these units, so that various management schemes or choices are tested on replicate subsets of units, has obvious scientific advantages over looking at the data from each substock in isolation. Overall environmental influences and trends can be detected sooner than from a single stock analysis, repeatable responses to management can be quantified sooner, and policy options can be compared in terms of performance measures that are directly and visibly evident to everyone concerned with management. Instead of asking the fishing industry to trust us about assessment model predictions, the management authority can ask the industry to think about its own experience with units managed in different ways.

Spatial structuring and replication also allow the management agency to isolate the costs and risks of experimental policy choices on a few units or substocks, without disrupting the entire fishing industry or risking the entire resource. In many cases an experimental design can be devised to test the effects of different fishing intensities, at no net loss to the industry: reduced fishing on some substocks is compensated by allowing increased fishing on others. The expected value of an experimental policy choice is often much larger when it can be performed on a few substocks than when it must be tested on the stock as a whole. If the policy fails (pessimistic model correct), the costs of failure (lost catches, monitoring costs, etc.) are measured only over the experimental substocks. If the policy works (on all or most substocks tested), it can then be applied to the rest of the substocks with only some loss in value associated with discounting the value to be obtained out to the time required to perform the policy test.

Time-treatment interactions and staircase designs

The main difficulty in using planned experiments where there are opportunities for spatial replication and control is a problem called time-treatment interaction (Walters et al. 1988a,b). Suppose that an experimental policy such as hatchery enhancement is applied to some substocks, whereas others are left as unenhanced controls. Suppose that the management policy seems to work at first (e.g., lots of hatchery fish in catches), but later declines without any corresponding decline in the control substocks. Should we interpret this as evidence that the policy has failed? A supporter of the experimental policy could defend it by claiming that it will still work in the long term but is suffering a temporary setback due to the treated substocks being more sensitive to environmental changes than the untreated substocks. In other words, he would argue that there is an interaction between the policy treatment and environmental effects over time, so that untreated substocks should not be used as controls for long-term response of the treated substocks. This is a perfectly valid argument for him to make, from a scientific viewpoint; there may well be good biological reasons for expecting treated substocks to respond differently from untreated ones to various environmental changes that cannot be controlled or prevented. For example, hatchery fish stocks may carry higher loads of disease organisms, whose impact on their hosts depends on environmental stress factors, such as water temperature; then the hatchery stock might decline during a period of several warm (or cold) years, while there is no corresponding reaction in natural stocks.

From a policy perspective, the trouble with time-treatment interactions is that rationalizations based on them can always be used as excuses for short-term policy failure, and to justify continuing any policy without reference to comparative data from areas or substocks managed differently. This has been a particularly acute problem in the management of hatchery programs for Pacific salmons, where declines in hatchery survival rates have been described by hatchery managers and promoters as being caused by differential responses of hatchery fish to changes in marine conditions (Walters 1988); just wait a few more years, they tell us, and the hatcheries will again perform as originally expected.

Scientific designs that can deal with time-treatment interaction are called staircase designs (Walters et al. 1988a,b). Such designs are costly. Essentially, it is necessary to start any policy treatment on different substocks at several points in time, so that it is eventually possible to determine which transient responses are in fact repeatable no matter when (in time) the experimental treatment is started. Thus staircase designs require a large number of similar substocks (at least six), which are all monitored well, and responses must be measured over a long period of time.

Optimum size of experiments

There are three main factors to consider in planning how large an experiment to conduct (how many replicate treatment and control areas to use):

(1) the cost of monitoring each experimental unit, (2) the effect of the number of experimental units on the time required to reliably determine whether the experimental policy represents an improvement on the control baseline, and (3) heterogeneity among experimental units and the increasing difficulty of finding similar units to use as the size of experiment is increased. Factors (1) and (3) work against large experiments, while factor (2) favors large experiments.

Considerable economies of scale are often possible in monitoring programs for replicated experiments. Many facilities for sample gathering and processing (measurement devices, aging equipment and staff, data processing, etc.) have the same basic cost whether the data are gathered from one or several spatial units. Field crews for onboard observation, enforcement of experimental regulations, and research sampling can usually be scheduled to rotate their activities among units in an efficient pattern. If fishermen become interested and involved in the experimental program, they may be co-opted (or provided various economic incentives) to do much of the sampling and enforcement work with relatively little training and supervision by the management agency.

The relationship between required duration of experiment and number of experimental replicates can be estimated by Monte Carlo simulation, where the statistical algorithms (estimation schemes, analysis of variance, etc.) are provided simulated data with expected levels of variability. This is essentially a special case of modelling learning as discussed in the previous section.

Potential experimental areas and substocks under field conditions are very different from laboratory experimental units such as laboratory animals, where it is possible in principle to supply a large number of essentially similar units. In the field, there is usually a very limited total number of potential units, and these units will differ from one another to various degrees in terms of size, similarity of environmental effects, ecosystem structure, and history of disturbance through fishing and management. Historical time-series data from candidate units can be used to measure similarity in behavior (and history of disturbance) among units, so that similar units can be paired or blocked for treatment-control comparisons. Figure 17.3 shows such a clustering for some sockeye salmon stocks in British Columbia. Experience to date in measuring similarity of behavior among substocks has been limited (see Walters et al. 1988b), and this experience indicates that larger experiments will have to deal with progressively greater heterogeneity among units until the heterogeneity (error variance) outweighs the benefits of increased replication. Based on our experience, the break-even point (error variance increasing faster than precision improvement due to replication) is likely to occur at an experiment size of approximately 6–8 units.

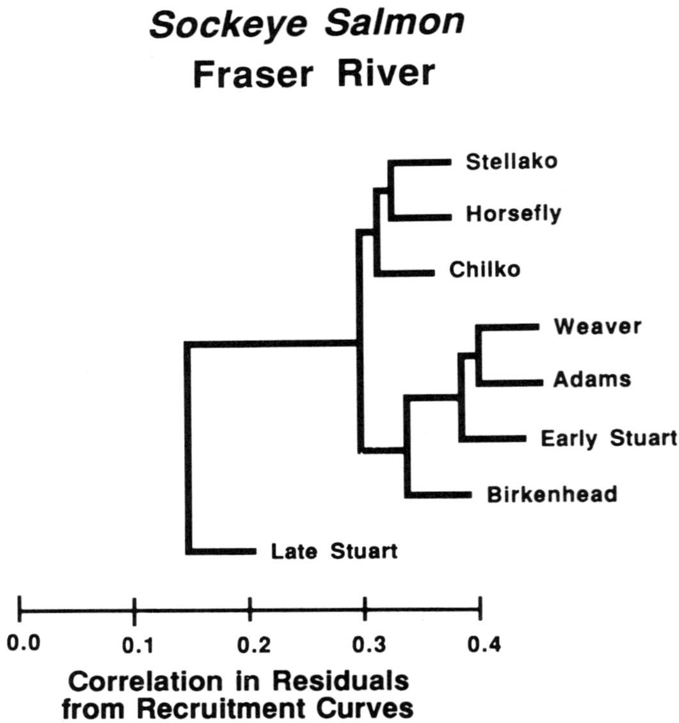

Figure 17.3. Similarities in residuals from stock-recruitment curves for different Fraser River salmon stocks. From Walters et al. (1988). Can. J. Fish. Aquat. Sci. 45: Figure 4.

17.6. Impediments to Adaptive Management

The idea of adaptive, deliberately experimental management has strong scientific appeal, and most fishery managers will openly admit that it is ultimately necessary to learn by doing in order to find improved management policies. But a number of serious difficulties have been encountered in practical attempts to justify and implement adaptive policies. This section reviews these difficulties, and offers suggestions about arguments to help overcome them.

Risk aversion and discounting

We noted earlier that strong risk aversion (to poor experimental outcomes, lost catches) and high discounting of future fishery benefits both act to reduce the expected benefits of adaptive management proposals. We noted that care must be taken not to confuse personal risk aversion (most biologist/managers are personally very risk averse) with the best interests of the fishing industry and the public in managing a fishery.

We have found it impossible in practice to convince working fishery managers that they should keep their personal views about risk out of their work places. Most people are simply not objective enough about their work; indeed, they would not do a good job if they were not personally committed to the fisheries and to the maintenance of various values associated with them. No responsible manager is going to accept the numerical results of models and decision tables uncritically (at face value), and act mechanically as these results prescribe; inevitably considerable personal judgement (including risk aversion, if only subconsciously) will be used in any final weighing of decision choices. Under these circumstances, the best that the adaptive policy analyst can do is to repeatedly point out and emphasize any instances where the management debate appears to include improper elements of personal risk aversion, and hope that the manager will listen and take some care.

Most fishermen operate with very high personal discount rates (10% per year or more) that reflect uncertainty about the future even in circumstances where their future interests are protected through license limitation, vessel quotas, and other measures. In many cases, government policies also call for high discount rates in project and policy evaluation (for example, the Canadian Salmonid Enhancement Program uses 10% or higher discount rates in evaluating hatchery and other salmon production projects). High discount rates favor experimental policies that give big immediate returns (e.g., experimental overfishing), but the longer term value of information from such experiments is not the reason they are favored. Information values are generally evident and important only when discount rates are less than approximately 4% per year.

To justify experiments that sacrifice short-term catch in favor of learning about long-term potential, it is generally necessary to appeal to arguments about resource husbandry and legacies of uncertainty for future generations of fishery users. Although there is often a legal mandate for the management agency to consider such arguments, that mandate is often either confused with short-term risk aversion or ignored completely in management debates.

Spatial scales and dependent economic communities

Adaptive management plans for stocks such as Pacific salmon have been most readily accepted by management agencies and fishermen when the experimental policy changes involve relatively small, isolated substocks. Such substocks are usually harvested by relatively mobile fishermen who can spread the risks and costs to them of the experimental policy by moving to other areas. Strong opposition to local experiments is likely to come mainly from dependent economic communities of fishermen, resort owners, and fish processors who cannot easily move to other fishing areas or stocks.

It is important to recognize that there are some large unit fish stocks that have correspondingly large dependent economic communities, for which it will probably never be economically or socially acceptable to engage in risky management experiments. A good example is the Pacific halibut off the coast of Canada and the American Pacific Northwest. As noted earlier, there is great uncertainty about whether fluctuations of this stock are due to environmental factors or to stock-recruit effects (Deriso et al. 1986), and the alternative hypotheses will not likely be resolved unless the International Pacific Halibut Commission elects to allow heavy fishing pressure to hold the halibut spawning stock at low levels. This experiment is very risky; if recruitment rates are depressed by some environmental effect in conjunction with spawning stock depression by fishing, the stock may collapse to such a low level that it will take decades to rebuild. Even under low discount rates (very long term view of the resource) and completely ignoring the present fishing industry, this risk is difficult to justify.

Trust and management credibility

Fisheries managers are often concerned about maintaining scientific credibility with the fishing industry, by showing that their policies and predictions can be trusted. In practice, the "safest" policies from this perspective are usually seen to be ones that involve only slow (or no) change in regulations over time (i.e., tried and true or steady as she goes policies). The design of adaptive policies begins with a deliberate confession to ignorance and uncertainty, through the formulation of a set of alternative hypotheses. If this confession to ignorance is seen as a threat to the credibility of the management agency or as a contradiction to past claims of understanding, it may provoke bitter arguments and reactions from fishery managers.

The key job of the adaptive analyst in such situations is to dispel any myth that the managers may hold about having great trust or credibility in the first place. Fishermen and other fishing industry spokesmen are rarely so foolish as put much trust in scientific assessment, and fisheries managers who pretend that they have achieved such trust are usually fooling themselves. Often the strongest support for innovative management plans comes from fishing industry representatives in management, and their assistance should be solicited in efforts to dispel misconceptions about credibility and trust.

Institutional impediments to learning

Adaptive management requires that institutions collect data, do analyses, and actually change their actions as a result of what they have learned. Even though there is little experience with implementation of adaptive management programs in fisheries agencies, the experience we do have indicates that there can be a great deal of trouble in maintaining the monitoring programs and in getting the managers to actually change their actions once the experiment is complete.

Sainsbury (1989) describes an experimental management program for a trawl fishery in Northwestern Australia. In many ways, Sainsbury's study is the best example designed and implemented using the procedures discussed in this chapter. Although he had considerable cooperation from different levels of government in the design and execution of this program, Sainsbury believes that the maintenance of the program is completely dependent on his own presence; if he were to leave his current position, the program would simply disappear.

We believe this may be a very common problem: no organizations we know of have adaptive management as a standard operating procedure. There are legal incentives or imperatives to develop adaptive management plans in some major systems such as the Columbia River Basin (Lee 1989) and the Florida Everglades marsh region, and the Canadian Department of Fisheries and Oceans has a planning team that is struggling to develop and coordinate a set of standard procedures for adaptive management of Pacific salmon stocks. But even in these cases, where there is strong commitment on paper, it appears that successful and persistent implementation of significant changes will require great personal commitment from one or several individuals.

Turnover of key personnel is only one possible barrier; adaptive programs almost always require substantial additional investments in monitoring, and such investments are highly vulnerable to the periodic retrenchments and budget crises that seem to afflict all organizations. It may be very difficult to maintain a critical monitoring program for the 5, 10, or even 20 years that are required.

Finally, once the experimental period is over, the agency must change its management. If this involves a departure from what it has done in the past, it may be impossible to make the change, even though the experimental period shows the change advisable. Where the experiment involves a deliberate large-scale change in management action, the agency may be prepared to stay with the change. But if the experimentation has been on a smaller scale, it may be almost impossible to get the required large-scale change implemented.

System stability

A final impediment to adaptive programs is temporal change in the underlying system. The purpose of most adaptive management programs is to identify which of several alternative models is correct. If the underlying system changes quickly, it may be impossible to identify the "real" model before the model has changed to something else. If the system structure is sufficiently variable, then it may be impossible to learn about it.

17.7. Summary and Critique

In this chapter, we have described some straightforward principles regarding how to deal with uncertainty and some complex computational procedures for evaluating alternative policies. We believe it is far more important to explicitly consider uncertainty than to apply sophisticated analytic techniques. Fisheries managers have generally not gone far enough in recognizing and acting on the very real possibility that their assessments are flawed.

The most important step in any fisheries policy design is to list alternative hypotheses on one side of a table, alternative management plans on another side and examine the consequences of the different plans under the different hypotheses! Do this in every policy evaluation! Although this may seem obvious, and something anyone with any training in decision analysis would naturally do as a first step, few fishery managers are trained in decision analysis.

If we accept that policy design should explicitly consider uncertainty, we still have a long way to go before knowing what methods work best. The concept of adaptive management was first proposed by Walters and Hilborn (1976), and, even though the basic principles have been widely accepted in many fisheries agencies, there have been few published examples where these methods have been applied.

Bibliographic Notes

The use of the term adaptive management for fisheries originated with Walters and Hilborn (1976), and we have remained the most verbose of the advocates in a series of papers including Walters (1985), Walters (1988) and Hilborn and Sibert (1988). Walters (1986) book is the most complete description of analytic techniques. Most of the techniques have been borrowed from optimal control theory, and a selection of the technical papers and books are Gelb (1974), Meinhold and Singpurwalla (1983), and Young (1974). The best published, worked example of an adaptive experimental design is Sainsbury (1989). An excellent discussion of the role of adaptive management in a complex situation involving many competing water users and fisheries problems is Lee (1989).

18

Making Stock Assessment and Management Work

18.1. Introduction

In this last chapter, we want to summarize how stock assessments are used to formulate management advice, and to emphasize where they typically go wrong. Our objective is not to be so pessimistic that you abandon fisheries for a more profitable profession, but rather to emphasize what we have been able to learn from past lessons. We begin the chapter with a quick look at three actual fisheries management systems, how they work, what data are used, and how management advice is formulated. We next summarize where failures usually occur, and what can be done about them.

We next consider some of the basic tenets of most fisheries management practice in developed countries and consider some alternative approaches to managing fisheries, including ideas about cooperative and experimental management. The final section examines basic directions for improving the science of fishery management.

18.2. Stock Assessment and Management in the Real World

In this section we review how the tools of fisheries stock assessment and management are put together in several developed fisheries. We review three actual stock assessment and management systems: (1) estimation of sustainable yields and setting annual quotas by using biomass dynamics models, (2) estimation of sustainable yield from analysis of catch-at-age data and implementation using annual quotas, and (3) yield-per-recruit maximization by regulating the opening date of a prawn fishery. Each of these examples shows how different tools can be combined to produce advice for managers and where the combination of the tools will frequently go wrong. These examples are intended to show how the techniques are currently used rather than how they should be used.

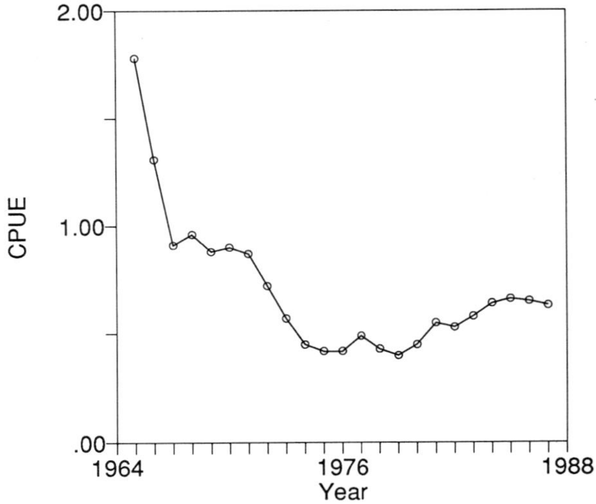

Figure 18.1. Spanish CPUE series for area 1.3 and 1.4 Cape hake.

Biomass dynamics models: the Cape hake

The International Commission for the Southeast Atlantic Fisheries (ICSEAF) managed the resources of Cape hake off Angola, Namibia, and South Africa prior to 1990. The following text describes the fishery as it was while ICSEAF was managing. The fishery is somewhat unique in that two very similar species are caught (*Merluccius capensis* and *M. paradoxus*) but because of identification difficulty, the data are combined for all analysis. Thus this is a multispecies fishery treated as an aggregate production unit.

Data

A number of countries, principally Spain, Poland, South Africa, and the Soviet Union, trawl for Cape hake. Each country collects its own catch and effort data. The effort data are standardized to a specific class of trawler, and the data are transmitted to the ICSEAF secretariat in Madrid. The ICSEAF area was divided into spatial divisions, and for purposes of analysis, some of the divisions were lumped together. The ICSEAF practice was to use the CPUE sequence of the Spanish fleet as the index of abundance. The trend in Spanish CPUE for divisions 1.3 and 1.4 is shown in Figure 18.1. The CPUE trend for the Polish fleet is quite different, and was ignored in ICSEAF analysis.

Methods

The catch and CPUE data series were analyzed using biomass dynamic models. The Spanish, South African, and Soviet Union delegations each proposed different models and estimation procedures. Each delegation did its own analysis and then the scientific committee of ICSEAF argued over the differences between the methods and produced a quota recommendation. Until about 1987, several of the delegations were using Fox's equilibrium fitting method, but in the late 1980s each delegation used some form of dynamic model. The South African delegation used a time-series fitting method to the basic Schaefer model (Butterworth and Andrew 1984), the delegation from the Soviet Union used a different transformation of the Schaefer model (Babayan and Kizner 1988) also fit using a time-series method, and the Spanish delegation used a different model (Lleonart et al. 1985, Lleonart and Salat 1989) that considers CPUE to be a quantity with inertia that is modified by fishing effort.

There has naturally been considerable discussion between delegations about the relative merits of the methods, and they agreed in 1989 to use an operating model (see Chapter 6) concept to test the reliability of each estimation procedure. An age structured simulation model, with all parameters specified, was designated as the true model, and each delegation used this model, and the historical catch and effort history, to see how well their own estimation procedures performed in Monte-Carlo trials.

Products

Each estimation procedure produces an estimate of MSY, the effort required for an $F_{0.1}$ policy (designated $E_{0.1}$), and the quota that would be taken if $E_{0.1}$ were applied (designated $TAC_{0.1}$). Table 18.1 shows the estimates derived in 1989. ICSEAF adopted $F_{0.1}$ as the harvest strategy, so once the scientific committee agreed on what TAC is appropriate under $F_{0.1}$, the political process of the commission then allocated the total quota among the national fleets.

Figure 18.2 summarizes the entire assessment and management process.

Where it goes wrong

ICSEAF was formed in the 1960s in response to conservation concerns stimulated by decline in CPUE. ICSEAF established quotas considerably lower than the historical catch, and CPUE (at least of the Spanish fleet) has responded. It appears, therefore, that the scientific and political process has actually worked! This apparent success depends critically of course on the assump-

Table 18.1. Estimates of management parameters by the Spanish, Soviet Union, and South African procedures for division 1.3 and 1.4 Cape hake in 1989. MSY and $TAC_{0.1}$ are in thousands of tons, $E_{0.1}$ is in thousands of standardized vessel days. Results from ICSEAF 1989, table 1.

Method	MSY	$E_{0.1}$	$TAC_{0.1}$
South Africa	257	376	255
Spain	301	730	408
Soviet Union	255	409	266

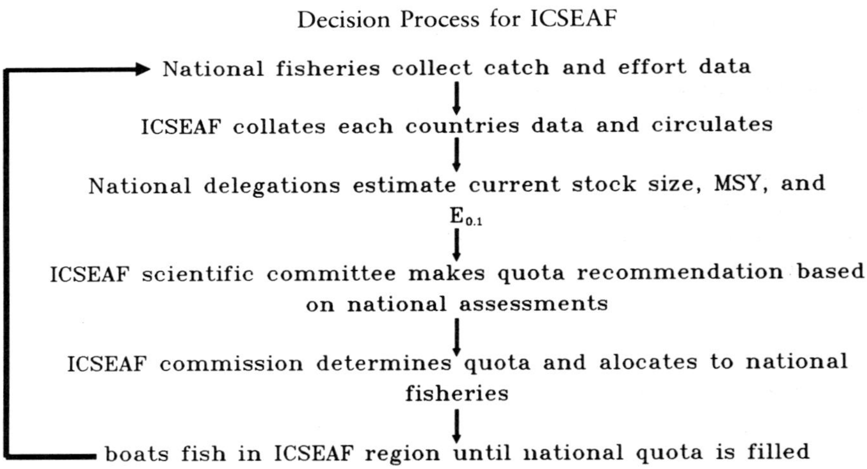

Figure 18.2. The data collection and decision making process for Cape hake.

tion that abundance is proportional to CPUE in the Spanish fleet, and that the response in CPUE was due to the reduction in catch, instead of natural events. In the absence of other, independent estimates of abundance, there is simply no way to judge whether ICSEAF has been as successful as the CPUE trend data might lead you to suspect.

Catch-at-age analysis: Pacific halibut

Pacific halibut (*Hippoglossus stenolepis*) is commercially fished on the west coasts of Canada and the United States, and is regulated under a bilateral treaty by the International Pacific Halibut Commission (IPHC). Stimulated by declining halibut catches in the first two decades of this century, the IPHC was formed in 1923. It has used a closed season and (after 1930) catch quotas by area (IPHC 1987).

Data

The IPHC staff have catch, catch-at-age, and commercial fishing effort data by detailed area. IPHC also conducts mark recapture studies and research cruises to calibrate changes in gear effectiveness. For instance, in the early 1980s the fishermen changed the types of hooks they use on the longline fishing gear, and the new hooks effectively doubled the catching power. IPHC has a near unrivaled 60 year data base on catch-at-age and catch-per-unit-effort. The major gap in the data has been incidental and unreported catches of halibut in other gears.

Methods

The scientific staff of the IPHC perform an assessment of the current stock size each year. The stock size (of fish vulnerable to the fishery) is estimated in three ways (Quinn et al. 1985): (1) a catch-at-age analysis using the computer program CAGEAN, (2) a catch-at-age analysis that includes movements between IPHC regulatory areas, and (3) an analysis based on CPUE. IPHC staff currently use the CAGEAN output and produce an estimate of the abundance of halibut in each of the regulatory areas.

Products

The IPHC has adopted a constant harvest rate policy of 35% of the exploitable biomass. Once the current stock abundances have been estimated from the assessment procedures described above, the scientific staff of the IPHC makes recommendations to the political commissioners of each country, who normally accept the recommendations and implement the area-specific quotas.

Where it goes wrong

As with any assessment method that is based on catch-at-age data, the IPHC assessment procedures are very good at determining what happened in the past, but not nearly as good at determining the current stock status. IPHC staff rely heavily on the CPUE data, to determine the terminal F's for the catch-at-age analysis. The major failure, that is now recognized, was the development of a major incidental fishing mortality problem in the 1960s and 1970s that was not detected and incorporated into the stock assessments until the early 1980s. Thus the stock was inexplicably decreasing, and reductions in the commercial harvest did not slow the decline as predicted.

Another "failure" of the halibut management system is that after 60 years of management and data collection, the IPHC still has very little idea what determines recruitment and net production. The constant harvest rate policy has been adopted as a management strategy that is likely to produce a viable and stable fishery, but to some extent the IPHC staff have abandoned any attempt at optimal yield. There continues to be a fundamental debate about whether long-term recruitment variation is due to density-dependent mechanisms or oceanographic factors and harvest rates could be increased somewhat if the density-dependence hypothesis is correct.

Yield-per-recruit maximization: Gulf of Carpentaria prawns

The commercial trawl fishery for prawns in the Gulf of Carpentaria, in the Northern Territories of Australia, began in the late 1960s, and by the mid 1970s concern was expressed about excess capacity (Bowen and Hancock 1985). In 1977 a limited entry regime was adopted. The fishery is different from those discussed earlier in this chapter, in that it has been assumed that recruitment overfishing is unlikely and that the fishery should be managed for economic return per recruit.

Data

The major data source available for management of these prawns is commercial and research vessel catch sampling. Since the prawns are highly aggregated, CPUE is not thought to be proportional to abundance. Sampling has focussed primarily on examination of the size and growth of the prawns, to identify the best times and places to fish so as to maximize yield-per-recruit.

Methods

The Australian scientists sample the growth of the prawns prior to the opening of the trawl fishery. Given the expected pattern of fishing effort, and historical data on growth of the prawns, they can estimate the total yield-per-recruit that will be obtained by opening the fishery on any date (Somers 1985). Most of the catch takes place in the first 2 weeks after the fishery is opened, so the optimum opening date is roughly when the prawns are at a size where growth and natural mortality are balanced.

Products

The total fleet size has been fixed for a number of years, and is roughly the same as when limited entry was adopted. There is no pretense that the fleet size is economically optimum. In practice, the opening dates tend to be set a little earlier than the economic optimum predicted by the yield-per-recruit models.

Where it goes wrong

The major weakness in any yield-per-recruit maximization is that it ignores recruitment overfishing. It remains unclear if this is a potential problem in the Gulf of Carpentaria, although it is a serious concern with similar species elsewhere in Australia (e.g., Exmouth Gulf, Penn and Caputi 1985). The other weakness in this system is that it is a mixed species fishery, with two species of tiger prawns caught as well as the banana prawn. There is concern that recruitment overfishing may be occurring in the banana prawn, so that total yields and value from the system may slowly deteriorate as this component of the catch is lost.

18.3. Does Stock Assessment Work?

In the four sections of this book we have (1) reviewed the role of stock assessment and the objectives of fisheries management, (2) examined how exploited populations and fishermen behave, (3) studied in great detail a number of methods for estimation of the parameters of population dynamics models from data, and (4) put it together in how to evaluate and formulate fisheries management plans. Despite the substantial length of the book, our treatment is at an elementary level; full-length books could be, and have been, written on many of the individual chapters. It is clear, therefore, that there is a very substantial body of fisheries theory, data, and experience.

Does stock assessment work? Can we do a reasonable job of managing a fishery to produce the biological, economic and social objectives of fisheries management?

The record is not very good!

There is, of course, the obvious litany of great fisheries failures: the great whales, California sardine, Peruvian anchoveta, North Sea herring, southern bluefin tuna etc. etc. etc. But there are many other subtle and yet more disturbing failures.

The use of mesh size in the northeast Atlantic and North Sea in particular, and the associated yield-per-recruit theory, is in many ways one of the cradles of civilization of modern fisheries theory. Yet recent work suggests that many if not most of the fish escaping the cod-ends are dying anyway (DeAlteris and Reifsteck, 1988; Main, 1988) and that most of the selectivity work was formulated using research cruises that exhibit very different selectivity relationships than commercial fishing gear. The showpiece of fisheries theory and practice may have been one gigantic mistake.

The most important fishery in Canada (the biggest fish exporter in the world) is the northern cod stock of Newfoundland and Labrador. The best minds in Canadian fisheries have been responsible for its assessment and management, and the rehabilitation of the northern cod following declaration of the 200 mile zone in the late 1970s was the showpiece of Canadian fisheries science and policy. In 1989 it was announced that the published stock assessments were wrong, that the stock was declining instead of building, and that quotas would have to be cut with thousands of jobs lost.

Do we really know what we are doing?

Fisheries stock assessment and management is a very difficult job. Before we can usefully address how to make it work better, we should take a realistic look at the major impediments to successful assessment and management. We see three major areas that impede progress:

Science. Successful science requires replication and control. If we really wanted to determine which assessment and management procedures worked, we would have to compare alternatives by using several replicates and controls, lined up like little test tubes in the laboratory. Most of the major failures of fisheries management have occurred in situations where replication and control are not possible. There is only one Peruvian anchoveta fishery, one southern bluefin tuna and so on. The large-scale marine fisheries of the world are not amenable to replication and control, and therefore it will be impossible in principle to ever determine with confidence whether the changes we have observed are due to the fisheries management actions that have been taken.

Social dynamics. Fisheries managers have traditionally ignored the dynamic response of fishermen to regulation. A fishery consists of fish and fishermen, and by concentrating almost all of our efforts on the fish, we have missed a great deal. The mistake of stock assessments in the northern cod fishery described later in this chapter appears to have been due to increased efficiency of fishermen, leading to increasing catch rates which were mistaken for increased abundance. The failure to understand the changing fishing behavior of the fishermen led to a failure of the stock assessment and management system.

Politics. The best stock assessment in the world does not do much good if the politicians do not implement it or if you cannot successfully control fishing effort. There are many examples where stock assessment teams appear to have known what was happening and made the correct recommendations to managers, but the advice was not accepted, was accepted so late, or was so dilute that it was ineffective (Saetersdal 1980). These are true failures of fisheries management: the best science is wasted if it is not translated into action.

It would be easy to be pessimistic; the impediments to successful stock assessment and management are formidable. However, we are optimistic; despite the obstacles listed above, there is a great potential to do good fisheries management, and we believe it can be done. Improvement in fisheries management will come from two directions: technical progress and reorientation of thinking.

There is certainly a lot of technical progress underway: in each of the areas reviewed in this book there are a great number of advances being made. But perhaps more important than technical progress is a reorientation of thinking. If assessment and management are to be successful, the basic assumptions about the process need changing: we need to recognize the limitations of the science, devote energies to understanding fishermens' behavior, and recognize the importance of breaking the political barriers to progress.

18.4. Integration of Data Collection and Fishery Development

One of the major impediments to making stock assessment work is the often antagonistic relationship between fishermen and managers. If we really want stock assessments to work, we are going to have to rethink this relationship. We challenge the classical European and North American presumption that fishery management must take a passive advisory position or an adversarial regulatory position with respect to the fishermen. In this classical model for management, the fishermen develop fishing locations and gears in whatever pattern they find to be most economic, while the management system scrambles to make sense of whatever data it can gather from the fishing pattern and then acts to impose regulations only in the face of obvious abuses such as catching fish when they are too small or after severe recruitment overfishing.

It should be obvious from the previous chapters on assessment methods that there are gross pitfalls in trying to make sense of the data that come from fisheries that are developing under the classical model for data gathering and management. These pitfalls involve assumptions about factors ranging from fishermen's search patterns to their gear selectivity, and some of these assumptions cannot be well tested even in the major fisheries where many years of data are available. In other words, better stock assessment models and methods of analysis are not going to eventually provide ways around the pitfalls; instead there is going to have to be a basic change in the quality of data being gathered. Notice that we say quality, not quantity: gathering more of the same information as it is now collected (i.e., getting more precise data) is not going to help matters much at all; most of the pitfalls have nothing to do with statistical errors due to imprecision in the data.

The classical European/North American answer to data-quality problems (e.g., nonrepresentative sampling of relative abundances and size structures) from commercial catch statistics has been to develop survey sampling systems based on gear and vessels that are owned or chartered by the management agencies. These systems are hideously expensive, especially considering the tiny sample sizes that can be obtained in comparison to the potential number of samples obtainable from the fishery itself (a 200-boat trawl fleet might make 50,000 trawls per year; a one-million-dollar research vessel might make a few hundred trawls over the same period.).

Beyond the problem of nonrepresentative sampling, the classical fisheries development pattern usually does not result in clear and properly contrasting tests of how the harvested ecosystem responds to alternative exploitation patterns. Fishery developments are usually "one way trips": they knock the stocks down over wide areas at once, without time to observe sustained responses to intermediate levels of development, and lead to temporal patterns of data where a whole range of factors are moving in the same or opposite directions and hence are fundamentally confounded in their effects. Any scientist can tell you that such an experimental design, without proper treatment and control arrangements, will not lead to defensible answers about various factor effects, no matter how much elaborate mathematical analysis is done with the data. In short, there is nothing truly scientific about classical approaches to fisheries management and stock assessment.

In this section, we will discuss the possibility of structuring fisheries development in ways that provide for more representative sampling patterns by the fishermen and for deliberate experimental contrasts among exploitation regimes. This possibility would require different fundamental attitudes and responsibilities by both fishermen and managers than in the classical European/North American model, involving open admission of uncertainty and a cooperative search for fishing regimes (places, times, information recording) that are likely to provide a sound scientific basis for resolving the uncertainty.

We will also challenge the traditional notion that management should be directed toward the establishment of a productive equilibrium with steady yields each year. Although this objective may be reasonable for large fisheries and fishing industries as a whole, it may often be better to manage local areas by rotational harvest policies involving periods of depletion and recovery.

Regulation versus cooperative management

The first and most fundamental step in the search for more informative fishery development patterns is to break away from the classical attitude that management is there to gather data and develop conservation regulations, that is, to become an adversary of short-sighted fishing interests. This adversarial attitude is not in fact present in the earliest development stages of most fisheries, when the management agencies are seen as assistants of the fishermen in gear development, exploratory fishing, and finding financial assistance. It appears later, when catch rates begin to fall, competition among fishermen becomes more intense, and suggestions begin to surface about regulations such as limited entry, size limits, and quotas.

A basic challenge in management is to find ways to prevent the shift toward adversarial positions. That prevention begins by breaking down the distinction between managers and fishermen as data gatherers, by trying to make data gathering for management a basic responsibility of the fishermen. The next step is to involve the fishermen right from the outset of development in planning for ways to reduce and mitigate the impacts of falling catch rates, through policies such as initially setting aside closed areas that are both sources of later catch and can act as experimental control areas against which to measure development effects. Finally, there must develop a clear consensus about how important it is to manage the fishery for sustained, long-term yields in the first place; in many cases it may be socially and economically better to plan right from the start on sequentially depleting local stocks and hoping that each will recover over time (as is done in forest management and agricultural crop rotation).

The role of fishermen in data collection

Suppose that a few essential needs for better stock assessment are spelled out to the fishermen from the outset: (1) information on the distribution and life history of key species, (2) systematic fishing patterns that can be used to map out spatial patterns and changes in relative abundance over time, and (3) information on the size/age structure of the stocks (not just the fish of preferred size for eating or sale) that can be used to assess growth and mortality rates. Then the fishermen are asked to help provide these needs as a basic part of their fishing activity.

Reactions to requests for cooperation

What might their reactions be? First, in many fisheries it is likely that they can already provide much valuable information on life histories and distribution. Second, they will likely ask for tools and incentives to provide assistance to the scientists. A common form of such assistance is voluntary logbook systems. Third, they will likely object to requests that they spend a significant amount of time fishing systematically in locations and for size/species that they would not normally choose. Here the challenge to fishery managers is to provide incentives, beyond vague promises that new fishing grounds might be found and better biological understanding obtained. One obvious and possibly foolish choice is to require them to do it, through regulations that make the systematic fishing a part of the "cost" for being allowed to participate (in effect, part of the "license cost" even when there is no formal licensing system).

The requirement that fishermen make their vessel available as a no-cost charter vessel has been used with success in South Australia. It has been tried with less success elsewhere. How well such a requirement works depends upon many factors including the relationship between the scientists and the fishermen, the legal structure of the licensing system, and the real opportunity cost to fishermen of research cruises. If a fishery is well managed, the fishermen should be able to afford the time and fuel required for research trips. It seems inefficient for government agencies to purchase or charter expensive vessels when the fishermen normally have similar vessels going unused due to season closures or quota limitations.

Incentives for cooperation

What incentives besides threats can the fishery manager offer? There are at least four general categories

1 *Ethical appeals.* Insist that data on changing spatial distributions will be essential to detect and prevent overfishing and hence losses to future fishermen. Here it is important that the arguments be precise and credible.

2 *Financial Incentives.* It may be far cheaper to pay for extra operating costs and lost catches during time spent in survey activities than for the management agency to operate its own survey vessels.

3 *Prestige and sport.* Offer rewards and honors of various kinds to those fishermen that provide the best information.

4 *Other material rewards.* Offer cooperating fishermen free gear, lottery tickets, preferential access to particularly good fishing spots, and so on.

It is essential to develop systems to efficiently store, summarize, and retrieve information supplied by fishermen, so that they can be provided regular feedback about how their information is being used and about stock status and trends. This means that some type of microcomputer-based data management and graphics (mapping, etc) system is a very important thing to have early in the development.

Experimental management in spatially structured systems

Many ecosystems are structured into relatively isolated spatial replicates, between which exchange rates of fish are expected to be relatively low (except possibly for species with long planktonic larval stages); examples are distinct coral reefs, seamounts, rocky reefs that produce abalone and spiny lobsters, and estuaries that produce barramundi and prawns. These situations offer a grand opportunity that is seldom utilized in fisheries management: to set up a range of alternative harvest policies for the various replicates. The alternative policies can be used both to test empirically for the best alternative much more quickly and safely than can be done with the system as a whole, and to provide buffers against the risk of overexploiting some areas (i.e., initially set aside a few areas as low-harvest-rate refuges, to support fishermen during periods of recovery of other areas that are overexploited).

In the absence of a spatial development plan, the fishermen would normally deplete the most accessible replicates first, then move on to others. Thus there might normally be a crude sort of spatial design in the pattern of fishery development. But this pattern would not provide any accessible buffer areas against overfishing, and the sequence of depletion responses would each follow the same relatively uninformative pattern (i.e., it is not much better to see 10 declines under local overfishing than to see one).

A designed management experiment would be essentially taking the empirical approach rightly advocated for dealing with complex ecologies and fishing practices by people like Marten and Polovina (1982). However, it would try to deliberately structure the application of different policy choices (effort levels, size limits, etc.) to avoid confusion about the relative importance of various factors.

The ingredients of experimental design

Good experimental designs have three essential features:

1. *Control units.* Areas that are left unfished or are managed to maintain historical abundance patterns;
2. *Replication.* having more than one area to which each type of treatment or policy is applied;
3. *Deliberate randomization or interspersion.* Picking the areas to receive each

treatment at random or in a systematic pattern to be representative of the whole range of units that constitute the managed system.

Control units provide a backdrop against which to measure large-scale environmental and ecological effects that would otherwise be confused (confounded) with effects of exploitation. For example, you can be sure that a cyclone or crown of thorns starfish outbreak will occur during any reef fishing experiment even if these phenomena have not occurred locally for decades. If there are no control areas, there will be no way of assessing the importance of such events in causing continued or exaggerated declines in catch rates in heavily fished treatment areas.

Replication is the only way to guarantee that measured control and treatment response patterns are representative of how other areas not initially included in the experiment are likely to behave if treated later. An apparent response pattern seen in only one area can always be attributed to, or explained away, by a whole variety of factors besides the deliberate treatment applied to the area; no amount of theoretical argument or statistical curve fitting can scientifically resolve which explanation is correct.

Deliberate randomization or interspersion is just a way of making sure that the treatment effects are not confused with other factors that might cause differences in responses among areas. For example, suppose a trawling experiment is arranged so that fishing pressure is highest in shallow water areas and lower pressure is applied in deeper offshore areas. If the offshore areas do decline substantially, could you tell whether this was because those areas were less productive, because recruitment from inshore areas was reduced due to the inshore treatments, because the trawls operate more efficiently offshore, or because of one of many other explanations?

A simple design to study exploitation rate effects

One obvious design would be to assist the fishermen to execute and monitor three levels of fishing effort (low, moderate, high enough to likely cause depletion) on different areas, and to maintain these effort differences over several years before rotating (so-called crossover design) the effort treatments among areas. The low effort areas would still be fished enough to provide basic abundance and composition data, yet would act as buffers against overfishing. The moderate effort areas would be managed at near the initial guesses at optimum harvest rates, based on yield/recruit and other biological assessments. The high effort areas would test for biological response and recovery potentials not suspected (or hoped for) in initial assessments.

The fishermen would certainly have to give up something in the short run in order to implement such a design, even if the low-, moderate-, and high-effort replicates were initially set up in reverse order of accessibility and preference for fishing. The fishermen would have to keep fishing in the high effort replicates after they would normally have moved on and to accept lower catch rates than they would have obtained by starting to fish in the low effort areas earlier.

Finding and maintaining workable designs

A key requirement in the design of spatial experiments is to develop a process for discussing various design options and constraints with the commercial fishermen, so as to seek out a compromise design that they will not find too uncomfortable and will support in terms of data collection and policing. If there are low fishing or refuge areas in the design, it is almost inevitable that some fishermen will try to cheat and fish in these areas while reporting that they have been in one of the other areas. It will generally not be economical for the management agency to undertake all enforcement activities; the fishermen must feel that the design is their plan and responsibility.

Even with the best atmosphere of cooperation, there will inevitably be some cheating, and, later in time, concerted efforts to change the experimental design and move from high-exploitation areas to the initial refuges sooner than originally planned. The cheating will be used as an argument in favor of the move or of giving up on the experiment entirely. It is important to recognize that fishermen's behavior and consensus in the fishing community can change very rapidly when conditions are perceived to be deteriorating. To counter and prevent such changes, managers and supportive fishermen have to remain ready to respond quickly to the arguments, by keeping up-to-date records and summaries of what is happening in the experimental areas. The best counter to arguments about cheating is to have catch and effort statistics showing how the areas are remaining different in spite of the cheating; if the data show otherwise, then probably the experiment should be terminated anyway.

An ideal experimental design from the biological point of view would be to hold exploitation rates constant in each area until it is clear that an equilibrium has been reached. However, it would likely take a decade or more for this to be achieved, especially if long-lived fish species are involved. It is politically and institutionally impractical for people (fishermen and managers alike) to maintain a plan over such long time scales. The best that can be hoped for is to maintain a 5–10 year time span between design "cross-overs" (switches of effort from high- to low-exploitation areas) unless most of the areas are subject only to moderate exploitation in the first place.

Rotational harvest strategies versus sustained yield management

A cherished concept in classical stock assessment is that management should seek to establish an equilibrium regime of sustained yields, with little variation from year to year. It is hard to understand how this concept got its start in fisheries, except that it involved mathematically simple ideas such as yield curves. Similar attitudes do not prevail in the management of other renewable resources, except on a large spatial scale. In local areas, some concept of rotation (with cycles of depletion and renewal) is a generally accepted feature of agricultural and forest management.

Why not apply a rotational concept in fisheries management, especially for long-lived species that occur in local aggregations or in fishing areas for which it is uneconomical or impractical to closely regulate fishing efforts? A rotational management strategy would not be the same as just letting fishermen deplete each area and then move on; it could involve a well-monitored and deliberate (i.e., experimental) plan for each area, with carefully chosen cutoff catch rates, below which the fishermen would have to move and with simple monitoring during the closed recovery period to help decide when the area should be harvested again.

For many ecosystems, a periodic or rotational harvest "disturbance" would be far more natural than any policy involving sustained fishing mortality. For example, coral reefs are periodically disturbed naturally by cyclones. Presumably, the fish community is preadapted (in terms of the life histories of the various species and their patterns of interaction) to cope with such disturbances. In other words, the species patterns we see and attempt to exploit are often already the result of disturbances similar to what would be introduced by rotational fishing.

The economics of fishing may also favor rotational policies, when (1) there is a high set-up cost for getting to, learning about, and taking fish out of each area; and (2) there are enough areas available so that a relatively steady overall yield can be taken to support processing and marketing systems.

The key idea in rotational fishing management is to manage for the recoveries, rather than to prevent the depletions. Thus the initial data requirements and monitoring requirements are much different, and generally simpler, than for sustained yield management.

Dealing with risk in the management of large unit stocks

While the ideas of experimental management and rotational fishing are attractive for systems that are spatially structured into substocks, they are obviously not applicable to large unit stocks that move extensively and are pursued by a variety of fisheries in different locations. Movement between fishing areas means that the results of any experimental policy in one fishing area are likely to depend on the treatments applied in other areas; the interpretation of such experiments would be hopelessly complex and ambiguous.

For the large unit stocks, there appears to be no alternative but to face long-term uncertainty about responses to development and to live with the risks of overfishing on the one hand and missed opportunity to take higher yields on the other hand. Provided the overall fishery development is a one-way trip, with steady increases in fishing pressure over time until some indications of overfishing become apparent, there will remain gross uncertainty during the development about how far the stock can be pushed. Even after an apparent state of overfishing has been reached, there will usually remain much uncertainty about how to interpret the signals that indicate overfishing: are catch rates just down temporarily due to some environmental factor? Is the fishery covering the whole range of the stock or is there some large, potentially exploitable fraction of it in areas or depths that have not yet been fished extensively? Are there slow responses still occurring in the ecosystem due to reduction of the stock, that will make potential yields higher/lower than appear possible at present?

Traditionally, fishermen and governments have demanded definitive predictions and answers from fishery managers; the usual description of this attitude is the statement "we don't care about your scientific qualms; give us a number!" Managers and stock assessment people have responded by agreeing to give the best numbers they can, since they have feared that to do otherwise (for example, deliberately provide several numbers or alternative hypotheses that appear equally probable given the data) would result in the politicians just looking elsewhere for advice and perhaps ending up using even worse estimates.

Stock-assessment people have sometimes tried to hedge against risks of overfishing by deliberately providing only conservative estimates. One problem with this approach is that other people may be willing to provide less conservative estimates. (You can always buy an expert who will tell you what you want to hear, with great confidence; ask the Peruvians how well this worked with their anchoveta fishery.) Another problem is that it represents a rather arrogant presumption by the assessment people that they know what is best for everyone in terms of risk avoidance, that is, how much potential yield it is worth giving up in order to be safe. It is questionable whether scientists, and especially biologists, have any special claim to good judgement about risk taking.

An important step in the development of better cooperative arrangements between fishermen and managers is to try to break that traditional insistence on pat answers, by insisting that uncertainties be clearly spelled out and alternative hypotheses regularly discussed in terms of their implications for further development. It is not important to provide precise ranges for parameter estimates or detailed collections of alternative predictions; in fact such presentations invite a return to insistence on single answers. What is important is to provide a clear and vivid picture of major alternative directions that further development might take, and to provide a sense of the odds that can be placed on these alternatives with the data available. In other words, the key need is not for more precise and elaborate calculations, but rather for better communication of the basic alternatives that emerge from even simple analyses.

18.5. Improving the Science of Fisheries Management

Is fisheries management scientific?

The key element in science is testing of hypotheses about the physical and natural world by experimentation and observation. At a superficial level, much of fisheries stock management is scientific: we estimate parameters from data. Each alternative parameter value is an alternative hypothesis, and we use data collected from the natural world to refine our understanding of alternative hypotheses.

There are many obvious failures of the scientific method: management actions are taken and not evaluated, and management agencies continue on their course without any use of data and analysis. However, in many fisheries a good "passive adaptive" scientific method is being used: each management action is recognized to be a new experiment, data are collected, and the experiment is evaluated.

Why then is fisheries in such trouble?

Why are there so many scientific failures? It is possible to proceed scientifically and yet make no progress, and this is unfortunately all too common in fisheries. As mentioned at the beginning of this chapter, large fisheries are not really amenable to the scientific method. For science to make progress, replication and control are necessary. The fields of science that one tends to think of as "hard" are precisely those fields characterized by the ability to do replicated measurements and experiments, and whose systems are not characterized by high variability.

Electrical engineers are incredibly successful at designing electronic circuits. We might be tempted to think this is because electricity is nice and simple, easy to predict and manipulate. But electrical engineers are not very successful at understanding large-scale power networks when these networks are perturbed. The great blackout on the east coast of the United States is just one of many examples of how electrical engineers do not really understand how to manage large-scale systems when they are severely perturbed. The physical science of large-scale networks is the same as circuit boards, but the difference is they cannot replicate and control large-scale power networks.

Fisheries scientists can replicate and control the incubation process in a hatchery quite well and make great scientific progress on how to turn eggs into fry. Similarly managers of small ponds and lakes can replicate and control their fisheries management actions quite successfully. Managers of blue whales, Peruvian anchoveta, and so on do not have the benefit of replication and control. We simply will never know if the collapse of the Peruvian anchoveta could have been prevented by alternative fisheries management actions. Fisheries, as a discipline, is going to have to accept that scientific progress on large unit-stocks will be slow, if not impossible. We believe there will always be great uncertainty about the importance of fisheries management actions in affecting the dynamics of these stocks.

There is the potential for temporal replication: we can learn from past experience in any given fishery, assuming that the production relationships are stationary. Similarly we can learn a lot about the behavior of clupeoid fisheries from experience with other clupeoid fisheries. Unfortunately, we will never be able to attain the level of understanding of these large unit stock fisheries that we can with fisheries that can be replicated.

Many fisheries are amenable to spatial replication and control. If we can break away from our tradition of allowing fishermen unlimited choice in areas to fish and move toward spatial variation in fishing policy, we can make great scientific progress. We believe that spatial variation in fishing policies provides perhaps the most powerful and underutilized tool available to fishery managers.

The northern Cod: the state of the art

As a final illustration of the difficulty, and yet potential, or fisheries management, let us now look at yet one more failure of fisheries management. The most important fishery in Canada is the northern cod, a fishery that peaked with a catch of about 800,000 tons (mostly foreign vessels) in 1969, and declined to less than 200,000 tons when Canada declared a 200-mile exclusive economic zone (EEZ) in 1977. Northern cod is particularly important for three reasons: (1) it is a very large and valuable fishery, (2) it takes place in the most economically depressed region of Canada where there are few alternative forms of employment, and (3) rehabilitation of

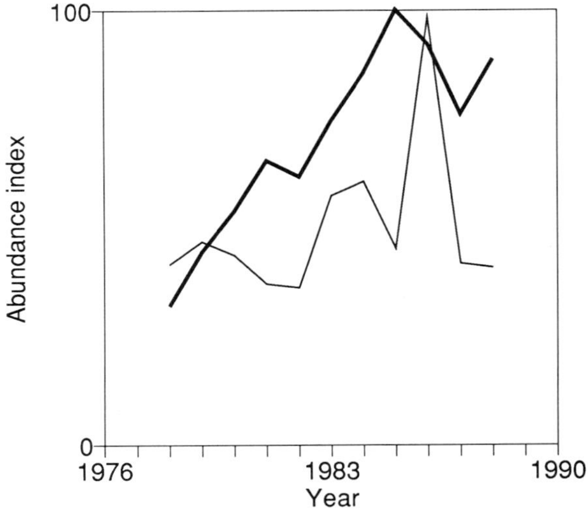

Figure 18.3. CPUE and research survey trends in the northern cod fishery. From CAFSAC 1989. Thick line is CPUE, thin line is survey.

northern cod, with increasing employment has been a showcase of Canadian fisheries policy and science. It came, therefore, as a major shock to fishermen and fisheries scientists when the Canadian government announced in 1989 that its stock assessments had been overly optimistic, and that quotas were to be greatly reduced. Some news reports indicated that layoffs of up to 20,000 workers might ensue.

The scientists working on northern cod have three types of data available: (1) catch-at-age data, (2) CPUE data of Canadian trawlers, and (3) research vessel surveys. Since cod are long lived, the catch-at-age data are only useful for estimating stock trends when combined with estimates of terminal fishing mortality rate, which in turn must be estimated from either the CPUE data or the surveys. Figure 18.3 shows the trends in research vessel index, and CPUE of Canadian trawlers.

The problem facing the scientists was the discrepancy between the surveys, which indicated that the stock had grown very little since 1978 (and therefore the fishing mortality rate was too high), and the CPUE which indicated that the stock was about three times higher in the late 1980s that it had been 10 years earlier. The objective of the management procedure was to implement $F_{0.1}$ policy, which means an annual exploitation rate of about 0.18.

Prior to 1986, the stock-assessment team had taken a midpoint between the two estimates of abundance in order to formulate the quota. The 1986 survey data were much higher than previous estimates, which caused the

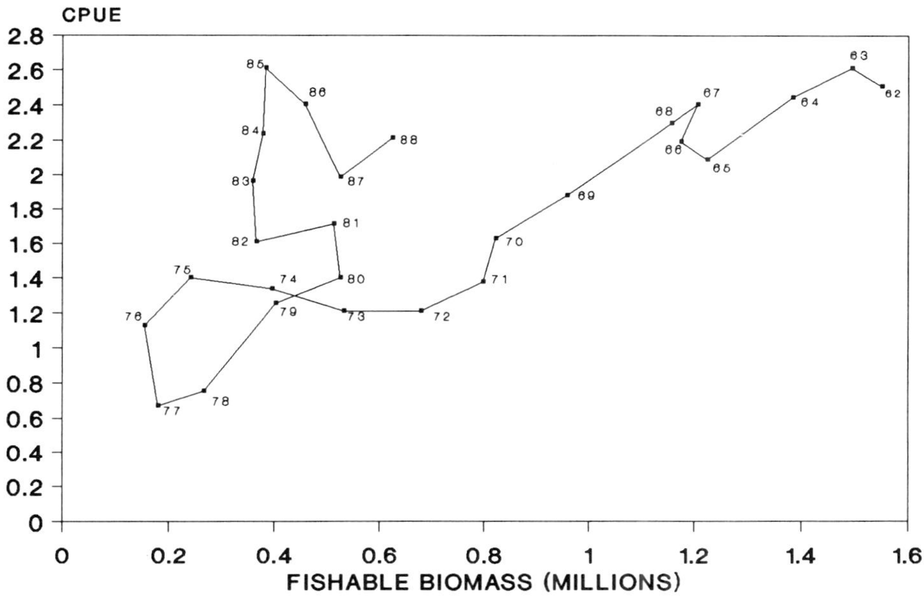

Figure 18.4. The relationship between CPUE and abundance for northern cod. From CAFSAC (1989).

midpoint to jump considerably. It appeared that the CPUE data had been correct all along. However, in 1987, and 1988, the survey data were again low and the CPUE had declined as well. The statistical analysis and quota recommendations are made by the Canadian Atlantic Fisheries Statistical Advisory Committee (CAFSAC), which in early 1989 said ". . . after intensive and extensive review of all possible factors, that the indications of stock status provided over the years by the research surveys have more accurately reflected historical trends in the cod populations." In other words, the cod biologists had been misled by growing commercial CPUE due to increased search and harvesting efficiency rather than by increased stock abundance! Figure 18.4 summarizes the new interpretation of the relationship between CPUE and abundance.

Integrating assessment and management

The assessment biologists working on the northern cod faced a common situation: they had several sources of data, providing different pictures of the status of the stock. Most stock assessments involve different types of information (catch rates, spatial distribution of catches, age or size fre-

Table 18.2 What a decision table for northern cod might have looked like in 1987.

Biological Hypothesis	Management Action Cautious quota	Aggressive quota
Stock declining or stable	modest yield	disaster
Stock increasing	modest yield	good yield

quency, tagging data, etc). Even within a single type of information such as catch rates, the trends may differ between vessel classes, types, or home ports. When your different sources of information provide the same basic picture, things are fine. Unfortunately, the pictures are often different. How do you make sense of conflicting information?

In Chapter 6 we discussed various approaches to utilizing multiple types of information using either Bayesian statistics or more ad hoc methods using auxiliary information. The current ultimate expression of this approach is the "stock synthesis" approach of Methot (1989), which can use an unlimited number of types of information, including (but not restricted to) catch rates, surveys, and catch-at-age data, to provide a single "synthetic" best estimate of stock status. The approach used for Northern Cod was a less formal "synthetic" method.

Warning: Beware of methods that average sources of information that are contradictory.

In the northern cod example, there were two types of information: catch rates that indicated the stock was growing, and surveys that indicated the stock was stable or declining. The DFO approach essentially averaged these two assessments, and then (and here is the problem) calculated the best management policy for the most likely stock status.

The problem with synthetic assessments is that assessment and management need to be integrated, not isolated. Fisheries management is a decision problem, not an assessment problem. Managers should be presented not with a single estimate but rather with a decision table, listing possible biological hypotheses (assessments), possible management actions, and the expected consequences of the management actions given different biological hypotheses. Table 18.2 shows qualitatively what such a decision table might have looked like in 1987 for the northern cod. The key element here is that the alternative hypotheses are explicitly listed, rather than "averaged" in a synthetic assessment. If the assessment process had said explicitly that the stock might actually be stable or was declining, the managers might have been more conservative in the quota, to try to avoid the possibility of having to reduce quotas drastically at a future date.

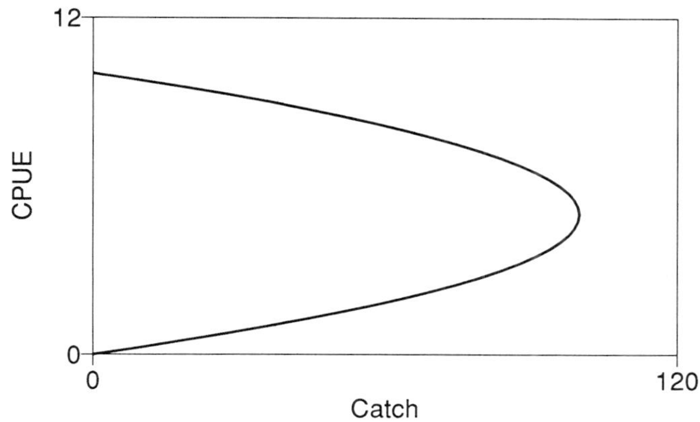

Figure 18.5. Equilibrium relationship between CPUE and catch for a simple biomass dynamics model.

The problem with synthetic assessments is that we are often faced with mutually incompatible hypotheses: the northern cod is either increasing or it is not. Synthetic assessment is simply a fancy way of taking the average between different types of information: yet it is possible (and perhaps likely) that the truth is not the average, but one of the extremes. Different hypotheses normally involve totally different assumptions, rather than simply being different parameter values in the same basic model.

The alternative hypotheses should not be merged in the assessment but should retain their separate identity and be carried through to the policy evaluation as true alternatives. An interesting note on the northern cod story is that the biologists doing the assessment have actually done analysis along the lines of Table 18.2, but the formal stock-assessment procedure does not allow these calculations to flow to the decision makers. Instead, the decision makers ask for the single best assessment!

18.6. Summary — What We Want You to Remember

In the many preceding chapters, and in the accompanying computer disk, we have shown you lots of techniques and given you lots of advice. In the final analysis, there are five things we want you to remember:

- You cannot predict how a stock will respond to exploitation. Your best hope is either spatial replication or qualitative experience with similar stocks elsewhere or the same stock in the past. You cannot predict MSY (or maximum economic yield) without exceeding it. You must weigh the risks of each possible management action against the potential benefits.
- Few stocks have the potential for stable yields. Irregularity of production,

cyclicity and cussedness are to be expected. In any fishery there is a tradeoff between average yield and variability of yield, and it is crucial to understand the appropriate management objective to decide whether average yield or lower variability is more important.

- There is almost always a tradeoff between catch rate and total catch. If you want to maximize catch, you will have to accept lower catch rates. The traditional opening picture of fisheries books, showing yield versus effort, should be replaced by Figure 18.5, showing catch rate versus catch. If a politician asks you what MSY is, you should ask what CPUE he is willing to accept.
- The biggest constraint on effective fisheries management is inability to change fishing mortality. Try to find some way to retain flexibility. Most fisheries disasters are caused by not being able to reduce fishing pressure once the biological or economic need is obvious. Devote research and management resources to understanding your fishermen and their gear.
- Always present managers with decision tables showing biological alternatives, and management alternatives. The output of a stock assessment should not be recommended quotas or fishing efforts — it should be the biological consequences of different actions. The people doing the stock assessment are not likely to be the right people to weigh the risks of alternative management actions.

Bibliography

Allen, K.R. 1966. Some methods for estimating exploited populations. J. Fish. Res. Bd. Can. 23:1553–1574.

Allen, P.M. and J.M. McGlade. 1986. Modelling complexity: the dynamics of discovery and exploitation in a fishery example. Ext. Mond. Dev. No 54–55.

Allen, P.M. and J.M. McGlade. 1987. Modelling and complex human systems: A fishery example. Eur. J. Op. Res. 30:147–167.

Allen, R. and R. Punsly. 1984. Catch rates as indices of abundance in yellowfin tuna, *Thunnus albacares,* in the eastern Pacific ocean. Inter-Am. Trop. Tuna Comm. Bull. 18(4).

Altukhov, Y.P. and E.A. Salmenkova. 1981. Applications of the stock concept to fish populations in the USSR. Can. J. Fish. Aquat. Sci. 38:1591–1600.

Alverson, D.L. and G.J. Paulik. 1973. Objectives and problems of managing aquatic living resources. J. Fish. Res. Bd. Can. 30:1936–1947.

Andrew, P.A. and D.S. Butterworth. 1987. Is $F_{0.1}$ an appropriate harvesting strategy for the cape hakes? S. Afr. J. Mar. Sci. 5:925–935.

Aranson R. 1984. Efficient harvesting of fish stocks: the case of Icelandic demersal fisheries. Ph.D. Thesis, Department of Economics, University of British Columbia.

Argue, A.W., R. Hilborn, R.M. Peterman, M.J. Staley, and C.J. Walters. 1983. The Strait of Georgia chinook and coho fishery. Can. J. Fish. Aquat. Sci. Bull. 211.

Armstrong, M.J. and P.A. Shelton. 1988. Bias in estimation of stock-recruit function parameters caused by non random environmental variability. Can. J. Fish. Aquat. Sci. 45:554–557.

Babayan, V.K. and Z.I. Kizner. 1988. Dynamic models for TAC assessment; logic, potentialities, development. Coll. Scient. Pap. Int. Comm. SE. Atl. Fish. 15(1):55–68.

Baltensweiler, W. 1964. *Zeiraphera griseana* Hubner (Lepidoperta: Tortricidae) in the European Alps. A contribution to the problem of cycles. Can. Entomol. 96:792–800.

Baranov, F.I. 1918. On the question of the biological basis of fisheries. Nauchn. Issled. Ikhtiologicheskii Inst. Izv. 1:81–128.

Bard, Y. 1974. *Nonlinear Parameter Estimation*. Academic Press, New York.

Bartoo, N.W. and K.R. Parker. 1983. Stochastic age-frequency estimation using the von Bertalanffy growth equation. Fish. Bull. 81:91–96.

Beamish, R.J. and G.A. McFarlane. 1983. The forgotten requirement for age validation in fisheries biology. Trans. Am. Fish. Soc. 112:735–743.

Beddington, J.R. 1978. On the risks associated with different harvesting strategies. Rep. Int. Whal. Comm. 28:165–167.

Beddington, J.R. and B. Grenfell. 1979. Risk and stability in whale harvesting. Rep. Int. Whal. Comm. 29:171–173.

Beddington, J.R. and R.M. May. 1977. Harvesting natural populations in a randomly fluctuating environment. Science 197:463–465.

Beddington, J.R. and D.B. Taylor. 1973. Optimum age specific harvesting of a population. Biometrics 29:801–809.

Bellman, R. 1961. *Adaptive Control Processes: A Guided Tour*. Princeton University Press, Princeton, NJ.

Berger, J.O. 1985. Statistical decision theory and Bayesian Analysis (second edition). Springer-Verlag, N.Y. 617 p.

Beverton, R.J.H. and S.J. Holt. 1956. A review of methods for estimating mortality rates in fish populations, with special references to sources of bias in catch sampling. Rapp. P.V. Reun. Cons. Int. Explor. Mer. 140:67–83.

Beverton, R.J.H. and S.J. Holt. 1957. *On the dynamics of exploited fish populations*. Fisheries Investment Series 2, Vol. 19 U.K. Ministry of Agriculture and Fisheries, London.

Bhattacharya. C.G. 1967. A simple method of resolution of a distribution into Gaussian components. Biometrics 23:115–135.

Bjorndal, T. and J. M. Conrad. 1987. Capital dynamics in the North Sea herring fishery. Mar. Res. Econ. 4:63–74.

Blackbourn, D.J. 1987. Sea surface temperature and the pre-season prediction of run timing in Fraser River sockeye salmon (*Oncorhynchus nerka*), In: H.D. Smith, L. Margolis, and C.C. Wood (Eds.) *Sockeye salmon* (Oncorhynchus nerka) *population biology and future management*. Canadian Special Publications on Fisheries and Aquatic Science, No. 96 pp. 296–306.

Bockstoce, J.R. and D.B. Botkin. 1983. The historical status and reduction of the Western Arctic bowhead whale *(Balaena mysticetus)* population by the pelagic whaling industry, 1848–1914. Rep. Int. Whal. Comm. (Special Issue) 5:107–141.

Bohlin, T. and B. Sundstrom. 1977. Influence of unequal catchability on population estimates using the Lincoln index and the removal method applied to electrofishing. Oikos 28:123–129.

Botsford, L.W. 1981. Optimal fishery policy for size-specific density-dependent population models. J. Math. Biol. 12:265–293.

Botsford, L.W., R.D. Method Jr., and W.E. Johnston. 1983. Effort dynamics of the north California Dungeness crab *(Cancer magister)* fishery. Can. J. Fish. Aquat. Sci. 40:337–346.

Bowen, B.K. and D.A. Hancock. 1985. Review of penaeid prawn fishery management regimes in Australia. *In* P.C. Rothlisberg, B.J. Hill, and D.J. Staples (Eds.) *Second Australian National Prawn Seminar*. NPS2, Cleveland, Queensland, Australia, pp. 247–265.

Box, G.E.P. and G.M. Jenkins. 1976. *Time Series Analysis: Forecasting and Control*. Holden Day, San Francisco.

Bradford, M.J. and R.M. Peterman. 1989. Incorrect parameter used in virtual population analysis (VPA) generate spurious time trends in reconstructed abundances. In: R.J. Beamish and S.A. MacFarlane (Eds.) *Effects of Ocean Variability on Recruitment and on Evaluation of Parameters Used in Stock Assessment*. Canadian Special Publications on Fisheries and Aquatic Science, No. 108. pp. 87–99.

Brouard, F. and R. Grandperrin. 1985. Deep-bottom fishes of the outer reef slope in Vanuatu. Working paper 12, Sevententh Regional Technical Meeting on Fisheries, South Pacific Commission, Noumea, New Caledonia.

Buckland, S.T. 1984. Monte Carlo confidence intervals. Biometrics 40:811–817.

Burkenroad, M.D. 1948. Fluctuations in abundance of Pacific halibut. Bull. Bingham Ocean-ogr. Coll. 11(4):81–129.

Burkenroad, M.D. 1950. Population dynamics in a regulated mareine fishery. [Book review of Thompson 1950] Texas J. Sci. 2(3):438–441.

Burkenroad, M.D. 1951. Some principles of marine fishery biology. Texas Inst. Mar. Sci. 2(1):177–212.

Burkenroad, M.D. 1953. Theory and practice of marine fishery management. J. Cons. Int. Explor. Mer 18:300–310.

Burnham, K.P., D.R. Anderson, G.C. White, C. Brownie, and K.H. Pollock. 1987. *Design and Analysis Methods for Fish Survival Experiments Based on Release–Recapture*. American Fisheries Society. Monograph 5. Bethesda, Maryland.

Butterworth, D.S. and P.A. Andrew. 1984. Dynamic catch-effort models for the hake stocks in ICSEAF Divisions 1.3 to 2.2. Coll. Scient. Pap. Int. Comm. S.E. Atl. Fish. 11:29–58.

Caddy, J.F. 1975. Spatial model for an exploited shellfish population, and its application to the Georges Bank scallop fishery. J. Fish. Res. Bd. Can. 32:1305–1328.

Caddy, J.F. and G.P. Bazigos. 1985. Practical guidelines for statistical monitoring of fisheries in manpower limited situations. FAO Fisheries Technical Paper 257.

Caddy, J.F. and J.A. Gulland. 1983. Historical patterns of fish stocks. Mar. Policy 7:267–278.

CAFSAC. 1989. Advice for 1989 on the management of cod in divisions 2J3KL. Canadian Atlantic Fisheries Scientific Advisory Committee Advisory Document 89/1.

Caputi, N. 1988. Factors affecting the time series bias in stock–recruitment relationships and the interaction between time series and measurement error bias. Can. J. Fish. Aquat. Sci. 45:178–184.

Cassie, R.M. 1954. Some uses of probability paper in the analysis of size frequency distibutions. Aust. J. Mar. Freshwater Res. 5:513–522.

Caswell, H. and D.E. Weeks. 1986. Two-sex models: chaos, extinction, and other dynamic consequences of sex. Am. Nat. 128:707–735.

Cauvin, D. 1980. The valuation of recreational fisheries. Can. J. Fish. Aquat. Sci. 37:1321–1327.

Chapman, D.G. 1974. Estimation of population size and sustainable yield of Sei whales in the Antarctic. Rep. Int. Whal. Comm. 24:82–90.

Charles, A.T. 1983a. Optimal fisheries investment: comparative dynamics for a deterministic seasonal fishery. Can. J. Fish. Aquat. Sci. 40:2069–2079.

Charles, A.T. 1983b. Optimal fisheries investment under uncertainty. Can. J. Fish. Aquat. Sci. 40:2080–2091.

Clark, C.W. 1976. *Mathematical Bioeconomics: The Optimal Management of Renewable Resources*. John Wiley. New York.

Clark, C.W. 1980. Towards a predictive model for the economic regulation of commercial fisheries. Can. J. Fish. Aquat. Sci. 37:1111–1129.

Clark, C.W. 1982. Concentration profiles and the production and management of marine fisheries. *In*: W. Eichorn, R. Henn, K. Neumann, and R.W. Shepard (Eds.) *Economic Theory of Natural Resources*. Physica-Verlag, Wurzburg-Wien, pp. 97–112.

Clark, C.W. 1985. *Bioeconomic Modelling and Fisheries Management*. Wiley, New York.

Clark, C., G. Edwards, and M. Friedlaender. 1973. Beverton-Holt model of commercial fisheries: optimal dynamics. J. Fish. Res. Bd. Can. 30:1629–1640.

Clark, C.W. and G.P. Kirkwood. 1979. Bioeconomic model of the Gulf of Carpentaria prawn fishery. J. Fish. Res. Bd. Can. 36:1304–1312.

Clark, C.W. and R.H. Lamberson. 1982. An economic history and analysis of pelagic whaling. Mar. Policy 6:103–120.

Clark, I. 1979. *Practical Geostatistics*. Elsevier, London.

Cochran, W.G. 1977. *Sampling Techniques*. Wiley, New York.

Collie, J.S. and M.P. Sissenwine. 1983. Estimating population size from relative abundance data measured with error. Can. J. Fish. Aquat. Sci. 40:1871–1879.

Cooke, J.G. and J.R. Beddington. 1985. The relationship between catch rates and abundance in fishes. IMA J. Math. Appl. Med. Biol. 2:1–15.

Cowx, I.G. 1983. Review of the methods for estimating fish population size from survey removal data. Fish. Mgmt. 14:67–82.

Crutchfield, J.A. (Ed.). 1965. *The Fisheries Problems in Resource Management*. University Washington Press, Seattle.

Crutchfield, J.A. 1979. Economic and social implications of the main policy alternatives for controlling fishing effort. J. Fish. Res. Bd. Can. 36:742–752.

Csirke, J. and G.D. Sharp. 1984. Reports of the expert consultation to examine changes in abundance and species composition of neritic fish resources. FAO Fisheries Report No. 291.

Cushing, D.H. 1971. The dependence of recruitment on parent stock in different groups of fishes. J. Cons. Int. Explor. Mer. 33:340–362.

Cushing, D.H. 1973. Dependence of recruitment on parent stock. J. Fish. Res. Bd. Can. 30:1965–1976.

Cushing, D. 1982. *Climate and Fisheries*. Academic Press, London.

Cushing, D.H. 1988. The study of stock and recruitment. *In*: J.A. Gulland (Ed.) *Fish Population Dynamics*. Wiley, New York. pp. 105–128.

Cushing, D.H. and J.G.K. Harris. 1973. Stock and recruitment and the problem of density-dependence. Rapp. P.-V. Reun. Cons. Explor. Perm. Int. Mer 164:142–155.

Daan, N. 1987. Multispecies versus single-species assessment of North Sea fish stocks. Can. J. Fish. Aquat. Sci. 44 (Suppl. 2):360–370.

Daniel, C. and F.S. Wood. 1974. *Fitting Equations to Data*. Wiley, N.Y.

Das Gupta, P. 1972. On two-sex models leading to stable populations. Theor. Pop. Biol. 3:358–375.

Davis, T.L.O. 1987. Biology of wildstock *Lates calcarifer* in northern Australia. *In*: J.W. Copland and D.L. Grey (Eds.) *Management of Wild and Cultured Sea Bass/Barramundi* (Lates calcarifer), Proceedings of an International Workshop Held at Darwin, N.T., Australia, 24–30 September 1986. ACIAR Proceedings No. 20, pp 22–29.

De Alteris, J.T. and D.M. Reifsteck. 1988. Selectivity of cod-ends on scup, *Stenotomus chrysops*, and the survivability of cod-end escapees: A preliminary report. *In*: *Proceedings of*

Square Mesh Workshop, Vol. 1, St. Johns, Newfoundland, November 25, 1988, pp. 38–53.

De Klerk, P. and M. Gatto. 1981. Some remarks on periodic harvesting of a fish population. Math. Biosci. 56:47–69.

de la Mare, W.K. 1986. Fitting population models to time series of abundance data. Rep. Int. Whal. Comm. 36:399–418.

De Lury, D.B. 1947. On the estimation of biological populations. Biometrics 3:145–167.

De Lury, D.B. 1951. On the planning of experiments for the estimation of fish populations. J. Fish. Res. Bd. Can. 8:281–307.

De Vries, T.J and W.G. Pearcy. 1982. Fish debris in sediments of the upwelling zone off central Peru; a late Quaternary record. Deep-Sea Res. 29(A):87–109.

Deriso, R.B. 1980. Harvesting strategies and parameter estimation for an age-structured model. Can. J. Fish. Aquat. Sci. 37:268–282.

Deriso, R.B. 1985a. Risk adverse harvesting strategies. *In*: M. Mangel (Ed.) *Resource Management*, Proceedings of the Second Ralf Yorque Workshop, Lecture Notes in Biomathematics No. 61. Springer-Verlag, Berlin. pp. 65–73

Deriso, R.B. 1985b. Stock assessment and new evidence of density-dependence. *In*: *Fisheries Dynamics: Harvest, Management, and Sampling*, Washington Sea Grant 85–1, Seattle, pp. 49–60.

Deriso, R.B. 1987. Optimal $F_{0.1}$ criteria and their relationship to maximum sustainable yield. Can. J. Fish. Aquat. Sci. 44(Suppl. 2):339–348.

Deriso, R.B., S.H. Hoag, and D.A. McCaughran. 1986. Two hypotheses about factors controlling production of Pacific halibut. Int. North Pacific Fish. Comm. Bull. 47:167–173.

Deriso, R.B. T.J. Quinn II, and P.R. Neal. 1985. Catch-age analysis with auxiliary information. Can. J. Fish. Aquat. Sci. 42:815–824.

Doubleday, W.G. 1976a. A least squares approach to analyzing catch at age data. Res. Bull. Int. Comm. NW Atl. Fish. 12:69–81.

Doubleday, W.G. 1976b. Environmental fluctuations and fisheries management. ICNAF. Sel. Pap. 1:141–150.

Doubleday, W.G. and D. Rivard (Eds.). 1981. *Bottom Trawl Surveys*. Canadian Special Publication on Fisheries and Aquatic Science, No. 58

Doubleday, W.G., D. Rivard, and W.D. McKone. 1984. Estimation of partial recruitment and yield per recruit for an otter trawl fishery for deepwater redfish. N. Am. J. Fish. Mgmt. 4:15–31.

Dragesund, O., J. Hamre, and O. Ulltang. 1980. Biology and population dynamics of the Norwegian spring-spawning herring. Rapp. P.-V. Reun. Cons. Int. Explor. Mer. 177:43–71.

Draper, N.R. and H. Smith. 1966. *Applied Regression Analysis*. Wiley, New York.

Edser, T. 1908. Note on the number of plaice at each length, in certain samples from the southern part of the North Sea, 1906. J.R. Stat. Soc. 71:686–690.

Edwards, A.W.F. 1972. *Likelihood; An Account of the Statistical Concept of Likelihood and its Applications to Scientific Inference*. Cambridge University Press, Cambridge.

Efron, B. 1979. Bootstrap methods: another look at the jackknife. Ann. Stat. 7:1–26.

Efron, B. 1981. Non parametric estimates of standard error: the jackknife, the bootstrap and other methods. Biometrika 68:589–599.

Efron, B. 1982. The jackknife, the bootstrap and other resampling plans. SIAM-CBMS 38:1–92.

Efron, B. 1985. Bootstrap confidence intervals for a class of parametric problems. Biometrika 72:45–58.

Efron, B. 1987. Better bootstrap confidence intervals. J. Am. Stat. Assoc. 82:171–185.

Fabens, A.J. 1965. Properties and fitting of the von Bertalanffy growth curve. Growth 29:265–289.

Fonteneau, A. 1984. Analyse de l'etat des stocks d'albacore au 30 Mai 1984. Int. Comm. Cons. Atl. Tunas 21(2):80–101.

Fournier, D.A. and C. Archibald. 1982. A general theory for analyzing catch at age data. Can. J. Fish. Aquat. Sci. 39:1195–1207.

Fournier, D.A. and P.A. Breen. 1983. Estimation of abalone mortality rates with growth analysis. Trans. Am. Fish. Soc. 112:403–411.

Fournier, D.A. and I.J. Doonan. 1987. A length-based stock assessment method utilizing a generalized delay-difference model. Can. J. Fish. Aquat. Sci. 44:422–437.

Fournier, D.A. and A.R. Warburton. 1989. Evaluating fisheries management models by simulated adaptive control-introducing the composite model. Can. J. Fish. Aquat. Sci. 46:1002–1012.

Fournier, D.A., J.R. Sibert, J. Majkowski, and J. Hampton. 1990. MULTIFAN a likelihood-based method for estimating growth parameters and age composition from multiple length frequency data sets illustrated by using data for southern bluefin tuna (*Thunnus maccoyii*). Can. J. Fish. Aquat. Sci. 47:301–317.

Fox, I.K. and O.C. Herfindahl. 1964. Attainment of efficiency in satisfying demands for water resources. Am. Econ. Rev. 54(3):198–206.

Fox, W.W. 1975. Fitting the generalized stock production model by least-squares and equilibrium approximation. U.S. Fish. Bull. 73:23–37.

Francis, R.I.C. 1988. Are growth parameters estimated from tagging and age-length data comparable? Can. J. Fish. Aquat. Sci. 45:936–942.

Fried, S.M. and R. Hilborn. 1988. Inseason forecasting of Bristol Bay, Alaska, sockeye salmon (*Oncorhynchus nerka*) abundance using Bayesian probability theory. Can. J. Fish. Aquat. Sci. 45:850–855.

García, S. 1984. The problems of unstable resource management. FAO publication CECAF/ECAF Series 84/28.

Garrod, D.J. 1969. Empirical assessments of catch/effort relationships in North Atlantic cod stocks. Int. Comm. NW Atl. Fish. Res. Bull. 6:26–34.

Garrod, D.J. and J.W. Horwood. 1979. Whale management:strategy and risk — a comment. Rep. Int. Whal. Comm. 29:215–218.

Gelb, A. (Ed). 1974. *Applied Optimal Estimation*. MIT Press, Cambridge, MA.

Getz., W. M. 1980. Harvesting models and stock recruitment curves in fisheries management. *In*: W. M. Getz (Ed.) *Mathematical Modeling in Biology and Ecology*. Lecture notes in biomathematics No 33. Springer-Verlag, Berlin, pp. 284–304.

Getz, W.M. and G.L. Swartzman. 1981. A probability transition matrix model for yield estimation in fisheries with highly variable recruitment. Can. J. Fish. Aquat. Sci. 38:847–855.

Getz, W.M., G.L. Swartzman, and R.C. Francis. 1985. A conceptual model for multispecies, multifleet fisheries. Lect. Notes Biomath. 61:49–63.

Glantz, M.H. 1979. Science, politics and economics of the Peruvian anchoveta fishery. Mar. Policy 3:201–210.

Goeden, G.B. 1978. A monograph of the coral trout, *Plectropomus leopardus* (Lacepede). Res. Bull. Fish. Serv. Queensl. 1:1–42

Gordon, H.S. 1953. An economic approach to the optimum utilization of fishery resources. J. Fish. Res. Bd. Can. 10:442–457.

Graham, M. 1935. Modern theory of exploiting a fishery and application to North Sea trawling. J. Cons. Int. Explor. Mer. 10:264–274.

Green, R.H. 1979. *Sampling Design and Statistical Methods for Environmental Biologists*. Wiley, New York.

Gulland, J.A. 1961. Fishing and the stocks of fish at Iceland. U.K. Ministry Agriculture and Fisheries Food, Fisheries Investment (Series 2) 23(4).

Gulland, J.A. 1965. Estimation of mortality rates. Annex to Arctic Fisheries Working Group Report (ICES, C.M. 1965. Doc. No. 3. mimeographed).

Gulland, J.A. 1971. Science and fishery management. J. Cons. Int. Explor. Mer. 33:471–477.

Gulland, J.A. 1973. Can a study of stock and recruitment aid management decisions. Rapp. P-V. Reun. Cons. Perm. Int. Explor. Mer. 164:368–372.

Gulland, J.A. 1974. Guidelines for fisheries management. FAO, Rome IOFC/DEV/ 74/36.

Gulland, J.A. 1983. *Fish Stock Assessment: A Manual of Basic Methods*. Wiley, New York. 223 p.

Hall, D.L. 1991. Growth, fecundity and recruitment response of stunted brook trout populations to density reduction. Ph.D. Thesis, Department of Zoology, University of British Columbia.

Hallier, J.P. 1985. Purse seining on debris-associated school in the western Indian Ocean. FAO TWS/85/30.

Hampton, J. and J. Majkowski. 1986. Computer simulations of future southern bluefin tuna parental biomass, recruitment, and catches under the 1982 regime. N. Am. J. Fish. Mgmt. 6:77–87.

Hannesson, R. 1975. Fishery dynamics: a North Atlantic cod fishery. Can. J. Econ. 8:151–173.

Harris, J.G.K. 1975. The effect of density-dependent mortality on the shape of the stock and recruitment curve. J. Cons. Int. Explor. Mer. 36:144–149.

Heincke, F. 1913. Investigations on the plaice. General report. 1. Plaice fishery and protective regulations. Part I. Rapp. P.-V. Reun. Cons. Perm. Int. Explor. Mer. 17A:1–153.

Helland, A. 1913–1914. Rovdyrene i Norge. Tidsskrift for Skogbruk 1913–1914. (cited from Hjort et al. 1933 in Ricker 1975).

Hightower, J.E. and W.H. Lenarz. 1989. Optimal harvesting policies for the widow rockfish fishery. Am. Fish. Soc. Symp. 6:83–91.

Hilborn, R. 1976. Optimal exploitation of multiple stocks by a common fishery: a new methodology. J. Fish. Res. Bd. Can. 33:1–5.

Hilborn, R. 1979. Comparison of fisheries control systems that utilize catch and effort data. J. Fish. Res. Bd. Can. 36:1477–1489.

Hilborn, R. 1985a. Fleet dynamics and individual variation: why some people catch more fish than others. Can. J. Fish. Aquat. Sci. 42:2–13.

Hilborn, R. 1985b. Apparent stock recruitment relationships in mixed stock fisheries. Can. J. Fish. Aquat. Sci. 42:718–723.

Hilborn, R. 1985c. Simplified calculation of optimum spawning stock size from Ricker's stock recruitment curve. Can. J. Fish. Aquat. Sci. 42:1833–1834.

Hilborn, R. 1985d. A comparison of harvest policies for mixed stock fisheries. *In*: M. Mangel (Ed.) *Resource Management: Proceedings of the Second Ralf Yorque Workshop*, Lecture Notes in Biomathematics No. 61. Springer-Verlag, Berlin. pp. 75–87

Hilborn, R. 1986. A comparison of alternative harvest tactics for invertebrate fisheries. *In*: G.S. Jamieson and N. Bourne (Eds.) Canadian Special Publications in Fisheries and Aquatic Sciences, No. 92, pp. 313–317.

Hilborn, R. 1987. Living with uncertainty in resource management. N. Am. J. Fish. Mgt. 7:1–5.

Hilborn, R. 1988. Determination of tag return from recaptured fish by sequential examination for tags. Trans. Am. Fish. Soc. 117:510–514.

Hilborn, R. 1990. Determination of fish movement patterns from tag recoveries using maximum likelihood estimators. Can. J. Fish. Aquat. Sci. 47:635–643.

Hilborn, R. and R.B. Kennedy. 1991. Spatial pattern in catch rates: a test of economic theory. J. Math. Biol. in press.

Hilborn, R. and M. Ledbetter. 1979. Analysis of the British Columbia salmon purse seine fleet: dynamics of movement. J. Fish. Res. Bd. Can. 36:384–391.

Hilborn, R. and M. Ledbetter. 1985. Determinants of catching power in the British Columbia salmon purse-seine fleet. Can. J. Fish. Aquat. Sci. 42:51–56

Hilborn, R. and J. Sibert. 1988. Adaptive management of developing fisheries. Mar. Policy 12:112–121.

Hilborn, R. and P.J. Starr. 1984. Making stock-recruitment work. *In*: P.E.K. Symons and M. Waldichuk (Eds.) *Proceedings of the Workshop on Stream Indexing for Salmon Escapement Estimation*. Canadian Technical Report on Fisheries and Aquatic Science, No. 1326, pp. 227–244.

Hilborn, R. and C.J. Walters. 1977. Differing goals of salmon management on the Skeena River. J. Fish. Res. Bd. Can. 34:64–72.

Hilborn, R. and C.J. Walters. 1987a. Microcomputer simulation for training and teaching. Environ. Softw. 1:156–163.

Hilborn, R. and C.J. Walters. 1987b. A general model for simulation of stock and fleet dynamics in spatially heterogeneous fisheries. Can. J. Fish. Aquat. Sci. 44:1366–1369.

Hjort, J., G. Jahn, and P. Ottestad. 1933. The optimum catch. Hvalradets Skr. 7:92–127.

Hoenig, J.M. and D.M. Heisey. 1987. Use of a log-linear model with the EM algorithm to correct estimates of stock composition and to convert length to age. Trans. Am. Fish. Soc. 116:232–243.

Hoenig, J.M., W.D. Lawing, and N.A. Hoenig. 1983. Using mean age, mean length, and median length data to estimate the total mortality rate. Int. Counc. Explor. Sea C.M. 23:1–11.

Holling, C.S. 1959. Some characteristics of simple types of predation and parasitism. Can. Entomol. 91:385–398.

Holling, C.S. 1973. Resilience and stability of ecological systems. Ann. Rev. Ecol. Syst. 4:1–23.

Holling, C.S. (Ed.). 1978. *Adaptive Environmental Assessment and Management*. Wiley, Chichester.

Holling, C.S. 1980. Forest insects, forest fires, and resilience. *In*: H. Mooney, J.M. Bonnicksen, N.L. Christensen, J.E. Lotan and W.A. Reiners (eds.), *Fire Regimes and Ecosystem Properties*. USDA Forest Service General Technical Report WO-26.

Hudson, D.J. 1971. Interval estimation from the likelihood function. Proc. Roy. Stat. Soc. 33:256–262.

Hunter, J.R., A.W. Argue, W.H. Bayliff, A.E. Dizon, A. Fonteneau, D. Goodman, and G.R. Seckel. 1986. The dynamics of tuna movement: an evaluation of past and future research. FAO Fisheries Technical Paper 277.

ICSEAF, 1989. Report of the ad hoc working group on stock assessment methodology. International Commission for the Southeast Atlantic fisheries. Report ICSEAF/89/METH/11.

ICES. 1988. Report of the workshop on methods of fish stock assessment. International Council for the Exploration of the Seas, Copenhagen, Denmark.

IFREMER. 1987. Assessment of technical interactions in mixed fisheries: Report of a workshop held at IFREMER in Nantes (France) under the auspices of EEC (DG XIV). IFREMER BP. 1049, 44037 Nantes CEDEX France.

International Whaling Commission. 1988. Comprehensive assessment workshop on management. Rep. No. 38.

IPHC. 1987. The Pacific halibut: Biology, fishery and management. International Pacific Halibut Commission Technical Report No. 22.

Ivlev, V.S. 1961. *Experimental Ecology of the Feeding of Fishes*. Yale University Press, New Haven.

Jacobson, D.L., W.R. MacCallum, and G.R. Spangler. 1987. Biomass dynamics of Lake Superior lake herring *(Coregonus artedii)*: application of Schnute's difference model. Can. J. Fish. Aquat. Sci. 44 (Suppl. 2):275–288.

Jakobsson, J. 1980. Exploitation of the Icelandic spring- and summer-spawning herring in relation to fisheries management, 1947–1977. Rapp. P-V. Reun. Cons. Perm. Int. Explor. Mer. 177:23–42.

Johannesson, K.A. and R.B. Mitson. 1983. Fisheries acoustics: a practical manual for aquatic biomass estimation. FAO Fisheries Technical Paper 240.

Johnson, B.L. and C. Walters. 1988. Estimation of abundance and survey catchability from catch-at-age and survey data. Can. J. Fish. Aquat. Sci. (submitted).

Jones, R. 1961. The assessment of long-term effects of changes in gear selectivity and fishing effort. Mar. Res. (Scotland) 1961(2):1–19.

Jones, R. 1981. The use of length composition data in fish stock assessment (with notes on VPA and cohort analysis). FAO Fisheries Circular No. 734.

Jones, R. 1984. Assessing the effects of changes in exploitation pattern using length composition data (with notes on VPA and cohort analysis). FAO Fisheries Technical Paper No. 256.

Kanwisher, J. K., K. Lawson, and G. Sundness. 1974. Acoustic telemetry from fish. Fish. Bull. 72:251–255.

Kearney, R.E. 1982. An assessment of the skipjack and baitfish resources of Fiji. Skipjack Survey and Assessment Programme Final Country Report No.1, South Pacific Commission, Noumea, New Caledonia.

Keeney, R.L. 1977. A utility function for examining policy affecting salmon in the Skeena River. J. Fish. Res. Bd. Can 34:49–63.

Keeney, R. and H. Raiffa, 1976. *Decisions with Multiple Objectives*. Wiley, New York.

Kendall, M.G. and A. Stuart. 1977. The Advanced Theory of Statistics. Hafner, New York.

Kerr, S.R. and R.A. Ryder. 1989. Current approaches to multispecies analysis of marine fisheries. Can. J. Fish. Aquat. Sci. 46:528–534.

Kimura, D.K. 1989. Variability, tuning, and simulation for the Doubleday-Deriso catch-at-age model. Can. J. Fish. Aquat. Sci. 46:941–949

Kimura, D.K., J.W. Balsinger, and D.H. Ito. 1984. Generalized stock reduction analysis. Can. J. Fish. Aquat. Sci. 41:1325–1333.

Kirkwood, G.P., and I.F. Somers. 1984. Growth of two species of tiger prawn, *Penaeus esculentus* and *P. semisulcatus*, in the western Gulf of Carpentaria. Aust. J. Mar. Freshw. Res. 35:703–712.

Kleiber, P. and E.F. Edwards. 1988. A model of tuna vessel and dolphin school movement in the eastern tropical Pacific ocean: technical description of the model. NMFS Southwest Fishery Center Administrative Report LJ-88–28.

Kope, R.G. 1987. Seperable virtual population analysis of Pacific salmon with application to marked chinook salmon *(Oncorhynchus tshawytscha)* from California's central valley. Can. J. Fish. Aquat. Sci. 44:1213–1220.

Krebs, C.J. 1989. *Ecological Methodology*. Harper and Row, New York.

Lack, D. 1954. *The Natural Regulation of Animal Numbers*. Oxford University Press, London.

Laevastu, T., F. Favorite, and H. A. Larkins. 1982. Resource assessment and evaluation of the dynamics of the fisheries resources in the northeastern Pacific with numerical ecosystem models. *In*: M.C. Mercer, (Ed.). *Multispecies Approaches to Fisheries Management Advice*. Canadian Special Publications in Fisheries and Aquatic Sciences, No. 59.

Lai, H.-L. and V.F. Gallucci. 1988. Effects of parameter variability on length-cohort analysis. J. Cons. Int. Explor. Mer 45:82–91.

Lane, D.E. 1988. Investment decision making by fishermen. Can. J. Fish. Aquat. Sci. 45:782–796.

Larkin, P.A. 1977. An epitaph for the concept of maximum sustained yield. Trans. Am. Fish. Soc. 106:1–11.

Larkin, P.A. and W.E. Ricker. 1964. Further information on sustained yields from fluctuating environments. J. Fish. Res. Bd. Can. 21:1–7.

LaPointe, M.F., R.M. Peterman, and A.D. MacCall. 1989. Trends in fishing mortality rate along with errors in natural mortality rate can cause spurious time trends in fish stock abundances estimated by virtual population analysis (VPA). Can. J. Fish. Aquat. Sci. 46:2129–2139.

Lasker, R. and A. MacCall. 1983. New ideas on the fluctuations of the clupeoid stocks off California. pp. 110–120 *In*: CNC/SCOR. *Proceedings of the Joint Oceanographic Assembly 1982 — General Symposia*. Canadian National Committee/Scientific Committee on Ocean Research. Ottawa, Ontario 189 p.

Laurec, A. and J.G. Shepherd. 1983. On the analysis of catch and effort data. J. Cons. Int. Explor. Mer 41:81–84.

Lawson, T.A. and R. Hilborn. 1985. Equilibrium yields and yield isopleths from a general age-structured model of harvested populations. Can. J. Fish. Aquat. Sci 42:1766–1771.

Ledbetter, M. 1986. Competition and information among British Columbia salmon purse seiners. Ph.D. Thesis, Department of Zoology, University of British Columbia, Vancouver.

Lee, K. 1989. The Columbia River Basin: experimenting with sustainability. Environment 31(6):7–11,30–33.

Leslie, P.H. and D.H.S. Davis. 1939. An attempt to determine the absolute number of rats on a given area. J. Anim. Ecol. 8:94–113.

Lewis, R.K. 1981. Southern rock lobster (*Jasus novaehollandiae*) — a review of the zone N fishery. SAFIC 5:3–16.

Lewis, R.K. 1983. Southern rock lobster (*Jasus novaehollandiae*) — a review of the zone S fishery. SAFIC 7:31–43.

Linhart, H. and W. Zucchini. 1986. *Model Selection*. Wiley, New York.

Lleonart, J. and J. Salat. 1989. A dynamic approach to catch- effort data analysis using a new concept: the inertia of the stock. Can. J. Fish. Aquat. Sci. 46:1367–1374.

Lleonart, J., J. Salat, and B. Roel. 1985. A dynamic production model. Colln. Scient. Pap. Int. Comm. SE. Atl. Fish 12(I): 119–146.

Lotka, A.J. 1925. *Elements of Physical Biology*. Williams and Wilkins, Baltimore.

Ludwig, D. 1981. Harvesting strategies for a randomly fluctuating population. J. Const Int. Explor. Mer. 39: 168–174.

Ludwig, D. and R. Hilborn. 1983. Adaptive probing strategies for age structured fish stocks. Can. J. Fish. Aquat. Sci. 40:559–569.

Ludwig, D. and C.J. Walters. 1981. Measurement errors and uncertainty in parameter estimates for stock and recruitment. Can. J. Fish. Aquat. Sci. 38:711–720.

Ludwig, D. and C.J. Walters. 1982. Optimal harvesting with imprecise parameter estimates. Ecol. Model. 14:273–292.

Ludwig, D. and C.J. Walters. 1985. Are age structured models appropriate for catch-effort data? Can. J. Fish. Aquat. Sci. 42:1066–1072.

Ludwig, D. and C.J. Walters. 1989. A robust method for parameter estimation from catch and effort data. Can. J. Fish. Aquat. Sci. 46:137–144.

Ludwig, D., C.J. Walters, and J. Cooke. 1988. Comparison of two models and two estimation methods for catch and effort data. Nat. Res. Model 2:457-498.

MacCall, A.D. 1984. Population models of habitat selection, with application to the northern anchovy. National Marine Fisheries Service, Southwest Fisheries Centre, Admin. Rept. No. LJ-84–01.

MacCall, A.D. 1990. *Dynamic Geography of Marine Fish Populations*. University of Washington Press, Seattle.

MacDonald, P.D.M. and T.J. Pitcher. 1979. Age groups from size-frequency data: a versatile and efficient method of analysing distribution mixtures. J. Fish. Res. Bd. Can. 36:987–1001.

Mace, P.M. and I.J. Doonan. 1988. A generalized bioeconomic simulation model for fish population dynamics. N.J. Fish. Ass. Res. Doc. 88/4.

MacLulich, D.A. 1937. Fluctuations in the numbers of the varying hare (*Lepus americanus*). University of Toronto Studies, Biol. Ser. No. 43.

Mahon, R. 1980. Accuracy of catch-effort methods for estimating fish density and biomass in streams. Env. Biol. Fish. 5:343–360.

Maier, H.N. 1906. Beiträge zur Alterbestimmung der Fische. I. Allgemeines. Die Alterbestimmung nach den Otolithen bei Scholle und Kabeljau. Wiss. Meeresuntersuchungen (Helgoland), N.F. 8:57–115 (cited in Ricker 1975).

Maine, J. 1988. Scale damage and survival of young gadoid fish escaping from the cod-end of a demersal trawl. *In*: J. DeAlteris (Ed.) *Proceedings of Selectivity and Survivability Workshop*. URI Sea Grant Marine Advisory Service, Nargansett, RI.

Mangel, M. 1985. *Decision and Control in Uncertain Resource Systems*. Academic Press, New York.

Mangel, M. and C.W. Clark. 1988. *Dynamic Modelling in Behavioral Ecology*. Princeton University Press, Princeton, NJ.

Marten, G.G. and J.J. Polovina. 1982. A comparative study of fish yields from various tropical ecosystems. *In*: D. Pauly and G.I. Murphy (Eds.) *Theory and Management of Tropical Fisheries*. ICLARM Contrib. 105, Manila, pp. 255–285.

Marten, G.G., P.M. Kleiber, and J.A.K. Reid. 1975. A computer program for fitting tracer kinetic and other differential equations to data. Ecology 56:752–754.

May, R.M. 1973. *Stability and Complexity in Model Ecosystems*. Princeton University Press, Princeton, NJ.

May, R.M. (Ed.). 1984. *Exploitation of Marine Communities. Report of the Dahlem Workshop on Exploitation of Marine Communities*. Berlin, April 1–6, 1984. Life Sciences Research Report 32. Springer-Verlag, Berlin.

May, R.M. J.R. Beddington, J.W. Horwood, and J.G. Shepherd. 1978. Exploiting natural populations in an uncertain world. Math. Biosci. 42:219–252.

McCullagh, P. and J.A. Nelder. 1983. *Generalized Linear Models*. Chapman and Hall, New York.

McGlade, J.M. and P.M. Allen. 1984. The fishing industry as a complex system. *In*: R. Mahon (Ed.) *Toward the Inclusion of Fishery Interactions in Management Advice*. Canadian Technical Report on Fisheries and Aquatic Science No. 1347, pp. 209–216.

McHugh, J.L. 1984. Fishery Management. Lecture Notes on Coastal and Estuary Studies No.10. Springer-Verlag, Berlin..

Megrey, B.A. 1989. Review and comparison of age-structured stock assessment models from theoretical and applied points of view. Am. Fish. Soc. Symp. 6:8–48.

Meinhold, R.J. and N.D. Singpurwalla. 1983. Understanding the Kalman filter. Am. Stat. 37:123–127.

Mendelssohn, R. 1980. A systematic approach to determining mean-variance tradeoffs when managing randomly varying populations. Math. Biosci. 50:75–84.

Mendelssohn, R. 1982. Discount factors and risk aversion in managing random fish populations. Can. J. Fish. Aquat. Sci. 39:1252–1257.

Mercer, M.C. (Ed.) 1982. Multispecies approaches to fisheries management advice. Canadian Special Publications on Fisheries and Aquatic Sciences, No. 59.

Methot, R.D. 1989. Synthetic estimates of historical abundance and mortality for northern anchovy. Am. Fish. Soc. Symp. 6:66–82.

Millington, P. 1984. The fishermen as predator: numerical responses of British Columbia gillnet fishermen to salmon abundance. MS Thesis. Department of Zoology, University of British Columbia, Vancouver.

Mittertreiner, A. and J. Schnute. 1985. Simplex: a manual and software package for easy nonlinear estimation and interpretation in fishery research. Canadian Technical Report on Fisheries and Aquatic Science, No. 1384.

Miyabe, N. and W. H. Bayliff. 1990. A review of the Japanese longline fishery for tunas and billfishes in the eastern Pacific Ocean, 1971–1980. Inter-Amer. Trop. Tuna Comm., Bull., 19:1–163.

Mohn, R.K. 1980. Bias and error propagation in logistic models. Can J. Fish. Aquat. Sci. 37:1276–1283.

Moloney, D. G. and P. H. Pearse. 1979. Quantitative rights as an instrument for regulating commercial fisheries. J. Fish. Res. Bd. Can. 36:859–866.

Moussalli, E. and R. Hilborn. 1986. Optimal stock size and harvest rate in multistage life history models. Can. J. Fish. Aquat. Sci. 43:135–141.

Munro, J.L. 1980. Stock assessment models: applicability and utility in tropical small-scale fisheries. *In*: S.B. Saila and P.M. Roedel (eds.) *Stock Assessment for Tropical Small-scale Fisheries*. International Center Marine Resource Development, University of Rhode Island, Kingston.

Munro, J.L. and R. Thompson. 1983. The Jamaican fishing industry. *In*: J.L. Munro, (Ed.) *Caribbean Coral Reef Fish Resources*. ICLARM, Manila, Philippines.

Murawski, S.A. 1984. Mixed-species yield-per-recruitment analysis accounting for technological interactions. Can. J. Fish. Aquat. Sci. 41:897–916.

Murawski, S.A. and J.S. Idoine. 1989. Yield sustainability under constant-catch policy and stochastic recruitment. Trans. Am. Fish. Soc. 118:349–367.

Murphy, G.I. 1965. A solution to the catch equation. J. Fish. Res. Bd. Can. 22:191–202.

Murphy, G.I. 1977. New understanding of southern bluefin tuna. Aust. Fish. 36:2–6.

Murphy, G.I. 1982. Recruitment of tropical fisheries. *In*: D. Pauly and G.I. Murphy (Eds.) *Theory and Management of Tropical Fisheries*. ICLARM Contribution, No. 105, Manila.

Nansen, F. 1924. *Hunting and Adventures in the Arctic*. Kristiania, Oslo.

Nelder, J.A. and R. Mead. 1965. A simplex method for function minimization. Comput. J. 7:308–313.

O'Neill, R.V. 1973. Error analysis of ecological models. Report presented at the Third National Symposium on Radioecology in 1971. Deciduous Forest Biome Memo Report 71–15.

Overholtz, W.J., M.P. Sissenwine, and S.H. Clark. 1986. Recruitment variability and its implications for managing and rebuilding the Georges Bank haddock stock. Can. J. Fish. Aquat. Sci. 43:748–753.

Pacific Salmon Commission. 1988. 1987 Annual report of the joint chinook technical committee. TCCHINOOK 88–2 Appendix II, Supplement B. Vancouver, Canada.

Paloheimo, J.E. 1980. Estimation of mortality rates in fish populations. Trans. Am. Fish. Soc. 1094:378–386.

Paloheimo, J.E. and L.M. Dickie. 1964. Abundance and fishing success. Rapp. P.-C. Reun. Cons. Perm. Int. Explor. Mer 155:152–163.

Pálsson, G. and P. Durrenberger. 1982. To dream of fish: the causes of Icelandic skippers' fishing success. J. Anthropol. Res. 38:227-242.

Parish, B.B. (Ed.). 1973. *Fish Stocks and Recruitment*. Rapp. P.-V. Reun. Vol. 164.

Parrish, B.B. and R.S. Keir. 1959. IV. The measurement of fishing power and its relation to the characteristics of vessels. ICNAF Ann. Proc. 9:106–112.

Paulik, G.J. 1961. Detection of incomplete reporting of tags. J. Fish. Res. Bd. Can. 18:817–832.

Paulik, G.J. 1973. Studies of the possible form of the stock and recruitment curve. Rapp. P-V. Reun. Cons. Perm. Int. Explor. Mer 164:302–315.

Paulik, G.J., A.S. Hourston, and P.A. Larkin. 1967. Exploitation of multiple stocks by a common fishery. J. Fish. Res. Bd. Can. 24:2527–2537.

Pauly, D. 1979. *Theory and Management of Tropical Multispecies Stocks: A Review, with Emphasis on the Southeast Asian Demersal Fisheries*. ICLARM Studies and Reviews No. 1. ICLARM, Manila.

Pauly, D. 1984. *Fish Population Dynamics in Tropical Waters: a Manual for Use with Programmable Calculators*. ICLARM, Manila.

Pauly, D. and G.R. Morgan (Eds.). 1987. *Length-Based Methods in Fisheries Research*. ICLARM Conference Proceedings, No. 13, ICLARM, Manila.

Pauly, D. and L. Tsukayama (Eds). 1987. *The Peruvian Anchoveta and its Upwelling Ecosystem: Three Decades of Change*. ICLARM Studies and Reviews, No. 15, ICLARM, Manila.

Pauly, D., M.L. Palomares, and F.C. Gayanilo. 1987. VPA estimates of monthly population length composition, recruitment, mortality, biomass and related statistics of Peruvian anchoveta, 1953–1981. *In*: D. Pauly and L. Tsukayama (Eds). 1987. *The Peruvian Anchoveta and its Upwelling Ecosystem: Three Decades of Change*. ICLARM Studies and Reviews, No. 15. ICLARM, Manila, pp. 142–166

Pella, J.J. and P.K. Tomlinson. 1969. A generalized stock production model. Bull. Inter-Am. Trop. Tuna Comm. 13:419–496.

Penn, J.W. and N. Caputi. 1984. Stock recruitment relationships for the tiger prawn, (*Penaeus esculentus*) fishery in Exmouth Gulf, Western Australia, and their implications for management. *In*: P.C. Rothlisberg, B.J. Hill and D.J. Staples (Eds.) *Second Australian Natural Prawn Seminar*, NPS2, Cleveland, Australia, pp. 165–173.

Peterman, R.M. 1975. New techniques for policy evaluation in ecological systems: methodologies for a case study of Pacific salmon fisheries. J. Fish. Res. Bd. Can. 32:2179–2188.

Peterman, R.M. 1977. Graphical evaluation of environmental management options: examples from a forest-insect pest system. Ecol. Model. 3:133–148.

Peterman, R.M. 1980. Dynamics of native indian food fisheries on salmon in British Columbia. Can. J. Fish. Aquat. Sci. 37:561–566.

Peterman, R.M. 1981. Form of random variation in salmon smolt- adult relations and its influence on production estimates. Can. J. Fish. Aquat. Sci. 38:1113–1119.

Peterman, R.M. and M. Gatto. 1978. Estimation of functional responses of predators on juvenile salmon. J. Fish. Res. Bd. Can. 35:797–808.

Peterman, R.M., W.C. Clark, and C.S. Holling. 1979. Dynamics of resilience: shifting stability domains in fish and insect systems. *In*: R.M. Anderson, B.D. Turner, and L.R. Taylor (Eds.). *Population Dynamics*. Blackwell Scientific, Oxford, pp. 321–341.

Petersen, C.G.J. 1892. Fiskensbiologiske forhold i Holboek Fjord, 1890–91. Beret. Danske Biol. Sta. 1890–91, 1:121–183.

Peterson, N.P. and C.J. Cederholm. 1984. A comparison of the removal and mark-recapture methods of population estimation for juvenile coho salmon in a small stream. No. Amer. J. Fish. Mgmt. 4:99–102.

Pikitch, E.K. 1987. Use of a mixed-species yield-per-recruit model to explore the consequences of various management policies for the Oregon flatfish fishery. Can. J. Fish. Aquat. Sci. 44 (Suppl II):349–359.

Pikitch, E.K. 1988. Objectives for biologically and technically interrelated fisheries. *In*: W.S. Wooster (Ed.) *Fisheries Science and Management: Objectives and Limitations,* Lecture Notes on Coastal and Estuarine Studies, No. 28 Springer-Verlag, New York, Chapter 5, pp. 107–136.

Pikitch, E.K., D.L. Erickson, and J.R. Wallace. 1988. An evaluation of the effectiveness of trip limits as a management tool. Northwest and Alaska Fisheries Center, National Marine Fisheries Service NWAFC Processed Report No. 88–27.

Polovina, J.J. 1989. A system of simultaneous dynamic production and forecast models for multispecies or multiarea applications. Can. J. Fish. Aquat. Sci. 46:961–963.

Pope, J.G. 1972. An investigation of the accuracy of virtual population analysis using cohort analysis. ICNAF Res. Bull. 9:65–74.

Pope, J.G. and B.J. Knight. 1982. Simple models of predation in multi-age multispecies fisheries for considering the estimation of fishing mortality and its effects. *In*: M.C. Mercer, (Ed.). *Multispecies Approaches to Fisheries Management Advice,* Canadian Special Publications on Fisheries and Aquatic Science, No. 59, pp. 64–69

Pope, J.G. and J.G. Shepherd. 1985. A comparison of the performance of various methods for tuning VPAs using effort data. J. Cons. Int. Explor. Mer 42:129–151.

Postel, E. 1955. Contribution a l'etude de la biologie de quelques Scombridae de l'Atlantique tropico-oriental. Ann. Stn. Oceangr. Salammbo, 10:1–167 (cited in Pauly, 1984).

Powers, J.E. and R.T. Lackey. 1976. A multiatribute utility function for management of a recreational resource. Va. J. Sci. 27:191–198.

Press, W.H., B.P. Flannery, S.A. Teukolsky, and William T. Vetterling. 1986. *Numerical Recipes: The Art of Scientific Computing*. Cambridge University Press, Cambridge.

Prince, J.D. 1989. The fisheries biology of the Tasmanian stocks of *Haliotis rubra*. Ph.D. Thesis, Department of Zoology, University of Tasmania.

Punt, A.E. 1988. Model selection for the dynamics of southern African hake resources. M.S. Thesis, Department of Applied Mathematics, University of Cape Town, South Africa.

Punt, A.E. and D.S. Butterworth. 1989. Results of first stage screening trials for a proposed whale stock management procedure. Rep. Int. Whal. Commn. (Special Issue II) 191–197.

Quinn, T.J. II. R.B. Deriso, and S.H. Hoag. 1985. Methods of population assessment of Pacific halibut. IPHC Scientific Report No. 72.

Ralston, S. 1976. Age determination of a tropical reef butterfly fish utilizing daily growth rings in otoliths. Fish. Bull. 74:990–994.

Ralston, S. and J.J. Polovina. 1982. A multispecies analysis of the commercial deep-sea handline fishery in Hawaii. Fish. Bull. 80:435–448.

Ratkowsky, D.A. 1983. *Nonlinear Regression Modelling, A Unified Practical Approach*. Marcel Dekker, New York.

Reed, W.J. 1978. The steady state of a stochastic harvesting model. Math. Biosci. 41:273–307.

Reed, W.J. 1979. Optimal escapement levels in stochastic and deterministic harvesting models. J. Env. Econ. Mgmt. 6:350–363.

Reed, W.J. 1980. Optimum age-specific harvesting in a nonlinear population model. Biometrics 36:579–593.

Ricker, W.E. 1948. Methods of estimating vital statistics of fish populations. Indiana University Publications in Science Series, No. 15.

Ricker, W.E. 1954. Stock and recruitment. J. Fish. Res. Bd. Can. 11:559–623.

Ricker, W.E. 1958a. *Handbook of Computations for Biological Statistics of Fish Populations*. Fisheries Research Board of Canada, Bulletin No. 119.

Ricker, W.E. 1958b. Maximum sustained yields from fluctuating environments and mixed stocks. J. Fish. Res. Bd. Can. 15:991–1006.

Ricker, W.E. 1973a. Linear regressions in fishery research. J. Fish. Res. Bd. Can. 30:409–434.

Ricker, W.E. 1973b. Critical statistics from two reproduction curves. Rapp. P-V. Reun. Cons. Perm. Int. Explor. Mer. 164:333–340.

Ricker, W.E. 1973c. Two mechanisms that make it impossible to maintain peak period yields from Pacific Salmon and other fishes. J. Fish. Res. Bd. Can. 30:1275–1286.

Ricker, W.E. 1975. *Computation and Interpretation of Biological Statistics of Fish Populations*. Fisheries Research Board of Canada, Bulletin No. 191.

Ricker, W.E. 1979. Growth rates and models. *Fish Physiology*, Vol. 8. Academic Press, New York, pp. 678–743.

Rivard, D. 1989. Overview of the systematic, structural and sampling errors in cohort analysis. Am. Fish. Soc. Symp. 6:49–65.

Robbins, H. and Monro, S. 1951. A stochastic approximation method. Ann. Math. Stat. 22:400–407.

Roedel, P.M. (Ed). 1975. *Optimum Sustainable Yield as a Concept in Fisheries Management*. Special Publication No. 9, American Fisheries Society.

Roff, D.A. 1983. Analysis of catch/effort data: a comparison of three methods. Can. J. Fish. Aquat. Sci. 40:1496–1506.

Roff, D.A. and D.J. Fairbairn. 1980. An evaluation of Gulland's method for fitting the Schaefer model. Can. J. Fish. Aquat. Sci. 37:1229–1235.

Rothschild. B.J. 1986. *Dynamics of Marine Fish Populations*. Harvard University Press, Cambridge, MA.

Royce, W.F. 1987. *Fishery Development*. Academic Press, New York. 248 pp.

Ruppert, D., R.L. Reish, R.B. Deriso, and R.J. Carroll. 1983. Development of a biological simulation model for Atlantic menhaden. Volume II: The stochastic model. NMFS Beaufort Laboratory, Report No. 2.

Ryder, R.A., S.R. Kerr, K.H. Loftus, and H.A. Regier. 1974. The morphoedaphic index, a fish yield estimator — review and evaluation. J. Fish. Res. Bd. Can. 31:663–688.

Saetersdal, G. 1980. A review of past management of some pelagic stocks and its effectiveness. Rapp. P.-V. Reun. Cons. Int. Explor. Mer 177:505–512.

Saila, S.B. 1983. Importance and assessment of discards in commercial fisheries. FAO Fisheries Circular No. 765.

Saila, S.B. and K. Erzini. 1987. Empirical approach to multispecies stock assessment. Trans. Am. Fish. Soc. 116:601–611.

Sainsbury, K. 1980. Effect of individual variability on the von Bertalanffy growth equation. Can. J. Fish. Aquat. Sci. 37:241–247.

Sainsbury, K. 1984. Optimal mesh size for tropical multispecies trawl fisheries. J. Cons. Int. Explor. Mer 41:129–139.

Sainsbury, K. 1988. The ecological basis of multispecies fisheries, and management of a demersal fishery in tropical Australia. *In*: J. Gulland (Ed.) *Fish Population Dynamics*. Wiley, Chichester. pp. 349–382

Saville, A. and R.S. Bailey. 1980. The assessment and management of the herring stocks in the North Sea and to the west of Scotland. Rapp. P.-V. Reun. Cons. Int. Explor. Mer 177:112–142.

Schaefer, M.B. 1954. Some aspects of the dynamics of populations important to the management of commercial marine fisheries. Bull. Inter-Am. Trop. Tuna Comm. 1:27–56.

Schaefer, M.B. 1957. A study of the dynamics of the fishery for yellowfin tuna in the eastern tropical Pacific Ocean. Inter-Am. Trop. Tuna Comm. Bull. 2:247–285.

Schnute, J. 1977. Improved estimates from the Schaefer production model theoretical considerations. J. Fish. Res. Bd. Can. 34:583–603.

Schnute, J. 1981. A versatile growth model with statistically stable parameters. Can. J. Fish. Aquat. Sci. 38:1128–1140.

Schnute, J. 1982. *A Manual for Easy Nonlinear Parameter Estimation in Fishery Research, With Interactive Microcomputer Programs*. Canadian Technical Report on Fisheries and Aquatic Science, No. 1140.

Schnute, J. 1985. A general theory for analysis of catch and effort data. Can. J. Fish. Aquat. Sci. 42:414–429.

Schnute, J. 1987. A general fishery model for a size-structured fish population. Can. J. Fish. Aquat. Sci. 44:924–940.

Schnute, J., L.J. Richards, and A.J. Cass. 1989. Fish survival and recruitment: investigations based on a size-structured model. Can. J. Fish. Aquat. Sci. 46:743–769.

Searle, S.R. 1966. *Matrix Algebra for the Biological Sciences*. Wiley, New York.

Seber, G.A.F. 1982. *The Estimation of Animal Abundance and Related Parameters*. Oxford University Press, New York. (Charles Griffin & Co. Ltd., High Wycombe Bucks, England).

Shackleton, L.Y. 1987. A comparative study of fossil fish scales from three upwelling regions. S. Afr. J. Mar. Sci. 5:79–84.

Shepherd, J.G. 1988. An exploratory method for the assessment of multispecies fisheries. J. Cons. Int. Explor. Mer 44:189–199.

Shepherd, J.G. and D.H. Cushing. 1980. A mechanism for density dependent survival of larval fish as the basis of a stock- recruitment relationship. J. Cons. Int. Explor. Mer 39:160–167.

Shepherd, J.G. and J.W. Horwood. 1977. The sensitivity of exploited populations to environmental "noise", and the implications for management. J. Const. Int. Explor. Mer, 38:318–323.

Shepard, M.P. and F.C. Withler. 1958. Spawning stock size and resultant production for Skeena sockeye. J. Fish. Res. Bd. Can. 15:1007–1025.

Sibert, J.R. 1984. A two-fishery tag attrition model for the analysis of mortality, recruitment, and fishery interaction. Tuna and Billfish Assessment Programme. South Pacific Commission, Noumea, New Caledonia Technical Report, No 13.

Sinclair, M., V.C. Anthony, T.D. Iles, and R. N. O'Boyle. 1985. Stock assessment problems in Atlantic Herring *(Clupea harengus)* in the Northwest Atlantic. Can. J. Fish. Aquat. Sci. 42:888–898.

Sissenwine, M.P. 1977. The effect of random fluctuations on a hypothetical fishery. Int. Comm. Northw. Atlant. Fish., Sel. Pap. 2:137–144.

Sissenwine, M.P. 1978. Is MSY an adequate foundation for optimum yield. Fisheries 3(6):22–24, 37:42.

Sissenwine, M.P. and J.E. Kirkley. 1982. Fishery management techniques practical aspects and limitations. Mar. Policy 6:43–57.

Sissenwine, M.P. and J.G. Shepherd. 1987. An alternative perspective on recruitment overfishing and biological reference points. Can. J. Fish. Aquat. Sci. 44:913–918.

Sissenwine, M.P., M.J. Fogarty, and W.J. Overholtz. 1988. Some fisheries management implications of recruitment variability. *In*: J.A. Gulland (Ed.). *Fish Population Dynamics*. Wiley, New York. pp. 129–152.

Skud, B.E. 1975. Revised estimates of halibut abundance and the Thompson-Burkenroad debate. Int. Pac. Halibut Comm. Scientific Report No 56.

Skud, B.E. 1977. Drift, migration, and the intermingling of Pacific halibut stocks. Int. Pac. Halibut Comm. Scientific Report No. 63.

Skud, B.E. 1982. Dominance in fishes: the relation between environment and abundance. Science 216:144–149.

Smith, G. G. M. 1940. Factors limiting distribution and size in the starfish. J. Fish. Res. Bd. Can. 5:84–103.

Smith, V. L. 1969. On models of commercial fishing. J. Polit. Econ 77:181-198.

Somers, I.F. 1985. Maximizing value per recruit in the fishery for banana prawns, *Penaeus merguiensis*, in the Gulf of Carpentaria. *In*: P.C. Rothlisberg, B.J. Hill and D.J. Staples (Eds.) *Second Australian National Prawn Seminar*. NPS2. Cleveland, Queensland, Australia. pp. 185–191.

Soutar, A. and J.D. Isaacs. 1969. A history of fish populations inferred from fish scales in anaerobic sediments off California. Calif. Mar. Res. Comm. CalCOFI, 13:63–70.

Sparre, P., E. Ursin, and S.C. Venema. 1989. *Introduction to Tropical Fish Stock Assessment,* Part 1 — Manual. FAO Fishery Technical Paper 306/1.

Steele, J.H. and E.W. Henderson. 1984. Modeling long-term fluctuations in fish stocks. Science 224:985–987.

Stocker, M. (Ed.). 1981. Groundfish stock assessments off the West Coast of Canada in 1981 and recommended total allowable catches for 1982. Canadian Manuscript Report on Fisheries and Aquatic Science, No. 1626.

Stocker, M., V. Haist, and D. Fournier. 1985. Environmental variation and recruitment of Pacific herring *(Clupea harengus pallasi)* in the Strait of Georgia. Can. J. Fish. Aquat. Sci. 42(Suppl. 1):174–180.

Sullivan, P.J., H.-L. Lai, and V.F. Gallucci. 1990. A catch-at-length analysis that incorporates a stochastic model for growth. Can. J. Fish. Aquat. Sci. 47:184–198.

Tautz, A., P.A. Larkin, and W.E. Ricker. 1969. Some effects of simulated long-term environmental fluctuations on maximum sustained yield. J. Fish. Res. Bd. Can. 26:2715–2726.

Thompson, W.F. 1950. *The Effect of Fishing on Stocks of Halibut in the Pacific*. University of Washington Press, Seattle.

Thompson, W.F., H.A. Dunlop, and F.H. Bell. 1931. Biological statistics of the Pacific halibut fishery. (1) Changes in yield of a standardized unit gear. International Fisheries Commission Report, No. 6, 119 p.

Tovar, H., V. Guillen and M.E. Nakama. 1987. Monthly population size of three guano bird species off Peru. In Pauly D. and I. Tsukayama (eds.). The Peruvian Anchoreta and its upwelling ecosystem: three decades of change. ICLARM Studies and Reviews, No. 15 ICLARM Morila. pp. 208–218.

Uhler, R. 1979. Least squares regression estimates of the Schaefer production model: some Monte Carlo simulation results. Can. J. Fish. Aquat. Sci. 37:1284–1294.

Varley, G.C. 1949. Population change in German forest pests. J. Anim. Ecol 18:117–122.

Vibe, C. 1967. *Arctic Animals in Relation to Climatic Fluctuations, Meddelelser om Grönland,* The Danish Zoogeographical Investigations in Greenland. Copenhagen.

von Bertalanffy, L. 1938. A quantitative theory of organic growth. Hum. Biol. 10:181–213.

Wallace, W. 1915. Report on the age, growth and sexual maturity of the plaice in certain parts of the North Sea. U.K. Board Agric. Fish., Fish. Invest. Ser. 2, 2(2):79 p.

Walters, C.J. 1969. A generalized computer simulation model for fish population studies. Trans. Am. Fish. Soc. 98:505–512.

Walters, C.J. 1975. Optimal harvest strategies for salmon in relation to environmental variability and uncertain production parameters. J. Fish. Res. Bd. Can. 32:1777–1784.

Walters. C.J. 1981. Optimum escapements in the face of alternative recruitment hypotheses. Can. J. Fish. Aquat. Sci. 38:678–689.

Walters, C.J. 1985a. Bias in the estimation of functional relationships from time series data. Can. J. Fish. Aquat. Sci. 42:147–149.

Walters, C.J. 1985b. Approaches to adaptive policy design for harvest management. *In*: Vincent, T.L., Y. Cohen, W.J. Grantham, G.P. Kirkwood, and J.M. Skowronski (Eds.), *Modelling and Management of Resources Under Uncertainty,* Lecture Notes in Biomathematics, Vol. 72. Springer-Verlag, Berlin. pp. 114–124.

Walters, C.J. 1986. *Adaptive Management of Renewable Resources.* Macmillian, New York.

Walters, C.J. 1987a. Nonstationarity of production relationships in exploited populations. Can. J. Fish. Aquat. Sci. 44 (Suppl. 2):156–165.

Walters, C.J. 1987b. Users guide for GENEST: microcomputer stock assessment using Schnute's generalization of the Deriso delay-difference model. University of British Columbia, Vancouver.

Walters, C.J. 1988. Mixed stock fisheries and the sustainability of enhancement production for chinook and coho salmon. *In*: W.J. McNeil (Ed.), *Salmon Production, Management, and Allocation.* Oregon State University Press, Corvallis, OR. pp. 109–116.

Walters, C.J. and P.J. Bandy. 1972. Periodic harvests as a method of increasing big game yields. J. Wildl. Mgmt. 36:128–134.

Walters, C.J. and J.S. Collie. 1988. Is research on environmental factors useful to fisheries management? Can. J. Fish. Aquat. Sci. 45:1848–1854.

Walters, C.J. and J.S. Collie. 1989. Design of an experimental program for groundfish management in the face of large uncertainty about stock size and production. *In* R.J. Beamish and G.A. MacFarlane (Ed.) *Effects of Ocean Variability on Recruitment and an Evaluation of Parameters Used in Stock Assessment Models.* Canadian Special Publications in Fisheries and Aquatic Sciences, No. 108. pp. 13–25.

Walters, C.J. and R. Hilborn. 1976. Adaptive control of fishing systems. J. Fish. Res. Bd. Can. 33:145–159.

Walters, C.J. and R. Hilborn. 1978. Ecological optimization and adaptive management. Ann. Rev. Ecol. Syst. 9:157–188.

Walters, C.J. and D. Ludwig. 1981. Effects of measurement errors on the assessment of stock-recruitment relationships. Can. J. Fish. Aquat. Sci. 38:704–710.

Walters, C.J. and D. Ludwig. 1987. Adaptive management of harvest rates in the presence of a risk averse utility function. Nat. Res. Model. 1:321–337.

Walters, C.J. and B. Riddell. 1986. Multiple objectives in salmon management: the chinook sport fishery in the Strait of Georgia. Nw. Env. J., 2:1–15.

Walters, C.J., M. Stocker, A.V. Tyler, and S.J. Westrheim. 1986. Interaction between pacific cod *(Gadus macrocephalus)* and herring *(Clupea harengus pallasi)* in the Hecate Strait, British Columbia. Can. J. Fish. Aquat. Sci. 43:830–837.

Walters, C.J., J.S. Collie, and T. Webb. 1988. Experimental designs for estimating transient responses to management disturbances. Can. J. Fish. Aquat. Sci. 45:530–538.

Watt, K.E.F. 1963. Dynamic programming, "Look ahead programming", and the strategy of insect pest control. Can. Entomol. 95:525–536.

Wetherall, J. A. 1982. Analysis of double-tagging experiments. Fish. Bull. 80(4):687–701.

Wilen, J.E. 1976. Common property resources and the dynamics of overexploitation; the case of the North Pacific fur seal. University of British Columbia Programme Natural Resource Management, Resource Paper No. 3.

Winters, G.H. and J.P. Wheeler. 1985. Interaction between stock area, stock abundance and catchability coefficient. Can. J. Fish. Aquat. Sci. 42:989–998.

Young, A.H., P. Tytler, and F.G.T. Holliday. 1976. New developments in ultrasonic fish tracking. Proc. R. Soc. Edin. 75(B):145–155.

Young, P. 1974. Recursive approaches to time series analysis. Inst. Math. Appl. 10:209–224.

Zippin, C. 1956. An evaluation of the removal method of estimating animal populations. Biometrics 12:163–189.

Zippin, C. 1958. The removal method of population estimation. J. Wildl. Mgmt. 22:82–90.

Author Index

Subject Index